Lecture Notes in Mathematics

Edited by A. Dold and B. Eckmann

T0214412

1103

Models and Sets

Proceedings of the Logic Colloquium
held in Aachen, July 18–23, 1983
Part I

Edited by G. H. Müller and M. M. Richter

Springer-Verlag
Berlin Heidelberg New York Tokyo 1984

Editors

Gert H. Müller
Mathematisches Institut, Universität Heidelberg
Im Neuenheimer Feld 294, 6900 Heidelberg, Federal Republic of Germany

Michael M. Richter
Lehrgebiet mathematische Grundlagen der Informatik, RWTH Aachen
Templergraben 64, 5100 Aachen, Federal Republic of Germany

AMS Subject Classification (1980): 03C, 03E, 03G, 03H

ISBN 3-540-13900-1 Springer-Verlag Berlin Heidelberg New York Tokyo
ISBN 0-387-13900-1 Springer-Verlag New York Heidelberg Berlin Tokyo

Printing and binding: Beltz Offsetdruck, Hemsbach/Bergstr.
2146/3140-543210

VORWORT

Dieser Band enthält einen Teil der Proceedings des Logic Colloquiums '83, welches vom 18. - 23. Juli 1983 in Aachen stattfand; es war dies gleichzeitig der Europäische Sommerkongreß der Association for Symbolic Logic. Ein weiterer Band der Proceedings erscheint unter dem Titel "Computation and Proof Theory" ebenfalls in den Lecture Notes in Mathematics des Springer-Verlages.

Insgesamt hatte das Logic Colloquium '83 189 Teilnehmer aus 26 Ländern. Zusätzlich zu den eingeladenen Hauptvorträgen wurden siebzig angemeldete Vorträge gehalten. Ein Teil davon fand in "Special Sessions" statt: Boole'sche Algebren (organisiert von S.Koppelberg), Topologische Modelltheorie (J.Flum), Nonstandard Analysis (K.H.Diener), Logic versus Computer Science (E.Börger). Abstracts aller angemeldeten Vorträge sowie eine vollständige Liste aller eingeladenen Vorträge werden im Bericht der Veranstalter im Journal of Symbolic Logic veröffentlicht.

Das Logic Colloquium '83 wurde ermöglicht durch großzügige finanzielle Unterstützung der Deutschen Forschungsgemeinschaft, des Landes Nordrhein-Westfalen, der Division of Logic, Methodology and Philosophy of Science, dem Deutschen Akademischen Austauschdienst, der Deutschen Stiftung für Internationale Entwicklung, der Maas-Rhein-Euregio, der Stadt Aachen, der RWTH Aachen und nicht zuletzt der deutschen Industrie. Ihnen allen sei herzlich gedankt!

Die alte Kaiserstadt Aachen gab einen würdigen Rahmen ab. Oberstadtdirektor Dr. H. Berger eröffnete den Kongreß als Schirmherr im Namen der Maas-Rhein-Euregio. Die erste Bürgermeisterin Frau Prof. Dr. W. Kruse lud ein zu einem Empfang im Krönungssaal des Rathauses, der für die Teilnehmer zu einem bleibenden Erlebnis wurde. Auch hierfür ein herzliches "Danke"!

Die Herausgeber

CONTENTS

* Invited Lecture

TABLE OF CONTENTS - PART II

(published in LNM vol.1104)

VORWORT

† Professor Dr. D. Rödding died on June 4, 1984

* An asterisk indicates a contributed paper.

FILTERS AND ULTRAFILTERS OVER DEFINABLE SUBSETS OF ADMISSIBLE ORDINALS

by Jos Baeten, Technische Hogeschool Delft, The Netherlands.

This article is part of the author's Ph.D. thesis, to appear under the same title at the University of Minnesota.

0. Abstract

The search for a recursive analogue of a measurable cardinal leads to a study of filters and ultrafilters over certain definable subsets of an admissible ordinal, using the hierarchy of constructible sets.

Connections with admissibility are explored in sections 2 and 3, and we find that the existence of a normal filter is stronger than the existence of the same type of filter (section 3). We look at the analogües of certain classical filters, namely the co-finite filter in section 2 and the normal filter of closed unbounded sets in section 3.

In section 4, we find that any filter (resp. normal filter) of a certain type can, on a <u>countable</u> ordinal, be extended to an ultrafilter (resp. normal ultrafilter) of the same type.

1. Preliminaries and notation

Lower case Greek letters represent ordinals, and lower case Latin letters represent non-negative integers. We work in the constructible hierarchy $L = \bigcup\{L_\kappa : \kappa \in \mathrm{Ord}\}$. The Lévy hierarchy of Σ_m, Π_m and Δ_m formulas is defined as usual. A relation R on L_κ is ΦL_κ (where Φ is a set of $\in$- formulas), if it is definable by a Φ-formula with parameters from L_κ. An ordinal κ is Σ_n-admissible if for all $\varphi \in \Sigma_n L_\kappa$ and all $\mu<\kappa$ $L_\kappa \models \forall\beta<\mu\ \exists\alpha\varphi(\alpha,\beta) \rightarrow \exists\gamma\forall\beta<\mu\exists\alpha<\gamma\ \varphi(\alpha,\beta)$.

$\mathrm{Pow}(\kappa)$ is the powerset of κ, $\Delta\{X_\beta : \beta<\kappa\} = \{\xi<\kappa : \forall\beta<\xi\ \xi \in X_\beta\}$,

$X\backslash Y = \{x \in X : x \notin Y\}$.

Note: to avoid unnecessary complications we will always assume κ is an ordinal with $\omega\kappa=\kappa$ and $n>0$.

2. Filters

2.1 Definition: Let $F \subseteq \mathrm{Pow}(\kappa)$ be a nonprincipal filter on κ, i.e.

i. $X, Y \in F \rightarrow X \cap Y \in F$

ii. $X \in F$ & $X \subseteq Y \rightarrow Y \in F$

iii. $\emptyset \notin F$

iv. $\forall\beta<\kappa\ \ \kappa\backslash\{\beta\} \in F$.

Let Φ be a set of $\in$-formulas (usually Σ_n, Π_n or Δ_n).

a. F is a **Φ-filter** if

$\forall\mu<\kappa\ \forall\langle X_\beta : \beta<\mu\rangle \in \Phi L_\kappa \cap {}^\mu F \quad \bigcap\{X_\beta : \beta<\mu\} \in F$.

b. F is a **Φ-normal filter** if

$\forall\langle X_\beta : \beta<\kappa\rangle \in \Phi L_\kappa \cap {}^\kappa F \quad \Delta\{X_\beta : \beta<\kappa\} \in F$.

c. F is **Φ-ultra** if $\forall X \in \Phi L_\kappa$ $X \in F$ or $\kappa\backslash X \in F$.

d. F is a **Φ-ultrafilter** if a. and c. hold;

F is a **Φ-normal ultrafilter** if b. and c. hold.

2.2 Note: a Δ_n-filter is often considered to be defined just on the Boolean algebra of $\Delta_n L_\kappa$ sets; a Π_n- or Σ_n-filter on the Boolean algebra generated by the $\Sigma_n L_\kappa$ sets (the so-called $B_n L_\kappa$ sets).

2.3 Definition: $H = \{X \subseteq \kappa : \kappa \setminus X$ is bounded in $\kappa\}$.
Note that if F is a Δ_1-filter, then $H \subseteq F$.

For the next theorem we need a lemma from Kranakis [1982a]:

2.4 Lemma: The following are equivalent:

i. κ is Σ_{n+1}-admissible

ii. For all $\mu<\kappa$ and all $\Delta_n L_\kappa$ $f : \kappa \longrightarrow \mu$ there is a $\beta<\mu$ such that $f^{-1}(\{\beta\})$ is cofinal in κ.

2.5 Theorem: The following are equivalent:

i. κ is Σ_{n+1}-admissible

ii. there is a Δ_n-filter on κ

iii. there is a Π_n-filter on κ

proof: iii $\Rightarrow$ ii: immediate.

ii $\Rightarrow$ i: we use 2.4. Let F be a Δ_n-filter on κ. Suppose $\mu<\kappa$, $f : \kappa \rightarrow \mu$ is $\Delta_n L_\kappa$ but for each $\beta<\mu$ $f^{-1}(\{\beta\})$ is bounded in κ. Then for each $\beta<\mu$ we have that $\kappa \setminus f^{-1}(\{\beta\}) \in H \subseteq F$ (by 2.3), so $\emptyset = \bigcap\{\kappa \setminus f^{-1}(\{\beta\}) : \beta<\mu\} \in F$, a contradiction.

i $\Rightarrow$ iii: we show H is a Π_n-filter on κ. It is easy to see that H is a nonprincipal proper filter on κ. Thus let $\mu<\kappa$ and $\langle X_\beta : \beta<\mu\rangle \in \Pi_n L_\kappa \cap {}^\mu H$. Take $\varphi \in \Pi_n L_\kappa$ such that $\xi \in X_\beta \Leftrightarrow L_\kappa \models \varphi(\beta,\xi)$ (for $\beta<\mu$). Then $L_\kappa \models \forall\beta<\mu\ \exists\alpha\ \forall\xi\geq\alpha\ \varphi(\beta,\xi)$. Since κ is Σ_{n+1}-admissible, there is a $\gamma<\kappa$ such that $L_\kappa \models \forall\beta<\mu\ \exists\alpha<\gamma\ \forall\xi\geq\alpha\ \varphi(\beta,\xi)$, so $L_\kappa \models \forall\xi\geq\gamma\ (\forall\beta<\mu\ \varphi(\beta,\xi))$, or $\bigcap\{X_\beta : \beta<\mu\} \in H$.

2.6 Remark: the case of the Σ_n-filter is harder. We have that H is a Σ_n-filter on κ iff κ is Σ_{n+2}-admissible, but there are κ below the least Σ_{n+2}-admissible which have a Σ_n-filter.

3. Normal filters

3.1 Definition:

i. $L_\beta \prec_n L_\kappa$ iff for all $\varphi \in \Sigma_n L_\beta$ $\quad L_\kappa \models \varphi \Rightarrow L_\beta \models \varphi$.

ii. $S_\kappa^n = \{\beta<\kappa : L_\beta \prec_n L_\kappa\}$. Note that S_κ^n is $\Pi_n L_\kappa$.

iii. Let Φ be a set of $\in$-formulas and $X \subseteq \kappa$.

κ is Φ-reflecting on X iff for all $\varphi \in \Phi L_\kappa$

$L_\kappa \models \varphi \quad \Rightarrow \quad \exists\sigma \in X \quad L_\sigma \models \varphi$.

For the next theorem we need a lemma from Kranakis [1982a]:

3.2 Lemma: The following are equivalent:

i. κ is Σ_n-admissible

ii. κ is Π_{n+1}-reflecting on S_κ^{n-1}.

3.3 Theorem: Let κ be Σ_{n+1}-admissible, $\langle C_\beta : \beta<\kappa\rangle \in \Sigma_{n+1} L_\kappa$ and C_β is closed unbounded in κ for $\beta<\kappa$. Then $\Delta\{C_\beta : \beta<\kappa\}$ is closed unbounded and $\Sigma_{n+1} L_\kappa$.

Proof: Take $\langle C_\beta : \beta<\kappa\rangle$ as stated, and $\varphi \in \Sigma_{n+1} L_\kappa$ such that $\xi \in C_\beta \Leftrightarrow L_\kappa \models \varphi(\beta,\xi)$. It is not hard to see that ΔC_β is closed and, using admissibility, that ΔC_β is $\Sigma_{n+1} L_\kappa$. Fix $\mu<\kappa$. We'll show $\Delta C_\beta \backslash \mu \neq \emptyset$, which finishes the proof. Since each C_β is unbounded, we have $L_\kappa \models \forall\beta\, \forall\alpha\, \exists\xi>\alpha\, \varphi(\beta,\xi)$. This sentence is $\Pi_{n+2} L_\kappa$, so using 3.2 there is a $\sigma \in S_\kappa^n$, $\sigma>\mu$ with $L_\sigma \models \forall\beta\, \forall\alpha\, \exists\xi>\alpha\, \varphi(\beta,\xi)$. This means $\forall\beta<\sigma\, \forall\alpha<\sigma\, \exists\xi<\sigma\, (\xi>\alpha \;\&\; L_\sigma \models \varphi(\beta,\xi))$.

Since $\sigma \in S_\kappa^n$ we have $\forall\beta<\sigma\, \forall\alpha<\sigma\, \exists\xi<\sigma\, (\xi>\alpha \;\&\; L_\kappa \models \varphi(\beta,\xi))$, so $\forall\beta<\sigma\, \forall\alpha<\sigma\, \exists\xi<\sigma\, (\xi>\alpha \;\&\; \xi \in C_\beta)$. Thus if $\beta<\sigma$, C_β is unbounded in σ, so $\sigma \in C_\beta$ and we get $\sigma \in \Delta C_\beta$.

3.4 <u>Definition</u>: $F_n = \{X \subseteq \kappa : \exists C \subseteq X \quad C$ is closed unbounded in κ and $\Sigma_{n+1}L_\kappa\}$. <u>Note</u>: if κ is Σ_{n+1}-admissible, then $S^n_\kappa \in F_n$.

3.5 <u>Note</u>: there are κ that are Σ_{n+1}-admissible, but that have two $\Pi_{n+1}L_\kappa$-definable closed unbounded subsets with empty intersection. Therefore, on a Σ_{n+1}-admissible ordinal we have that F_n should be the analogue of the (classical) normal filter of the cubsets. However, we will find in 3.11 that F_n is not a Δ_1-normal filter.

3.6 <u>Definition</u>: $X \subseteq \kappa$ is Σ_m-**stationary** iff for all $C \subseteq \kappa$ that are cub and $\Sigma_m L_\kappa$-definable we have $C \cap X \neq \emptyset$.

3.7 <u>Theorem</u>: Let κ be Σ_{n+1}-admissible, $X \subseteq \kappa$.
κ is Π_{n+2}-reflecting on $S^n_\kappa \cap X \Leftrightarrow X$ is Σ_{n+1}-stationary.
<u>Proof</u>: $\Rightarrow$: Let C be cub in κ and $\varphi \in \Sigma_{n+1}L$ such that $\xi \in C \Leftrightarrow$ $\Leftrightarrow L_\kappa \models \varphi(\xi)$. Then $L_\kappa \models \forall\beta\ \exists\xi>\beta\ \varphi(\xi)$, so there is a $\sigma \in S^n_\kappa \cap X$ with $L_\sigma \models \forall\beta\ \exists\xi>\beta\ \varphi(\xi)$. As in 3.3 we see that C is unbounded in σ, so $\sigma \in C$. But also $\sigma \in X$.

$\Leftarrow$: Let $\varphi \in \Pi_{n+2}L_\kappa$ and $L_\kappa \models \varphi$. Put $C = \{\beta \in S^n_\kappa : L_\beta \models \varphi\}$. Since κ is Σ_{n+1}-admissible, we have by 3.2 that C is unbounded in κ. Also C is $\Pi_n L_\kappa$. It is not hard to see that C is closed, since φ is Π_{n+2}. Thus $C \cap X \neq \emptyset$.

3.8 <u>Corollary</u>: i. $X \in F_n \Leftrightarrow \kappa$ is not Π_{n+2}-reflecting on $S^n_\kappa \setminus X$.

ii. each $\Sigma_{n+1}L_\kappa$-definable cubset C contains a $\Pi_n L_\kappa$-definable subset that is cub, namely $\{\beta \in S^n_\kappa : L_\beta \models \forall\alpha\ \exists\xi>\alpha$ "$\xi \in C$"$\}$.

3.9 <u>Theorem</u>: Let F be a Π_n-normal filter on κ. Then $F_n \subseteq F$.
<u>Proof</u>: Note κ is Σ_{n+1}-admissible by 2.5. Let $X \in F_n$. By 3.8.ii there is a $\Pi_n L_\kappa$-definable cub $C \subseteq X$. For $\beta<\kappa$, define
$\xi \in X_\beta \Leftrightarrow \exists\gamma<\xi\ (\gamma>\beta\ \&\ \gamma \in C)$.

Then $\langle X_\beta : \beta<\kappa\rangle \in \Pi_n L_\kappa$ and since C is unbounded, $X_\beta \in H$ for each $\beta<\kappa$. Thus $\Delta\{X_\beta : \beta<\kappa\} \in F$. But we see $\Delta X_\beta \subseteq C$, using the closedness of C.

3.10 Theorem: Let F be a Δ_1-normal filter on κ with $S^n_\kappa \in F$. Let $\varphi \in \Pi_{n+3}L_\kappa$ and $L_\kappa \models \varphi$. Then $\{\beta \in S^n_\kappa : L_\beta \models \varphi\} \in F$.

Proof: Write φ as $\forall\xi\psi(\xi)$ with $\psi \in \Sigma_{n+2}L_\kappa$.
Suppose $X = \{\beta \in S^n_\kappa : L_\beta \models \forall\xi\psi(\xi)\} \notin F$.
Define, for $\xi<\kappa$, $X_\xi = \{\beta<\kappa : L_\beta \models \psi(\xi)\}$. Then $\langle X_\xi : \xi<\kappa\rangle \in \Delta_1 L_\kappa$ and $S^n_\kappa \cap \Delta\{X_\xi : \xi<\kappa\} = X$, so we can take $\xi_o<\kappa$ with $X_{\xi_o} \notin F$ (since $S^n_\kappa \in F$). But then $S^n_\kappa\backslash X_{\xi_o}$ is cofinal in κ. Also $S^n_\kappa\backslash X_{\xi_o} = \{\beta \in S^n_\kappa : L_\beta \models \neg\psi(\xi_o)\}$, so $L_\kappa \models \neg\psi(\xi_o)$, a contradiction.

3.11 Corollary: i. If there is a Π_n-normal filter on κ, then κ is Π_{n+3}-reflecting on S^n_κ, so, since the property of being Σ_{n+1}-admissible is expressible by a Π_{n+3}-sentence, κ is a limit of Σ_{n+1}-admissibles.

ii. F_n is not a Δ_1-normal filter on κ, since $\{\beta \in S^n_\kappa : \beta \text{ is } \Sigma_{n+1}\text{-admissible}\} \notin F_n$ (use 3.7).

4. Extending filters to ultrafilters

Kranakis [1982b] proved corollary 4.2. His ideas are used in the proof of the following theorem. He credits the basic idea to MacDowell & Specker [1961].

4.1 Theorem: Let κ be countable. Let Φ be a set of $\in$-formulas closed under disjunction and bounded universal quantification, and such that $\{\kappa\backslash\{\beta\} : \beta<\kappa\} \subseteq \Phi L_\kappa$ (e.g. Π_n, and if κ is Σ_n-admissible also Σ_n and Δ_n).
Let G be a Φ-filter on κ (resp. a Φ-normal filter).
Let $X \subseteq \kappa$ be such that $\kappa\backslash X$ is ΦL_κ and $\kappa\backslash X \notin G$.
Then there is a Φ-ultrafilter F on κ (resp. a Φ-normal ultrafilter F) with $X \in F$ and $G \cap \Phi L_\kappa \subseteq F$.

<u>Proof</u>: I will prove only one case. Other cases are similar.
Let G be a Π_n-normal filter on κ and $X \in \Sigma_n L_\kappa$ with $\kappa\setminus X \notin G$.
Enumerate all $\varphi \in \Pi_n L_\kappa$ with two free variables in a sequence $\langle\varphi_m : 1 \leq m < \omega\rangle$ such that each formula occurs infinitely many times in the list (this is the only place we use the countability of κ).
By induction, we will define a sequence $\langle Z_m : m < \omega\rangle$ of $\Sigma_n L_\kappa$ sets such that $Z_0 \supsetneq Z_1 \supsetneq \ldots$ and $\kappa\setminus Z_m \notin G$ for $m < \omega$.
Put $Z_0 = X$. Now suppose $Z_0 \supsetneq Z_1 \supsetneq \ldots \supsetneq Z_m$ have been defined, $Z_m \in \Sigma_n L_\kappa$ and $\kappa\setminus Z_m \notin G$. To define Z_{m+1} we look at φ_{m+1}.
For $\beta<\kappa$ define $X_\beta = \{\xi<\kappa : L_\kappa \models \varphi_{m+1}(\xi,\beta)\}$.
There are two cases: if $Z_m \cap \triangle\{X_\beta : \beta<\kappa\} \neq \emptyset$, put $Z_{m+1} = Z_m$.
Otherwise $Z_m \cap \triangle X_\beta = \emptyset$, and we

<u>claim</u>: $\exists \beta<\kappa \quad (\kappa\setminus Z_m) \cup X_\beta \notin G$.

<u>proof</u>: otherwise $\langle(\kappa\setminus Z_m)\cup X_\beta : \beta<\kappa\rangle \in {}^\kappa G \cap \Pi_n L_\kappa$, so $\kappa\setminus Z_m = (\kappa\setminus Z_m)\cup\triangle X_\beta = \triangle\{(\kappa\setminus Z_m)\cup X_\beta : \beta<\kappa\} \in G$, contradiction.

In the second case we put $Z_{m+1} = Z_m\setminus X_\beta$, with β as in the claim.
Then $Z_m \supsetneq Z_{m+1}$, $Z_{m+1} \in \Sigma_n L_\kappa$ and $\kappa\setminus Z_{m+1} \notin G$.
To define F, it is enough to define membership for $\Pi_n L_\kappa$ and $\Sigma_n L_\kappa$ sets (see 2.2).
If $A \in \Sigma_n L_\kappa$, put $A \in F$ iff $\exists m<\omega \quad Z_m \subseteq A$;
if $A \in \Pi_n L_\kappa$, put $A \in F$ iff $\forall m<\omega \quad Z_m \not\subseteq \kappa\setminus A$.
It is not hard to check that F is a proper nonprincipal filter on κ, is Π_n-ultra, that $X \in F$ and $G\cap\Pi_n L_\kappa \subseteq F$.
To prove Π_n-normality, take $\langle X_\beta : \beta<\kappa\rangle \in \Pi_n L_\kappa$ and suppose $\triangle\{X_\beta : \beta<\kappa\} \notin F$. We have to find a $\beta<\kappa$ with $X_\beta \notin F$.
Since $\triangle X_\beta \notin F$ and is $\Pi_n L_\kappa$, there is a $m_0<\omega$ such that $Z_{m_0} \subseteq \kappa\setminus\triangle X_\beta$.
Take $\varphi \in \Pi_n L_\kappa$ such that $\xi \in X_\beta \Leftrightarrow L_\kappa \models \varphi(\xi,\beta)$. Since this φ occurs infinitely many times in the list $\langle\varphi_m : 1\leq m<\omega\rangle$, it occurs with index $k+1>m_0$. Then $Z_k \subseteq Z_{m_0} \subset \kappa\setminus\triangle X_\beta$, so there is $\beta<\kappa$ with $Z_{k+1} = Z_k\setminus X_\beta$. Thus $Z_{k+1} \subseteq \kappa\setminus X_\beta$, so $\kappa\setminus X_\beta \in F$ and $X_\beta \notin F$.

4.2 Corollary: Let κ be countable. The following are equivalent:

i. κ is Σ_{n+1}-admissible

ii. there is a Δ_n-ultrafilter on κ

iii. there is a Π_n-ultrafilter on κ.

4.3 Corollary: Let κ be countable and Σ_{n+1}-admissible. Then $\bigcap\{F : F \text{ is a } \Delta_n\text{-ultrafilter on } \kappa\} = H$.

4.4 Note: By passing to a "ultrapower", connections can be made with the existence of certain kinds of end extensions.

References

E. Kranakis, 1982a: Reflection and partition properties of admissible ordinals, Annals of Math. Logic, 1982, pp. 213 - 242.

E. Kranakis, 1982b: Definable ultrafilters and end extensions of constructible sets, Zeitschrift für Math. Logik und Grundlagen der Math., Band 28, 1982, pp. 395 - 412.

R. MacDowell & E. Specker, Modelle der Arithmetik, in Infinitistic Methods, Pergamon Press, London, 1961, pp. 257 - 263.

SUPERINFINITESIMALS AND THE CALCULUS OF THE GENERALIZED RIEMANN INTEGRAL

Benjamin Benninghofen*
RWTH Aachen, Germany

§1. Introduction

In [Hen 2] Henstock describes a generalized Riemann integral which looks very similar to the Riemann integral but it contains the Lebesgue integral and the Denjoy-Perron integral. This integral was first introduced by Kurzweil [Ku 1] in 1957 and independently by Henstock [Hen 1] in 1961.

In 1980 there appeared a whole book on this integral ([McL]). In this book an integration theory based on Henstock's integral is developed. For many important theorems from Lebesgue integration one finds a corresponding theorem for the generalized Riemann integral. In the last chapter of that book it is shown that a function f is Lebesgue integrable iff f and |f| have a generalized Riemann integral.

In this paper we present a nonstandard approach to the generalized Riemann integral. We will prove some of the results in [McL] in our technique. Furthermore we will show, that the function

$$\Phi(x) := \int_a^x \varphi(t)\, dt \quad \text{(indefinite generalized Riemann integral)}$$

has almost everywhere (in the Lebesgue sense) a derivative and we have $\Phi' = \varphi$ a.e.. Usually this is proved by using the fact that this holds for the Denjoy-Perron integral and the fact that the generalized Riemann integral is equivalent to the Denjoy-Perron integral (see also the remark following Theorem 2.5.). Here this result will be proved directly.

Furthermore we prove that if φ has a generalized Riemann integral, then φ is locally Lebesgue integrable on a dense open set, which is also known for the Denjoy-Perron integral.

Using this we will be able to show that the special Denjoy integral is

*The author was supported by the Deutsche Forschungsgemeinschaft (Ri 327/5-1).

more general than the generalized Riemann integral. In particular the proofs for 2.5. and 7.13. are another proof for the Looman-Alexandrow theorem, which states that the special Denjoy integral is more general than the Perron integral.
We are working in internal set theory (cf. [Ne] or [Ri]) and use the terminology and results of the theory of superinfinitesimals developped in [Ben] and [Ben-Ri], but will include below some of the basic material. Throughout this paper we consider natural numbers as ordinals, i.e. $n = \{0,\dots,n-1\}$.

1.1. Def.

Suppose $a < b$ are standard real numbers.

(a) $W_\infty(a,b) := \{(n,x,\xi) \mid n \in {}^*\mathbb{N},\ x \in {}^*\mathbb{R}^{n+1},\ \xi \in {}^*\mathbb{R}^n \wedge (\forall r \in n)$
$(x_r < x_{r+1},\ \xi_r \in [x_r,x_{r+1}],\ x_0 = a \wedge x_n = b)\}$;
$(n,x,\xi) \in W_\infty(a,b)$ is called a partition of $[a,b]$.

(b) $W_\approx(a,b) := {}^E\{(n,x,\xi) \in W_\infty(a,b) \mid (\forall r \in n)(x_r \approx x_{r+1})\}$;
$(n,x,\xi) \in W_\approx(a,b)$ is called an infinitesimal partition of $[a,b]$.

We observe that in this case n has to be nonstandard.

The upper index "E" is only a reminder that $W_\approx(a,b)$ is an external set. $W_\approx(a,b)$ is the monad of some filter; the easiest way to see this is observe that it is defined by an $\forall^{st}$-formula and use the theorem on monads (cf. [Ben]).
A standard bounded function $f: [a,b] \to {}^*\mathbb{R}$ is Riemann integrable iff there is a standard number $c \in {}^*\mathbb{R}$ s. t.:

$$(n,x,\xi) \in W_\approx(a,b) \Rightarrow c \approx \sum_{r\in n} (x_{r+1} - x_r)\cdot f(\xi_r)$$

In this case we have $c = \int_a^b f(x)\,dx$.

1.2. Def.

$\mathfrak{R}(a,b) := \{f: [a,b] \to {}^*\mathbb{R} \mid f$ is Riemann integrable on $[a,b]\}$.

In order to obtain a wider class of integrable functions we consider two types of changes:

(1) We replace $W_\approx(a,b)$ by a subset $W(a,b)$ of "superinfinitesimal" partitions.
(2) It is no longer required that f is bounded.

Next we recall some concepts concerning families of monads from [Ben],

[Ben-Ri].

We assume that we have a standard familiy of sets $(X_i)_{i\in I}$ and for each standard $i \in I$ we have a monad $M_i \subseteq X_i$. In this case we say that we have a family of monads. We then find a standard family $(\mathcal{F}_i)_{i\in I}$ of filters, s. t. $(\forall^{st} i \in I)(\mu(\mathcal{F}_i) = M_i)$. Now define for each $i \in I$ (standard or nonstandard)

$$^{\pi}M_i := {}^{E}\{x \in X_i \mid (\forall^{st}\tilde{\mathcal{F}} \in \prod_{i\in I} \mathcal{F}_i)(x \in \tilde{\mathcal{F}}(i)\}.$$

Note that this definition depends on the whole family $(\mathcal{F}_i)_{i\in I}$. For standard $i \in I$ we have $^{\pi}M_i = M_i = \mu(\mathcal{F}_i)$, i.e. the usual monad. If for all standard $i \in I$ $M_i = M = \mu(\mathcal{F})$, then we write $^{\pi}M[i]$ for $^{\pi}M_i$.

An important special case is given if the index set I is a topological space E and the family of monads is given by $M_a = \mu(a)$, where $\mu(a)$ denotes the usual monad of the point a; for arbitrary $a \in E$ we write $^{\pi}\mu(a)$ for $^{\pi}M_a$.

In order to describe the π-monad by formulas we consider formulas $\varphi(x_1 \dots x_n; Z_1 \dots Z_m; Y_1 \dots Y_l; X_1 \dots X_k)$ s. t. all external quantifiers are of the form

$\forall^{st} u \in Z_\mu$ or $\exists^{st} u \in Z_\mu$ and the Y_λ occur only in the form $z \in Y_\lambda$.

If we have families of sets $(A_{1i})_{i\in I}, \dots, (A_{ki})_{i\in I}$,
families of lattices $(U_{1i})_{i\in I}, \dots, (U_{ni})_{i\in I}$
and families of monads $(M_{1i})_{i\in I}, \dots, (M_{li})_{i\in I}$
then we may form for each standard $i \in I$ a new formula

$$\varphi_i(x_1 \dots x_n) = \varphi(x_1 \dots x_n; U_{1i}, \dots, U_{mi}; M_{1i}, \dots, M_{li}; A_{1i}, \dots, A_{ki}).$$

Form these formulas $\varphi_i(x_1 \dots x_n)$ we now construct one new formula $\hat{\Phi}(i, x_1 \dots x_n)$. In [Ben-Ri] we call φ the base formula and $\hat{\Phi}$ the product form for the formulas φ_i. This construction is illustrated in the following example:

Suppose that (E,d) is a standard metric space; for standard $i \in E$ we take

$$\varphi_i := \varphi(;]0,\infty[;\mu(i);e,d,<,i)$$
$$:= (\forall x \in E)[(x \in \mu(i)) \longleftrightarrow (\forall^{st}\varepsilon \in]0,\infty[)(d(x,i) < \varepsilon)].$$

Here we have $n = 0$ (i.e. there are no variables $x_1 \dots x_n$), and most of the families $(U_{\mu i})_{i\in I}$, $(A_{\kappa i})_{i\in I}$ are constant.

Now we construct the formulas $\hat{\Phi}(i)$ as follows:

$(Q^{st} u \in U_{\gamma i})\Phi(\dots u \dots)$ is replaced by
$(Q^{st}\tilde{u} \in \prod_{i\in I} U_{\gamma i})\Phi(\dots \tilde{u}(i) \dots)$, $Q \in \{\forall, \exists\}$,

and $z \in M_{\gamma i}$ is replaced by $z \in {}^{\pi}M_{\gamma i}$.
In the above example we obtain:

$$\hat{\Phi}(i) = (\forall x \in E)[x \in {}^{\pi}\mu(i) \leftrightarrow (\forall^{st}\tilde{\varepsilon} \in]0,\infty[^{E})\ \ d(x,a) < \tilde{\varepsilon}(i)].$$

Then the following transfer theorem holds:

1.3. Theorem

Suppose that the φ_i have no free variables and are simply reducible (this is defined in [Ben], [Ben-Ri]).
Then we have:

$$(\forall^{st} i \in I)(\varphi_i) \leftrightarrow (\forall i \in I)(\hat{\Phi}(i)).$$

Now we get a useful description for the π-monads of a point x in a metric space E.

1.4. Corollary

In a standard metric space (E,d)

$$(\forall a \in E)({}^{\pi}\mu(a) = {}^{E}\{x \in E \mid (\forall^{st}\tilde{\varepsilon} \in]0,\infty[^{E})\ \ d(x,a) < \tilde{\varepsilon}(i)\}) \text{ holds.}$$

For standard $a \in E$ the usual monad of a is ${}^{\pi}\mu(a)$; for nonstandard $a \in E$ ${}^{\pi}\mu(a)$ contains the points which are "superinfinitesimally close to a".
Now we observe that $\approx\ \subseteq {}^*\mathbb{R} \times {}^*\mathbb{R}$ is a monad. If we take as index set $I = \mathcal{R}(a,b)$ we have a constant family of filters which define all have $\approx$ as their monad; for arbitrary $f \in \mathcal{R}(a,b)$ we have the π-monad ${}^{\pi}\!\approx_{[f]}$.
Similar as above we have:

(I) $\forall^{st}(a,b) \in {}^*\mathbb{R} \times {}^*\mathbb{R}\ [a < b \rightarrow (\forall f \in \mathcal{R}(a,b))(\forall(n,x,\zeta) \in {}^{\pi}W_{\approx}(a,b)[f])$

$$(\int_a^b f(x)\,dx\ \ {}^{\pi}\!\approx_{[f]} \sum_{r \in n} (x_{r+1} - x_r)\, f(\zeta_r))];$$

(II) $\forall(a,b) \in {}^*\mathbb{R} \times {}^*\mathbb{R}\ [a < b \rightarrow (\forall f \in \mathcal{R}(a,b))(\forall(n,x,\zeta) \in {}^{\pi\pi}W_{\approx}(a,b)[f])$

$$(\int_a^b f(t)\,dt\ \ {}^{\pi\pi}\!\approx_{[f][(a,b)]} \sum_{r \in n} (x_{r+1} - x_r)\, f(\zeta_r))];$$

(III) ${}^{\pi}W_{\approx}(a,b) = {}^{E}\{(n,x,\zeta) \in W_{\infty}(a,b) \mid (\forall r \in n)(x_r\ {}^{\pi}\!\approx_{[(a,b)]}\ x_{r+1})\}$.

In order to get an idea of the type of integral we obtain by weakening the conditions for integrability as indicated we consider a function $g\colon [a,b] \rightarrow {}^*\mathbb{R}$ with a derivative $f = g'$.
Taking $\dot{\mu}(x) := \mu(x)\setminus\{x\}$ we get

(*) $(\forall^{st} x \in [a,b])(\forall t \in [a,b])(t \in \dot{\mu}(x) \rightarrow f(x) \approx \frac{g(t) - g(x)}{t - x})$.

If we try to integrate f we have to consider the Riemann sums

$\sum_{r\in n} (x_{r+1} - x_r)\cdot f(\zeta_r)$ for some $(n,x,\zeta) \in W_{\approx}(a,b)$, where we try to replace $f(\zeta_r)$ by $\dfrac{g(x_{r+1}) - g(x_r)}{x_{r+1} - x_r}$; the last term equals $f(\eta_r)$ for some $\eta_r \in [x_r, x_{r+1}]$. But because f need not be continuous, it is possible that $f(\zeta_r) \not\approx f(\eta_r)$; hence we may fail to integrate f this way. In fact, there are examples of derivatives of functions g s. t. $g' \notin \mathfrak{R}(a,b)$ (cf e.g. [Ru], [McL]).

We may however enforce $f(\zeta_r) \approx \dfrac{g(x_{r+1}) - g(x_r)}{x_{r+1} - x_r}$ if we put two restrictions on the infinitesimal partition:

(i) $(\forall r \in n)(\zeta_r = x_r \vee \zeta_r = x_{r+1})$

(ii) $(\forall r \in n)([x_r, x_{r+1}] \subseteq {}^{\pi}\mu(\zeta_r))$.

By (*) and the transfer theorem we obtain $f(\zeta_r) \approx \dfrac{g(x_{r+1}) - g(x_r)}{x_{r+1} - x_r}$.

Now an easy calculation shows

$$\sum_{r\in n} (x_{r+1} - x_r)\cdot f(\zeta_r) \approx g(b) - g(a).$$

§2. Elementary properties of the generalized Riemann Integral

According to the motivation in §1. we define for standard $a < b$ the following monad:

2.1. Def.

$$W(a,b) := {}^{E}\{(n,x,\zeta) \in W_{\infty}(a,b) \mid (\forall r \in n)((\zeta_r = x_r \vee \zeta_r = x_{r+1}) \wedge [x_r, x_{r+1}] \subseteq {}^{\pi}\mu(\zeta_r))\}.$$

In order to show $W(a,b) \neq \emptyset$ we put for $\tilde{\varepsilon}: [a,b] \to]0,\infty[$:

$$W_{\tilde{\varepsilon}}(a,b) := \{(n,x,\zeta) \in W_{\infty}(a,b) \mid (\forall r \in n)((\zeta_r = x_r \vee \zeta_r = x_{r+1}) \wedge |x_{r+1} - x_r| < \tilde{\varepsilon}(\zeta_r))\}.$$

Then we have

$$W(a,b) = {}^{E}\{(n,x,\zeta) \in W_{\infty}(a,b) \mid (\forall^{st}\tilde{\varepsilon} \in]0,\infty[^{[a,b]})(n,x,\zeta) \in W_{\tilde{\varepsilon}}(a,b)\} .$$

By the axiom of the ideal point it suffices to show $W_{\tilde{\varepsilon}}(a,b) \neq \emptyset$, which follows from the compactness of $[a,b]$.

Now we can give a nonstandard definition of the generalized Riemann integral.

2.2. Def.

Let $a < b$, $f: [a,b] \longrightarrow {}^*\mathbb{R}$ be standard.

(i) f is called integrable iff

$$(\exists^{st} c \in {}^*\mathbb{R})(\forall(n,x,\zeta) \in W(a,b))(c \approx \sum_{r \in n} (x_{r+1} - x_r) \cdot f(\zeta_r)).$$

In that case we put $\int_a^b f(t)\,dt := c$.

(ii) $\mathbb{D}(a,b) := {}^S\{f: [a,b] \to {}^*\mathbb{R} \mid f \text{ is integrable}\}$.

Here ${}^S\{x \in A \mid \varphi(x)\}$ denotes the (unique) standard set the standard elements of which satisfy $\varphi(x)$.

Remark

The definition of the generalized Riemann integral given here is equivalent to the definition given in [McL].

We put $W'(a,b) = \{(n,x,\zeta) \in W_\infty(a,b) \mid (\forall r \in n)([x_r, x_{r+1}] \subseteq {}^\pi\mu(\zeta_r))\}$.

If we replace in the above definition $W(a,b)$ by $W'(a,b)$, then we obtain the same integral. In order to see this, we take some $(n,x,\zeta) \in W'(a,b)$ and construct $(m,y,\eta) \in W(a,b)$ as follows:

If $\zeta_r = x_r \vee \zeta_r = x_{r+1}$, then for some $\mu \in m$

$\eta_\mu = \zeta_r$, $y_\mu = x_r$, $y_{\mu+1} = x_{r+1}$.

If $x_r < \zeta_r < x_{r+1}$, then for some $\mu \in m$

$x_r = y_{\mu-1}$, $\zeta_r = y_\mu = \eta_{\mu-1} = \eta_\mu$, $x_{r+1} = y_{\mu+1}$.

From this we get

$$\sum_{r \in n} (x_{r+1} - x_r) \cdot f(\zeta_r) = \sum_{\mu \in m} (y_{\mu+1} - y_\mu) \cdot f(\eta_\mu).$$

Applying Nelson's Reduction Algorithm (see [Ne]) to the formula $(\exists^{st} c)(\forall(n,x,\zeta) \in W'(a,b))(c \approx \sum_{r \in n} (x_{r+1} - x_r) \cdot f(\zeta_r))$ we obtain the formula

$$(\exists c)(\forall \varepsilon > 0)(\exists \gamma \in \,]0,\infty[^{[a,b]})(\forall(n,x,\zeta) \in W_\infty(a,b))$$

$$[(\forall r \in n)([x_r, x_{r+1}] \subseteq K(\zeta_r; \gamma(\zeta_r))) \to (|c - \sum_{r \in n} (x_{r+1} - x_r) \cdot f(\zeta_r)| < \varepsilon]$$

where $K(z;r) = \,]z-r, z+r[$, i.e. the standard definition of the generalized Riemann integral as given in [McL].

Next we integrate derivatives of differentiable functions.

2.3. Theorem

Suppose $g: [a,b] \to {}^*\mathbb{R}$ is continuous and $g'(x)$ exists for $x \in [a,b]\setminus H$ where $H \subseteq [a,b]$ is countable; put $f(x) = g'(x)$ for $x \notin H$ and $f(x) = 0$ otherwise. Then $f \in W(a,b)$ and

$$\int_a^b f(t)\,dt = g(b) - g(a).$$

Proof: We may assume that all entries are standard.
For $(n,x,\zeta) \in W(a,b)$ we put

$$N := \{r \in n \mid \zeta_r \notin H\} \text{ and } M := n \setminus N.$$

By Robinsons Lemma (see [Ben-Ri]) there is a positive infinitesimal r s. t. $(n,x,\zeta) \in {}^{\pi}W(a,b)[r]$. We also find a standard function $e: H \to]0,\infty[$ satisfying $\sum_{x\in H} e(x) < 1$, from where we get

$$\left|\sum_{r\in M} g(x_{r+1}) - g(x_r)\right| \leq \sum_{r\in M} |g(x_{r+1}) - g(x_r)| \leq \sum_{r\in M} r\cdot e(\zeta_r) \leq r \approx 0$$

(because $x_{r+1}, x_r \in {}^{\pi\pi}\mu(\zeta_r)[r]$);
this implies

$$\sum_{r\in n} (x_{r+1} - x_r)\cdot f(\zeta_r) = \sum_{r\in N} (x_{r+1} - x_r)\cdot f(\zeta_r) \approx$$

$$\approx \sum_{r\in N} (x_{r+1} - x_r).\frac{g(x_{r+1}) - g(x_r)}{x_{r+1} - x_r} = \sum_{r\in N} g(x_{r+1}) - g(x_r) \approx$$

$$\approx \sum_{r\in n} g(x_{r+1}) - g(x_r) = g(b) - g(a).$$

Remark: The theorem becomes false if the exceptional set H is only required to have Lebesgue measure 0 (see [Ru; p. 179]).

Another integral which integrates all derivatives is the Perron integral. In order to compare the generalized Riemann integral with the Perron integral, we describe shortly the latter (cf. e.g. [Na]).
For functions $f,F: [a,b] \to {}^*\mathbb{R}$ we recall the notation

$$\underline{D}\,F(x) = \liminf_{t\to x} \frac{F(t) - F(x)}{t - x} \quad (\in [-\infty,\infty])$$

$$\overline{D}\,F(x) = \limsup_{t\to x} \frac{F(t) - F(x)}{t - x} \quad (\in [-\infty,\infty])$$

$$T(f) = \{F \in \mathcal{C}(a,b) \mid F(a) = 0 \wedge (\forall x \in [a,b])(\underline{D}\,F(x) \geq f(x) > -\infty)\}$$

$$\perp(f) = \{F \in \mathcal{C}(a,b) \mid F(a) = 0 \wedge (\forall x \in [a,b])(\overline{D}\,F(x) \leq f(x) < +\infty)\}.$$

If $T(f) \neq \emptyset \wedge \perp(f) \neq \emptyset$ and if $c = \inf_{F\in T(f)} F(b) = \sup_{G\in\perp(f)} G(b)$

then $(P)\int_a^b f(t)\,dt := c$ is the Perron integral of f.

$\mathcal{P}(a,b)$ and $\mathcal{L}^1(a,b)$ denote Perron integrable resp. Lebesgue integrable functions on $[a,b]$; the Lebesgue integral is denoted by
$(L) \int_a^b f(t)\,dt$.

Without proof we quote (cf. [Na]):

2.4. Theorem
$\mathcal{L}^1(a,b) \subsetneqq \mathcal{P}(a,b)$ and $(L) \int_a^b f(t)\,dt = (P) \int_a^b f(t)\,dt$ for all $f \in \mathcal{L}^1(a,b)$.

The next theorem compares the generalized Riemann and the Perron integral.

2.5. Theorem
(i) $\mathcal{P}(a,b) \subseteq \mathcal{D}(a,b)$;
(ii) $(\forall f \in \mathcal{P}(a,b))\ (P) \int_a^b f(t)\,dt = \int_a^b f(t)\,dt$;
(iii) $\mathcal{L}^1(a,b) \subseteq \mathcal{D}(a,b)$.

Proof: We have to show:

$$(\forall^{st} F \in T(f))(\forall^{st} G \in \bot(f))(\forall (n,x,\zeta) \in W(a,b)) \left[G(b) \lesssim \sum_{r \in n} (x_{r+1} - x_r)\cdot f(\zeta_r) \lesssim F(b)\right]$$

(where $u \lesssim v$ means $u \le v \vee u \approx v$).
First we get

$$(\forall^{st} x \in [a,b])(\forall t \in [a,b])\left[t \in \dot{\mu}(x) \to \frac{G(t) - G(x)}{t - x} \lesssim f(x)\right]$$

because if $\overline{D}\, G(x) > -\infty$ then $\frac{G(t) - G(x)}{t - x} \lesssim \overline{D}\, G(x) \le f(x) < \infty$

and if $\overline{D}\, G(x) = -\infty$ then $\frac{G(t) - G(x)}{t - x} \in \mu_{\mathbb{R}}(-\infty)$

and hence $\frac{G(t) - G(x)}{t - x} < f(x)$.

With the transfer theorem we obtain

$$(\forall x \in [a,b])(\forall t \in [a,b] \cap {}^{\pi}\dot{\mu}(x))\left[\frac{G(t) - G(x)}{t - x} \lesssim f(x) \lesssim \frac{F(t) - F(x)}{t - x}\right]$$

and hence

$$(\forall r \in n)\left[\frac{G(x_{r+1}) - G(x_r)}{x_{r+1} - x_r} \lesssim f(\zeta_r) \lesssim \frac{F(x_{r+1}) - F(x_r)}{x_{r+1} - x_r}\right].$$

By taking the sums we obtain

$$G(b) = G(b) - G(a) = \sum_{r\in n} G(x_{r+1}) - G(x_r) \lesssim \sum_{r\in n} (x_{r+1} - x_r\, f(\zeta_r) \lesssim$$

$$\lesssim \sum_{r\in n} F(x_{r+1}) - F(x_r) = F(b) - F(a) = F(b).$$

From this follow (i), (ii) and also (iii).

Remark

It was already mentioned in the introduction, that $\mathbb{D}(a,b) \subseteq \mathbb{P}(a,b)$ is also true. The difficulty of the proof for this depends on the definition of $\mathbb{P}(a,b)$. We might for example modify the definition of $\mathbb{P}(a,b)$ such that we only require that the functions in $T(f)$ and $\perp(f)$ are real valued but not necessarily continuous. We denote the set of all functions integrable in that sense by $\mathbb{P}'(a,b)$. Obviously we have $\mathbb{P}(a,b) \subseteq \mathbb{P}'(a,b)$. Because the proof of 2.5. did not need the fact that $F \in T(f)$, $G \in \perp(f)$ are continuous we have $\mathbb{P}(a,b) \subseteq \mathbb{P}'(a,b) \subseteq \mathbb{D}(a,b)$. The proof for $\mathbb{D}(a,b) \subseteq \mathbb{P}'(a,b)$ is as easy as the proof of 2.5. and it can be found in [Ku 1]. Then it remains to show that $\mathbb{P}'(a,b) = \mathbb{P}(a,b)$. This is done in chapters VI, VII, VIII of [Saks]. That $\mathbb{D}(a,b) \subseteq \mathbb{P}(a,b)$ also follows from 7.13. and the theorem of Hake.

§3. Some generalizations

In this chapter we will generalize the inclusion $\mathfrak{L}^1(a,b) \subseteq \mathbb{D}(a,b)$ not referring to $\mathfrak{L}^1(a,b) \subseteq \mathbb{P}(a,b)$.

Throughout this section X will denote a standard compact metric space.

Some notation:

$\mathfrak{m}(X) = \{\Gamma \mid \Gamma \text{ a real Borel measure on } X\}$

$\mathfrak{m}_+(X) = \{\Gamma \in \mathfrak{m}(X) \mid \Gamma \geq 0\}$.

We may identify (by the Riesz representation theorem) $\mathfrak{m}(X)$ with $(\mathfrak{C}(X))'$, the dual space of real valued continuous functions on X.

3.1. Def.

$$\hat{W}(X) = {}^{E}\{(n,(I_r)_{r\in n},(\zeta_r)_{r\in n}) \mid n \in {}^*\mathbb{N},\ \zeta_r \in X,\ (I_r)_{r\in n} \text{ a disjoint family of Borel sets of } X,\ \bigcup_{r\in n} I_r = X,\ (\forall r \in n)(I_r \subseteq {}^{\pi}\mu(\zeta_r))\}.$$

In a similar way as for $W(a,b)$, one shows that $\hat{W}(X) \neq \emptyset$. Now let $f\colon X \to {}^*\mathbb{R}$, $c \in {}^*\mathbb{R}$ and $\Gamma \in \mathfrak{m}(X)$ be standard.

3.2. Def.

(i) f is called $\hat{\mathfrak{W}}$-integrable iff
$(\exists^{st} c \in {}^*\mathbb{R})(\forall (n,I,\zeta) \in \hat{\mathfrak{W}}(X))\ (c \approx \sum_{r\in n} \Gamma(I_r)\cdot f(\zeta_r))$.

In that case we put $⨍ f\ d\Gamma := ⨍ f(x)\ d\Gamma(x) := c$.

(ii) $\hat{\mathfrak{L}}(X,\Gamma) := {}^S\{f : X \to {}^*\mathbb{R} \mid f \text{ is } \hat{\mathfrak{W}}\text{-integrable}\}$.

Remark: Similar integrals were also considered by McShane.

In order to compare the new integral with the Lebesgue integral we consider lower semi-continuous (lsc) functions h: $X \to]-\infty,\infty]$. These are in our terminology described by:

$(\forall^{st} z \in X)(\forall x \in \mu(z))[(h(z) < \infty \Rightarrow h(x) \gtrsim h(z)) \wedge$

$\wedge\ (h(z) = \infty \Rightarrow h(x) \in \mu_{\mathbb{R}}(\infty))]$,

where $\mu_{\mathbb{R}}(\infty) = {}^E\{x \in {}^*\mathbb{R} \mid (\forall^{st} a \in {}^*\mathbb{R})(x > a)\}$.

3.3. Lemma

Suppose f: $X \to {}^*\mathbb{R}$, h: $X \to]-\infty,\infty]$ lsc, $h \geq f$, $z \in X$, $\Gamma \in \mathfrak{M}_+(X)$ are all standard.

If $I \subseteq \mu(z)$ a Borel set s. t. $\Gamma(I) > 0$, then $\frac{1}{\Gamma(I)} \int_I h\ d\Gamma \gtrsim f(z)$.

Proof:

Case 1: $h(z) = \infty$.

Then $(\exists r \in \mu_{\mathbb{R}}(\infty))(\forall x \in I)(h(x) \geq r)$ and $\frac{1}{\Gamma(I)} \int_I h\ d\Gamma \geq r > f(z)$.

Case 2: $h(z) < \infty$.

Then $(\exists \eta > 0)(\eta \approx 0 \wedge (\forall x \in I)\ h(x) \geq h(z) - \eta)$, and

$$\frac{1}{\Gamma(I)} \int_I h\ d\Gamma \geq h(z) - \eta \approx h(z) \geq f(z).$$

Next we consider the upper integral ${}^*\!\int f\ d\Gamma$.

3.4. Lemma

Suppose f: $X \to {}^*\mathbb{R}$ and $\Gamma \in \mathfrak{M}_+(X)$ are standard, $(n,I,\zeta) \in \hat{\mathfrak{W}}(X)$.

Then we have:

If ${}^*\!\int f\ d\Gamma > -\infty$ then ${}^*\!\int f\ d\Gamma \approx \sum_{r\in n} (I_r)\cdot f(\zeta_r)$

if ${}^*\!\int f\ d\Gamma = -\infty$ then $\sum_{r\in n} \Gamma(I_r)\ f(\zeta_r) \in \mu_{\mathbb{R}}(-\infty)$.

Proof:

By definition we have:

$$^*\!\int f\,d\Gamma = \inf\{\int h\,d\Gamma \mid h: X \to]-\infty,\infty],\ h \text{ lsc},\ h \geq f\}\ .$$

Therefore it is sufficient to show:

If $h: X \to]-\infty,\infty]$ is lsc and $h \geq f$ then $\int h\,d\Gamma \gtrsim \sum_{r\in n} \Gamma(I_r)\,f(\zeta_r)$.

By the above lemma and the transfer theorem we have for $r \in n$:

$$\Gamma(I_r) > 0 \Rightarrow \frac{1}{\Gamma(I_r)} \int_{I_r} h\,d\Gamma \gtrsim f(\zeta_r).$$

Because $\sum_{r\in n} |\Gamma(I_r)|$ is finite we have

$$\sum_{r\in n} \Gamma(I_r)\,f(\zeta_r) = \sum_{\substack{r\in n\\ \Gamma(I_r)>0}} \Gamma(I_r)\cdot f(\zeta_r) \lesssim \sum_{\substack{r\in n\\ \Gamma(I_r)>0}} \Gamma(I_r)\cdot\frac{1}{\Gamma(I_r)}\cdot\int_{I_r} h\,d\Gamma$$

$$= \sum_{\substack{r\in n\\ \Gamma(I_r)>0}} \int_{I_r} h\,d\Gamma = \sum_{r\in n} h\,d\Gamma = \int h\,d\Gamma.$$

Remark

An analogous result holds for the lower integral.

If $x,y \in [-\infty,\infty]$ then we write $x \simeq y$ for $x \approx y \vee x,y \in \mu_{\mathbb{R}}(-\infty) \cup \{-\infty\} \vee$ $\vee\ x,y \in \mu_{\mathbb{R}}(\infty) \cup \{\infty\}$. (This is exactly the uniformity on the two point compactification of $^*\mathbb{R}$).

3.5. Lemma

Assume $^*\!\int f\,d\Gamma = {}_*\!\int f\,d\Gamma$ for standard $f: X \to {}^*\mathbb{R}$ and $\Gamma \in \mathbb{M}_+(X)$.
Then $(\forall(n,I,\zeta) \in \hat{\mathfrak{A}}(X))(\sum_{r\in n} \Gamma(I_r)\,f(\zeta_r) \simeq {}^*\!\int f\,d\Gamma)$ holds.

Proof:

If $^*\!\int f\,d\Gamma = -\infty$, then we have by the above lemma $\sum_{r\in n} \Gamma(I_r)\cdot f(\zeta_r) \in \mu_{\mathbb{R}}(-\infty)$; the case $^*\!\int f\,d\Gamma = +\infty$ is treated similarly.

For $^*\!\int f\,d\Gamma \in {}^*\mathbb{R}$ we get

$$^*\!\int f\,d\Gamma \gtrsim \sum_{r\in n} \Gamma(I_r)\cdot f(\zeta_r) = -\sum_{r\in n} \Gamma(I_r)\cdot(-f)(\zeta_r) \gtrsim -{}^*\!\int(-f)\,d\Gamma = {}_*\!\int f\,d\Gamma.$$

3.6. Theorem

If $\Gamma \in \mathfrak{M}_+(X)$ then $\mathcal{L}^1(X,\Gamma) \subseteq \mathfrak{D}(X,\Gamma)$ and $(\forall f \in \mathcal{L}^1(X,\Gamma))(\int f\, d\Gamma = ⨍ f\, d\Gamma)$.

Proof: This follows at once from the above lemma.

3.7. Theorem

If $\Gamma \in \mathfrak{M}_+(X)$ and $f: X \to [0,\infty[$ are both standard and if f is Γ-measurable, then

$(\forall (n,\Gamma,\zeta) \in \hat{\mathfrak{W}}(X))(\sum_{r\in n} \Gamma(I_r)\cdot f(\zeta_r) \simeq {}^*\!\int f\, d\Gamma)$ holds.

Proof:

It is known that for such functions on a compact space ${}^*\!\int f\, d\Gamma = {}_*\!\int f\, d\Gamma$ holds; (cf. e.g. [Flo: Ü8.15]); therefore the theorem follows from lemma 3.5..

Now we consider again integrals on intervals of ${}^*\mathbb{R}$. We denote $\mathfrak{M}([a,b])$ by $\mathfrak{M}(a,b)$ and $\mathfrak{M}_+([a,b])$ by $\mathfrak{M}_+(a,b)$. When we calculated the generalized Riemann integral we considered sums of the form

$\sum_{r\in n} (x_{r+1} - x_r)\cdot f(\zeta_r)$ which can also be written as

$\sum_{r\in n} \lambda^1([x_r,x_{r+1}])\cdot f(\zeta_r)$.

Calculating $\int f\, d\Gamma$ for any $\Gamma \in \mathfrak{M}(a,b)$ by the sums $\sum_{r\in n} \Gamma([x_r,x_{r+1}])\cdot f(\zeta_r)$

would yield an incorrect result (take e.g. $\Gamma = \varepsilon_c$, the Dirac-measure for some $c \in]a,b[$).

To enable us to neglect endpoints of intervals we consider only diffus measures:

$\mathfrak{M}_o(a,b) = \{\Gamma \in \mathfrak{M}(a,b) \mid \forall x \in [a,b] : \Gamma(\{x\}) = 0\}$

$\mathfrak{M}_o^+(a,b) = \mathfrak{M}_o(a,b) \cap \mathfrak{M}_+(a,b)$;

for $\Gamma \in \mathfrak{M}_o(a,b)$ we put $\Gamma(x,y) := \Gamma([x,y]) = \Gamma(]x,y[)$.

3.8. Def.

(i) For $\Gamma \in \mathfrak{M}_o(a,b)$, $f: [a,b] \to {}^*\mathbb{R}$, both standard, f is called W-Γ-integrable iff:

$(\exists^{st} c \in {}^*\mathbb{R})(\forall (n,x,\zeta) \in W(a,b))(c \approx \sum_{r\in n} \Gamma(x_r,x_{r+1})\cdot f(\zeta_r))$.

In this case we put $\int_a^b f\, d\Gamma := \int_a^b f(x)\, d\Gamma(x) := c$.

(ii) $\mathbb{D}(a,b;\Gamma) := {}^S\{f: [a,b] \to {}^*\mathbb{R} \mid f \text{ is W-}\Gamma\text{-integrable}\}$

3.9. Lemma

If $\Gamma \in \mathbb{M}_o^+(a,b)$ then $\mathcal{L}^1([a,b],\Gamma) \subseteq \hat{\mathbb{D}}([a,b];\Gamma) \subseteq \mathbb{D}(a,b;\Gamma)$ and in particular:
$\mathcal{L}^1(a,b) \subseteq \hat{\mathbb{D}}([a,b];\lambda^1) \subseteq \mathbb{D}(a,b;\lambda^1) = \mathbb{D}(a,b)$.

Proof:

We only have to prove that $\hat{\mathbb{D}}([a,b],\Gamma) \subseteq \mathbb{D}(a,b;\Gamma)$ holds for $(n,x,\zeta) \in W(a,b)$.
Taking

$$I_r := \begin{cases} [x_r,x_{r+1}[& \text{if } r+1 < n \\ [x_r,x_{r+1}] & \text{if } r+1 = n \end{cases}$$

we have $(n,(I_r)_{r\in n},\zeta) \in \hat{\mathbb{N}}([a,b])$ and
$\sum_{r\in n} (x_r,x_{r+1})\ f(\zeta_r) = \sum_{r\in n} \Gamma(I_r)\ f(\zeta_r) \approx \mathop{{\int\!\!\!\!\!\!-}} f\ d\Gamma$.

Remark

If $\Gamma \in \mathbb{M}_o(a,b)$ then there is a continuous function $\alpha: [a,b] \to {}^*\mathbb{R}$ s.t. α is of bounded variation and
$(\forall u,v \in [a,b])(u \leq v \to \Gamma(u,v) = \alpha(v) - \alpha(u))$ is true.
Therefore the W-Γ-integral is exactly the Riemann-Stieltjes integral (g) $\int f\ d\Gamma$ which is also described in [McL]; here we only consider the case that α is of bounded variation and continuous.

We close the section with some additivity and continuity properties of the W-Γ-integral.

3.10. Proposition

Let $a < c < b$, $\Gamma \in \mathbb{M}_o(a,b)$, $f \in \mathbb{D}(a,b;\Gamma)$.
Then $\int_a^b f\ d\Gamma = \int_a^c f\ d\Gamma + \int_c^b f\ d\Gamma$ and in particular the right side integrals exist.

Proof:

First we show $f \in \mathbb{D}(a,c)$. We take $(n,x,\zeta) \in W(c,b)$ and assume $(m,y,\eta),(l,z,\xi) \in W(a,c)$.
Then we get

$$\int_a^b f\,d\Gamma \approx \sum_{\mu\in m} \Gamma(y_\mu,y_{\mu+1})\cdot f(\eta_\mu) + \sum_{r\in n} \Gamma(x_r,x_{r+1})\cdot f(\zeta_r)$$

$$\int_a^b f\,d\Gamma \approx \sum_{\lambda\in l} \Gamma(z_\lambda,z_{\lambda+1})\, f(\xi_\lambda) + \sum_{r\in n} \Gamma(x_r,x_{r+1})\cdot f(\zeta_r)$$

and $\sum_{\lambda\in l} \Gamma(z_\lambda,z_{\lambda+1})\, f(\xi_\lambda) \approx \sum_{\mu\in m} \Gamma(y_\mu,y_{\mu+1})\cdot f(\eta_\mu)$;

hence ${}^E\{\sum_{\mu\in m} \Gamma(y_\mu,y_{\mu+1})\cdot f(\eta_\mu) \mid (m,y,\mu) \in W(a,c)\}$ is a Cauchy monad, i.e.

$f \in \mathbb{D}(a,c)$ and also $f \in \mathbb{D}(c,b)$.

Taking $(n,x,\zeta) \in W(a,b)$ and $m \in n$ s.t. $x_m = c$ we see

$$\int_a^c f\,d\Gamma + \int_c^b f\,d\Gamma \approx \sum_{r=0}^{m-1} \Gamma(x_r,x_{r+1})\cdot f(\zeta_r) + \sum_{r=m}^{n-1} \Gamma(x_r,x_{r+1})\cdot f(\zeta_r)$$

$$= \sum_{r\in n} \Gamma(x_r,x_{r+1})\cdot f(\zeta_r) \approx \int_a^b f\,d\Gamma, \text{ i.e. } \int_a^c f\,d\Gamma + \int_c^b f\,d\Gamma = \int_a^b f\,d\Gamma.$$

3.11. Corollary

$a \le u < v \le b$, $\Gamma \in \mathbb{M}_o(a,b)$, $f \in \mathbb{D}(a,b;\Gamma)$ implies $f \in \mathbb{D}(u,v;\Gamma)$.

3.12. Lemma

Assume $\Gamma \in \mathbb{M}_o(a,b)$ and $f \in \mathbb{D}(a,b;\Gamma)$ are standard. Then for any $(n,x,\zeta) \in W(a,b)$ and any internal $N \subseteq n = \{r \mid r < n\}$:

$$\sum_{r\in N} \Gamma(x_r,x_{r+1})\cdot f(\zeta_r) \approx \sum_{r\in N} \int_{x_r}^{x_{r+1}} f\,d\Gamma.$$

Proof:

If $(n,x,\zeta) \in W(a,b)$, $Z \subseteq n$ we take some $(m,y,\zeta) \in {}^{\pi\pi}W(a,b)[(n,x,\zeta)][Z]$. We know that $(\forall^{st} u \in [a,b])(\exists r \in n)(u = x_r)$ because $u \in [x_r,x_{r+1}] \subseteq {}^{\pi}\mu(\zeta_r)$, and therefore $u = \zeta_r$.
By the transfer theorem 1.3. it follows that
$(\forall r \in (n+1))(\exists \mu \in (m+1))(x_r = y_\mu)$ holds;
then we find an internal function $\tilde{m} : \{0,\ldots,n\} \to \{0,\ldots,m\}$ s.t.
$(\forall r \in (n+1))(x_r = y_{\tilde{m}(r)})$.
Now we define $(l,z,\xi) \in W(a,b)$ by omitting from (m,y,η) those y_j for which $y_{\tilde{m}(r)} < y_j < y_{\tilde{m}(r+1)} \wedge x_r \notin Z$ holds. Now we have:

$$\sum_{r\in n} \Gamma(x_r,x_{r+1})\cdot f(\zeta_r) \approx \sum_{\lambda\in l} \Gamma(z_\lambda,z_{\lambda+1})\cdot f(\xi_\lambda)$$

$$= \sum_{r\in n\setminus Z} \Gamma(x_r,x_{r+1})\cdot f(\zeta_r) + \sum_{r\in Z} \sum_{\mu=\tilde{m}(r)}^{\tilde{m}(r+1)-1} \Gamma(y_\mu,y_{\mu+1})\cdot f(\eta_\mu)$$

$$\approx \sum_{r\in n\setminus Z} \Gamma(x_r,x_{r+1})\cdot f(\zeta_r) + \sum_{r\in Z} \int_{x_r}^{x_{r+1}} f\, d\Gamma$$

$$\Rightarrow \sum_{r\in Z} \Gamma(x_r,x_{r+1})\cdot f(\zeta_r) \approx \sum_{r\in Z} \int_{x_r}^{x_{r+1}} f\, d\Gamma.$$

3.13. Corollary (Henstock's lemma)

If $\Gamma \in \mathbb{M}_o(a,b)$ and $f \in \mathbb{D}(a,b;\Gamma)$ are standard, then

$$(\forall(n,x,\zeta) \in W(a,b))(\sum_{r\in n} |\Gamma(x_r,x_{r+1})\, f(\zeta_r) - \int_{x_r}^{x_{r+1}} f\, d\Gamma| \approx 0).$$

Proof:

Taking $Z := \{r \in n \mid \Gamma(x_r,x_{r+1})\cdot f(\zeta_r) \geq \int_{x_r}^{x_{r+1}} f\, d\Gamma\}$ we get

$$\sum_{r\in n} |\Gamma(x_r,x_{r+1})\cdot f(\zeta_r) - \int_{x_r}^{x_{r+1}} f\, d\Gamma|$$

$$= \sum_{r\in Z} \Gamma(x_r,x_{r+1})\, f(\zeta_r) - \int_{x_r}^{x_{r+1}} f\, d\Gamma - \sum_{r\in n\setminus Z} \Gamma(x_r,x_{r+1})\cdot f(\zeta_r) - \int_{x_r}^{x_{r+1}} f\, d\Gamma$$

$\approx 0.$

Remark

3.13. is weaker than Henstock's lemma shich says:
If $(\forall(n,x,\zeta) \in W_{\tilde{\delta}}(a,b))(|\sum_{r\in n} \Gamma(x_r,x_{r+1})\cdot f(\zeta_r) - \int_0^r f\, d\Gamma| < \varepsilon)$ then
$(\forall(n,x,\zeta) \in W_{\tilde{\delta}}(a,b))(\sum_{r\in n}|\Gamma(x_r,x_{r+1})\cdot f(\zeta_r) - \int_{x_r}^{x_{r+1}} f\, d\Gamma| < 2\varepsilon)$; but
3.13. will be sufficient for our applications.

3.14. Theorem

Suppose $\Gamma \in \mathbb{M}_o(a,b)$ and $\varphi \in \mathbb{D}(a,b;\Gamma)$. Then Φ: $[a,b] \to {}^*\mathbb{R}$ defined by
$\Phi(x) := \int_a^x \varphi\, d\Gamma$ is continuous.

Proof:

We put $\Gamma(x,y) := -\Gamma(y,x)$ if $x > y$ and $\int_x^y \varphi\, d\Gamma = -\int_y^x \varphi\, d\Gamma$.

If $u_o \in [a,b]$ is standard and $u \in [a,b] \cap \mu(u_o)$, then we find
$(n,x,\zeta) \in W(a,b)$ and $r \in n$ s.t. $\{x_r,x_{r+1}\} = \{u,u_o\}$ and $u_o = \zeta_r$.
Then $\Phi(u_o) - \Phi(u) = \int_u^{u_o} \varphi\, d\Gamma \approx \Gamma(u,u_o) - \varphi(u_o)$ (by 3.13. with $Z = \{r\}$).

We have $0 \leq |\Gamma(u,u_o)| \lesssim \lim_{m\to\infty} \Gamma([u_o - \frac{1}{m}, u_o + \frac{1}{m}] \cap [a,b]) = 0$, and hence $\Phi(u_o) \approx \Phi(u)$.

§4. Some Applications

For a mapping $\varphi: X \to Y$ into some Hausdorff space Y and a filter $\mathfrak{D}$ on X the limit b of φ along $\mathfrak{D}$ is denoted by $\lim_{x\in\mu(\mathfrak{D})} \varphi(x)$ iff $\varphi(x) \in \mu_Y(b)$ for all $x \in \mu(\mathfrak{D})$.
Taking the filter $\mathfrak{F}$ the monad of which are the nonzero infinitesimals of ${}^*\mathbb{R}$, i.e. $\mu(\mathfrak{F}) = \dot{\mu}_{\mathbb{R}}(0) = \mu_{\mathbb{R}}(0)\setminus\{0\}$ the monad of the $\mathfrak{G} = \mathfrak{F} \times \mathfrak{F}$ can be described by $\mu(\mathfrak{G}) = \{(h,k) \in {}^*\mathbb{R}^2 \mid h \in \dot{\mu}_{\mathbb{R}}(0),\ k \in {}^{\pi}\dot{\mu}_{\mathbb{R}}(0)[h]\}$.

For $f: {}^*\mathbb{R} \to {}^*\mathbb{R}$ and $h \neq 0$, $k \neq 0$ we put

$$f^{[1]}(h)(x) := \frac{f(x+h) - f(x)}{h},\quad f^{[2]}(h,k)(x) := \left(f^{[1]}(h)\right)^{[1]}(k)(x)$$ and

observe $f^{[2]}(h,k)(x) = f^{[2]}(k,h)(x)$.

4.1. Lemma

If $f: {}^*\mathbb{R} \to {}^*\mathbb{R}$ is twice differentiable then

$$\lim_{(h,k)\in\mu(\mathfrak{G})} f^{[2]}(h,k)(x) = f''(x) \text{ for } x \in {}^*\mathbb{R}.$$

Proof:

If $h \in \mu_{\mathbb{R}}(0)$ and $k \in {}^{\pi}\dot{\mu}_{\mathbb{R}}(0)$ then for all standard $x \in {}^*\mathbb{R}$

$$f^{[2]}(h,k)(x) = \frac{f^{[1]}(k)(x+h) - f^{[1]}(k)(x)}{h} \ {}^{\pi}{\approx}_{[h]}\ \frac{f'(x+h) - f'(x)}{h} \approx f''(x).$$

The existence of $\lim_{(h,k)\in\mu(\mathfrak{G})} f^{[2]}(h,k)(x)$ does not imply the continuity of f because $f^{[2]}(h,k)(x) = 0$ for each $\mathbb{Q}$-linear f. We observe also that f is $\mathbb{Q}$-linear iff $f(x) - f(y) = f(x-y)$ for $x,y \in {}^*\mathbb{R}$.

4.2. Theorem

Suppose $\lim_{(h,k)\in\mu(\mathfrak{G})} f^{[2]}(h,k)(x)$ exists for $x \in {}^*\mathbb{R}$.

Then $f = f_o + l$, where l is $\mathbb{Q}$-linear, f_o is twice differentiable and $f_o''(x) = \lim_{(h,k)\in\mu(\mathfrak{G})} f^{[2]}(h,k)(x)$.

Proof:

W.l.o.g. we may assume $f(0) = 0$ and put
$f_2(x) := \lim_{(h,k)\in\mu(\mathbb{G})} f^{[2]}(h,k)(x)$ for $x \in {}^*\mathbb{R}$ and get

$$(*)\ (\forall^{st} x \in {}^*\mathbb{R})(\forall h \in \dot{\mu}_{\mathbb{R}}(0))(\forall k \in {}^{\pi}\dot{\mu}_{\mathbb{R}}(0)[h])(f_2(x) \approx f^{[2]}(h,k)(x)).$$

Now we take $b > 0$ and standard, $(n,x,\zeta) \in \mathbb{D}(0,b)$ and
$k \in {}^{\pi}\dot{\mu}_{\mathbb{R}}(0)[(n,x,\zeta_r)]$.
For $r \in n$ we put

$$h_r := \begin{cases} x_{r+1} - x_r & \text{if } \zeta_r = x_r \\ x_r - x_{r+1} & \text{if } \zeta_r = x_{r+1} \end{cases} .$$

Then we have $h_r \in {}^{\pi}\dot{\mu}_{\mathbb{R}}(0)[\zeta_r]$ and $k \in {}^{\pi\pi}\dot{\mu}_{\mathbb{R}}(0)[h_r][\zeta_r]$ for $r \in n$ and hence $(h_r,k) \in {}^{\pi}\mu(\mathbb{G})[\zeta_r]$.
By (*) and the transfer theorem 1.3. we obtain:
$f_2(\zeta_r) \approx f^{[2]}(h_r,k)(\zeta_r)$ for $r \in n$.
Taking sums we get

$$\sum_{r\in n} (x_{r+1} - x_r)\cdot f_2(\zeta_r) \approx \sum_{r\in n} (x_{r+1} - x_r)\cdot f^{[2]}(h_r,k)(\zeta_r)$$

$$= \sum_{r\in n} f^{[1]}(k)(x_{r+1}) - f^{[1]}(k)(x_r) = f^{[1]}(k)(b) - f^{[1]}(k)(0).$$

Now we put $M := {}^{E}\{f^{[1]}(k)(b) - f^{[1]}(k)(0) \mid k \in \dot{\mu}_{\mathbb{R}}(0)\}$ and want to show that M is a Cauchy monad.
Therefore we assume $u_1,u_2 \in M$ and take $(k_1,k_2) \in \dot{\mu}_{\mathbb{R}}(0) \times \dot{\mu}_{\mathbb{R}}(0)$ s.t.
$u_i = f^{[1]}(k_i)(b) - f^{[1]}(k_i)(0)$, $i = 1,2$.
By Robinson's lemma (see [Ben-Ri]) there is $(n,x,\zeta) \in W(a,b)$ s.t.
$(k_1,k_2) \in {}^{\pi}(\dot{\mu}_{\mathbb{R}}(0) \times \dot{\mu}_{\mathbb{R}}(0))[(n,x,\zeta)]$.
Hence we have $k_1,k_2 \in {}^{\pi}\dot{\mu}_{\mathbb{R}}(0)[(n,x,\zeta)]$ and get by the calculation from above

$$f^{[1]}(k_1)(h) - f^{[1]}(k_1)(0) \approx \sum_{r\in n} (x_{r+1} - x_r)\cdot f_2(\zeta_r)$$

$$\approx f^{[1]}(k_2)(b) - f^{[1]}(k_2)(0).$$

Hence M is a Cauchy monad and therefore $(\exists^{st} d \in {}^*\mathbb{R})(\forall u \in M)(u \approx d)$.
We also have for each $(n,x,\zeta) \in W(0,b)$
$\sum_{r\in n} (x_{r+1} - x_r)\cdot f_2(\zeta_r) \approx d$ and $f_2 \in \mathbb{D}(0,b)$ follows.
A similar argument for $b < 0$ yields $f_2 \in \mathbb{D}(a,b)$ whenever $a < b$.
Furthermore $(\forall^{st} x \in {}^*\mathbb{R})(\forall k \in \dot{\mu}_{\mathbb{R}}(0))(\int_0^x f_2(t)\,dt \approx f^{[1]}(k)(x) - f^{[1]}(k)(0))$ holds.

Now we put $f_1(x) := \int_0^x f_2(t)\,dt$ and $f_o(x) := \int_0^x f_1(t)\,dt$.
By theorem 3.14. f_1 is continuous and therefore $f_o' = f_1$, $f_o(0) = 0$.
Defining $l(x) := f(x) - f_o(x)$ gives $f = f_o + l$, $f_o(0) = l(0) = f(0) = 0$.
For standard x and $h \in \dot{\mu}_{\mathbb{R}}(0)$ we obtain

$$f^{[1]}(h)(x) = f_o^{[1]}(h)(x) + l^{[1]}(h)(x) \approx f_1(x) + l^{[1]}(h)(x)$$

$\approx f^{[1]}(h)(x) - f^{[1]}(h)(0) + l^{[1]}(h)(x)$ and hence

$l^{[1]}(h)(x) \approx f^{[1]}(h)(0)$ and $(\forall^{st} x \in {}^*\mathbb{R})(\forall h \in \dot{\mu}_{\mathbb{R}}(0))(l^{[1]}(h)(x) \approx l^{[1]}(h)(0))$.
For each $a \in {}^*\mathbb{R}$ we define $g_a(x) := l(x+a) - l(x-a)$.
For standard a and for $h \in \dot{\mu}_{\mathbb{R}}(0)$ we see

$$g_a^{[1]}(h)(x) = \frac{[l(x+h+a) - l(x+h-a)] - [l(x+a) - l(x-a)]}{h}$$

$= l^{[1]}(h)(a+x) - l^{[1]}(h)(x-a) \approx 0;$

which shows that g_a is differentiable and $g_a'(x) = 0$.
Therefore $\gamma(a) := l(x+a) - l(x-a)$ is well-defined. In particular for $a := \frac{u-v}{2}$ and $x := \frac{u+v}{2}$ we get $l(u) - l(v) = \gamma(\frac{u-v}{2})$; the choice of $v = 0$ shows $\gamma(\frac{u}{2}) = l(u)$ which implies $l(u) - l(v) = l(u-v)$.

Therefore l is Q-linear. In order to show $f_o'' = f_2$ it remains to prove $f_1' = f_1$. For standard x and $h \in \dot{\mu}_{\mathbb{R}}(0)$ we take some $k \in {}^{\pi}\mu_{\mathbb{R}}(0)[h]$ and obtain the assertion from

$$\frac{f_1(x+h) - f_1(x)}{h} \approx \frac{f_o^{[1]}(k)(x+h) - f_o^{[1]}(k)(x)}{h} = f_o^{[2]}(h,k)(x)$$

$= f^{[2]}(h,k)(x) - l^{[2]}(h,k)(x) = f^{[2]}(h,k)(x) \approx f_2(x).$

Next we consider the theorem on bounded convergence. The theorem says that if a sequence $(f_n)_{n\in{}^*\mathbb{N}}$ of Lebesgue-integrable functions is bounded by some integrable function g and converges a.e. to f then f is integrable and $\int f\,d\lambda^1 = \lim_{n\to\infty} \int f_n\,d\lambda^1$. Here the convergence is taken along the Frechet filter on ${}^*\mathbb{N}$; the theorem remains true if one takes convergence along a filter with a countable base. Without any restriction on the filter the theorem is false, however. In terms of monads $\mu(\mathcal{F})$ rather than of filters $\mathcal{F}$ the usual counterexample is:

$\mathfrak{C} := \{f\colon [0,1] \to \{0,1\} \mid f^{-1}(\{0\}) \text{ finite}\} \subseteq \mathcal{L}^1(0,1)$,
$\mathfrak{m} := {}^E\{f \in \mathfrak{C} \mid (\forall^{st} x \in [0,1])\ f(x) = 0\} = \mu(\mathcal{F})$.

Then $\lim_{f\in\mu(\mathcal{F})} f(x) = 0$ for $x \in [0,1]$ and $\lim_{f\in\mu(\mathcal{F})} \int_0^1 f\,d\lambda^1 = 1$.

A superficial view might suggest that the uncountability of $\mathfrak{C}$ is responsible for this effect. For a sequence $(f_n)_{n\in {}^*\mathbb{N}}$ of functions in $\mathcal{L}^1(0,1)$, $f: [0,1] \to {}^*\mathbb{R}$, $Y \subseteq [0,1]$ s. t. $\lambda^1(Y) = 0$ and a filter $\mathfrak{U}$ on ${}^*\mathbb{N}$ we therefore consider the hypothesis

(H) (i) $|f_n(x)| \leq 1$ for $x \in [0,1]$ and $n \in {}^*\mathbb{N}$;

(ii) $\lim_{n\in\mu(\mathfrak{U})} f_n(x) = f(x)$ for $x \in [0,1]\setminus Y$.

We will show that (H) does not imply $f \in \mathcal{L}^1(0,1)$ and $\lim_{n\in\mu(\mathfrak{U})} ||f_n - f||_1 = 0$, i.e. the theorem of dominated convergence relies more on the countability of the filter base than on the cardinality of the underlying set.

4.3. Theorem

There is some sequence $(f_n)_{n\in {}^*\mathbb{N}}$ of functions from $\mathcal{L}^1(0,1)$ s. t. for all free ultrafilters $\mathfrak{U}$ there is some function $f_{\mathfrak{U}} \notin \mathfrak{M}(0,1)$ satisfying (H) with $Y = \emptyset$; in particular f is not measurable. The construction will in addition guarantee ran $f_{\mathfrak{U}} \subseteq \{0,\frac{1}{2},1\}$ and $\mathfrak{U} \neq \mathfrak{V} \to f_{\mathfrak{U}} \neq f_{\mathfrak{V}}$.

Proof: We restrict ourselves to the standard case and construct first the sequence $(f_n)_{n\in {}^*\mathbb{N}}$. We put $X := \{x \in [0,1] \mid x$ has only one binary representation$\}$;

then $Y := [0,1]\setminus X = \{\frac{m}{2^n} \mid m,n \in {}^*\mathbb{N},\ 0 \leq m \leq 2^n\}$ is countable and hence $\lambda^1(Y) = 0$.

For $x \in X$ we define $\tilde{x}$ by $\sum_{n=1}^{\infty} \tilde{x}(n)\cdot 2^{-n} = x$, $\tilde{x}(n) \in \{0,1\}$.

Now we define $f_n(x)$ for $x \in [0,1]$ by

$$f_n(x) := \begin{cases} \tilde{x}(n) & \text{if } x \in X \\ \frac{1}{2} & \text{if } x \in Y. \end{cases}$$

Next we put $f_{\mathfrak{U}}(x) := \lim_{n\in\mu(\mathfrak{U})} f_n(x)$ for each (standard) free ultrafilter $\mathfrak{U}$ on ${}^*\mathbb{N}$.

This limit exists because $f_n(x) = f_m(x)$ for all $n,m \in \mu(\mathfrak{U})$.

Let $\mathfrak{U},\mathfrak{V}$ be in $\beta\mathbb{N}\setminus\mathbb{N}$, $A \in \mathfrak{U}$, $f_{\mathfrak{U}} = f_{\mathfrak{V}}$. Now suppose $f_{\mathfrak{U}} = f_{\mathfrak{V}}$; take $A \in \mathfrak{U}$ standard and put $x := \sum_{n\in A} 2^{-n}$.

Because $\mu(\mathfrak{U}) \subseteq A$ we have $1 = f_{\mathfrak{U}}(x) = f_{\mathfrak{V}}(x)$; hence $f_n(x) = 1$ for $n \in \mu(\mathfrak{U})$ and therefore $\mu(\mathfrak{V}) \subseteq A$ and $A \in \mathfrak{V}$, i.e. $\mathfrak{U} \subseteq \mathfrak{V}$ and also $\mathfrak{U} = \mathfrak{V}$.

Now we proceed indirectly and assume $f_{\mathfrak{U}} \in \mathfrak{M}(a,b)$. For $(n,x,\zeta) \in W(0,1)$ we define $(n,y,\eta) \in W(0,1)$ by

$y_r := 1 - x_{n-r}$ and $\eta_r := 1 - \zeta_{n-1-r}$.
From $f_{\mathfrak{U}}(1-x) = 1 - f_{\mathfrak{U}}(x)$ we get $\sum_{r \in n} (x_{r+1} - x_r) \cdot f_{\mathfrak{U}}(1-\zeta_r)$

$= \sum_{r \in n} (y_{r+1} - y_r) \cdot f_{\mathfrak{U}}(\eta_r) \approx \int_0^1 f_{\mathfrak{U}}\, d\lambda^1$, and
$\int_0^1 f_{\mathfrak{U}}(x)\, dx = 1 - \int_0^1 f_{\mathfrak{U}}(x)\, dx$; therfore $\int_0^1 f_{\mathfrak{U}}(x)\, dx = \frac{1}{2}$.

On the other hand $f_{\mathfrak{U}}(x + r2^{-n}) = f_{\mathfrak{U}}(x)$ for $x \in [0, 2^{-n}]$, $(r+1) \cdot 2^{-n} \leq 1$ implies

$$\frac{1}{2} = \int_0^1 f_{\mathfrak{U}}(x)\, dx = \sum_{r=0}^{2^n - 1} \int_{r2^{-n}}^{(r+1)\cdot 2^{-n}} f_{\mathfrak{U}}(x)\, dx = 2^n \cdot \int_0^{2^{-n}} f_{\mathfrak{U}}(x)\, dx.$$

Therefore

$$\int_0^{2^{-n}} f_{\mathfrak{U}}(x)\, dx = \frac{1}{2^{n+1}} \quad \text{and } (\forall a,b \in Y) \int_a^b f_{\mathfrak{U}}(x)\, dx = \frac{b-a}{2}.$$

Because $\overline{Y} = [0,1]$ we get from 3.14. $\int_a^b f_{\mathfrak{U}}(x)\, dx = \frac{b-a}{2}$ for all $a,b \in [0,1]$. Finally we define h by

$$h(x) := \begin{cases} \frac{1}{2} & \text{if } x \in Y \\ 0 & \text{if } x \in X \end{cases} \qquad \text{and put } g_{\mathfrak{U}} = f_{\mathfrak{U}} - h.$$

Then $h \in \mathfrak{L}^1(0,1) \subseteq \mathfrak{D}(0,1)$ and $h(x) = 0$ a. e. holds; therefore
$\int_0^1 h\, d\lambda^1 = 0$ and $\int_a^b g_{\mathfrak{U}}(x)\, dx = \int_a^b f_{\mathfrak{U}}(x)\, dx = \frac{b-a}{2}$ for $a,b \in [0,1]$.

For $(n,x,\zeta) \in W(0,1)$ the following calculation leads to a contradiction:

$$\frac{1}{2} = \int_0^1 g_{\mathfrak{U}}(x)\, dx \approx \sum_{\substack{r \in n \\ g_{\mathfrak{U}}(\zeta_r)=1}} (x_{r+1} - x_r)\, g_{\mathfrak{U}}(\zeta_r) \approx \sum_{\substack{r \in n \\ g_{\mathfrak{U}}(\zeta_r)=1}} \cdot \int_{x_r}^{x_{r+1}} g_{\mathfrak{U}}(x)\, dx$$

$$\approx \frac{1}{2} \cdot \int_0^1 f_{\mathfrak{U}}(x)\, dx = \frac{1}{4}.$$ This shows $f_{\mathfrak{U}} \notin \mathfrak{D}(0,1)$.

Remark

This theorem also shows that not every bounded function has a generalized Riemann integral.

§5 Measurability and Differentiation

In this section we will show that indefinite generalized Riemann integrals have a.e. a derivative.

We put $\mu_+ := \mu_{\mathbb{R}}(0) \cap]0,\infty[$.
If $\tilde{\varepsilon} = [a,b] \longrightarrow]0,\infty[$ s.t. $(\forall x \in [a,b])(\tilde{\varepsilon}(x) \in {}^{\pi}\mu_+[x])$ then
$W_{\tilde{\varepsilon}}(a,b) \subseteq W(a,b)$.
For $A \subseteq [a,b]$ we denote its characteristic function by 1_A.

5.1. Theorem

Suppose $\Gamma \in \mathfrak{M}_0^+(a,b)$, $A \subseteq [a,b]$ are standard, $1_A \in \mathfrak{D}(a,b;\Gamma)$ and
$\int_a^b 1_A \, d\Gamma = 0$.
Then A is Lebesgue integrable and $\Gamma(A) = 0$.

Proof:
It suffices to find an open set $U \supseteq A$ and $\Gamma(U) \approx 0$. We start with some $\tilde{\varepsilon}: [a,b] \longrightarrow]0,\infty[$ s.t. $\tilde{\varepsilon}(x) \in {}^{\pi}\mu_+[x]$ for $x \in [a,b]$.
We put $E_n := \tilde{\varepsilon}^{-1}(]\frac{1}{n+1},\frac{1}{n}[)$. W.l.o.g. we may assume that none of the endpoints of $[a,b]$ belongs to A, i.e. $A \subseteq]a,b[$.
Now we will define inductively a function $\tilde{m}(n)$, a sequence $((L_{n,r})_{r=1}^{\tilde{m}(n)})_{n\in {}^*\mathbb{N}}$ with $L_{n,r}$ a closed subinterval of $[a,b]$ and a sequence $((1^n,z^n,\zeta^n))_{n\in {}^*\mathbb{N}}$ of partitions from $W_{\tilde{\varepsilon}}(a,b)$. From this we define

$$L_n = \cup(L_{n,r} \mid 1 \le r \le \tilde{m}(n)),$$

$$V_l = L_{n,r} \text{ for } l = \sum_{i=1}^{l-1} \tilde{m}(i) + r \text{ and } 1 \le r \le \tilde{m}(n), \text{ and}$$

$K_n = \cup(L_n \mid 1 \le r \le n)$. With E_n and A from above the following will be true:

1) $E_n \cap A \subseteq K_n$, 2) $\partial K_n \subseteq \{z_\lambda^n \mid 0 \le \lambda \le 1^n\}$, 3) if $[z_\lambda^n, z_{\lambda+1}^n] \subseteq K_n$ then $\zeta_\lambda^n \in A$.

n = 1:
We put $\tilde{m}(1) = 0$ and choose an arbitrary $(1^1,z^1,\zeta^1) \in W_{\tilde{\varepsilon}}(a,b)$.
Because $E_1 = \emptyset$ the conditions 1) - 3) are trivially true.
Induction step:
We assume the conditions for n-1 and distinguish two cases.
Case 1: $(E_n \cap A) \setminus K_{n-1} = \emptyset$.

We put $\tilde{m}(x) = 0$, $(1^n,z^n,\zeta^n) = (1^{n-1},z^{n-1},\zeta^{n-1})$. Again the conditions are trivially true.

Case 2: $(E_n \cap A)\setminus K_{n-1} \neq \emptyset$.

$K_0' := K_{n-1}$.

Using the notation $\overline{K}(x,r) := [a,b] \cap [x-r,x+r]$ we choose $x_1 \in (E_n \cap A)\setminus K_0'$ and put

$L_{n,1} := \overline{K}(x_1,\tilde{\varepsilon}(x_1))$, $K_1' := K_0' \cup L_{n-1}$; if $(A \cap E_n)\setminus K_1' \neq \emptyset$ we choose $x_2 \in (A \cap E_n)\setminus K_1'$ and take $L_{n,2} := \overline{K}(x_2,\tilde{\varepsilon}(x_2))$, $K_2' := K_1' \cup L_{n,2}$.

For some m we get $(A \cap E_n)\setminus K_m' = \emptyset$ because $|x_j - x_k| \geq \frac{1}{n+1}$ for $j \neq k$;

We put $\tilde{m}(n) := m$.

Let $L_{n,\mu} = [a_\mu,b_\mu]$ for $1 \leq \mu \leq \tilde{m}(n)$; we put $L_{n,\mu}' := \overline{L_{n,\mu}\setminus K_{\mu-1}'}$ and write $L_{n,\mu}' = [a_\mu',b_\mu']$ with $a_\mu \leq a_\mu' < x_\mu < b_\mu' \leq b_\mu$ (always we have $a_\mu < x_\mu < b_\mu$ because $x_\mu \in A$ and hence $a < x_\mu < b$).

Now we construct $(1^n,z^n,\zeta^n)$. If $[z_k^{n-1},z_{k+1}^{n-1}] \subseteq K_{n-1}$, we have for some $\lambda \in 1^n$ s.t. $z_k^{n-1} = z_\lambda^n$, $z_{k+1}^{n-1} = z_{\lambda+1}^n$, $\zeta_k^{n-1} = \zeta_\lambda^n$ (i.e. $(1^n,z^n,\zeta^n)$ and $(1^{n-1},z^{n-1},\zeta^{n-1})$ are identical on K_{n-1}).

If $1 \leq \mu \leq \tilde{m}(n)$ we therefore have some $\lambda \in 1^n$ satisfying

$a_\mu' = z_{\lambda-1}^n < x_\mu = z_\lambda^n = \zeta_{\lambda-1}^n = \zeta_\lambda^n < b_\mu' = z_{\lambda+1}^n$. Having defined $(1^n,z^n,\zeta^n)$ on

$K_n = K_{n-1} \cup \bigcup_{\mu=1}^{\tilde{m}(n)} L_{n,\mu} = K_{n-1} \cup \bigcup_{\mu=1}^{\tilde{m}(n)} L_{n,\mu}'$ we conclude the definition by chosing an arbitrary extension on $[a,b]\setminus K_n$.

Our conditions are satisfied:

1) $\emptyset = (A \cap E_n)\setminus K_{\tilde{m}(n)}' = (A \cap E_n)\setminus K_n \Rightarrow A \cap E_n \subseteq K_n$.

2) From the construction it is obvious that

$$\partial K_n \subseteq \partial K_{n-1} \cup \{a_\mu' \mid 1 \leq \mu \leq \tilde{m}(n)\} \cup \{b_\mu' \mid 1 \leq \mu \leq \tilde{m}(n)\}$$
$$\subseteq \{z_\lambda^n \mid 0 \leq \lambda \leq 1^n\}.$$

3) Suppose that $[z_\lambda^n,z_{\lambda+1}^n] \subseteq K_n$ then

(i) $[z_\lambda^n,z_{\lambda+1}^n] \subseteq K_{n-1}$ or

(ii) $[z_\lambda^n,z_{\lambda+1}^n] \subseteq [a',b']$.

In case (i) $\zeta_\lambda^n \in A$ follows, because 3) holds for n - 1, in case (ii) we have $\zeta_\lambda^n = x_\mu \in A$.

Using the above data we obtain $0 = \int_a^b 1_A \, d\Gamma \approx \sum_{\lambda \in 1^n} \Gamma(z_\lambda^n,z_{\lambda+1}^n)\cdot 1_A(\zeta_\lambda^n)$

$$= \sum_{\substack{\lambda \in 1^n \\ \zeta_\lambda^n \in A}} \Gamma(z_\lambda^n, z_{\lambda+1}^n) \geq \sum_{\substack{\lambda \in 1^n \\ [z_\lambda^n, z_{\lambda+1}^n] \subseteq K_n}} \Gamma(z_\lambda^n, z_{\lambda+1}^n) = \Gamma(K_n).$$

Hence $\Gamma(K_n) \approx 0$ for all $n \in {}^*\mathbb{N}$. Now we cover the sompact sets K_n by open sets $U_n \subseteq [a,b]$ s.t. $K_n \subseteq U_n$, $\Gamma(U_n) \approx 0$ and $U_n \subseteq U_{n+1}$.

For $U := \bigcup_{n=1}^{\infty} U_n$ we have $\Gamma(U) = \sup_{1 \leq n < \infty} \Gamma(U_n) \approx 0$.

Finally we obtain $A = \bigcup_{n=1}^{\infty} A \cap E_n \subseteq \bigcup_{n=1}^{\infty} K_n \subseteq \bigcup_{n=1}^{\infty} U_n = U$ from which we get $\Gamma(A) = 0$.

For the main result of this section we need the following technical lemma.

5.2. Lemma

Suppose $a = x_0 < x_1 < \ldots < x_{n-1} < x_n = b$, $I_\nu := [x_\nu, x_{\nu+1}]$ for $\nu \in n$ is a partition of $[a,b]$, $N \subseteq n$, $\Gamma \in \mathfrak{m}_0^+(a,b)$, and for each $\nu \in N$ there is some interval I'_ν, $I_\nu \subseteq I'_\nu \subseteq [a,b]$.

Then there is some $N' \subseteq N$ satisfying

(i) $I'_\nu \cap I'_\mu = \emptyset$ for $\nu, \mu \in N'$, $\nu \neq \mu$;

(ii) $\sum_{\nu \in N'} \Gamma(I'_\nu) \geq \frac{1}{2} \sum_{\nu \in N} \Gamma(I_\nu)$.

Proof: We proceed by induction on $|N|$; the case $N = \emptyset$ is obvious. For $N \neq \emptyset$ we put $\alpha = \min N$ and define inductively a sequence $(\nu_0, \ldots, \nu_k)$. $\nu_0 \in N$ is chosen s.t. $I_\alpha \subseteq I'_{\nu_0}$ and $I_\alpha \subseteq I'_\nu$ implies $\sup I'_\nu \leq \sup I'_{\nu_0}$ for $\nu \in N$. Assume $\nu_0, \ldots, \nu_j$ has already been defined. If there is some $\nu_k \in N$ s.t. $I'_{\nu_k} \cap I'_{\nu_{k+1}} \neq \emptyset$, $(\forall \nu \in N)(I'_{\nu_k} \cap I'_{\nu_{k+1}} \neq \emptyset \Rightarrow \sup I'_\nu \leq \sup I'_{\nu_{k+1}})$ and $\sup I'_{\nu_{k+1}} > \sup I'_{\nu_k}$ then ν_{j+1} is the last such ν, otherwise ν_{j+1} is undefined and the sequence terminates, i.e. $\nu_{j+1} = \nu_j$.

Then $I = \bigcup_{\kappa=0}^{k} I'_\kappa$ is an interval. Putting

$$\beta := \begin{cases} \max \{\nu \in N \mid I_\nu \cap I \neq \emptyset & \text{if this exists} \\ n & \text{otherwise} \end{cases}$$

$N_- := \{\nu \in N \mid \nu \leq \beta\} = \{\nu \in N \mid \nu < \beta\}$

$N_+ := \{\nu \in N \mid \nu > \beta\}$

and

$c := \sup I_\beta \geq \sup I$ we observe $\beta \notin N_+$ and apply the induction hypothesis to $c = x_{\beta+1} < \ldots < x_n = b$ and the intervals $(I'_\nu \cap [c,b])_{\nu \in N_+}$.

Because $|N_+| < |N|$ there is $N'_+ \subseteq N_+$ satisfying $\sum_{\nu \in N'_+} \Gamma(I'_\nu) \geq \sum_{\nu \in N'_+} \Gamma(I'_\nu \cap [c,b]) \geq \frac{1}{2} \sum_{\nu \in N_+} \Gamma(I_\nu)$ and $I'_\nu \cap I'_\mu = \emptyset$ for $\nu \in N'_+$. In addition $I'_\nu \cap I_\nu = \emptyset$ holds for $\nu \in N_+$ and $I'_{\nu_j} \leq \inf I'_{\nu_{j+2}}$ for $0 \leq j \leq j+2 \leq k$.

Taking $N_i := \{\nu_{2\kappa+i} \mid 0 \leq 2\kappa+1 \leq k\} \subseteq N_-$ for $i = 0,1$ we obtain

$$(\forall \nu,\mu \in N_i \cup N'_+)(\nu \neq \mu \rightarrow I'_\nu \cap I'_\mu = \emptyset) \qquad (*)$$

From $\sum_{\nu \in N_-} \Gamma(I_\nu) \leq \Gamma(I) \leq (\dot{\bigcup}_{\nu \in N_0} I'_\nu) + (\dot{\bigcup}_{\nu \in N_1} I'_\nu)$ we get for either $i = 0$ or $i = 1$ $\sum_{\nu \in N_i} \Gamma(I'_\nu) \geq \frac{1}{2} \sum_{\nu \in N_-} \Gamma(I_\nu)$.

We finally put $N'_- := N_i$ and $N' := N'_- \cup N'_+$. The assertion follows from

$$\sum_{\nu \in N'} \Gamma(I'_\nu) = \sum_{\nu \in N'_-} \Gamma(I'_\nu) + \sum_{\nu \in N'_+} \Gamma(I'_\nu) \geq \frac{1}{2} \sum_{\nu \in N_-} \Gamma(I_\nu) + \frac{1}{2} \sum_{\nu \in N_+} \Gamma(I_\nu)$$

$= \frac{1}{2} \sum_{\nu \in N} \Gamma(I_\nu)$ and the condition $(*)$.

For $\varphi \in \mathbb{D}(a,b)$ we have seen that $\Phi(x) = \int_a^x \varphi(t)\, dt$ is continuous. In order to show that φ is measurable it would suffice to prove that Φ has a derivative a.e. and $\Phi' = \varphi$ a.e.. This argument will work for the Lebesgue measure. For arbitrary $\Gamma \in \mathfrak{m}_0^+(a,b)$ some modification is necessary, however. This essentially amounts in replacing $\frac{f(t) - f(x)}{\lambda^1([x,t])}$ by $\frac{f(t) - f(x)}{\Gamma([x,t])}$ in the notion of differentiability.

5.3. Definition

For $\Gamma \in \mathfrak{m}(a,b)$ the support of Γ is

$\underline{\mathrm{supp}}(\Gamma) := [a,b] \setminus \{x \mid (\exists \varepsilon > 0)\ \Gamma([x-\varepsilon, x+\varepsilon] \cap [a,b]) = 0\}$.

5.4 Definition

Let $f: [a,b] \longrightarrow {}^*\mathbb{R}$ be continuous and $\Gamma \in \mathfrak{m}_0^+(a,b)$.

$\overline{D\Gamma} f, \underline{D\Gamma} f : [a,b] \longrightarrow [-\infty,\infty]$ are defined by:

$$\overline{D\Gamma}\, f(x) := \begin{cases} \limsup\limits_{\substack{s\to x- \\ t\to x+}} \dfrac{f(t) - f(s)}{\Gamma(s,t)} & \text{if } x \in \,]a,b[\, \cap \underline{\text{supp}}(\Gamma) \\ 0 & \text{otherwise} \end{cases}$$

$$\underline{D\Gamma}\, f(x) := \begin{cases} \liminf\limits_{\substack{s\to x- \\ t\to x+}} \dfrac{f(t) - f(s)}{\Gamma(s,t)} & \text{if } x \in \,]a,b[\, \cap \underline{\text{supp}}(\Gamma) \\ 0 & \text{otherwise} \end{cases}$$

If $x \in \,]a,b[\, \cap \underline{\text{supp}}(\Gamma)$ and $a \leq s < x < t \leq b$ then $\Gamma(s,t) > 0$.

5.5. Definition

Let $f : [a,b] \to {}^*\mathbb{R}$ be continuous, $\Gamma \in \mathfrak{M}_o^+(a,b)$ and ;
$G := \{x \in [a,b] \mid \overline{D\Gamma}\, f(x) = \underline{D\Gamma}\, f(x) \in {}^*\mathbb{R}\}$; f is called Γ-differentiable iff $\Gamma([a,b] \setminus G) = 0$.
In this case we define $D\Gamma\, f : [a,b] \to {}^*\mathbb{R}$ by: $D\Gamma\, f(x) = \overline{D\Gamma}\,(x)$ for $x \in G$ and 0 otherwise.

Remark:

f ist λ^1-differentiable iff f' exists λ^1-a.e. and f is continuous.

5.6. Lemma

If $\Gamma \in \mathfrak{M}_o^+(a,b)$ and $f : [a,b] \longrightarrow {}^*\mathbb{R}$ is Γ-differentiable, then $D\Gamma\, f$ is Γ-measurable.

Proof:

For $X := \,]a,b[\, \cap \underline{\text{supp}}(\Gamma) \cap G$ we have

$$D\Gamma\, f(x) = \lim_{\substack{s\to x- \\ t\to x+}} \frac{f(t) - f(s)}{\Gamma(s,t)} \quad \text{if } x \in X \text{ and o otherwise.}$$

We consider the continuous function $\gamma_n(x) := \Gamma([a,b] \cap [x-\frac{1}{n}, x+\frac{1}{n}])$; we have $\gamma_n(x) > 0$ for $x \in X$ and $\gamma_n(x) = \Gamma(x-\frac{1}{n}, x+\frac{1}{n})$ for standard $x \in \,]a,b[$ and nonstandard $n \in {}^*\mathbb{N}$.

If $g_n(x) := \dfrac{f(x+\frac{1}{n}) - f(x-\frac{1}{n})}{\gamma_n(x)}$ for $x \in X$ and 0 otherwise, then g_n is Γ-measurable and $D\Gamma\, f = \lim\limits_{n\to\infty} g_n$ (pointwise). Therefore $D\Gamma\, f$ is Γ-measurable.

5.7. Theorem

Suppose $\Gamma \in \mathbb{M}_o^+(a,b)$, $\varphi \in \mathbb{D}(a,b)$ and $\Phi(x) := \int_a^x \Phi\, d\Gamma$.
Then Φ is Γ-differentiable and $D\Gamma\ \Phi = \varphi$ a.e. In particular Γ is measurable.

Proof: We may w.l.o.g. assume that $\varphi = \varphi\, 1_X$ with $X =]a,b[\ \cap\ \underline{supp}(\Gamma)$ because $\Gamma(\{a,b\} \cup [a,b] \setminus \underline{supp}(\Gamma)) = 0$. For $m \in {}^*\mathbb{N}$ we take

$$A_m := \{x \in [a,b] \mid |\varphi(x) - \overline{D\Gamma}\ \Phi(x)| > \frac{2}{m} \text{ or } |\varphi(x) - \underline{D\Gamma}\ \Phi(x)| > \frac{2}{m}\}$$

$\subseteq]a,b[\ \cap\ \underline{supp}(\Gamma)$, and we define a family of monads by putting fixed m $M_X := {}^E\{(s,t) \in [a,b]^2 \mid x \in \underline{supp}(\Gamma),\ s < x < t,\ [s,t] \subseteq \mu(x)$ and

$|\varphi(x) - \dfrac{\Phi(t) - \Phi(s)}{\Gamma(s,t)}| > \frac{1}{m}\}$ for each standard $x \in [a,b]$.

If $x \in \{a,b\} \cup [a,b] \setminus \underline{supp}(\Gamma)$ then $M_X = \emptyset$, but $M_X \neq \emptyset$ for $x \in A_m$.
Now we will use theorem 5.1. in order to show that $\Gamma(A_m) = 0$. Therefore we assume that m is standard and $(n,x,\zeta) \in W(a,b)$. If $I_\nu := [x_\nu, x_{\nu+1}]$ for $\nu \in n$ and $N := \{\nu \in n \mid \zeta_\nu \in A_m\}$ then $(\forall^{st} x \in A_M)\ (M_X \neq \emptyset)$.
Applying Robinson's lemma (cf. [Ben-Ri]) then yields
$(\forall^{st} x \in A_m)(\exists h \in \mu_+)(\exists (s,t) \in M_X)(h \in {}^\pi\mu_+[(s,t)])$ and in particular

$(\forall^{st} x \in A_m)(\exists h \in \mu_+)(\exists (s,t) \in M_X)([x-h,x+h] \subseteq [s,t])$.

Now we apply the transfer theorem (1.3.) and obtain
$(\forall x \in A_m)(\exists h \in {}^\pi\mu_+[x])(\exists (s,t) \in {}^\pi M_X)([x-h,x+h] \subseteq [s,t])$.
In particicluar for $\nu \in N$, $x = \zeta_\nu$ and $h = x_{\nu+1} - x_\nu$ we get
$(\forall \nu \in N)(\exists (s,t) \in {}^\pi M_{(\zeta_\nu)})(I_\nu \subseteq [s,t])$.
Then we also can find internal sequences $(s_\nu)_{\nu \in N}$, $(t_\nu)_{\nu \in N}$ s.t.:
$(\forall \nu \in N)\left[(s_\nu,t_\nu) \in {}^\pi M_{(\zeta_\nu)} \wedge I_\nu \subseteq [s_\nu,t_\nu]\right]$ holds.
We put $I'_\nu := [s_\nu,t_\nu] \subseteq {}^\pi\mu(\zeta_\nu)$; by lemma 5.1. we find $N' \subseteq N$ s.t.:
$I'_\nu \cap I'_\mu = \emptyset$, $\nu,\mu \in N'$, $\nu \neq \mu$, $\sum_{\nu \in N'} \Gamma(I'_\nu) \geq \frac{1}{2} \sum_{\nu \in N} \Gamma(I_\nu)$.
Now we take $(m,y,\eta) \in W(a,b)$ and $\tilde{m}: N' \to m$ s.t.:

$$(\forall \nu \in N')(s_\nu = y_{\tilde{m}(\nu)-1} < y_{\tilde{m}(\nu)} = \zeta_\nu = \eta_{\tilde{m}(\nu)-1} = \eta_{\tilde{m}(\nu)} < y_{\tilde{m}(\nu)+1} = t_\nu)$$

and put

$K := \{k \in m \mid (\exists \nu \in N')\ k = \tilde{m}(\nu) \text{ or } k = \tilde{m}(\nu) - 1\}$.

Then we can conclude

$$\sum_{\nu \in n} \Gamma(x_\nu, x_{\nu+1})\ 1_{A_m}(\zeta_\nu) = \sum_{\nu \in N} \Gamma(I_\nu) \leq 2 \sum_{\nu \in N'} \Gamma(I'_\nu) = 2m \sum_{\nu \in N'} \Gamma(s_\nu, t_\nu)\ \frac{1}{m}$$

$$\leq 2m \sum_{\nu \in N'} \Gamma(s_\nu, t_\nu) \left| \varphi(\zeta_\nu) - \frac{\Phi(t_\nu) - \Phi(s_\nu)}{\Gamma(s_\nu, t_\nu)} \right|$$

$$= 2m \sum_{k \in K} \left| \Gamma(y_k, y_{k+1})\ \varphi(\eta_k) - \int_{y_k}^{y_{k+1}} \varphi\, d\Gamma \right. \approx 0 \text{; here we used}$$

$(s_\nu, t_\nu) \in {}^\pi M_{\zeta_\nu b}$ and for the last step Henstock's lemma 3.13. This gives $1_{A_m} \in \mathbb{D}(a,b)$, ${}^\nu\!\int_a 1_{A_m}\, d\Gamma = 0$ and by theorem 5.1. $\Gamma(A_m) = 0$.

We put $B := \{x \in [a,b] \mid \overline{D\Gamma}\ \Phi(x) = \underline{D\Gamma}\ \Phi(x) = \varphi(x)\}$ and $G := \{x \in [a,b] \mid \overline{D\Gamma}\ \Phi(x) = \underline{D\Gamma}\ \Phi(x) \in R\}$.

Then $B = [a,b]\ \ \cup(A_m \mid m \in {}^*N) \subseteq G$ and $\Gamma([a,b]\ \ G) \leq \Gamma([a,b]\ \ B)$ $= \Gamma(\cup(A_m \mid m \in {}^*N))$. Therefore Φ is Γ-differentiable and $D\Gamma\ \Phi(x) = \varphi(x)$ for all $x \in B$.

§6. Comparison with the Lebesgue integral

Let $\mu_{\mathbb{R}}$ denote the ideal monad of finite real numbers.

6.1. Lemma

Suppose $\Gamma \in \mathbb{M}_0^+(a,b)$, $f\colon [a,b] \longrightarrow [0,\infty[$ is Γ-measurable and both are standard. Then the following are equivalent:

(i) $\exists (n,x,\zeta) \in W(a,b)$ s.t. $\sum_{\nu \in n} \Gamma(x_\nu, x_{\nu+1})\ f(\zeta_\nu) \in \mu_{\mathbb{R}}$

(ii) $f \in \mathbb{D}(a,b\,;\zeta)$

(iii) $f \in \mathcal{L}^1(a,b\,;\zeta)$.

Proof: (iii) → (ii) → (i) is obvious.

(i) → (iii): By (3.7.) we have for such an (n,x,ζ):

$${}^*\!\!\int_a^b f\, d\Gamma = \int_{*a}^b f\, d\Gamma \simeq \sum_{\nu \in n} \Gamma(x_\nu, x_{\nu+1})\ f(\zeta_\nu) \in \mu_R \text{ thus } f \in \mathcal{L}^1(a,b\,;\zeta).$$

6.2. Corollary

Under the assumption of 6.1. $f \in \mathcal{L}^1(a,b\,;\zeta)$ implies

$(\forall(n,x,\zeta) \in W(a,b))(\sum_{\nu\in n} \Gamma(x_\nu,x_{\nu+1})\ f(\zeta_\nu) \in \mu_R(+\infty))$.

6.3. Corollary

If $\Gamma \in \mathbb{M}_o^+(a,b)$, $f \in \mathbb{D}(a,b;\Gamma)$ and $f \geq 0$ then $f \in \mathcal{L}^1(a,b;\Gamma)$.

Proof:

By (5.7.) f is measurable and by (6.1.) proves $f \in \mathcal{L}^1(a,b;\Gamma)$.

6.4. Definition

For $p \in [1,\infty[$ and $\Gamma \in \mathbb{M}_o^+(a,b)$ we put

$\mathbb{D}^p(a,b;\Gamma) := \{f \in \mathbb{D}(a,b;\Gamma) \mid |f|^p \in \mathbb{D}(a,b;\Gamma)\}$.

6.5. Theorem

If $\Gamma \in \mathbb{M}_o^+(a,b)$ then $\mathbb{D}^p(a,b;\Gamma) = \mathcal{L}^p(a,b;\Gamma)$.

Proof:

The inclusion "$\supseteq$" is clear and for the other direction we take $f \in \mathbb{D}^p(a,b;\Gamma)$; then f is Γ-measurable by (5.7.) and $|f|^p \in \mathcal{L}^1(a,b;\Gamma)$ by (6.3.); thus $f \in \mathcal{L}^p(a,b;\Gamma)$ holds.

Now we will use our methods to show that the (1-dimensional) integral which McShane introduced in [McS] is exactly the Lebesgue integral.

6.6. Definition

For $a < b$ and $\Gamma \in \mathbb{M}_o^+(a,b)$ we put

$$\overline{W}(a,b) := {}^E\{(n,x,\zeta) \mid n \in {}^*\mathbb{N},\ x \in [a,b]^{n+1}, \zeta \in [a,b]^n,\ a = x_o < x_1 < \dots < x_n = b,\ (\forall\nu \in n)([x_\nu,x_{\nu+1}] \subseteq {}^\pi\mu(\zeta_\nu))\};$$

$$\overline{\mathbb{D}}(a,b;\Gamma) := {}^S\{f\colon [a,b] \to {}^*\mathbb{R} \mid (\exists^{st} c \in {}^*\mathbb{R})(\forall(n,x,\zeta) \in \overline{W}(a,b))(\sum_{\nu\in n} \Gamma(x_\nu,x_{\nu+1})\ f(\zeta_\nu) \approx c)\}.$$

The corresponding integral is denoted by $\displaystyle\oint_a^b f\, d\Gamma$; if $f \in \overline{\mathbb{D}}(a,b;\Gamma)$ then we call f $\overline{W}$-integrable.

Remark

(i) If $\Gamma = \lambda^1$ then the $(\overline{W})$-integral is exactly the integral described in [McS] and [McL 8.3.].

(ii) Because we do not require for $(n,x,\zeta) \in \overline{W}(a,b)$ that $\zeta_\nu \in [x_\nu, x_{\nu+1}]$ we have $\overline{W}(a,b) \not\subseteq W_{\approx}(a,b)$; therefore $\mathfrak{R}(a,b) \subseteq \overline{\mathfrak{W}}(a,b\,;\Gamma)$ is not obviously clear.

6.7. Lemma

If $\Gamma \in \mathfrak{m}_o^+(a,b)$ then $\mathfrak{L}^1(1,L\,;\Gamma) \subseteq \hat{\mathfrak{D}}([a,b]\,;\Gamma) \subseteq \overline{\mathfrak{W}}(a,b\,;\Gamma) \subseteq \mathfrak{W}(a,b\,;\Gamma)$.

Proof: (3.6.) and the definitions of $\hat{W}([a,b])$, $\overline{W}(a,b)$, $W(a,b)$.

6.8. Theorem

If $\Gamma \in \mathfrak{m}_o^+(a,b)$ then $\mathfrak{L}^1(a,b\,;\Gamma) = \hat{\mathfrak{D}}([a,b]\,;\Gamma) = \overline{\mathfrak{W}}(a,b\,;\Gamma)$.

Proof:

It remains to show $\overline{\mathfrak{W}}(a,b\,;\Gamma) \subseteq \mathfrak{L}^1(a,b\,;\Gamma)$; thus we take $f \in \overline{\mathfrak{W}}(a,b\,;\Gamma)$. By (6.7.), (5.7.) f is Γ-measurable, and hence it is sufficient to show that $f_+ := \frac{|f| + f}{2} \in \mathfrak{L}^1(a,b\,;\Gamma)$.

By (6.1.) we have ot find some $(n,x,\zeta) \in W(a,b)$ such that

$$\sum_{\nu \in n} \Gamma(x_\nu, x_{\nu+1})\, f_+(\zeta_\nu) \in \eta_{\mathbb{R}}.$$

We assume that this is not possible and will derive a contradiction. We take some $i := (n,x,\zeta) \in W(a,b)$ and $(m,y,\eta) \in {}^\pi W(a,b)[i]$. This gives $A := \sum_{k \in m} \Gamma(y_k, y_{k+1})\, f_+(\eta_k) \in {}^\pi\mu_{\mathbb{R}}(+\infty)[i]$. Now we take

$\tilde{m}: (n+1) \to (m+1)$ s.t. $(\forall \nu \in (n+1))(x_\nu = y_{\tilde{m}(\nu)})$ holds and put:

$I_\nu := [x_\nu, x_{\nu+1}]$, $\nu \in n$, $J_k := [y_k, y_{k+1}]$, $k \in m$.

We define $(m,y,\zeta) \in \overline{W}(a,b)$ by:

$$\zeta_k : \begin{cases} \zeta_\nu & \text{if } f(\eta_k) < 0 \wedge J_k \subseteq I_\nu \\ \eta_k & \text{if } f(\eta_k) \geq 0 \end{cases}$$

If $B := \sum_{\substack{k \in m \\ f(\eta_k)<0}} \Gamma(y_k, y_{k+1}) \cdot f(\zeta_k)$ then $|B| \leq \sum_{\substack{k \in m \\ f(\eta_k)<0}} \Gamma(y_k, y_{k+1}) \cdot |f(\zeta_k)|$

$\leq \sum_{\nu \in n} \Gamma(x_\nu, x_{\nu+1}) \cdot |f(\zeta_\nu)| \in {}^\pi\eta_{\mathbb{R}}[i]$.

Therefore we get $\displaystyle\fint_a^b f\, d\Gamma \approx \sum_{\mu \in m} \Gamma(y_\mu, y_{\mu+1}) \cdot f(\zeta_\mu) =$

$= A + B \in {}^{\pi}\mu_R(+\infty)[i] + {}^{\pi}\eta_R[i] \subseteq \mu_R(+\infty)$ and $\oint_a^b f\, d\Gamma \in \mu_R(\infty)$, a contradiction.

§7. Local Lebesque integrability and comparison with the special Denjoy integral

7.1. Theorem

Suppose $\Gamma \in \mathfrak{M}_o^+(a,b)$, $\varphi \in \mathfrak{D}(a,b;\Gamma)$, $S \subseteq [a,b]$, S closed and $S \cap]a,b[\neq \emptyset$, Then there are c, d such that $a \leq c < d \leq b$, $S \cap]c,d[\neq \emptyset$ and $\varphi \cdot 1_S \in \mathcal{L}^1(c,d;\Gamma)$.

Proof

We know from (5.7) that $\varphi \cdot 1_S$ is measurable. First see that it is sufficient to prove $\varphi_+ \cdot 1_S \in \mathcal{L}^1(c, d; \Gamma)$ for some c and d. Assuming that we can apply it again to $-\varphi$ and the intervall $[c, d]$. Thus we get c', d' $\in [c,d]$ s.t. $]c', d'[\cap S \neq \emptyset$ and $\varphi_- \cdot 1_S \in \mathcal{L}^1(c',d';\Gamma)$. Combining this fact with $\varphi_+ \cdot 1_S \in \mathcal{L}^1(c',d';\Gamma)$ gives $\varphi \cdot 1_S \in \mathcal{L}^1(c',d';\Gamma)$. It therefor remains to show $.(\exists\, c,d \in [a,b])\ (c<d,\ S\cap]c,d[\neq \emptyset$ and $\varphi_+ \cdot 1_S \in \mathcal{L}^1(c, d;\Gamma)$. We proceed indirectley and assume the contrary.

Using (6.1) and the fact $(\forall^{st}\omega \in [o,\infty[)\ (\forall(n,x,\zeta) \in W(a,b))\ (\sum_{\substack{\nu\in n \\ |\varphi(\zeta_\nu)|\leq\omega}} \Gamma(x_\nu, x_{\nu+1})\ |\varphi(S_\nu)| \in \eta_R)$.

We obtain $(\forall^{st}c,d \in [a,b])\ (\forall^{st}\omega \in [o,\infty[)\ (\forall(n,x,\zeta) \in W(a,b))$

$$[\,(c<d \wedge]c,d[\cap S \neq \emptyset) \rightarrow \sum_{\substack{\nu\in n \\ \zeta_\nu\in S\cap]c,d[\\ \varphi(\zeta_\nu)\geq\omega}} \Gamma(x_\nu,x_{\nu+1})\varphi(\zeta_\nu) \in \mu_R(+\infty)\,]$$

In the sequel $\overset{o}{X}$ devotes the interior of $X \subseteq [a,b]$ and we put $\Phi(x) = \int_a^x \varphi(t)dt$. We are going to define inductively two standard sequences with members $(n^l, x^l, \zeta^l) \in W_{2^{-l}}(a,b)$ and $\omega^l \in [o,\infty[$ resp. for $l \geq 0$

From the sequences three other sequences will be obtained by putting

(a) $I_\nu^l = [x_\nu^l, x_{\nu+1}^l] \subseteq [a,b]$ for $\nu \in n_l$;

(b) $N^l = \{\nu \in n^l \mid I_\nu^l \subseteq \overset{o}{G}{}^{l-1},\ \varphi(\zeta_\nu^l) \geq \omega^l,\ \overset{o}{I}{}_\nu^l \cap S \neq \emptyset\}$

(c) $G^{-1} = [a,b]$, $G^l = \cup(I_\nu^l \mid \nu \in N^l)\ \ (\subseteq \overset{o}{G}{}^{l-1})$

The definition of N^l and G^l is again by (simultaneous) induction. Parallel to these definitions we will show that the following conditions hold:

(i) $\sum_{\nu \in N^o} \Phi(x^o_{\nu+1}) - \Phi(x^o_\nu) \geq 5$, and for $l>0$, $x \in N^{l-1}$

$\sum_{\nu \in N^l} \Phi(x^l_{\nu+1}) - \Phi(x^l_\nu) \geq \max\{5, \Phi(x^{l-1}_{k+1}) - \Phi(x^{l-1}_k)\}$.

$I^l_\nu \subseteq \overset{o}{I}{}^{l-1}_k$

(ii) $N^l \neq \emptyset$; $\{\nu \in N^l \mid I^l_\nu \subseteq I^{l-1}_k\} \neq \emptyset$ for $l<0$ and $k \in N^{l-1}$.

(iii) $\Gamma(G_l) \leq \frac{1}{2}\Gamma(G^{l-1})$

From these conditions we will finally derive a contradiction. Assume that our sequences have been defined for all $\lambda \in l$ s. t. (i) - (iii) hold; for the induction step it is now sufficient to assume that these sequences are standard. By transfer it is also sufficient to extend the sequences by taking $\omega^l \in \mu_N$ and $(n^l, x^l, \zeta^l) \in {}^\pi W(a,b)\,[\omega]$. Proceeding this way we first note that (i) implies (ii). From (i) we also get (iii) by observing $G^{l-1} \neq \emptyset$, $\Gamma(G^{l-1})>0$ and

$$\Gamma(G^l) = \sum_{\nu \in N^l} \Gamma(x^l_\nu, x^l_{\nu+1}) \leq \sum_{\nu \in n^l, \varphi(\zeta^l_\nu) \geq \omega^l} \Gamma(x^l_\nu, x^l_{\nu+1})$$

$${}^\pi\!\approx_{[\omega^l]} \Gamma(\{x \in [a,b] \mid \varphi(x) \geq \omega^l\}) \approx 0.$$

Hence it remains to prove (i).

Let $I = [c,d] \subseteq G^{l-1}$ be any standard interval s. t. $\overset{o}{I} \cap S \neq \emptyset$.

We put $A: \{\nu \in N^l \mid \zeta^l_\nu \in S \wedge S \cap \overset{o}{I}{}^l_\nu = \emptyset\}$, $B: = n^l \setminus A$.

We take $(m,y,\eta) \in {}^{\pi\pi}W(a,b)\,[\omega^l]\,[(n^l,x^l,\zeta^l)]$ and define $(k,z,\zeta) \in {}^\pi W(a,b)\,[\omega^l]$ by requiring that the restriction to I^l_ν is identical with (m,y,ζ) for $\nu \in A$ and otherwise coincides with (n^l,x^l,ζ^l).

We take $\tilde{m} : (n^l+1) \to (m+1)$ with $x^l_\nu = y_{\tilde{m}(\nu)}$ for $\nu \in n$ and get by our indirect assumption (where "<<" means infinitely smaller):

$$1 \ll \sum_{c_1} \Gamma(z_\kappa, z_{\kappa+1}) \cdot \varphi(\zeta_\kappa) = \sum_{\nu \in n^l} \sum_{c_2} \Gamma(z_\kappa, z_{\kappa+1})\, \varphi(\zeta_\kappa)$$

$$= \sum_{\nu \in A} \sum_{c_3} \Gamma(y_\mu, y_{\mu+1})\, \varphi(\eta_\mu) + \sum_{c_4} \Gamma(x^l_\nu, x^l_{\nu+1})\, \varphi(\zeta^l_\nu)$$

Here the conditions for the sums are as follows:

c_1: $\kappa \in K$, $\zeta_\kappa \in S \cap I$, $\varphi(\zeta_\kappa) \geq \omega^l$;

c_2: $\kappa \in K$, $[z_\kappa, z_{\kappa+1}] \subseteq I^l_\nu$, $\zeta_\kappa \in S_o \cap \overset{o}{I}$, $\varphi(\zeta_\kappa) \geq \omega^l$;

c_3: $\tilde{m}(\nu) \leq \mu \leq \tilde{m}(\nu+1)-1$, $\eta_\mu \in S \cap \overset{o}{I}$, $\varphi(\zeta_\kappa) \geq \omega^l$;

c_4: $\nu \in B$, $\zeta^l_\nu \in S \cap I$, $\varphi(\zeta^l_\nu) \geq \omega^l$.

If $\nu \in A$, $\tilde{m}(\nu) \leq \mu < \tilde{m}(\nu+1)$, $\eta_\mu \in S$ then $\mu = \tilde{m}(\nu) \vee \mu = \tilde{m}(\nu+1)-1$, because all other η_μ are in $\overset{o}{I}{}^1_\nu$ and $\overset{o}{I}{}^1_\nu \cap S = \emptyset$. Therefor the first sum consists of at most $2n^1$ terms and each term is in ${}^\pi\mu_R(o)\ [(n^1,x^1,\zeta^1)] \subseteq {}^\pi\mu_R(o)[n^1]$. Hence the first sum is infinitesimal.

Therefore we obtain

$$1 \ll \sum_{D_1} \Gamma(x^1_\nu, x^1_\nu+1)\ \varphi(\zeta^1_\nu) \leq \sum_{D_2} \Gamma(x^1_\nu, x^1_{\nu+1})\ \varphi(\zeta^1_\nu) \approx \sum_{D_2} \Phi(x^1_{\nu+1}) - \Phi(x^1_\nu)$$

where D_1: $\nu \in B,\ \zeta^1_\nu \in S \cap \overset{o}{I},\ \varphi(\zeta^1_\nu) \geq \omega^1$

D_2: $\nu \in n^1,\ I^1_\nu \subseteq \overset{o}{I},\ \varphi(\zeta^1_\nu) \geq \omega^1,\ S \cap \overset{o}{I}{}^1_\nu \neq \emptyset$.

Here we used that $S \cap \overset{o}{I}{}^1_\nu \neq \emptyset$ implies $S \cap {}^\pi\mu(\zeta^1_\nu) \neq \emptyset$ and applied Henstock's lemma for the last relation. Treating the case $l=0$ we chose $I = [a,b] = G^{-1}$ and get

$$1 \ll \sum_{D_3} \Phi(x^o_{\nu+1}) - \Phi(x^o_\nu) = \sum_{\nu \in N^o} \Phi(x^o_{\nu+1}) - \Phi(x^o_\nu)$$ where D_3 is D_1 with $l=0$ which shows (i).

If $l \geq 1$ and $\kappa \in N^{l-1}$ we choose $I = I^{l-1}_\kappa \subseteq G^{l-1}$ and obtain

$$1 \ll \sum_{D_4} \Phi(x^1_{\nu+1}) - \Phi(x^1_\nu) = \sum_{D_5} \Phi(x^1_{\nu+1}) - \Phi(x^1_\nu)$$

replacing $\overset{o}{I}$ by $\overset{o}{I}{}^{l-1}_\kappa$ and $\overset{o}{I}{}^1_\nu$ by I^1_ν in D_2 gives D_4 and D_5 is $\nu \in N^1$, $I^1_\nu \subseteq \overset{o}{I}{}^{l-1}_\kappa$. This again shows condition (i). Hence our sequences are defined and (i), (ii) and (iii) are true.

Next we define

$$\kappa = \bigcap_{l=0}^{\infty} G^l = \bigcap_{l=0}^{\infty} \overset{o}{G}{}^l$$

Then $\kappa = \bar{\kappa} \neq \emptyset$ and because of $G^l \subseteq G^{l-1}$ and $\Gamma(G^l) \leq \frac{1}{2}\Gamma(G^{l-1})$ we have $\Gamma(\kappa) \simeq 0$. For $x \in \kappa$ and $l \geq 0$ there is exactly one interval $I^1_\nu \subseteq G^1$ with $x \in \overset{o}{I}{}^1_\nu$; we put $G^1_x := I^1_\nu$.

Taking $\tilde{\varepsilon} : [a,b] \to]o,\infty[$ s.t.: $(\forall x \in [a,b])(\tilde{\varepsilon}(x) \in {}^\pi\mu_+[x])$ we find a function $\tilde{l} : \kappa \to N$ satisfying

$$\kappa \subseteq \bigcup_{x \in \kappa} \overset{o}{G}{}^{\tilde{l}(x)}_x$$

There is also some finite $H \subseteq K$ s.t. $\kappa \subseteq \bigcup_{x \in H} \overset{o}{G}{}^{\tilde{l}(x)}_x$ and

$\kappa \nsubseteq \bigcup (\overset{o}{G}{}^{l(x)}_x | x \in H')$ for all $H \subseteq \kappa$ of smaller cardinality.

Then $(\kappa \cap \overset{o}{G}{}^{\tilde{l}(x)}_x)_{x \in H}$ is partition of κ.

Defining $\mu_x := \underline{\inf}\, G_x^{\tilde{I}(x)}$, $\nu_x := \underline{\sup}\, G_x^{\tilde{I}(x)}$ there is $(m, y, \eta) \in W_{\tilde{\varepsilon}}(a,b) \subseteq W(a,b)$ s.t. $(\forall x \in H)(\exists \mu \in m)(\mu_x = y_{\mu-1} \wedge x = y_\mu = \eta_{\mu-1} = \eta_\mu \wedge \nu_x = y_{\mu+1}$.

Now $\Gamma(\kappa) = 0$ and hence $\varphi \cdot 1_\kappa$ is a zero function. This yields, using 3.12., $0 \approx \sum_{\substack{\mu\in m\\ \eta_\mu\in\kappa}} \Gamma(y_\mu, y_{\mu+1})\, \varphi(\eta_\mu) \approx \sum_{\mu\in m, \eta_\mu\in\kappa} \int_{y_\mu}^{y_{\mu+1}} \varphi d\Gamma$

$= \sum_{\mu\in m,\eta_\mu\in\kappa} \Phi(y_{\mu+1}) - \Phi(y_\mu) = \sum_{x\in H} \Phi(\nu_x) - \Phi(\mu_x)$ and we thus get

$$\sum_{x\in H} \Phi(\nu_x) - \Phi(\mu_x) \leq 1 .$$

If $l := \max_{x\in H} \tilde{I}(x)$, then $l = \tilde{I}(z)$ for some $z \in H$.

Putting $H' := (H \setminus G_z^{l-1}) \cup \{z\}$ and $\tilde{I}'(x) = \tilde{I}(x)$ for $x \in H' \setminus \{z\}$ and $l-1$ for $x=z$ we get by $(\kappa \cap \overset{\circ}{G}{}_x^{l'(x)})_{x\in H}$ a partition of K in equivalence classes. (obtained from $(\kappa \cap \overset{\circ}{G}{}_x^{\tilde{I}(x)})_{x\in H}$ by replacing the classes $(\kappa \cap \overset{\circ}{G}{}_x^{\tilde{I}(x)})_{x\in H_o}$ by the new class $\kappa \cap \overset{\circ}{G}{}_z^{l-1}$ where $H_o = H \cap G_z^{l-1}$).

For $\mu'_x = \inf \overset{\circ}{G}{}_x^{\tilde{I}'(x)}$ and $\nu'_x = \sup \overset{\circ}{G}{}_x^{\tilde{I}(x)}$ we now want to show

$$\sum_{x\in H'} \Phi(\nu'_x) - \Phi(\mu'_x) \leq \sum_{x\in H} \Phi(\nu_x) - \Phi(\mu_x);$$

It is sufficient to show $\Phi(\nu'_x) - \Phi(\mu'_x) \leq \sum_{x\in H_o} \Phi(\nu_x) - \Phi(\mu_x)$.

Taking $\kappa \in N^{l-1}$ s.t. $G_z^{l-1} = I_\kappa^{l-1}$ and put

$X = \{G_x^l \mid x \in H_o\}$; $Y = \{I_\nu^l \mid I_\nu^l \subseteq \overset{\circ}{I}{}_\kappa^{l-1} \wedge \nu \in N^l\}$. Then we get

a) $X \subseteq Y$: By definition $G_x^l = I_\nu^l$ for some $\nu \in N^l$ for $x \in H_o$ we have $I_\nu^l = G_x^l \subseteq \overset{\circ}{G}{}_z^{l-1} = \overset{\circ}{I}{}_\kappa^{l-1}$

b) $X \supseteq Y$: If we would have $\supsetneq$ then we could find some $y \in \kappa$ and some I_ν^l, $\nu \in N^l$ s.t.: $y \in I_\nu^l$. But then $y \notin \bigcup_{x\in H} G_x^{\tilde{I}(x)}$ leads to a contradiction. Therefore $X=Y$ holds.

Using condition (ii) we get $\Phi(\nu'_z) - \Phi(\mu'_z) = \Phi(x_{\kappa+1}^{l-1}) - \Phi(x_\kappa^{l-1})$

$$\leq \sum_{\nu\in N^1, I^1_\nu \subseteq \overset{o}{I}{}^{1-1}_\kappa} \Phi(x^1_{\nu+1}) - \Phi(x^1_\nu) = \sum_{x\in H_o} \Phi(\nu_x) - (\mu_x) \text{ and}$$

$$1 \geq \sum_{x\in H} \Phi(x^1_{\nu+1}) - \Phi(x^1_\nu) = \sum_{x\in H'} \Phi(\nu'_x) - \Phi(\mu'_x).$$

Continuing this way in constructing partitions we eventually get a partition $(\kappa \cap \overset{o}{G}{}^{1*(x)}_x)_{x\in H^*}$ for which $\tilde{1}^*(x) = 0$ holds for all $x \in H^*$.

Then we have $1 \geq \sum_{x\in H} \Phi(\nu_x) - \Phi(\mu_x) \geq \sum_{x\in H'} \Phi(\nu'_x) - \Phi(\mu'_x)$

$$\geq \sum_{x\in H^*} \Phi(\nu^*_x) - \Phi(\mu^*_x) .$$

For each $x \in H^*$ we find $\nu \in N^o$ s.t.: $G_x^{\tilde{1}^*(x)} = G^o_x = I^o_\nu$, and if $\nu \in N^o$ then $I^o_\nu = G^o_x$ for some $x \in H^*$, because otherwise there would be some $y \in \kappa \cap I^o_\nu$ s.t. $y \notin \bigcup_{x\in H^*} G^o_x$. Hence $\{G^o_x \mid x \in H^*\} = \{I^o_\nu \mid \nu \in N^o\}$

and therefore $1 \geq \sum_{x\in H^*} \Phi(\nu^*_x) - \Phi(\mu^*_x) = \sum_{\nu\in N^o} \Phi(x^o_{\nu\pm 1}) - \Phi(x^o_\nu) \geq 5.$

This is the desired contradiction which finishes the proof of theorem 7.1..

7.2. Corollary

If $\Gamma \in \mathbb{D}^+_o(a,b)$ and $\varphi \in \mathbb{D}(a,b;\Gamma)$ then there are c,d such that $a\leq c<d\leq b$ and $\varphi \in \sim\mathcal{L}^1(c,d;\Gamma)$.

Proof

S = [a,b] in the last theorem gives the result.

7.3. Lemma

For $S = \overline{S} \subseteq R$ and $f: S \to]o,1[$

put $g(x) := \limsup_{t\in\mu_S(x)} f(t)$ there are $a,b,\nu \in {}^*R$ such that $a < b$, $S \cap]a,b[\neq \emptyset$ and $g(x) \geq \nu$ for all $x\in S\cap]a,b[$.

Proof

We define $A_n = \{x \in S \mid g(x) \geq \frac{1}{n}\}$ for $n \geq 1$. In order to see that the A_n are closed we take $a \in \overline{A}_n$. For such a we first choose some $x, a\approx x\in A_n$ and then $t \in {}^{\pi}\mu(x) \cap S$. Then $t \in \mu(a)$ and $g(a) \gtrsim f(t) \approx g(x) \geq \frac{1}{n}$ holds and we get $a \in A_n$.

Now $g(x) \geq f(x) > 0$ for $x \in S$ implies $S = \bigcup_{n\in N} A_n$; therefor $\overset{\circ}{A}_n \neq \emptyset$ for some $n \geq 1$, where the interior $\overset{\circ}{A}_n$ is taken in the complete metric space S.

7.4. Theorem

Assume the following: $\Gamma \in \mathbb{D}_o^+(a,b)$, $\varphi \in \mathbb{D}(a,b)$, $S \subseteq [a,b]$, S closed and $S \cap]a,b[\neq \emptyset$, $G := [a,b] \setminus S$, $\varphi(x) = 0$ for $x \in S$;
Let $(G_n)_{n\in N}$ be the connected components of G (where $N \in {}^*N$ or $N = {}^*N$), $a_n := \inf G_n$, $b_n := \sup G_n$ and $\gamma_n := \sup_{a_n \leq d < \beta \leq b_n} | \int_d^\beta \varphi \, d\Gamma |$.

Then there are $c,d \in [a,b]$ such that $c < d$, $S \cap]c,d[\neq \emptyset$ and

$$\sum_{n\in N, G_n \subseteq]a,b[} \gamma_n < \infty$$

Proof

There is a standard function $\tilde{\varepsilon}$ defined on [a,b] with values in]0,1[s.t. for (m,y,η), $(\kappa,z,\zeta) \in W_{\tilde{\varepsilon}}(a,b)$ we have

$$|\sum_{\mu\in m} \Gamma(y_\mu,y_{\mu+1}) \varphi(\eta_\mu) - \sum_{\kappa\in K} \Gamma(z_\kappa,z_{\kappa+1}) \varphi(\zeta_\kappa)| < 1$$

we define $\bar{\varepsilon}: S \to]0,1]$; $\bar{\varepsilon}(x) = \limsup_{t\in\mu_S(x)} \tilde{\varepsilon}(x)$.

Lemma 7.3. guarantees ν, c', $d' \in {}^*R$, $a \leq c' < d' \leq b$, $\nu > 0$, $S \cap]c',d'[\neq \emptyset$ satisfying $\bar{\varepsilon}(x) > \nu$ for $x \in S \cap]c',d'[$;
w. l. o. g. we may also assume $|d' - c'| < \nu$. We proceed indirectly and will arrive at a contradiction from the assumption that

$\sum_{n\in N,\ G_n \subseteq]c',d'[} \gamma_n$ diverges. The first implies $G_{n_i} \subseteq]c',d'[$ for infinitely many $n_i \in N$. Take two such indices, say n_1 and n_2, where G_{n_1} is left of G_{n_2} and choose $c \in G_{n_1}$ and $d \in G_{n_2}$.

Again we assume $\sum_{n\in N, G_n \subseteq]c,d[} \gamma_n = +\infty$

Now we take a_n', b_n' s.t. $a_n \leq a_n' < b_n' \leq b_n$ and $\gamma_n = |\int_{a_n'}^{b_n'} \varphi \, d\Gamma|$ holds and we put $\gamma_n' := \int_{a_n'}^{b_n'} \varphi \, d\Gamma$, $\gamma_n^+ := \underline{\max} \{\gamma_n', 0\}$, $\gamma_n^- := \underline{\min} \{\gamma_n', 0\}$.

W. l. o. g. we may assume $\sum_{n\in N, G_n \subseteq]c,d[} \gamma_n^+ = +\infty$ (otherwise we consider the function $-\varphi$)

Next we pick (standard) finitely many $G_{\sigma(o)}, \ldots, G_{\sigma(1+1)}$ satisfying

(i) $a_{\nu(0)} < c < b_{\nu(0)}$,

(ii) $a_{\sigma(1+1)} < d < b_{\sigma(1+1)}$,

(iii) $b_{\sigma(\lambda)} < a_{\sigma(\lambda+1)}$ for $0 \leq \lambda \leq 1$,

(iv) $\sum_{\lambda=1}^{1} \gamma_{\sigma(\lambda)}^+ > 5$,

(v) $\gamma_{\sigma(\lambda)}' = \gamma_{\sigma(\lambda)}^+ > 0$ for $1 \leq \lambda \leq 1$

Furthermore for $0 \leq \lambda \leq 1$ we put $H_\lambda := S \cap [b_{\sigma(\lambda)}, a_{\sigma(\lambda+1)}]$ and get $(\forall \lambda \in 1+1)(\forall x \in H_\lambda)(\nu < \tilde{\varepsilon}(x) = \limsup_{t\in\mu_S(x)} \tilde{\varepsilon}(t))$;

therefore there is $(W_\lambda)_{\lambda=0}^{1} \in \prod_{\lambda=0}^{1} H_\lambda$ for which $\tilde{\varepsilon}(W_\lambda) > \nu$ holds for $\lambda \in 1+1$

In particular we have $[c,d] \subseteq [W_\lambda - \tilde{\varepsilon}(W_\lambda), W_\lambda + \tilde{\varepsilon}(W_\lambda)]$. We choose some $(n,x,\zeta) \in W(a,b)$; two partitions (m,y,η), $(\kappa,z,\zeta) \in W_{\tilde{\varepsilon}}(a,b)$ will be defined by stating how they look on certain intervals.

Firstly, on $[a,c] \cup [d,b]$ both partitions coincide with (n,x,ζ).

Next we describe (m,y,η) as follows:

a) On $[c, a_{\sigma(1)}']$ we put $y_\mu = c$, $y_{\mu+1} = W_o = \eta_\mu = \eta_{\mu+1}$ and $y_{\mu+2} = a_{\sigma(1)}'$;

b) On $[b_{\sigma'(1)}, d]$ we let $y_\mu = b_{\sigma(1)}'$, $y_{\mu+1} = W_1 = \eta_\mu = \eta_{\mu+1}$ and $y_{\mu+2} = d$;

c) for $1 \leq \lambda < 1$ we take on $[b'_{\sigma(\lambda)}, a_{\sigma(\lambda+1)}']$, $y_\lambda = b_{\sigma(\lambda)}'$, $y_{\mu+1} = W_\lambda = \eta_\mu = \eta_{\mu+1}$ and $y_{\mu+2} = a_{\sigma(\lambda+1)}'$;

d) for $1 \leq \lambda < l$ we require that on $[a'_{\sigma(\lambda)}, b'_{\sigma(\lambda)}]$ (m,y,η) coincides with (n,x,ζ).

Secondly, (κ,z,ζ) is defined similarly by:

a') On $[c, b_{\sigma(1)}]$ we put $z_\kappa = c$, $z_{+1} = W_o = \zeta_\kappa = \zeta_{\kappa+1}$ and $z_{\kappa+2} = b'_{\sigma(1)}$,

b') on $[b'_{\sigma(1)}, d]$ we let (k,z,ζ) coincide with (m,y,η);

c') for $1 \leq \lambda < l$ we put on $[b'_{\sigma(\lambda)}, b'_{\sigma(\lambda+1)}]$: $z_\kappa = b'_{\sigma(\lambda)}$ and $z_{\kappa+1} = W_\lambda = \zeta_\kappa = \zeta_{\kappa+1}$ and $z_{\kappa+2} = b'_{\sigma(\lambda+1)}$.

From our choice of $\tilde{\epsilon}$ we now obtain (using in particular (iv) from above):
$$1 > \sum_{\mu\in m} \Gamma(y_\mu,y_{\mu+1})\ \varphi(\eta_\mu) - \sum_{\kappa\in K} \Gamma(z_\kappa,z_{\kappa+1})\ \varphi(\zeta_\kappa)$$

$$= \sum_{\mu\in m}{}' \Gamma(y_\mu,y_{\mu+1})\ \varphi(\eta_\mu) - \sum_{\kappa\in K}{}'' \Gamma(z_\kappa,z_{\kappa+1})\ \varphi(\eta_\mu)$$

(where Σ' is restricted to μ with $[y_\mu,y_{\mu+1}] \subseteq [c,d]$ and $\eta_\mu \neq W_i$; $0 \leq i \leq l$ and Σ'' is restricted to $[z_\kappa,z_{\kappa+1}] \subseteq [c,d]$ and $\eta_\mu \neq W_i$; $0 \leq i \leq l$.).

$$= \sum_{\lambda=1}^{l} \sum_{\nu\in n}{}''' \Gamma(x_\nu,x_{\nu+1})\ \varphi(\zeta_\nu) \quad \text{(where } \Sigma''' \text{ is restricted}$$

to $[x_\nu,x_{\nu+1}] \subseteq [a'_{\sigma(\lambda)},b'_{\sigma(\lambda)}]$);

$$\approx \sum_{\lambda=1}^{l} \int_{a'_{\sigma(\lambda)}}^{b'_{\sigma(\lambda)}} \varphi\, d\Gamma = \sum_{\lambda=1}^{l} \gamma'_{\sigma(\lambda)} = \sum_{\lambda=1}^{l} \gamma'_{\sigma(\lambda)} > 5$$

Hence $1 \geq 5$ is the desired contradiction which finishes the proof.

The next theorem says that a kind of "improper W-integral" is the W-integral.

7.5. Theorem

Suppose $\Gamma \in \mathfrak{M}_o(a,b)$, φ is real valued $[a,b]$, $\varphi \in \mathfrak{D}(\alpha,\beta;\Gamma)$ for $a < \alpha < \beta < b$ and $c := \lim\limits_{\substack{\alpha\to a+ \\ \beta\to b-}} \int_\alpha^\beta \varphi\, d\Gamma$ exists.

Then $\varphi \in \mathfrak{D}(a,b)$ and
$$c = \int_a^b \varphi\, d\Gamma$$

Proof

W.l.o.g. we may assume $a < \alpha < \beta$ and $\varphi \in \mathbb{D}(\alpha,b;\Gamma)$. Let $(a_m)_{m=0}^{\infty}$ be a standard sequence decreasing monotonically from b to a :
$b = a_0 > a_1 > a_2 > \ldots > a$ and $\lim_{m\to\infty} a_m = a$; define $\tilde{m}$ on $]a,b]$ by

$x \in]a_{\tilde{m}(x)+1}, a_{\tilde{m}(x)}]$ for $x \in]a,b]$.

For $(n,x,\zeta) \in W(a,b)$, $0 < \nu < n$ and $\zeta_\nu \in]a_{m+1},a_m]$ then
$[x_\nu,x_{\nu+1}] \subseteq {}^{\pi}\mu(\zeta_\nu) = {}^{\pi\pi}\mu(\zeta_\nu)[\tilde{m}(\zeta_\nu)] = {}^{\pi\pi}\mu(\zeta_\nu)[m]$ follows.
Putting $\omega := \tilde{m}(\zeta_1)$ then for $0 \leq m < \omega$ the restriction of (n,x,ζ) to $[a_{m+1},a_m]$ is in ${}^{\pi}W(a_{m+1},a_m)$; furthermore the restriction of (n,x,ζ) to $[a_{\omega+1},a_\omega]$ can be expanded to some $(\kappa,y,\eta) \in {}^{\pi}W(a_{\omega+1},a_\omega)$.

This implies for $I_\nu = [x_\nu,x_{\nu+1}]$ and $m \in \omega$ the relation

$$\Rightarrow \forall\ m \in \omega \quad \sum_{\nu\in n, I_\nu \subseteq [a_{m+1},a_m]} \Gamma(x_\nu,x_{\nu+1})\ \varphi(\zeta_\nu) \quad {}^{\pi}\!\approx_{[m]} \quad \int_{a_{m+1}}^{a_\omega} \varphi\, d\Gamma$$

as well as
$$\sum_{\nu\in n, I_\nu \subseteq [a_{\omega+1},a_\omega]} \Gamma(x_\nu,x_{\nu+1})\ \varphi(\zeta_\nu) \quad {}^{\pi}\!\approx_{[\omega]} \quad \int_{x_1}^{a_\omega} \varphi\, d\Gamma$$

Now we are able to compute our integral:

$$\sum_{\nu\in n} \Gamma(x_\nu,x_{\nu+1})\ \varphi(\zeta_\nu) \ :$$
$$= \Gamma(a,x_1)\ \varphi(a) + \sum_{\nu\in n, I_\nu \subseteq [a_{\omega+1},a_\omega]} \Gamma(x_\nu,x_{\nu+1})\ \varphi(\zeta_\nu) +$$
$$\sum_{m\in\omega}\ \sum_{\nu\in n, I_\nu \subseteq [a_{m+1},a_m]} \Gamma(x_\nu,x_{\nu+1})\ \varphi(\zeta_\nu) \approx 0 + \int_{x_1}^{a_\omega} \varphi\, d\Gamma + \sum_{m\in\omega} \int_{a_{m+1}}^{a_m} \varphi\, d\Gamma$$
$$= \int_{x_1}^{b} \varphi\, d\Gamma \approx c.$$

Finally we consider the special Denjoy integral (cf.e.g.[Na.:XVI § 7]). It is defined by a limit process over ordinal numbers which we shortly recall. First we have $\mathcal{D}_0(a,b) = \mathcal{L}^1(a,b)$, $(0)\int_a^b f(t)\,dt := \int_a^b f\, d\lambda^1$.

Suppose $\mathcal{D}_\alpha(a,b)$ and $(\alpha)\int_a^b f(t)\,dt$ have been defined for all $\alpha \in \gamma$.

For γ a limit ordinal one simply puts: $\mathcal{D}_\gamma(a,b) := \bigcup_{\alpha\in\gamma} \mathcal{D}_\alpha(a,b)$

and $(\gamma) \int_a^b f(t)\,dt := (\alpha) \int_a^b f(t)\,dt$, where $\alpha \in \gamma$ and $f \in \mathcal{D}_\alpha(a,b)$

If $\gamma = \beta+1$ is a sucessor ordinal we first take the closed set

$S_\beta(f;\, a,b) := \{x \in [a,b] \mid (\forall\, \varepsilon > 0)\ (f \notin \mathcal{D}_\beta(a,b) \cap [x-\varepsilon, x+\varepsilon]\}$;

We say that $f \in \mathcal{D}_{\beta+1}(a,b)$ if the following conditions are fulfilled:

(i) $\overset{o}{S}_\beta(f;a,b) = \emptyset$ and $f\cdot 1_{S_\beta(f;a,b)} \in \mathcal{L}^1(a,b)$

(ii) If $(R_n \mid n\geq 0)$ are the connected components of $R_\beta(f;a,b)$,

$a_n := \inf R_n$, $b_n := \sup R_n$ then $\theta_n := \lim_{\substack{a'\to a_n^+ \\ b'\to b_n^-}} (\beta) \int_{a'}^{b'} f(t)\,dt \in {}^*\mathbb{R}$

(iii) $\upsilon_n := \sup_{a_n<a'<b'<b_n} |(\beta) \int_{a'}^{b'} f(t)\,dt| < \infty$ and $\sum_n \upsilon_n < \infty$.

For $f \in \mathcal{D}_{\beta+1}(a,b)$ we put $(\beta+1) \int_a^b f(t)\,dt = \int_{S_\beta(f;a,b)} f\,d\lambda^1 + \sum_n \theta_n$.

We have for $\alpha \geq \beta \geq \Omega$ where Ω is the first uncountable ordinal (see [Na.: XVI, §7]).

We put $\mathcal{D}(a,b) := \mathcal{D}_\Omega(a,b)$ and $(\mathcal{D}) \int_a^b f(t)\,dt := (\Omega) \int_a^b f(t)\,dt$

This is called the special Denjoy integral.

7.6. Proposition (H. Hake 1921)

$\mathcal{D}(a,b) \subseteq \mathcal{P}(a,b)$ and $\forall\, f \in \mathcal{D}(a,b)$ $\quad (\mathcal{D})\int_a^b f(t)\,dt = (p) \int_a^b f(t)\,dt$

Proof See [Na.: XVI §8].

7.7. Corollary

$\mathcal{D}(a,b) \subseteq \mathbb{D}(a,b)$

Proof

We have $\mathcal{D}(a,b) \subseteq \rho(a,b) \subseteq \mathbb{D}(a,b)$ by (7.6.) and (2.5.).

7.8. Lemma

If $\varphi \in \mathbb{D}(a,b)_o$ then $\overset{o}{S}_o(\varphi;a,b) = \emptyset$

Proof

If $\overset{\circ}{S}_0(\varphi;a,b) \neq \emptyset$ then there are $a',b' \in [a,b]$, $a' < b'$ and $[a',b'] \subseteq S_0(\varphi;a,b)$. By (7.2.) for some $c,d \in [a',b']$ $c < d$ and $\varphi \in \mathcal{L}^1(c,d)$ holds, therefore $]c,d[\subseteq R_0(\varphi;a,b)$ and $]c,d[\subseteq [a',b'] \subseteq S_0(\varphi;a,b)$, a contradiction.

7.9. Corollary

If $\varphi \in \mathcal{D}(a,b)$ for $\gamma \in \Omega$ then $S_\gamma(\varphi;a,b) = \emptyset$

Proof

$\mathcal{D}_0(a',b') \subseteq \mathcal{D}_\gamma(a',b')$ for all $a' < b'$ implies $R_0(\varphi;a,b) \subseteq R(\varphi;a,b)$ and $S_\gamma(\varphi;a,b) \subseteq S_0(\varphi;a,b)$, hence $\overset{\circ}{S}_\gamma(\varphi;a,b) \subseteq \overset{\circ}{S}_0(\varphi;a,b) = \emptyset$.
From [Na.: XVI §7, Thm 2] we quote

7.10. Proposition

For realvalued f defined on $[a,b]$ one has $f \in \mathcal{D}(a,b)$ iff $_{\gamma\in\Omega}(S_\gamma(f;a,b) \mid \gamma \in \Omega) = \emptyset$.

Next we will see that the S_γ are strictly decreasing.

7.11. Theorem

For $\varphi \in \mathcal{D}(a,b)$, $\gamma \in \Omega$ and $S_\gamma(\varphi;a,b) \neq \emptyset$ one has $S_{\gamma+1}(\varphi;a,b) \subsetneqq S_\gamma(\varphi;a,b)$.

Proof

Take $S := S_\gamma(\varphi;a,b)$; the easy case is $S \cap\,]a,b[\, = \emptyset$. Then $S \subseteq [a,b]$ and hence $\varphi \cdot 1_S \in \mathcal{L}^1(a,b)$; furthermore we have:

$$\int_a^b \varphi(t)\,dt = \lim_{\substack{a'\to a+\\ b'\to b-}} \int_{a'}^{b'} \varphi(t)\,dt = \lim_{\substack{a'\to a+\\ b'\to b-}} (\gamma)\int_{a'}^{b'} \varphi(t)\,dt.$$

(Note that by (7.7.) the (W)-integral is an extension of the $\mathcal{D}_\gamma$-integral).

We get $S_{\gamma+1}(\varphi;a,b) = \emptyset \subsetneqq S_\gamma(\varphi;a,b)$. The second case is $S \cap\,]a,b[\, \neq \emptyset$.

Using (7.1.) we take $a',b' \in [a,b]$, $a' < b'$, $S \cap]a',b'[\neq \emptyset$ and $\varphi \cdot 1_S \in \mathcal{L}^1(a',b')$.

We put $\Psi := (\varphi - \varphi \cdot 1_S) \upharpoonright [a',b']$, $S' := [a',b'] \cap S$; then $\Psi \in \mathcal{D}(a',b')$ and $\Psi \upharpoonright S' = 0$ holds.

Let $(G_n)_{n \in N}$ be the connected components of $G := [a',b'] \setminus S$, $a_n := \inf G_n$ and $b_n := \sup G_n$. By (7.7.), (7.5.) we get

$$\lim_{\substack{a'' \to a_n + \\ b'' \to b_n -}} (\gamma) \int_{a''}^{b''} \varphi(t)\, dt = \int_{a_n}^{b_n} \varphi(t)\, dt = \int_{a_n}^{b_n} \Psi(t)\, dt = \theta(n)$$

and

$$\upsilon_n := \sup_{a_n < a'' < b'' < b_n} \left| (\gamma) \int_{a''}^{b''} \varphi(t)\, dt \right| = \left| \int_{a_n'}^{b_n'} \varphi(t)\, dt \right| = \left| \int_{a_n'}^{b_n'} \Psi(t)\, dt \right|$$

for some $a_n', b_n' \in [a_n,b_n]$, $a_n' < b_n'$.

Now we apply (7.4.) to Ψ, S', a',b' and get $c,d \in [a',b']$, $c < d$, $S' \cap]c,d[\neq \emptyset$ and $\sum_{G_n \subseteq]c,d[} \upsilon_n < \infty$.

Now we choose $x_o \in S' \cap]c,d[\subseteq S_\gamma(\varphi;a,b)$ and have to show $x_o \notin S_{\gamma+1}(\varphi;a,b)$; it is sufficient to prove $\varphi \in \mathcal{D}_{\gamma+1}(c,d)$. We check the conditions in the definition of $\mathcal{D}_{\gamma+1}(\varphi;a,b)$:

(i) $\overset{o}{S}_\gamma(\varphi;c,d) \subseteq \overset{o}{S}_\gamma(\varphi;a,b) = \emptyset$ comes from (7.9.) and we have

$$\varphi \upharpoonright [c,d] \cdot 1_{S_\gamma(\varphi;c,d)} \underset{a.e.}{=} \varphi \cdot 1_S \upharpoonright [c,d] \in \mathcal{L}^1(c,d).$$

(ii) The R_n (connected components of $R_\beta(\varphi;c,d)$) are those G_m satisfying $G_m \subseteq]c,d[$ and perhaps intervals of the form $[c,b_m[$ or $]a_m,d]$. As we have already seen, the limits

$$\theta_m = \lim_{\substack{a' \to a_m^+ \\ b' \to b_m^-}} (\gamma) \int_{a'}^{b'} (t)\, dt$$

always exist; the remarks from above also yield (iii).

This shows $\varphi \in \mathcal{D}_{\gamma+1}(c,d)$ and proves the theorem.

From [Na:XV §3] we quote the Cantor - Baire Stationary Principle:

<u>7.12. Proposition</u>

If $(F_\alpha)_{\alpha<\Omega}$ is a family of closed subsets of $[a,b]$ satisfying $F_\beta \subseteq F_\alpha$

for $\alpha \leq \beta$. Then $F_\alpha = F_\gamma$ for some $\gamma < \Omega$ and all $\alpha, \gamma \leq \alpha < \Omega$.

An application gives our final comparison theorem.

__7.13. Theorem__

$\mathbb{D}(a,b) \subseteq \mathfrak{D}(a,b)$.

__Proof__

For $\varphi \in \mathbb{D}(a,b)$ and $\alpha < \Omega$ we put $F_\alpha := S_\alpha(\varphi;a,b)$; the stationary principle shows $F_\gamma = F_{\gamma+1}$ for some γ and (7.11.) proves $F_\gamma = \emptyset$. Using (7.10.) and (7.6.) we obtain $\varphi \in \mathfrak{D}(a,b)$.

We summarize the results in 7.6 and 7.13. and obtain $\mathbb{D}(a,b) = \mathfrak{D}(a,b)$ $= \mathfrak{P}(a,b) = \mathfrak{P}'(a,b)$.

References

[Ben] B. Benninghofen: Infinitesimalien und Superinfinitesimalien, Dissertation (1982) RWTH Aachen

[BenRi] B. Benninghofen, M. M. Richter: General Theory of Superinfinitesimals (1983), preprint.

[D] J. Dieudonné: Grundzüge der modernen Analysis 1, 2; Vieweg (1975).

[Flo] Floret: Maß- und Integrationstheorie; Teubner (1981).

[Hen 1] R. Henstock: Definitions of Riemann type of the variational integrals; Proc. London Math. Soc. (3), 11(1961), 402 - 418

[Hen 2] R. Henstock: A Riemann type integral of Lebesque power, Canad. J. Math. 20(1968), 79 - 87.

[Hen 3] R. Henstock: The equivalence of generalized forms of the Ward, variational, Denjoy Stieltjes and Penon - Stieltjes integrals; Proc. London Math. Soc. 10(1960), 281 - 303.

[Ku 1] J. Kurzweil: Generalized ordinary differential equations and continuous dependence on a parameter; Czechoslovak. Math. J. 7(82) (1957) 418 - 446

[Ku 2] J. Kurzweil: Nicht absolut konvergente Integrale; Teubner - Texte zur Mathematik, Band 26 (1980).

[McL] R. M. McLeod: The generalized Riemann integral; The Carus Mathematical Monographs (1980).

[McS] E. J. McShane: A unified theory of integration; Amer. Math. Monthly, 80(1973) 349 -359.

[Na] I. P. Natanson: Theory of the functions of a real variable.

[Ne] E. Nelson: Internal Set Theory; Bull. Amer. Math. Soc. 83, 1165 - 1198.

[Ri] M. M. Richter: Ideale Punkte, Monaden und Nichtstandardmethoden; Vieweg 1982

[Ru] W. Rudin: Real and Complex Analysis; Mc Graw - Hill Series in Higher Mathematics

[Saks] S. Saks: Theory of the Integral; Monografie Matematyczne VIII, 1937.

[StrBa] K. D. Stroyan, J.M. Bayod: Foundations of infinitesimal stochastic Analysis, to appear.

[StrLu] K. D. Stroyan, W. A. J. Luxemburg: Introduction to the Theory of Infinitesimals; Academic Press.

POINT-PICKING GAMES AND HFC'S

Andrew J. Berner*
University of Dallas

István Juhász
Mathematical Institute of the
Hungarian Academy of Sciences

1. INTRODUCTION AND DEFINITIONS

In this paper, we consider winning strategies for the following topological game.

Definition 1.1. If X is a topological space, α an ordinal and P a property of subsets of X, the game $G_\alpha^P(X)$ is played in the following manner:

Two players take turns playing. A round consists of Player I choosing an open set $U \subset X$ and Player II choosing a point $x \in U$. A round is played for each ordinal less than α. Player I wins the game if the set of points Player II played has property P. Otherwise, Player II wins.

In this paper, property P will either be dense, dense in itself, somewhere dense, or non-discrete. These properties will be abbreviated D, DI, SD, and ND, respectively.

The set of all well-ordered sequences of points of X of order-type less than α (including the null sequence) will be notated by $X^{<\alpha}$

Definition 1.2. a) A strategy for Player I in $G_\alpha^P(X)$ is a function $s: X^{<\alpha} \to \tau(X)$ (where $\tau(X)$ is the collection of non-empty open subsets of X).

b) A sequence $\langle x_\beta : \beta < \alpha \rangle$ of points of X is a play for s if $x_\gamma \in s(\langle x_\beta : \beta < \gamma \rangle)$ for all $\gamma < \alpha$.

c) A winning strategy for Player I in $G_\alpha^P(X)$ is a strategy s such that for every play $\langle x_\beta : \beta < \alpha \rangle$ for s, the set

* Research partially supported by the O'Hara Chemical Sciences Institute of the University of Dallas

$\{x_\beta : \beta < \alpha\}$ has property P.

Definition 1.3. a) A strategy for Player II in G_α^P is a function $s: \tau(X)^{<\alpha} \times \tau(X) \to X$ such that $s(f,U) \in U$ for all $(f,U) \in \tau(X)^{<\alpha} \times \tau(X)$.

b) A sequence $\langle (U_\beta, x_\beta): \beta < \alpha \rangle$, where $U_\beta \in \tau(X)$ and $x_\beta \in U_\beta$, is a play for s if $x_\gamma = s((\langle U_\beta : \beta < \gamma \rangle, U_\gamma))$ for all $\gamma < \alpha$.

c) A winning strategy for Player II in $G_\alpha^P(X)$ is a strategy s such that for every play $\langle (U_\beta, x_\beta): \beta < \alpha \rangle$ for s, the set $\{x_\beta : \beta < \alpha\}$ does not have property P.

A strategy for either player is thus a prescription for that player that tells him what to move in response to all the previous moves of his opponent.

Definition 1.4. a) We say $I \uparrow G_\alpha^P(X)$ (read Player I wins $G_\alpha^P(X)$) if there is a winning strategy for Player I in $G_\alpha^P(X)$. $II \uparrow G_\alpha^P(X)$ is defined similarly.

b) We say $I \not\uparrow G_\alpha^P(X)$ if there is no winning strategy for Player I in $G_\alpha^P(X)$; similarly we define $II \not\uparrow G_\alpha^P(X)$.

Definition 1.5. $G_\alpha^P(X)$ is neutral if $I \not\uparrow G_\alpha^P(X)$ and $II \not\uparrow G_\alpha^P(X)$.

In later sections we will need a different notion of winning strategy for Player II, which is equivalent to the notion in definition 1.3:

Definition 1.6. a) A point-strategy for Player II in $G_\alpha^P(X)$ is a function $s: X^{<\alpha} \times \tau(X) \to X$ such that $s(g,U) \in U$ for all $(g,U) \in X^{<\alpha} \times \tau(X)$.

b) A sequence $\langle (U_\beta, x_\beta): \beta < \alpha \rangle$, where $U_\beta \in \tau(X)$ and $x_\beta \in U_\beta$ is a play for s if $x_\gamma = s((\langle x_\beta : \beta < \gamma \rangle, U_\gamma))$ for all $\gamma < \alpha$.

c) A winning point-strategy for Player II in $G_\alpha^P(X)$ is a point-strategy s such that for every play $\langle (U_\beta, x_\beta): \beta < \alpha \rangle$

for s, the set $\{x_\beta : \beta < \alpha\}$ does not have property P.

Lemma 1.7. $II \uparrow G^P_\alpha(X)$ if and only if there is a winning point-strategy for Player II in $G^P_\alpha(X)$.

Proof. Suppose $s: \tau(X)^{<\alpha} \times \tau(X) \to X$ is a winning strategy for Player II in $G^P_\alpha(X)$. Inductively define functions $T_\beta: X^\beta \to \tau(X)^\beta$ for each $\beta < \alpha$ with the following properties:

i) if $T_\beta(\langle x_\delta : \delta < \beta\rangle) = \langle U_\delta : \delta < \beta\rangle$ then $x_\delta \in U_\delta$ for each

ii) if $\gamma < \beta$, $\langle x_\delta : \delta < \beta\rangle \in X^\beta$ and

$T_\beta(\langle x_\delta : \delta < \beta\rangle) = \langle U_\delta : \delta < \beta\rangle$ then

$T_\gamma(\langle x_\delta : \delta < \gamma\rangle) = \langle U_\delta : \delta < \gamma\rangle$

iii) suppose $\beta = \gamma + 1$, $\langle x_\delta : \delta < \beta\rangle \in X^\beta$ and there is a

$U \in \tau(X)$ such that $x_\gamma = s(T_\gamma(\langle x_\delta : \delta < \gamma\rangle), U)$;

if $T_\beta(\langle x_\delta : \delta < \beta\rangle) = \langle U_\delta : \delta < \beta\rangle$ then

$x_\gamma = s(\langle U_\delta : \delta < \gamma\rangle, U_\gamma)$ (Note that from property (ii),

$\langle U_\delta : \delta < \gamma\rangle = T_\gamma(\langle x_\delta : \delta < \gamma\rangle)$.)

If we are at stage β and β is a limit ordinal, property (ii) determines the choice of T_β. If $\beta = \gamma + 1$, then we can define $T_\beta(\langle x_\delta : \delta < \beta\rangle)$ consonant with property (iii) if $x_\gamma = s(T_\gamma(\langle x_\delta : \delta < \gamma\rangle), U)$ for some U, which will automatically satisfy (i) and (ii); and we can arbitrarily define $T_\beta(\langle x_\delta : \delta < \beta\rangle)$ otherwise, making sure (i) and (ii) are satisfied.

Property (ii) of the above inductive construction lets us define $T: X^\alpha \to \tau(X)^\alpha$ by the formula

$$T(\langle x_\delta : \delta < \alpha\rangle) = \bigcup\{T_\beta(\langle x_\delta : \delta < \beta\rangle) : \beta < \alpha\}$$

Finally, define the point-strategy $s': X^{<\alpha} \times \tau(X) \to X$ by

$$s'(\langle x_\delta : \delta < \beta\rangle, U) = s(T_\beta(\langle x_\delta : \delta < \beta\rangle, U)$$

Suppose $\langle (V_\beta, x_\beta) : \beta < \alpha\rangle$ is a play for s'. Let $T(\langle x_\beta : \beta < \alpha\rangle) = \langle U_\beta : \beta < \alpha\rangle$. Suppose $\gamma < \alpha$. Then $x_\gamma = s'(\langle x_\beta : \beta < \gamma\rangle, V_\gamma) = s(T_\gamma(\langle x_\beta : \beta < \gamma\rangle), V_\gamma)$ $= s(\langle U_\beta : \beta < \gamma\rangle, V_\gamma)$. Thus, since $T_{\gamma+1}(\langle x_\beta : \beta < \gamma + 1\rangle) = \langle U_\beta : \beta < \gamma + 1\rangle$, condition (iii) of the inductive construction shows that $x_\gamma = s(\langle U_\beta : \beta < \gamma \rangle, U_\gamma)$. Thus $\langle (U_\beta, x_\beta) : \beta < \alpha \rangle$ is a play for s, and therefore $\{x_\beta : \beta < \alpha\}$ does not have property P. This shows that s' is a winning point-strategy in $G^P_\alpha(X)$.

Conversely, suppose we have a winning point-strategy $s' : X^{<\alpha} \times \tau(X) \to X$. Inductively define functions $t_\beta : \tau(X)^\beta \to X^\beta$ for $\beta < \alpha$ with the following properties:

i) if $t_\beta(\langle U_\delta : \delta < \beta \rangle) = \langle x_\delta : \delta < \beta \rangle$ then $x_\delta \in U_\delta$

ii) if $\gamma < \beta$, then $t_\beta(\langle U_\delta : \delta < \beta \rangle)$ extends $t_\gamma(\langle U_\delta : \delta < \gamma \rangle)$

iii) if $\beta = \gamma + 1$, and if $t_\beta(\langle U_\delta : \delta < \beta \rangle) = \langle x_\delta : \delta < \beta \rangle$ then $x_\gamma = s'(\langle x_\delta : \delta < \gamma \rangle, U_\gamma)$

As in the previous construction, property (ii) determines t_β at limit stages β. In this constuction, property (iii) along with property (ii) determine t_β at successor stages. If Player I begins playing $G^P_\alpha(X)$ with the sequence $\langle U_\delta : \delta < \beta \rangle$, then the sequence $t_\beta(\langle U_\delta : \delta < \beta \rangle)$ is simply Player II's repsponses by using the point-strategy s'.

Now define $s : \tau(X)^{<\alpha} \times \tau(X) \to X$ by the formula $s(\langle U_\beta : \beta < \gamma \rangle, U) = s'(t_\gamma(\langle U_\beta : \beta < \gamma \rangle), U)$. The remark above shows that a play for s is also a play for the winning point-strategy s'. Thus s is a winning strategy.

<u>Definition 1.8.</u> A set V is <u>somewhere separable</u> if there is a separable open $U \subset V$.

<u>Definition 1.9.</u> [J.] An infinite subspace $X \subset 2^{\omega_1}$ (where $2 = \{0,1\}$ with the discrete topology) is called an <u>HFD</u> (short for hereditarily finally dense) if the following condition is satisfied:

For every $A \in [X]^\omega$ there exists a $\nu_A < \omega_1$ such that whenever ε is any finite partial function from $\omega_1 - \nu_A$ into 2, there is $f \in A$ with $\varepsilon \subset f$.

<u>Notation</u>: If X is a set, then <u>H(X)</u> is the set of all finite partial functions from X to $\{0,1\}$. If $f \in H(\omega_1)$ then

$\langle f\rangle = \{p \in 2^{\omega_1}: p \text{ extends } f\}$. Thus $\{\langle f\rangle: f \in H(\omega_1)\}$ is a basis for 2^{ω_1}.

All spaces considered in this paper will be T_3 spaces with no isolated points. Thus, if $T \subset X$ then

T is discrete $\Rightarrow$ T is nowhere dense $\Rightarrow$ T is not dense.

An immediate consequence of this is that a strategy for Player I in $G^D_\alpha(X)$ is a strategy for $G^{SD}_\alpha(X)$, and a strategy for $G^{SD}_\alpha(X)$ is a strategy for $G^{ND}_\alpha(X)$. For Player II, a strategy for $G^{ND}_\alpha(X)$ is a strategy for $G^{SD}_\alpha(X)$ which in turn is a strategy for $G^D_\alpha(X)$.

2. STRATEGIES FOR PLAYER I

Theorem 2.1. If $\varkappa^{\aleph} = \varkappa$, then $I \uparrow G^D_\varkappa(X) \Leftrightarrow \pi(X) \le \varkappa$.

Proof. If $\pi(X) \le \varkappa$, then let $\{U_\alpha: \alpha < \varkappa\}$ be a π-base for X and define $s: X^{<\varkappa} \to \tau(X)$ by $s(\langle x_\beta: \beta < \alpha\rangle) = U_\alpha$. If $\langle (V_\alpha, x_\alpha): \alpha < \varkappa\rangle$ is a play for s, then $V_\alpha = U_\alpha$ for all $\alpha < \varkappa$, and so $\{x_\alpha: \alpha < \varkappa\}$ is dense. Thus s is a winning strategy.

Suppose now $I \uparrow G^D_\varkappa(X)$ and $s: X^{<\varkappa} \to \tau(X)$ is a winning strategy for Player I in $G^D_\varkappa(X)$. Clearly $d(X) \le \varkappa$, since Player I can win the game! Let D be a dense subset of X with $|D| \le \varkappa$. Then $\mathcal{B} = s(D^{<\varkappa})$ is a π-base for X, for suppose, on the contrary, there is a non-empty open set U such that $B - \overline{U} \ne \emptyset$ for all $B \in \mathcal{B}$ (recall X is T_3). Since $s(\emptyset) \in \mathcal{B}$, we can inductively define a play $\langle (U_\alpha, x_\alpha): \alpha < \varkappa\rangle$ for s with $U_\alpha \in \mathcal{B}$ and $x_\alpha \in (U_\alpha - \overline{U}) \cap D$ for each $\alpha < \varkappa$. But then $\{x_\alpha: \alpha < \varkappa\} \cap U = \emptyset$, contradicting the fact that s was a winning strategy for Player I in $G^D_\varkappa(X)$. Thus $\mathcal{B}$ is indeed a π-base, and $|\mathcal{B}| \le \varkappa^{\aleph} = \varkappa$, so $\pi(X) \le \varkappa$.

Theorem 2.2. If X has no left separated sequence of length α then $I \uparrow G^D_\alpha(X)$.

Proof. Let $s: X^{<\alpha} \to \tau(X)$ be defined by $s(\langle x_\gamma: \gamma < \beta\rangle) = X - \overline{\{x_\gamma: \gamma < \beta\}}$ if this is not empty, and X otherwise. If a play for s were not dense, Player II would have played a left separated sequence of length α. Thus s is a winning strategy in $G^D_\alpha(X)$.

Corollary 2.3. a) If $|X| < \varkappa$, then I $\uparrow$ $G^D_\varkappa(X)$ (note that Player II is not required to play a "new" point in each round).

b) If $z(X) = \varkappa$ and $\alpha > \varkappa$, then I $\uparrow$ $G^D_\alpha(X)$.

c) If X is hereditarily separable, (e.g. an HFD), then I $\uparrow$ $G^D_{\omega_1}(X)$.

There are countable spaces for which II $\uparrow$ $G^{ND}_\alpha(X)$ for every $\alpha < \omega_1$, thus Corollary 2.3c is best possible. See also Example 2.6.

Since there are countable spaces X with $\pi(X) = c$, we cannot drop the hypothesis $\varkappa^{\aleph_0} = \varkappa$ from Theorem 2.1.

Our main result in this section uses the following lemma, first observed by van Douwen. It is the key to the proof that $\pi_d(X) = \omega$ iff $\pi_o(X) = \omega$, as noted in [J-W].

Recall that $\pi_o(X)$, as defined in [J-W], is $\min \{\pi(U) : U \text{ is a non-empty open subset of } X\}$.

Lemma 2.4. Suppose $\pi_o(X) > \omega$ and $\{U_i : i \in \omega\}$ is a countable collection of non-empty open sets in X. Then there is a collection $\{V_i : i \in \omega\}$ of pairwise disjoint non-empty open sets such that $V_i \subset U_i$ for every $i \in \omega$.

Main idea of the proof. Inductively define the sequence $\langle V_i : i \in \omega \rangle$ with V_i open and $\overline{V_i} \subset U_i - \bigcup\{\overline{V_j} : j < i\}$, using the fact that $\{U_k - \bigcup\{\overline{V_j} : j < i\} : k > i\}$ does not form a π-base for $U_i - \bigcup\{\overline{V_j} : j < i\}$.

Theorem 2.5. I $\uparrow$ $G^{SD}_\omega(X)$ if and only if $\pi_o(X) = \omega$.

Proof. If $\pi_o(X) = \omega$, Player I has an obvious strategy. So assume $\pi_o(X) > \omega$ and $s : X^{<\omega} \to \tau(X)$ is a strategy for Player I in $G^{SD}_\omega(X)$. We inductively construct a tree, as follows.

A node of the tree will be of the form $(O, U, C, \langle x_i : i < j \rangle)$ where

(1) $O = s(\langle x_i: i < j\rangle)$

(2) If O is somewhere separable, then U is a separable open subset of O and C is a countable dense subset of U. Otherwise, U = O and C is a singleton subset of O.

Note that given a finite sequence $\langle x_i: i < j\rangle \in X^{<\omega}$, we can define a node $(O, U, C, \langle x_i: i < j\rangle)$.

Let $(O_0, U_0, C_0, \langle\,\rangle)$ be the root of the tree (thus $O_0 = s(\langle\,\rangle)$). We will continue inductively by levels of the tree. Suppose we have defined a node $N = (O, U, C, \langle x_0 \ldots x_{k-1}\rangle)$ on level $k < \omega$. For each $x \in C$, define a node $N_x = (O_x, U_x, C_x, \langle x_0 \ldots x_{k-1}, x\rangle)$ which will be a child of N on level k + 1. Make x a label on the edge from N to N_x. This inductively defines a tree of countable height and countable width.

Notice that if $\langle x_i: i \in \omega\rangle$ is the sequence of labels of the edges of a branch through the tree, then $\langle x_i: i \in \omega\rangle$ is a play for s. Thus Player II's goal is to find a branch through the tree where the set of labels forms a nowhere dense set. We will now show that there is guaranteed to be such a branch.

Let $\mathcal{S} = \{O: (O, U, C, f)$ is a node of the tree and O is somewhere separable$\}$. Index $\mathcal{S}$ as $\{O_i: i \in A \subset \omega\}$ (Note: $\mathcal{S}$ may be countably infinite, finite or even empty; thus the use of $A \subset \omega$ as the index set.) If $O_i \in \mathcal{S}$, let (O_i, U_i, C_i, f_i) be the corresponding node of the tree (thus if two nodes have the same first coordinate, the indexing of $\mathcal{S}$ will not be one-to-one) Apply Lemma 2.4 to $\{U_i: i \in A\}$ to obtain $\{V_i: i \in A\}$ and choose a discrete set $D = \{d_i: d_i \in C_i \cap V_i\}$. Now inductively define a branch through the tree as follows. Start, of course, at the root. Suppose at level k we are at a node (O, U, C, f). If $O = O_i \in \mathcal{S}$, continue the branch by picking the edge labeled $d_i \in D \cap C_i$. If $O \notin \mathcal{S}$, then C is a singleton and there is a unique way to continue the branch.

Let P be the set of labels of the edges of the branch. $P \cap D$ is discrete. We will show that $P - D$ is nowhere dense. Suppose not, and let V be a non-empty open subset of $\overline{P - D}$. Let $x \in (P - D) \cap V$. Then $x \in O$ for some node $(O, O, \{x\}, f)$ with $O \notin \mathcal{S}$. But then $V \cap O$ is a separable open subset of O, contradicting the fact that $O \notin \mathcal{S}$

Thus P is nowhere dense, so s is not a winning strategy for Player I.

The proof shows that if every open subset of X is somewhere separable, then $I \uparrow G^{ND}_{\omega}(X)$ if and only if $\pi_o(X) = \omega$. The following example shows this is not true in general.

Example 2.6. $I \uparrow G^{DI}_{\omega}(\Sigma(2^{\omega_1}))$, where

$\Sigma(2^{\omega_1}) = \{f \in 2^{\omega_1} : \exists\, \alpha < \omega_1 \text{ s.t. } f(\beta) = 0 \text{ for all } \beta > \alpha\}$.

Proof. If $f \in (2^{\omega_1})$, let $\alpha(f) = \min\{\alpha : f(\beta) = 0 \text{ for all } \beta > \alpha\}$. For each $\alpha \in \omega_1$, index $H(\alpha) = \{\varepsilon_{\alpha i} : i \in \omega\}$. Let $\{A_i : i \in \omega\}$ be a partition of $\omega - \{0\}$ with $|A_i| = \omega$ for each i and $A_i \subset \omega - i{+}1 = \{j \in \omega : j > i\}$. Index $A_i = \{a_{ij} : j \in \omega\}$.

Define $s : \Sigma(2^{\omega_1})^{<\omega} \to \tau(\Sigma(2^{\omega_1}))$ as follows. Let $s(\langle\rangle) = \Sigma(2^{\omega_1})$. Suppose $\langle f_0 \ldots f_n\rangle \in \Sigma(2^{\omega_1})^{<\omega}$ and $n + 1 \in A_i$ for some i. Then $n + 1 = a_{ij}$ for some j. Let $\alpha = \alpha(f_i)$ and define $s(\langle f_0 \ldots f_n\rangle) = \langle \varepsilon_{\alpha j}\rangle \cap \Sigma(2^{\omega_1})$. Then s is a winning strategy for Player I in $G^{DI}_{\omega}(\Sigma(2^{\omega_1}))$.

Notation. If $X \subset 2^{\omega_1}$, let $\underline{H_X(\alpha)} = \{f \in H(\alpha) : \langle f\rangle \cap X \neq \emptyset\}$.

Theorem 2.7. If X is an HFD, then $I \uparrow G^{D}_{\omega\cdot\omega}(X)$.

Proof. If S is an infinite subset of an HFD X, note that

(*) there is an $\alpha(S) < \omega_1$ such that if $f \in H(\alpha(S))$ and $|S \cap \langle f\rangle| \geq \omega$, then $S \cap \langle f\rangle$ is dense past $\alpha(S)$ (i.e. $\{p|\omega_1 - \alpha(S) : p \in S \cap \langle f\rangle\}$ is dense in $2^{\omega_1 - \alpha(S)}$).

For each $\alpha < \omega_1$, index $H_X(\alpha)$ as $\{f_{\alpha,i} : i < \omega\}$.

Define $s : X^{<\omega\cdot\omega} \to \dot{\tau}(X)$ as follows:

Choose $\nu = \alpha(X)$, as defined by (*), and let $\mu = \nu + \omega$.

If $i < \omega$ and $\langle x_0 \dots x_{i-1}\rangle \in X^{<\omega} \subset X^{<\omega\cdot\omega}$, let $s(\langle x_0 \dots x_{i-1}\rangle) = \langle f_{\mu,i}\rangle \cap X$.

If $\beta = \omega\cdot n + i$ with $n, i < \omega$ and $n \geq 1$, and $\langle x_\gamma : \gamma < \beta\rangle \in X^{\omega\cdot\omega}$, let $S = \{x_\gamma : \gamma < \omega\cdot n\}$. If S is infinite, let $\alpha(S)$ be as defined by (*) above. (The cas^ that S is finite is irrelevant, and we can arbitrarily define $\alpha(S)$ in that case.) Now let $s(\langle x_\gamma : \gamma < \beta \rangle) = \langle f_{\alpha(S),i}\rangle \cap X$.

To see that s is a winning strategy for Player I, consider a play $\langle x_\beta : \beta < \omega\cdot\omega\rangle$ and let $f \in H_X(\omega_1)$. Let $S(n) = \{x_\beta : \beta < \omega\cdot n\}$. Note that for $n \geq 1$, $|S(n)| = \omega$. Let $\alpha = \sup\{\alpha(S(n)) : 1 \leq n < \omega\}$ and let $f = f_1 \cup f_2$ where $f_1 \in H(\alpha)$ and $\mathrm{dom}(f_2) \subset \omega_1 - \alpha$. Since $f_1 \in H(\alpha(S(n)))$ for some n, $D = \{x_\beta : \omega\cdot n \leq \beta < \omega\cdot(n+1) \text{ and } x_\beta \in \langle f_1\rangle\}$ is infinite, thus dense past α, so $D \cap \langle f\rangle \cap X \neq \emptyset$.

3. AN HFD WHERE PLAYER II WINS

<u>Theorem 3.1.</u> (CH ⇒) There is an HFD X with $|X| = \omega_1$ such that $II \uparrow G_\omega^{ND}(X)$.

We assume the reader is familiar with the standard construction of an HFD from CH. In particular, at stage $\alpha < \omega_1$ in the inductive construction, we define functions $f_{\beta\alpha} : \alpha + 1 \to 2$ for each $\beta < \omega_1$ that extend those defined at the earlier stages. The HFD will then be $\{f_\beta = \bigcup\{f_{\beta\alpha} : \alpha < \omega_1\} : \beta < \omega_1\}$. To do this, we have, at stage α, a countable collection $Z(\alpha)$ of countable subsets of ω_1. We find a set $B(\alpha) \subset \omega_1$ such that for each $A \in Z(\alpha)$ both $A \cap B(\alpha)$ and $A - B(\alpha)$ are infinite. We will say the set $B(\alpha)$ splits $Z(\alpha)$. Each $S \in [\omega_1]^\omega$ is included in $Z(\alpha)$ for some α, and if $S \in Z(\alpha)$, then the construction guarantees that $\{f_\beta : \beta \in S\}$ is dense after α.

The following two lemmas show how to modify the standard construction to construct HFD's with some pre-defined values.

<u>Lemma 3.2.</u> Suppose Z is a countable collection of countably infinite

sets and $\mathcal{S}$ is a collection of pairwise almost disjoint sets such that for every $A \in Z$ and every finite collection $\mathcal{F} \subset \mathcal{S}$, $|A - \cup\mathcal{F}| = \omega$. Then there is a set B such that for every $A \in Z$ and finite collection $\mathcal{F} \subset \mathcal{S}$, $(A \cap B) - \cup\mathcal{F}$ and $(A - B) - \cup\mathcal{F}$ are both infinite.

Proof. For each $A \in Z$, let $\mathcal{S}(A) = \{S \in \mathcal{S} : |A \cap S| = \omega\}$. We say A is light if $|\mathcal{S}(A)| < \omega$, and heavy otherwise. The hypothesis of the lemma guarantees that if A is light, $|A - \cup\mathcal{S}(A)| = \omega$. If A is heavy, let $\mathcal{S}'(A)$ be a countably infinite subset of $\mathcal{S}(A)$. Define

$Z' = \cup\{\{A \cap S: S \in \mathcal{S}'(A)\}: A \text{ is heavy}\} \cup \{A - \cup\mathcal{S}(A): A \text{ is light}\}$.

Then Z' is also a countable collection of countably infinite sets, so there is a set B which splits Z'. Suppose $\mathcal{F}$ is a finite subcollection of $\mathcal{S}$, and $A \in Z$. If A is light, then for each $S \in \mathcal{F}$, $|(A - \cup\mathcal{S}(A)) \cap S| < \omega$. Therefore, since $A - \cup\mathcal{S}(A) \in Z'$, $|((A - \cup\mathcal{S}(A)) \cap B) - \cup\mathcal{F}| = \omega$ and $|((A - \cup\mathcal{S}(A)) - B) - \cup\mathcal{F}| = \omega$, hence $(A \cap B) - \cup\mathcal{F}$ and $(A - B) - \cup\mathcal{F}$ are both infinite. If A is heavy, choose $S \in \mathcal{S}'(A) - \mathcal{F}$. Then $A \cap S \in Z'$ and, since $\mathcal{S}$ is pairwise almost disjoint, $|\cup\mathcal{F} \cap (A \cap S)| < \omega$. Since $(A \cap S) \cap B$ and $(A \cap S) - B$ are both infinite, so are $((A \cap S) \cap B) - \cup\mathcal{F}$ and $((A \cap S) - B) - \cup\mathcal{F}$, and thus

$(A \cap B) - \cup\mathcal{F}$ and $(A - B) - \cup\mathcal{F}$ are both infinite.

Lemma 3.3. (CH) Suppose $L \subset \omega_1$ consists of limits and $\mathcal{S} = \{S_\lambda: \lambda \in L\} \subset [\omega_1]^\omega$ is a pairwise almost disjoint collection such that $S_\lambda \subset \lambda$ for each $\lambda \in L$. Suppose for each $\beta < \omega_1$ there is a function $g_\beta \in 2^\beta$ and for each $\lambda \in L$ and $\beta \in S_\lambda$, $h^\lambda_\beta \in 2^{\lambda+\omega-\lambda}$. Then there is an HFD $X = \{f_\beta: \beta \in \omega_1\} \subset 2^{\omega_1}$ such that

a) If $\lambda \in L$ and $\beta \in S_\lambda$ then $h^\lambda_\beta = f_\beta$ and

b) For every $\beta < \omega_1$, $f_\beta|\beta = g_\beta$

Proof. First, for each $\beta < \omega_1$ define $p_\beta = \cup\{h^\lambda_\beta: \beta \in S_\lambda\} \cup g_\beta$. As in the standard construction, we will inductively define $f_{\beta\alpha}: \alpha+1 \to 2$

but we will be sure that $p_\beta | \alpha + 1 \subset f_{\beta\alpha}$. We will define $f_\beta = \cup\{f_{\beta\alpha} : \alpha < \omega_1\}$; f_β will therefore be an extension of p_β, and thus (a) and (b) will be satisfied.

Now for the inductive construction. Suppose $A \in [\omega_1]^\omega$ and L' is a finite subset of L such that $|A - \cup\{S_\lambda : \lambda \in L'\}| < \omega$. Then, since distinct elements of $\mathcal{S}$ are almost disjoint, if $\lambda \notin L'$, $|A \cap S_\lambda| < \omega$. Thus we can index $[\omega_1]^\omega$ as $\{A(\alpha) : \alpha \in I \subset \omega_1\}$ where

(1) $\alpha \geq \sup(A(\alpha))$ for every $\alpha \in I$ and

(2) If L' is a finite subset of L and for all $\lambda \in L'$, $\lambda + \omega \geq \alpha$, then $|A(\alpha) - \cup\{S_\lambda : \lambda \in L'\}| = \omega$.

We will carry out the usual inductive construction of an HFD with the proviso that if $Z(\gamma)$ is the countable collection of subsets of γ which was split at stage γ, we want to ensure that if $A \in Z(\gamma)$ and $\mathcal{F}$ is a finite subset of $\{S_\lambda : \lambda \in L$ and $\lambda + \omega > \gamma\}$ then $|A - \cup\mathcal{F}| = \omega$. Suppose we have done this for all $\gamma < \alpha$ and we are now at stage α.

As usual, first let

$$Z_1(\alpha) = \begin{cases} \cup\{Z(\gamma) : \gamma < \alpha\} & \text{if } \alpha \notin I \\ \cup\{Z(\gamma) : \gamma < \alpha\} \cup \{A(\alpha)\} & \text{if } \alpha \in I \end{cases}$$

If $\alpha \neq \beta + i$ for any $\beta \in L$, $i < \omega$, then let $Z_2(\alpha) = Z_1(\alpha)$. If $\alpha = \beta + i$ for some $\beta \in L$ and $i < \omega$, then

$$A \in Z_1(\alpha) \Rightarrow |A - S_\beta| = \omega.$$

In that case, let $Z_2(\alpha) = \{A - S_\beta : A \in Z_1(\alpha)\}$.

By applying Lemma 3.2, we can find $B(\alpha) \subset \omega_1$ which splits $Z_2(\alpha)$ such that for every $A \in Z_2(\alpha)$ and finite subset $\mathcal{F}$ of $\{S_\lambda : \lambda \in L$ and $\lambda + \omega > \alpha\}$, $(A \cap B(\alpha)) - \cup\mathcal{F}$ and $(A - B(\alpha)) - \cup\mathcal{F}$ are both infinite.

It follows from the inductive assumption and the care with which the indexing has been done that if $\delta \in A \in Z_2(\alpha)$ then $\alpha \notin \operatorname{dom}(p_\delta)$.

For all $\delta < \omega_1$ define $f_{\delta\alpha}: \alpha + 1 \to 2$ to extend $p_\delta | \alpha + 1$ and $f_{\delta\gamma}$ for all $\gamma < \alpha$ such that if $\delta \in A \in Z_2(\alpha)$ then

$$f_{\delta\alpha}(\alpha) = \begin{cases} 1 & \text{if } \delta \in B(\alpha) \\ 0 & \text{if } \delta \notin B(\alpha) \end{cases}$$

Finally, let

$Z(\alpha) = Z_1(\alpha) \cup \{A \cap B(\alpha): A \in Z_2(\alpha)\} \cup \{A - B(\alpha): A \in Z_2(\alpha)\}$.

This completes the inductive definition of an HFD

$X = \{f_\beta: \beta < \omega_1\}$.

Proof of theorem 3.1. Let $\{C_\alpha: \alpha \in \omega_1\}$ be a collection of pairwise disjoint uncountable subsets of ω_1 with $C_\alpha \subset \omega_1 - (\alpha + 1)$. For each $\alpha \in \omega_1$, index $H(\omega_1)$ as $\{h_{\alpha\beta}: \beta \in C_\alpha\}$ with dom $(h_{\alpha\beta}) \subset \beta$.

We say a sequence $\langle \alpha(i): i \in \omega \rangle$ is a strategic sequence if $\alpha(0) \in C_0$ and for each $i < \omega$, $\alpha(i+1) \in C_{\alpha(i)}$.

Note that a strategic sequence is increasing. We will identify a strategic sequence with its range. Let $\mathcal{S}$ be the set of all strategic sequences and index $\mathcal{S}$ as $\{S(\alpha) = \langle s(\alpha,i): i \in \omega \rangle: \alpha \in L \subset \omega_1\}$, where for each $\alpha \in L$, α is a limit ordinal and $\alpha > \sup(S(\alpha))$. Note that $\mathcal{S}$ is a pairwise almost disjoint collection of countable sets. For each $\alpha \in \omega_1$ and $\beta \in C(\alpha)$, define $g_\beta: \beta \to 2$ to agree with $h_{\alpha\beta}$. Next, suppose $\alpha \in L$ and $\beta = s(\alpha, i)$. Define $h^\alpha_\beta \in 2^{\alpha+\omega - \alpha}$ by $h^\alpha_\beta(\alpha + i) = 1$ and $h^\alpha_\beta(\alpha + j) = 0$ for all $j \neq i$, $j \in \omega$.

Using Lemma 3.3, construct an HFD $X = \{f_\beta: \beta \in \omega_1\}$. Notice that this will assure that for each $\alpha \in \omega_1$, $\{f_\beta: \beta \in C_\alpha\}$ is dense in X, and for $\alpha \in L$, $\{f_\beta: \beta \in S(\alpha)\}$ forms a discrete set in X.

Player II will have a winning point-strategy $s: X^{<\omega} \times \tau(X) \to X$ in $G^{ND}_\omega(X)$, defined as follows.

If $(\langle f_{\alpha(i)}: i < k \rangle, U) \in X^{<\omega} \times \tau(X)$ choose $h \in H(\omega_1)$ such that $\langle h \rangle \cap X \subset U$. Then $h = h_{\alpha(k-1)\beta}$ for some $\beta \in C_{\alpha(k-1)}$ (use $\alpha(k-1) = 0$ if $k = 0$).

Let $s(\langle f_{\alpha(i)}: i < k \rangle, U) = f_\beta$. The construction of X assures that f_β extends h, and thus $f_\beta \in U$.

If $\langle (U_i, f_{\alpha(i)}) : i < \omega\rangle$ is a play for s, then $\langle \alpha(i) : i < \omega\rangle$ is a strategic sequence, and thus $\{f_{\alpha(i)} : i < \omega\}$ is discrete in X.

Since s is a winning point-strategy, Lemma 1.7 shows II $\uparrow G^{ND}_{\omega}(X)$.

Theorem 3.1 and Theorem 2.7 show that Player I can have a winning strategy for $G^D_\alpha(X)$ while Player II has one for $G^D_\beta(X)$ for some $\beta < \alpha < \omega_1$

Notice that if $\alpha < \omega\cdot\omega$, then Player I has a strategy for $G^{SD}_\alpha(X)$ iff $\pi_o(X) = \omega$. To see this, note that for $n \in \omega$, Player II can treat the rounds $\{\omega\cdot n + i : i < \omega\}$ as the game $G^{SD}_\omega(X)$. If Player I announces a strategy for $G^{SD}_\alpha(X)$, Player II could then use the methods of Theorem 2.2 to play a set which is the union of a finite number of nowhere dense sets. Similarly, if Player II has a winning strategy for $G^{SD}_\omega(X)$, it extends to a strategy for $G^{SD}_\alpha(X)$ for each $\alpha < \omega\cdot\omega$. Thus Theorem 2.7 is best possible.

Question 3.4. Is there a space X in ZFC such that II $\uparrow G^D_\omega(X)$ but I $\uparrow G^D_{\omega\cdot\omega}(X)$?

Question 3.5. Is there (even consistently) a countable ordinal $\alpha \neq \omega\cdot\omega$ for which there is a space X such that I $\uparrow G^D_\alpha(X)$ but for all $\beta < \alpha$, II $\uparrow G^D_\beta(X)$, or even I $\not\uparrow G^D_\beta(X)$?

4. A NEUTRAL GAME

Theorem 4.1. ($\diamondsuit \Rightarrow$) There is an HFD X such that II $\not\uparrow G^D_\omega(X)$. Thus, in light of Theorem 2.5, $G^D_\omega(X)$ is neutral.

Proof. We will again modify the standard construction of an HFD from CH by modifying the countable collection $Z(\alpha)$ at stage α.

From $\diamondsuit$, for each $\alpha < \omega_1$, there is a function

$$s_\alpha : \alpha^{<\omega} \times H(\alpha) \to \alpha$$

such that if we have a function $s : \omega_1^{<\omega} \times H(\omega_1) \to \omega_1$ then $\{\alpha : s|\alpha^{<\omega} \times H(\alpha) = s_\alpha\}$ is stationary, thus non-empty!

Suppose we are at stage α in the construction of an HFD and suppose $Z_1(\alpha)$ is the countable collection of countable subsets of ω_1 which we need to split, as in the proof of Lemma 3.3. Let $H(\alpha)$ be indexed as $\{h_{\alpha,i} : i < \omega\}$. Inductively define a sequence $\langle \beta(\alpha,i) : i < \omega\rangle$ by $\beta(\alpha,i) = s_\alpha(\langle \beta(\alpha,j) : j < i\rangle, h_{\alpha,i})$.

and let $S_\alpha = \{\beta(\alpha,i): i < \omega\}$. In constructing the HFD, we have already constructed partial functions $f_\gamma: \alpha \to 2$ for each $\gamma < \omega_1$ ($f_\gamma = \cup\{f_{\gamma\delta}: \delta < \alpha\}$ as in the proof of Lemma 3.3). Let $S_{\alpha i} = \{\beta \in S_\alpha \ f_\beta$ extends $h_{\alpha,i}\}$.

Let $$\mathcal{S}(\alpha) = \begin{cases} \{S_\alpha\} \cup \{S_{\alpha i}: |S_{\alpha i}| = \omega\}, \text{if } |S_\alpha| = \omega \\ \emptyset \text{ if } S_\zeta \text{ is finite} \end{cases}$$

Finally, let $Z_2(\alpha) = Z_1(\alpha) \cup \mathcal{S}(\alpha)$. Choose $B(\alpha) \subset \omega_1$ which splits $Z_2(\alpha)$. Extend f_γ for each γ as usual and define $Z(\alpha) = Z_2(\alpha) \cup \{A \cap B(\alpha): A \in Z_2(\alpha)\} \cup \{A - B(\alpha): A \in Z_2(\alpha)\}$. This will inductively define an HFD $X = \{f_\beta: \beta < \omega_1\}$.

If $S_\alpha = \{\beta(\alpha,i): i \in \omega\}$ is the set defined at stage α, and $f_{\beta(\alpha,i)} \in \langle h_{\alpha,i}\rangle$ for each $i \in \omega$, then $\{f_{\beta(\alpha,i)}: i \in \omega\}$ is dense in X, since if $h \in H(\omega_1)$, then there is an $i \in \omega$ and $h' \in 2^{\omega_1 - \alpha}$ such that $h = h_{\alpha,i} \cup h'$. Then $S_{\alpha i}$ will be infinite. Since $S_{\alpha i} \in Z_2(\alpha)$, $\{f_\beta: \beta \in S_{\alpha i}\}$ is dense past α, and thus intersects $\langle h'\rangle$ Thus $\{f_\beta: \beta \in S_\alpha\} \cap \langle h\rangle \neq \emptyset$.

Suppose $s: X^{<\omega} \times \tau(X) \to X$ is a strategy for Player II in $G^D_\omega(X)$. Let $s': \omega_1^{<\omega} \times H(\omega_1) \to \omega_1$ be defined by $s'(\langle\alpha(i): i < j\rangle, h) = \beta$ iff $s(\langle f_{\alpha(i)}: i < j\rangle, \langle h\rangle \cap X) = f_\beta$. Then, using our $\diamondsuit$ sequence, there is an α such that $s'|\alpha^{<\omega} \times H(\alpha) = s_\alpha$.

If, in a play of the game, $\langle h_{\alpha,i}\rangle \cap X$ is the open set Player I plays on round i for each $i < \omega$, then Player II will play the set $\{f_\beta: \beta \in S_\alpha\}$ which, as remarked above, will be dense in X. Therefore s is not a winning strategy for Player II.

Question 4.2. Is there a space X in ZFC for which $G^D_\omega(X)$ is neutral?

REFERENCES

[H-J]. Hajnal, A. and Juhász, I., On Hereditarily α-Lindelof and α-Separable Spaces II, Fund. Math., 81 (1974), 147-158.

[J]. Juhász, I. Consistency results in topology, in Handbook of Mathematical Logic, Barwise, J. ed., North Holland, 1977, 503-522.

[J-W]. Juhász, I. and Weiss, W., Nowhere Dense Choices and π-weight, Prace Matematyczne Universytetu Slaskiego, to appear.

ON HOMOMORPHISM TYPES OF SUPERATOMIC INTERVAL BOOLEAN ALGEBRAS

Robert BONNET

Department of Mathematics
University Claude Bernard
69622 - VILLEURBANNE Cedex (France)

ABSTRACT. Let D_κ be the class of superatomic interval Boolean algebras of cardinality $< > \omega_1$. For $\kappa < \alpha < \kappa^+$ and for $m,n < \omega$, $m+n > 1$, let $B_{\alpha,m,n}$ be the superatomic interval algebra generated by the chain $\omega^\alpha . m + (\omega^\alpha + (\omega^\alpha)^*) . n$. Let N_κ be the subset of D_κ consisting of all $B_{\alpha,m,n}$. In the first part, we consider the following relation in D_κ : $B' < B''$ iff B' is embeddable in B''. We prove that for every B in D_κ, there is an unique $B_{\alpha,m,n}$ such that $B < B_{\alpha,m,n} < B$. We describe completely $< N_\kappa , < >$: this is a well-founded distributive lattice with the property that for every $B_{\alpha,m,n}$, there are only finitely many incomparable elements to $B_{\alpha,m,n}$ in N_κ. In the second part, we introduce other quasi-orderings $\leq$ on D_κ : for instance the relations being elementary embeddable, being a homomorphic image, being a dense homomorphic image. In contrast to the first part, for these relations $\leq$, the quasi-ordered class $< D_< , < >$ is very complicated : to each subset I of κ, we can associate a member B_I of B_κ, such that $I \subset J$ if $B_I \leq B_J$.

We thank the referees, I.ROSENBERG and S.KOPPELBERG for their comments, in particular concerning the proof of the theorem in § I, and S.SHELAH for his helpful comments and improvements of results in § II.

0 - NOTATIONS AND DEFINITIONS.

0.0. In the following, an algebra always means a Boolean algebra. We use the notations $\vee, \wedge$ for the lattice operations, and a' is the complement of a. For an algebra B, and an element a, we let $B \upharpoonright a$ be the algebra with underlying the set $\{t \in B : 0 < t < a\}$, whose lattice operations are those of B restricted to $[0,a]$, the complement of x being $x' \wedge a$.

For a chain C (i.e. a total ordering), we denote by C^* the reverse chain, i.e. $x < y$ in C^* iff $x > y$ in C. We denote by $\alpha, \beta, \ldots$ ordinals and $\kappa, \lambda, \ldots$ cardinals. ω is the set of non-negative integers and ω^* is order-isomorphic to the non-positive integers. We denote by $\delta + \gamma$, ω^α, $\omega^\alpha . p$ the ordinal operations ($2.\omega = \omega$ and $\omega.2 = \omega + \omega$).

Let $< P , \prec >$ be a quasi-ordered set, i.e. $\prec$ is a reflexive and transitive relation on P. Then recall that $x \equiv y$ iff $x \prec y$ and $y \prec x$, for x,y in P, and the quotient $< P/\equiv , \prec >$ is called the ordering associated with $< P , \prec >$.

A topological space T is called an *interval* space, whenever there is a linear ordering $\prec$ on T such that the topology of T is the interval topology of $< T , \prec >$, i.e. the open sets of T are the unions of open intervals of $< T , \prec >$.

0.1. Interval algebras.

Let C be a chain.

We put $C^o = C \cup \{-\infty\}$ and $C^+ = C \cup \{-\infty, +\infty\}$, where $-\infty < x < +\infty$ for all $x \in C$. We denote by $\mathbf{B} < C >$ the algebra of all subsets of C^o which are finite unions of intervals $[u,v[= \{t \in C^+ : u \prec t < v\}$, for $u < v$ in C^+ ; note that $+\infty \notin [u,v[$. Such an algebra $\mathbf{B} < C >$ is called the *interval algebra* generated by the set C. An element a of $\mathbf{B} < C >$ has an unique canonical decomposition : $a = \bigcup \{[a_{2i}, a_{2i+1}[: i \prec n\}$ where $-\infty \prec a_o < a_1 < \ldots < a_{2n+1} \prec +\infty$ and $a_i \in C^+$ $(i = 0 , \ldots , 2n+1)$. We put $\sigma(a) = \{a_o , a_1 , \ldots , a_{2n+1}\} \subset C^+$ and a_{2i} (resp. a_{2i+1}) is called a *left* (resp. *right*) *end* point of a.

Examples. $\mathbf{B} < Q >$ is the free countable algebra and $\mathbf{B} < \omega >$ is the algebra of finite or cofinite subsets of ω.

Now, let I(C), be the set, ordered by the inclusion relation, of initial intervals of C (we recall that $I \subset C$ is an initial interval of C iff $b \in I$ and $a \prec b$ implie $a \in I$). *The set* I(C), *endowed with the induced topology of* 2^C, *which is also the interval topology on* I(C) *is the Boolean space associated with* $\mathbf{B} < C >$.

0.2. Superatomic algebras.

PROPOSITION (DAY). *Let* $\mathbf{B}$ *be an algebra. The following properties are equivalent :*

(i). every subalgebra of $\mathbf{B}$ *is atomic*

(ii). every non-trivial quotient algebra of $\mathbf{B}$ *is atomic*

(iii). there is no one-to-one homomorphism from $\mathbf{B} < Q >$ *into* $\mathbf{B}$.

(iv). there is no one-to-one increasing function from the rational chain Q *into* $\mathbf{B}$.

An algebra satisfying one of the above conditions is said to be *superatomic*. It is known that infinite countable superatomic algebras are the algebras isomorphic to exactly one $B < \omega^{\alpha} . p >$, where $1 \leq p < \omega$ and $1 \leq \alpha < \omega_1$.

Now, let us recall that a compact space S is said to be *scattered* iff every non-empty subset F of S has an isolated point in its subspace topology. For a space X, we denote by X^{ξ} the ξth - CANTOR-BENDIXON derivative of X ($X^{\xi+1}$ is the subspace of non-isolated points of X^{ξ}). Let δ be the least ordinal such that $X^{\delta} = X^{\delta+1}$. The ordinal $\delta = rk(X)$ is called the *rank* of X. For a scattered compact space S, we have $S^{rk(S)} = \emptyset$. Also $rk(S) = \delta = \alpha+1$ for some ordinal α and $|S^{\alpha}| = p \geq 1$ is finite ; $\rho(S) = (\alpha,p)$ is called the *characteristic type* of S. For instance the characteristic type of the interval spaces $\omega^{\alpha} + 1$ and $\omega^{\alpha} + 1 + (\omega^{\alpha})^*$ are $(\alpha,1)$.

Let B be an algebra and S be its Boolean space. Obviously, according to (ii) of the above proposition, B is a superatomic algebra iff S is a scattered space.

A chain C is said to be *scattered* iff C does not contain a chain order--isomorphic to the rational chain Q. We can remark that a lexicographic sum of scattered chains, indexed by a scattered chain, is scattered too.

Remark. Let $B = B < C >$ be an interval algebra (its Boolean space $I(C)$ is an interval space).

The following properties are equivalent :

(i). C is a scattered chain

(ii). $B < C >$ is a superatomic algebra

(iii). $I(C)$ is a scattered chain

(iv). $I(C)$ is a scattered space.

For $\kappa \geq \omega$, we denote by D_{κ} the class of superatomic interval algebras of cardinality κ. So if $B \in D_{\omega}$, then B is isomorphic to $B_{\alpha,p} = B < \omega^{\alpha} . p >$ for unique $1 \leq p < \omega$ and $1 \leq \alpha < \omega_1$. We put, for $B', B'' \in D_{\omega}$, $B' \leq B''$ iff B' is isomorphic to a subalgebra of B''. Then $B_{\alpha,p} \leq B_{\beta,q}$ iff $\alpha < \beta$, or $\alpha = \beta$ and $p \leq q$.

Consequently $\langle \mathcal{D}_{\omega}, \preccurlyeq \rangle$ is order-isomorphic to the chain ω_1. In § 1, we will study the partially ordered set associated with the quasi-order $\langle \mathcal{D}_{\kappa}, \preccurlyeq \rangle$ for $\kappa \geqslant \omega_1$.

§ I. EMBEDDING RELATION BETWEEN SUPERATOMIC INTERVAL ALGEBRAS.

Let α be an ordinal and $1 \leqslant p < \omega$. Chains which are obtained as lexicographic sums of p chains, each order-isomorphic to $\omega^{\alpha}+1$, $1+(\omega^{\alpha})^*$, or $\omega^{\alpha}+1+(\omega^{\alpha})^*$, are scattered compact interval spaces of characteristic type (α,p). We shall call them *natural examples of type* (α,p).

I-1. LEMMA. *Let* S *be a compact scattered interval space of characteristic type* (α,p). *Then there is a natural example* S_o *of characteristic type* (α,p) *such that*

(1) S *is a quotient space of* S_o *and* S_o *is a quotient space of* S.

(2) S *is a quotient space of the interval space* $\omega^{\alpha}.2p+1$ *and the interval space* $\omega^{\alpha}.p+1$ *is a quotient of* S.

Proof. We shall prove (1) and (2) simultaneously by induction on α (these claims are obvious for $\alpha < \omega_1$, since an S of type (α,p) is homeomorphic to the interval space $\omega^{\alpha}.p+1$). We can assume without loss of generality that $p = 1$. Indeed if S is of type (α,p), then S is a finite direct topological sum of p spaces of type $(\alpha,1)$. We shall prove (1) and obtain, as a consequence, (2). Let ∞ be the unique element of S . Let $\preccurlyeq$ be a linear ordering on S, which generates the topology. Now let I -resp. F- be the set of $x \in S$ such that $x < \infty$ -resp. $x > \infty$-. If I is non-empty and has no greatest element, then we choose a strictly increasing sequence $(x_{\nu})_{\nu<\varphi}$ of elements of I such that :

(i) $(x_{\nu})_{\nu<\varphi}$ is cofinal in I

(ii) φ is a regular cardinal

(iii) $x_{\mu} = \sup\{x_{\nu} : \nu < \mu\}$ for every limit ordinal $\mu < \varphi$.

Moreover if F is non-empty and has no smallest element, then we choose a strictly decreasing sequence $(y_{\theta})_{\theta<\psi}$ satisfying the dual properties.

Now, for each $\nu < \varphi$, let I_{ν} be the set of all $x \in S$ such that $x_{\nu} \preccurlyeq x \preccurlyeq x_{\nu+1}$. We denote by I_{-1} the set of $x \in S$ such that $x \preccurlyeq x_o$ (if the sequence $(x_{\nu})_{\nu<\varphi}$ does not exist, then we put $I_{-1} = I$). Similarly, we define the F_θ's. For every $\nu < \varphi$ (or $\nu = -1$) there exist β_ν and p_ν such that $\rho(I_\nu) = (\beta_\nu, p_\nu)$, and for every $\theta < \psi$ (or $\theta = -1$) there are γ_θ and q_θ such that $\rho(F_\theta) = (\gamma_\theta, q_\theta)$. From $S^\alpha = \{\infty\}$ and $\infty \notin I_\nu$ for every ν and $\infty \notin F_\theta$ for every θ, it follows that I_ν^α and F_θ^α are empty sets and thus $\beta_\nu < \alpha$ for every ν and $\gamma_\theta < \alpha$ for every θ.

Moreover, if the sequence $(x_\nu)_{\nu < \varphi}$ exists, then we can assume that :

(iv) if $\mu < \nu < \alpha$, then $\beta_\mu \leq \beta_\nu$

(and the dual statement about γ_θ's).

By induction hypothesis, the compact scattered interval spaces I_ν and F_θ verify the claims (1) and (2).

Before going on, we begin to prove :

<u>Claim 1</u>. Let S and T be compact scattered spaces of characteristic types (β,s) and (γ,t), respectively. If there is a continuous function f from S onto T, then $\gamma < \beta$, or $\gamma = \beta$ and $t \leq s$.

<u>Proof</u>. By induction, we prove that $T^\xi \subset f(S^\xi)$, this will prove the claim. First, $T^0 = T = f(S) = f(S^0)$. Let us suppose $T^n \subset f(S^n)$. From compactness of S^n it follows that $T^{n+1} \subset f(S^{n+1})$. Let λ be a limit ordinal, and suppose $T^\xi \subset f(S^\xi)$ for $\xi < \lambda$. It is sufficient to prove that $\bigcap f(S^\xi) \subset f(\bigcap S^\xi)$. Let $y \in \bigcap f(S^\xi)$. For each $\xi < \lambda$, let $x_\xi \in S^\xi$ such that $f(x_\xi) = y$. Let x be a cluster point of the generalized sequence $(x_\xi)_\xi$. We have $x_\xi \in S_\xi \subset S_\zeta$ for $\zeta < \xi < \lambda$, and so $x \in \bigcap S_\xi$. Since $f(x) = y$, we are done. □

a) Suppose (x_ν) and (y_θ) exist. By induction hypothesis $\omega^{\beta_\nu} . p_\nu + 1$ is a quotient space of I_ν and I_ν is a quotient space of $\omega^{\beta_\nu} . 2p_\nu + 1$. Using the lexicographic sum, we have :

$\omega^\beta = \sum_{\nu < \varphi} (\omega^{\beta_\nu} . p_\nu + 1) = \sum_{\nu < \varphi} (\omega^{\beta_\nu} . 2p_\nu + 1)$ and thus I is a quotient space of ω^β and ω^β is a quotient space of I. By analogy, if we put $\omega^\gamma = \sum_{\theta < \psi} (\omega^{\gamma_\theta} + 1)$, then the space F is a quotient of the interval space $(\omega^\gamma)^*$ and $(\omega^\gamma)^*$ is a quotient space of F. It follows that S is a quotient space of the interval space

$T = \omega^{\beta} + 1 + (\omega^{\gamma})^*$ and T is a quotient space of S. Now, according to claim 1, we have $\beta = \alpha$ or $\gamma = \alpha$. If $\gamma < \beta = \alpha$, then clearly T is a quotient space of the interval space $S_o = \omega^{\alpha} + 1$ and $\omega^{\alpha} + 1$ is a quotient space of T, concluding the proof of (1), in this case. If $\gamma = \beta = \alpha$, then S is a quotient space of the interval space $S_o = \omega^{\alpha} + 1 + (\omega^{\alpha})^*$ and S_o is a quotient space of S, concluding the proof of (1), in this case.

b) Suppose F has a smallest element. So, F is a clopen subspace of S, of characteristic type (γ, q), with $\gamma < \alpha$, and I has no greatest element. In this case, as before, $I \cup \{\infty\}$ is a quotient space of $\omega^{\alpha} + 1$ and $\omega^{\alpha} + 1$ is a quotient of $I \cup \{\infty\}$. Moreover, S is the direct topological sum of $I \cup \{\infty\}$ and F. By induction hypothesis, F is a quotient space of the interval space $\omega^{\gamma} . 2q+1$ and $\omega^{\gamma} . q+1$ is a quotient space of F, since $rk(F) = \gamma < \alpha$.
From $\omega^{\gamma} . 2q+1 + \omega^{\alpha} + 1 = \omega^{\alpha} + 1$, we conclude that S is a quotient of $\omega^{\alpha} + 1$ and $\omega^{\alpha} + 1$ is a quotient of S, and thus we have proved (1).

c) Suppose I has a greatest element. We proceed as in case b).

Now we shall prove (2), as a consequence of (1). Let S_o be a natural example of type $(\alpha, 1)$. So S_o is the interval space $\omega^{\alpha} + 1$, $1 + (\omega^{\alpha})^*$ or $\omega^{\alpha} + 1 + (\omega^{\alpha})^*$. Then obviously the interval space $\omega^{\alpha} + 1$ is a quotient of S_o and S_o is a quotient space of the interval space $\omega^{\alpha} . 2 + 1 = \omega^{\alpha} + 1 + \omega^{\alpha} + 1$. Since each of S and S_o is a quotient of the other one, (2) follows. □

<u>Example</u>. Let $S = (1 + \omega^*) . \omega_1 + 1$. Obviously $S_o = \omega_1 + 1$ is a quotient space of the interval space S, and S is a quotient space of $S_o = \omega_1 + 1 = (\omega+1) . \omega_1 + 1$.

I-2. Notations.

Let $\alpha \geq \omega_1$ and m,n with $m+n \geq 1$ be integers. We denote by $\mathbf{B}_{\alpha,m,n}$ the superatomic interval algebra $\mathbf{B} < \omega^{\alpha} . m + (\omega^{\alpha} + (\omega^{\alpha})^*) . n >$. Let $\kappa \geq \omega_1$ be a cardinal. We denote by $\mathfrak{N}_{\kappa}$ be the set of $\mathbf{B}_{\alpha,m,n}$, for $\kappa \leq \alpha < \kappa^+$ and $m,n < \omega$, $m+n \geq 1$.

For two algebras $\mathbf{B}'$ and $\mathbf{B}''$, we put $\mathbf{B}' \leq \mathbf{B}''$ iff $\mathbf{B}'$ is isomorphic to a subalgebra of $\mathbf{B}''$, and we put $\mathbf{B}' \equiv \mathbf{B}''$ iff $\mathbf{B}' \leq \mathbf{B}'' \leq \mathbf{B}'$.

The algebra associated with natural examples of characteristic type (α,p) are exactly those isomorphic to some $B_{\alpha,m,p-m}$ (for $0 \leq m \leq p$).

The *characteristic type* (α,p) *of a superatomic algebra* is the characteristic type of its Boolean space.

I-3. THEOREM. *Let* B *be a superatomic interval algebra of cardinality* $\kappa \geq \omega_1$, *i.e.* $B \in D_\kappa$, *and of characteristic type* (α,p).

(1) (i) if $cf(\omega^\alpha) = \omega$, *then there is exactly one algebra* $B_{\alpha,m,0}$ *in* N_κ *such that* $B \equiv B_{\alpha,m,0}$.

(ii) if $cf(\omega^\alpha) = \omega$, *then there is exactly one algebra* $B_{\alpha,m,n}$ in N_κ *such that* $B \equiv B_{\alpha,m,n}$.

(2) Let $B_{\alpha,m,n}$ *and* $B_{\beta,p,q}$ *in* N_κ. *We have* $B_{\alpha,m,n} \leq B_{\beta,p,q}$ *iff*

- either $\alpha < \beta$

- or $\alpha = \beta$, $cf(\omega^\alpha) = \omega$ *and* $m+n \leq p+q$; *moreover if* $m+n = p+q$, *then* $B_{\alpha,m,n}$, $B_{\alpha,p,q}$ *and* $B_{\alpha,m+n,0}$ *are isomorphic.*

- or $\alpha = \beta$, $cf(\omega^\alpha) > \omega$ *and* $m+n \leq p+q$ *and* $m + 2n \leq p + 2q$.

(3) N_κ, *with the ordering* $\leq$, *is a distributive lattice such that for every element* $B_{\alpha,m,n}$ *in* N_κ, *there are only finitely many elements in* N_κ, *incomparable with* $B_{\alpha,m,n}$.

Remark. We denote by N_κ^α the set of all algebras $B_{\alpha,m,n}$ such that $m+n \geq 1$. N_κ^α has a smallest element $B_{\alpha,1,0} = B < \omega^\alpha >$, and $B_{\alpha+1,1,0}$ is the supremum of N_κ^α in N_κ.

Obviously, according to (2), N_κ is the lexicographic sum of N_κ^α, for $\kappa \leq \alpha < \kappa^+$. If $cf(\omega^\alpha) = \omega$, then N_κ^α is a chain of order type ω_1. The description of the lattice N_κ^α, for $cf(\omega^\alpha) > \omega$, is given in figure 1.

Proof. From (2), follows that if $cf(\omega^\alpha) = \omega$, then N_κ^α is order-isomorphic to ω_1 and if $cf(\omega^\alpha) > \omega$, then N_κ^α is a distributive lattice and that, for every $B_{\alpha,m,n}$, there are at most $\frac{1}{4}(m^2 + 2n^2)$ algebras $B_{\alpha,p,q}$ incomparable to $B_{\alpha,m,n}$. Therefore (3) is a consequence of (2) and of the obvious fact that the lexicographic sum of the distributive lattice N_κ^α, for $\kappa \leq \alpha < \kappa^+$ is a distributive lattice. So, it is

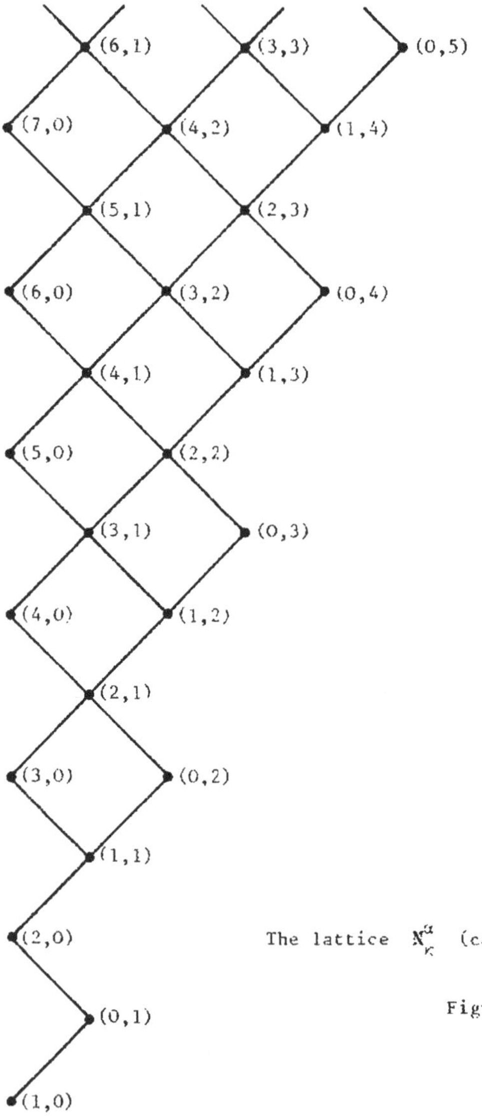

The lattice $\aleph_{\kappa}^{\alpha}$ (case where $cf(\omega^{\alpha}) > \omega$)

Figure 1.

sufficient to prove (1) and to study N^{α}_{κ}.

For integers m and n, we consider the direct topological sum A(m,n) of m copies $A_i(m,n)$ of $\omega^{\alpha}+1$ and the direct topological sum B(m,n) of n copies $B_i(m,n)$ of $\omega^{\alpha}+1+(\omega^{\alpha})^*$. Let X(m,n) be the interval space which is the topological sum of A(m,n) and B(m,n). So, N^{α}_{κ} is identified with the set $\underset{\sim}{L}$ of X(m,n), for $m+n \geq 1$. Let $\prec$ be the quasi-ordering on $\underset{\sim}{L}$, defined by $Y \prec Z$ iff Y is a quotient space of Z. Clearly $< N^{\alpha}_{\kappa}, \prec >$ and $< \underset{\sim}{L}, \prec >$ are order-isomorphic. We denote by + the operation of disjoint sum ; if $Y \prec Z$, then $Y + T \prec Z + T$ for every Y,Z and T in $\underset{\sim}{L}$.

<u>Proof of point (1).</u> If $cf(\omega^{\alpha}) > \omega_1$, then (1) is the dual statement of the above lemma I-1 (1). Now, if $cf(\omega^{\alpha}) = \omega$, then (1) follows from I-1 (1) and from the following obvious result :

<u>Claim 2.</u> Let us suppose $cf(\omega^{\alpha}) = \omega$. Then X(1,0) and X(0,1) are homeomorphic spaces.

So, if $cf(\omega^{\alpha}) = \omega$, then X(p,q) and X(p+q,0) are homeomorphic, and by claim 1, $X(m,0) \prec X(p,0)$ iff $m \leq p$.

Now, we conclude the proof with the study of $< N^{\alpha}_{\kappa}, \prec >$.

<u>Study of $< N^{\alpha}_{\kappa}, \prec >$.</u> According to the claim 2, if $cf(\omega^{\alpha}) = \omega$, then $< N^{\alpha}_{\kappa}, \prec >$ is order-isomorphic to ω. So, subsequently, we assume $cf(\omega^{\alpha}) > \omega$.

<u>Claim 3.</u> If $cf(\omega^{\alpha}) > \omega$, then $X(0,1) \not\prec X(1,0)$.

<u>Proof.</u> By contradiction. Let f be a continuous function from $X(1,0) = \omega^{\alpha}+1$ onto $X(0,1) = \omega^{\alpha}+1+(\omega^{\alpha})^*$. With obvious notation, we set $\omega^{\alpha}+1 = U + \{a\}$, with $U = \omega^{\alpha}$, so U is an open subset of X(1,0), we also set $X(0,1) = V + \{b\} + W$, with $V = \omega^{\alpha}$, $W = (\omega^{\alpha})^*$, thus V and W are disjoint open subsets of X(0,1). Consequently $f^{-1}(V)$ and $f^{-1}(W)$ are disjoint open subsets of X(1,0).

<u>Fact.</u> $f^{-1}(V) \subset U$ and $f^{-1}(W) \subset U$.

<u>Proof.</u> By contradiction. For instance, let us assume $a \in f^{-1}(V)$, i.e. $f(a) \in V$.

Consequently $f^{-1}(V)$ is a neighborhood of a. We set $L = X(1,0) \setminus f^{-1}(V)$. So $L = f^{-1}(\{a\} + W)$ is a compact space of rank $rk(L) = \beta < \alpha$ and f is a continuous function from L onto $\{a\} + W$, which is homeomorphic to $X(1,0)$, and of rank α. This contradicts claim 1. □

Fact. Considering U as a chain, $f^{-1}(V)$ and $f^{-1}(W)$ are cofinal subsets of U.

Proof. By contradiction. Let us suppose $f^{-1}(V)$ is not cofinal in U, i.e. $f^{-1}(V)$ is bounded in U. Consequently its topological closure $F = \overline{f^{-1}(V)}$ is a compact space of rank $\beta < \alpha$. Moreover $f(F)$ is a compact subspace of $X(0,1)$ such that $V \subset f(F)$. Consequently $b \in f(F)$ and thus $f(F)$ is a compact space of rank α. This contradict the claim 1. □

Now we finish the proof of claim 3. Let $(x_n)_{n<\omega}$ be a sequence in U such that $x_o < x_1 < x_2 < \ldots$ and $x_o, x_2, x_4, \ldots \in f^{-1}(V)$; $x_1, x_3, x_5, \ldots \in f^{-1}(W)$; let x be the supremum of $(x_n)_{n<\omega}$ in $X(1,0)$. Then, by $f(x) = \lim_k f(x_{2k})$ and $cf(\omega^\alpha) \geq \omega_1$, $f(x) \in V$. Similarly $f(x) \in W$. Contradiction. □

Claim 4. If $m+n \leq p+q$ and $m+2n \leq p+2q$, then $X(m,n) \leq X(p,q)$.

Proof. First, we have $X(1,0) \leq X(0,1) \leq X(2,0)$. Consequently :

- assuming $m > 0$, we have

$X(m,n) = X(m-1,n) + X(1,0) \leq X(m-1,n) + X(0,1) = X(m-1,n+1)$

- assuming $n > 0$, we have :

$X(m,n) = X(m,n-1) + X(0,1) \leq X(m,n-1) + X(2,0) = X(m+2,n-1)$.

Now define $(m,n) \leq (p,q)$ in $\mathbb{N} \times \mathbb{N}$ iff $X(m,n) \leq X(p,q)$. So, for instance $(m,n) \leq (m-1,n+1)$, provided $m > 0$ and $(m,n) \leq (m+2,n-1)$, provided $n \geq 0$. By induction, if $m+n = p+q$ and $0 \leq p \leq m$, then $(m,n) \leq (p,q)$, and if $m+2n = p+2q$ and $0 \leq q \leq n$, then $(m,n) \leq (p,q)$. Now, let $m+n \leq p+q$ and $m+2n \leq p+2q$. We set $s = 2(m+n) - (p+2q)$ and $t = p+2q - (m+n)$. We have the following properties :

(*) $s \leq m$ and $m+n = s+t$

(**) $q \leq t$ and $p+2q = s+2t$.

So, according to (*) and (**), if $s \geq 0$, then $X(m,n) \leq X(s,t) \leq X(p,q)$, and thus $X(m,n) \prec X(p,q)$.

Now, let us suppose $s < 0$. According to the definition of s, we have $2(m+n) < p+2q$. First, let us assume p is even. We have :
$X(m,n) \prec X(0,m+n) \prec X(0, \frac{p}{2}+q) \prec X(p,q)$ and thus $X(m,n) \prec X(p,q)$. If p is odd, i.e. $p = 2k+1$, then $2(m+n) \leq 2k+2q$, and thus :
$X(m,n) \prec X(0,m+n) \prec X(0,k+q) \prec X(2k,q) \prec X(p,q)$ and so $X(m,n) \prec X(p,q)$. □

Before going on, let us recall that $X(m,n) = A(m,n) + B(m,n)$, where $A(m,n)$ is the topological sum of m copies $A_i(m,n)$ of $\omega^\alpha + 1$ and $B(m,n)$ is the sum of n copies $B_j(m,n)$ of $\omega^\alpha + 1 + (\omega^\alpha)^*$. Consequently under CANTOR-BENDIXON derivation, the space $X^\alpha(m,n) = F(m,n)$ has $m+n \geq 1$ elements.
Moreover $F(m,n) = A^\alpha(m,n) + B^\alpha(m,n)$, and the sets $A^\alpha(m,n)$ and $B^\alpha(m,n)$ have m and n elements respectively.

Claim 5. Let f be a continuous function from $X(p,q)$ onto $X(m,n)$.
(i) for each $t \in A^\alpha(m,n)$, there is $u \in F(p,q) = A^\alpha(p,q) + B^\alpha(p,q)$ such that $f(u) = t$,
(ii) for each $t \in B^\alpha(m,n)$, either there is $u \in B^\alpha(p,q)$ such that $f(u) = t$, or there are $v_1 \neq v_2$ in $A^\alpha(p,q)$ such that $f(v_1) = f(v_2) = t$.

Proof. (i) and (ii) are consequences of the proof of claim 1 and of $X(1,0) < X(0,1) < X(2,0)$. □

Using the notation of claim 5, we set $\Sigma(f) = F(p,q) \cap f^{-1}(F(m,n))$, $\Sigma_A(f) = \Sigma(f) \cap A^\alpha(p,q)$ and $\Sigma_B(f) = \Sigma(f) \cap B^\alpha(p,q)$. So, f induces a function $\tilde{f}$ from $\Sigma(f)$ onto $F(m,n)$.

We denote by $|Z|$ the cardinality of a set Z. Clearly we have :

Claim 6. Let H be a clopen subspace of $X(p,q)$. We put $p_H = |H \cap A^\alpha(p,q)|$ and $q_H = |H \cap B^\alpha(p,q)|$. If $p_H + q_H \geq 1$, then H is homeomorphic to $X(p_H, q_H)$.

Claim 7. Let us suppose $X(m,n) \prec X(p,q)$. Then $m+n \leq p+q$ and $m+2n \leq p+2q$.

Proof. By induction on $m+n$. More precisely, let $k \geq 1$ be given. We denote by $H(k)$

the following statement :

Let m and n be integers such that $1 \leq m+n \leq k$.

Let $X(p,q)$ be given. If $X(m,n) \leq X(p,q)$, then $m+n \leq p+q$ and $m+2n \leq p+2q$.

<u>First step.</u> $H(1)$ is satisfied. We have either $m=1$ or $n=1$.

<u>Case 1.</u> $X(1,0) \leq X(p,q)$. Since $p+q \geq 1$, we obviously have $1 \leq p+q$ and $1 \leq p+2q$

<u>Case 2.</u> $X(0,1) \leq X(p,q)$. From claim 2, it follows that $X(0,1) \not\leq X(1,0)$. Consequently $p \geq 2$ or $q \geq 1$, and thus $1 \leq p+q$ and $2 \leq p+2q$.

<u>Second step.</u> Let us suppose $H(k-1)$ is satisfied. We shall prove $H(k)$. Let m,n be such that $m+n=k$. Let f be a continuous function from $X(p,q)$ onto $X(m,n)$, i.e. $X(m,n) \leq X(p,q)$.

<u>Case 1.</u> Let us assume $m > 0$. Let $A_o(m,n)$ be a clopen subspace of $X(m,n)$, homeomorphic to $\omega^\alpha + 1 = X(1,0)$. Let $U = f^{-1}(A_o(m,n))$, and $V = X(p,q) \setminus U$. So $X(p,q) = U + V$. According to claim 6, U is homeomorphic to $X(p_1, q_1)$ and V is homeomorphic to $X(p_2, q_2)$. Moreover $X(m,n) = A_o(m,n) + (X(m,n) \setminus A_o(m,n))$, i.e. $X(m,n) = X(1,0) + X(m-1,n)$. We have $X(1,0) \leq X(p_1, q_1)$, $X(m-1,n) \leq X(p_2, q_2)$ and $X(p,q) = U+V = X(p_1, q_1) + X(p_2, q_2) = X(p_1+p_2, q_1+q_2)$. From $H(k-1)$, it follows that $1 \leq p_1+q_1$, $1 \leq p_1+2q_1$, $m-1+n \leq p_2+q_2$ and $m-1+2n \leq p_2+2q_2$. Consequently $m+n \leq p+q$ and $m+2n \leq p+2q$.

<u>Case 2.</u> Let us assume $m=0$. Let $B_o(m,n)$ be a clopen subspace of $X(m,n)$, homeomorphic to $\omega^\alpha + 1 + (\omega^\alpha)^*$, $U = f^{-1}(B_o(m,n))$, and $V = X(p,q) \setminus U$. From the claim 6 follows :

$U = X(p_1, q_1)$, $V = X(p_2, q_2)$, $X(0,n) = X(0,1) + X(0,n-1)$, $X(0,1) \leq X(p_1, q_1)$ and $X(0,n-1) \leq X(p_2, q_2)$.

By induction hypothesis, we have $1 \leq p_1+q_1$, $2 \leq p_1+2q_1$, $n-1 \leq p_2+q_2$, $2(n-1) \leq p_2+2q_2$. From $p_1+p_2=p$ and $q_1+q_2=q$, follows that $n \leq p+q$ and $2n \leq p+2q$. □

I-4. Comment.

From $cf(\omega^{\alpha}) = \omega$ for $1 < \alpha < \omega_1$, and from the above theorem, it follows that $< \aleph_{\omega}, < >$ is order-isomorphic to ω_1.

§ II. OTHER RELATIONS BETWEEN SUPERATOMIC INTERVAL ALGEBRAS.

II-0. In the following, each algebra is an atomic algebra. For such an algebra B, we denote by $At(B)$ the set for all atoms of B. On the class of Boolean algebras, we define various quasi-orderings as follows :

(i) we put $B_1 \underset{i}{<} B_2$ iff B_1 is elementarily embeddable in B_2 (for a definition, see CHANG and KEISLER [1978]). It follows, from a well-known theorem of TARSKI [1949] (see CHANG, KEISLER [1978]), characterizing elementary equivalence for Boolean algebras, that for atomic Boolean algebras $B_1 \underset{i}{<} B_2$ iff there is an embedding f from B_1 into B_2 satisfying : for any $a \in B_1$, a is an atom of B_1 iff $f(a)$ is an atom of B_2.

(ii) We put $B_1 \underset{ii}{<} B_2$ iff there is an elementary embedding f from B_1 into B_2 such that f is an one-to-one function from $At(B_1)$ onto $At(B_2)$; we emphasize that such an f is not necessarily onto.

(iii) We put $B_1 \underset{iii}{<} B_2$ iff B_1 is a quotient algebra of B_2.

(iv) We put $B_1 \underset{iv}{<} B_2$ iff there is an homomorphism f from B_2 into B_1 such that $f(B_2)$ is a dense subalgebra of B_1, that means $At(B_1) \subset f(B_2)$.

(v) we put $B_1 \underset{v}{<} B_2$ iff there is an homomorphism f from B_2 into B_1 with the following property : for every atom a_1 of B_1, there is an unique atom a_2 of B_2 such that $f(a_2) = a_1$.

II-1. We have the following properties

(a) if $B_1 \underset{ii}{<} B_2$, then $B_1 \underset{i}{<} B_2$; and if $B_1 \underset{i}{<} B_2$, then B_1 is embeddable into B_2, i.e. $B_1 < B_2$.

(b) If $B_1 \underset{iii}{<} B_2$, then $B_1 \underset{iv}{<} B_2$.

(c) If $B_1 \underset{v}{<} B_2$, then $B_1 \underset{iv}{<} B_2$. Also, in the case B_1 and B_2 are interval algebras, then $B_1 \underset{iv}{<} B_2$ iff $B_1 \underset{v}{<} B_2$ (see BONNET and SI-KADDOUR [1984]).

(d) For two countable algebras, we have $B_1 \underset{iv}{<} B_2 \underset{iv}{<} B_1$ iff B_1 and B_2 are both superatomic, or both non-superatomic. Moreover if B_1 is superatomic and if B_2 is non superatomic, then $B_1 \underset{iv}{<} B_2$ and $B_2 \underset{iv}{\nless} B_1$ (see BONNET and SI-KADDOUR [1984]). For instance, let $B_1 = B < \omega^{\alpha} . p >$ and $B_2 = B < \omega >$. Obviously B_2 is a quotient algebra of B_1 and thus $B_2 \underset{iii}{<} B_1$. Moreover a one-to-one function φ from $At(B_2)$ onto $At(B_1)$, can be extended to an homomorphism from B_2 into B_1, that proves $B_1 \underset{iv}{<} B_2$.

(e) Now, let D_{ω} be the set of $B < \omega^{\alpha} . p >$ for $1 \leq p < \omega$ and $1 \leq \alpha < \omega_1$, that is the set of representatives, up to isomorphism, of the class of countable superatomic (interval) algebras. Let $\underset{\sim}{D}_k$ be the ordered set associated with the quasi-ordered set $< D_{\omega} , <_k >$, for $k = i , ii , \ldots ,$ or v.

(e.1) $\underset{\sim}{D}_i$, $\underset{\sim}{D}_{ii}$ and $\underset{\sim}{D}_{iii}$ are isomorphic to ω_1

(e.2) $\underset{\sim}{D}_{iv}$ and $\underset{\sim}{D}_v$ have an unique element.

The last point is a consequence of (d), and of the second part of (c).

II-2. The situation is quite different in D_{κ}, for $\kappa \geq \omega_1$.

<u>THEOREM</u>. *Let* $\kappa \geq \omega$ *be a cardinal. Let* $\underset{k}{<}$ *be one of the quasi-orderings* $\underset{i}{<}, \underset{ii}{<}, \ldots,$ or $\underset{v}{<}$. *To each subset* I *of* κ, *we can associate a superatomic interval algebra* B_I, *of cardinality* κ *such that* $I \subset J$ *iff* $B_I \underset{k}{<} B_J$.

The proof of the above theorem use a refinement of a technique due to S.SHELAH [1971], (see BONNET and SI-KADDOUR [1984]). For $\underset{1}{<}$, there is a proof due to S.SHELAH by the model-theoretic methods of S.SHELAH [1978].

II-3. We conclude, with a new result due to S.SHELAH. Let $\kappa \geq \omega_1$ be a regular cardinal. Let $P(\kappa)$ be the complete algebra of all subsets of κ. Let $NS(\kappa)$ be the ideal of non-stationary subsets of κ (i.e. subsets of κ whose complements contain some closed cofinal subsets ; see KUNEN [1980] or JECH [1978]). Let $A(\kappa)$ be the quotient algebra $P(\kappa)/NS(\kappa)$.

<u>THEOREM</u> (SHELAH). *Let* $\kappa \geq \omega_1$ *be a regular cardinal. To each element* I *of* $A(\kappa)$ *we can associate an atomic interval algebra* B_I *such that* $I \leq J$ *in* $A(\kappa)$ *iff* B_I *is elementary embeddable into* B_J.

<u>Remark</u>. In the above theorem, the algebras are not superatomic.

REFERENCES.

BAKER J.W. : *Compact spaces homeomorphic to a ray of ordinals,* Fund. Math. 76 (1972), p.19-27.

BIRKHOFF G. : *Lattice Theory,* A.M.S. Coll. Pub. Vol.25, New-York (1967).

BONNET R. - SI-KADDOUR H. : *The number of scattered interval algebras,* submitted to Order (1983).

CARPINTERO O.P. : *The number of different types of Boolean algebras of infinite cardinality* m *that posses* 2^{m} *primes ideals,* Rev. Math. Hisp. Am. 4, 31, (1951), p.93-97.

CHANG C.C. - KEISLER G. : *Model Theory,* Studies in Logic (1973), North-Holland.

DAY G.W. : *Superatomic Boolean algebras,* Pacific J. of Math. 23 (1967), p.479.

FODOR G. : *Eine Bemerking zür theorie der regressive functionen,* Acta. Sci. Math. 17, (1956), p.139-142.

FRAISSE R. : *La comparaison des types d'ordres,* C.R.A.S. Paris, 226 (1948), p.1330-1331.

HALMOS P. : *Lectures on Boolean algebras,* (Van Nostrand).

JECH T. : *Set Theory,* Academic Press (1978).

KUNEN K. : *Set Theory,* Studies in Logic, North-Holland 102 (1980).

LAVER R. : *On Fraissé order type conjecture,* Annals of Math. 93 (1971), p.89-111.

MAYER R.D. - PIERCE R.S. : *Boolean algebras with ordered basis,* Pacific J. of Math. 10 (1960), p.925-942.

MOSTOWSKI A. - TARSKI A. : *Boolesche ringe mit goerdnete basis,* Fund. Math. 32 (1939) p.69-86.

ROTMAN R. : *Boolean algebras with ordered basis,* Fund. Math. (1971), p.187-197.

SHELAH S. : *The number of non-isomorphic models of an unstable first order theory,* Israël J. of Math. 9, 4 (1971), p.473-487.

SHELAH S. : *Classification theory,* Studies in Logic, 92 (1978), North-Holland.

SHELAH S. : *Private communication* (1984).

SIKORSKI R. : *Boolean algebras,* Ergebnisse Mathematik 25, Springer-Verlag (1964).

SI-KADDOUR R. : *Sur la classe des algèbres de Boole d'intervalles,* (Thèse de 3° Cycle 1984, Lyon).

SOLOVAY R. : *Real-valued measurable cardinals,* A.M.S. Proc. of Symposia in Pure Math. 13, 1 (A.M.S. Providence R.I.), p.397-428.

TARSKI A. : *Arithmetical classes and types of Boolean algebras,* Bull. Amer. Math. Soc., 55 (1949), p.64-1192.

DECIDABLE THEORIES OF PSEUDO-ALGEBRAICALLY CLOSED FIELDS

Gregory L. Cherlin
Rutgers University

The work of James Ax on the first order properties of finite fields is mathematically one of the most attractive of the applications of model theoretic methods to algebra. On this foundation a systematic analysis of the model theory of a remarkably large class of fields--the pseudo-algebraically closed (also called regularly closed, or Ax) fields--has been erected over the last fifteen years, in the first instance by Jarden and his collaborators, and latterly by a number of others. My goal, or more precisely: my assignment, is to summarize our present understanding of this class of fields. I will attack the subject from three angles: a semihistorical approach at the outset, followed by a detailed outline of a specific decidability result of Haran and Lubotzky, at the level of nuts and bolts, and finally, a dicussion of what is variously called the model theory or the comodel theory of profinite groups, boiled down from the most recent draft of [12].

For a thorough treatment of the area, starting more or less from first principles, one will very soon be able, I believe, to consult the book of Fried and Jarden [15].

P.A.C. Fields

We will call a field pseudofinite if it is infinite, but is a model of the full first order theory of finite fields. Good axioms for this theory were found by Ax [1], who proved the decidability of the theory of all such fields. His axioms are obtained by formalizing the following three properties of a field K.

(1) K is perfect.

(2) K has a unique extension of each degree.

(3) Every absolutely irreducible variety defined over K has a

point in K.

This last condition is called the Lang-Weil theorem. Recall that a variety defined by polynomials over the field K is said to be *absolutely* irreducible if it remains irreducible over the algebraic closure of K. For example the polynomial $x^2 + 1$ is irreducible over the reals, but not absolutely, whereas the conic $x^2 + y^2 + 1 = 0$ is absolutely irreducible over the reals--so they do not satisfy the Lang-Weil theorem. Any field which *does* satisfy the Lang-Weil theorem is said to be *pseudo-algebraically closed*, or p.a.c.

What is the significance of the second property (or axiom scheme)? If we look at the absolute Galois group G(K), defined as the Galois group Gal(K/K) with K the algebraic closure of K, then (2) determines G(K) up to isomorphism as a *profinite* group. That is, following Krull, we notice that G(K) is the *inverse limit* of the finite groups Gal(L/K) with L/K finite and Galois. From this point of view property (2) can be phrased very neatly; it says that G(K) is the free profinite group on one generator. More generally many or most of the significant properties of various p.a.c. fields can be expressed very efficiently in the language and notation of profinite groups, though of course everything one needs to say can be expressed, in principle, in terms of the existence of various algebraic extensions of K. But I am getting ahead of myself.

Ax proceeds to develop a theory based on an explicit criterion for the elementary equivalence of two pseudofinite fields K,K', in terms of their subfields Abs K, Abs K' of absolutely algebraic numbers (elements algebraic over the prime subfield $\mathbb{F}_p$ or $\mathbb{Q}$). Indeed for such fields the following are equivalent:

(i) $K \equiv K'$

(ii) Abs K $\cong$ Abs K'

(iii) The sets of all polynomials $p \in \mathbb{Z}[x]$ having a root in K or in K' respectively are identical.

From this last condition one obtains very explicit elementary invariants for pseudofinite fields. Given such invariants, the decision problem for the theory of pseudofinite fields becomes very concrete:

which combinations of invariants actually occur?

The answer is quite simple.

Proposition.

Let the field A be algebraic over its prime subfield. Then the following two conditions are equivalent:

(i) A is the field of absolutely algebraic elements of some pseudofinite field.

(ii) The absolute Galois group G(A) is procyclic (that is, A has at most one extension of each degree).

I suppose this way of putting things has the drawback that it doesn't make the decidability of the theory very obvious, but it seems both elegant and informative.

Subsequent efforts were devoted to generalizing Ax's results, and to making them more explicit (for example, primitive recursive) via a "Galois stratification" procedure initiated by Fried-Sacerdote [2]. I will follow up the former line here, in Jarden's footsteps.

In the first place, Jarden [3,4] found a wealth of perfect p.a.c. fields K with various absolute Galois groups G(K). He begins by considering some n elements $\sigma_1,\ldots,\sigma_n$ of G(Q), or more generally of the absolute Galois group G(F) of any countable Hilbertian field F. Let Fix ($\overline{\sigma}$) denote the fixed field of $\sigma_1,\ldots,\sigma_n$ in $\tilde{Q}$ or $\tilde{F}$ respectively. Then with probability one, Fix ($\overline{\sigma}$) is a perfect p.a.c. field. To make sense of the notion of probability here, one notices that any profinite group carries a probability measure which is more or less the inverse limit of counting measures on its finite quotients. More formally, a profinite group carries a natural compact topology, and normalized Haar measure provides a suitable probability measure.

These perfect p.a.c. fields found in such profusion by Jarden have an absolute Galois group which is isomorphic (again with probability one) to the free profinite group on n generators; this group will be denoted $\hat{F}_n$, where the hat reminds us that we remain in the profinite category. In the case $n = 1$ we see that we have pseudofinite fields which are algebraic over $\mathbb{Q}$, while for n greater than one we have something entirely new. Can Ax's results be extended to this larger context? Can one perhaps treat perfect p.a.c. fields from a perfectly general point of view, with no restriction whatsoever on the Galois group? A large part of the answer to such questions is found in work of Jarden and Kiehne [5].

They give at the outset a very general <u>Embedding Lemma</u> which may be used in connection with back-and-forth constructions aimed at establishing an isomorphism between a pair of saturated perfect p.a.c. fields- a familiar approach to the determination of elementary invariants. If we imagine that K,K' are saturated perfect p.a.c. fields, and that $\phi:A \simeq A'$ is an isomorphism between suitable small subfields of K and K' respectively, and if we then wish to extend ϕ to a certain subfield E of K, we arrive at the following:

<u>Problem</u>.

Suppose that $A \leq E$, $A' \leq K'$ are four fields, where E is perfect, K' is card $(E)^+$-saturated, perfect, and p.a.c., that $\phi:A \simeq A'$ is a fixed isomorphism, and in addition that the extensions E/A and K'/A' are regular (that is, separable, and with the bottom field relatively algebraically closed in the top one). <u>Is there then</u> an embedding $\psi:E \to K'$ extending ϕ, so that $K'/\chi[E]$ is again a regular extension?

Evidently some further conditions will be required in general.

<u>Embedding Lemma</u>.

With the notation and hypotheses taken as above, suppose that we have also:

$$\text{an isomorphism } \Phi:\tilde{A}\simeq\tilde{A}' \text{ inducing } \phi, \text{ and:}$$

$$\text{a surjection } \gamma:G(K') \to G(E),$$

satisfying a compatibility condition, namely that the diagram:

$$\begin{array}{ccc} G(E) & \xrightarrow{\gamma} & G(K) \\ \downarrow & \hat{\Phi} & \downarrow \\ G(A) & \longrightarrow & G(A') \end{array}$$

commutes, where Φ induces $\hat{\Phi}$ and the vertical maps are given by restriction. Then we can solve the preceding problem, and indeed we can arrange to have $\psi:E\simeq E' \leq K'$ be the restriction of a map $\psi:\tilde{E}\simeq\tilde{E}'$ chosen so that $\psi:G(E')\simeq G(E)$ also satisfies the condition of compatibility with γ.

Remark.

The same result applies to imperfect fields, under the hypothesis that $[E:E^P] \leq [K':K'^P]$. It is not obvious that the p.a.c. hypothesis is adequate for this generalization, but a lemma given by Tamagawa [7] shows that is . Such remarks could be applied systematically throughout this report. Full details are given in [11].

Let us see where this lemma leaves us. If K,K' are both p.a.c., perfect, and saturated of the same cardinality, and if we take $A = \text{Abs}(K)$, $A' = \text{Abs}(K')$, then using the Jarden-Kiehne Embedding Lemma we will get $K\simeq K'$ if we can get maps:

$$\tilde{A}\simeq\tilde{A}' \text{ inducing } A\simeq A' \text{ and } G(K')\simeq G(K)$$

satisfying the compatibility condition that the diagram:

$$\begin{array}{ccc} G(K) & \simeq & G(K') \\ \downarrow & & \downarrow \\ G(A) & \simeq & G(A') \end{array}$$

commutes.

This suggests that the theory of K should be determined by statements solely about the map:

$$G(K) \to G(A).$$

In particular if for two perfect p.a.c. fields K, K' with the same field A of absolute numbers we happen to have a commutative diagram:

$$\begin{array}{ccc} G(K) & \simeq & G(K') \\ & \searrow \quad \swarrow & \\ & G(A) & \end{array}$$

then $K \equiv K'$ (here there is no hypothesis of saturation).

From these general considerations Jarden and Kiehne were able to derive the decidability of the theory of perfect p.a.c. fields K satisfying:

$$G(K) \simeq \hat{F}_n.$$

This however requires three additional facts:

(1) A profinite group is isomorphic to $\hat{F}_n$ iff the set $\mathrm{Quot}(G)$ of the isomorphism types of its finite quotients (with respect to continuous surjections) coincides with the set of isomorphism types of all finite, n-generated groups.

(2) Any two (continuous) surjections $I, II: \hat{F}_n \to G$ are compatible in the sense that the following diagram can be completed by an isomorphism (Gaschutz [8]):

$$\begin{array}{ccc} \hat{F}_n & \cdots\cdots> & \hat{F}_n \\ I \searrow & & \swarrow II \\ & G & \end{array}$$

(3) For A algebraic over the prime field, there will be a p.a.c. field of the type under consideration satisfying

$$A \simeq \mathrm{Abs}(K)$$

just in case the Galois group $G(A)$ is n-generated (as a profinite group).

From (1) we see easily that the class under consideration is axiomatizable.

From (2) elementary invariants can be obtained using the Embedding Lemma.

Then (3) supplies the information needed to conclude that the theory is decidable.

At this point, we might as well consider an arbitrary class $\mathcal{G}$ of profinite groups, and associate to this class the theory $T(\mathcal{G})$ of perfect p.a.c. fields with absolute Galois group in $\mathcal{G}$. It then remains to isolate the abstract content of facts (1-3) above in the general case. I think that the extension of facts (1) and (3) to the general context can be considered completely satisfactory (more on this later). Since the obvious generalization of (2) is for the most part false, we can adopt the usual strategem of making it a definition rather than a theorem. This leads to the notion of an "Iwasawa" group (a bit of terminology which evolved rather haphazardly, and is by no means fixed; anyone for "Gaschütz group"?). The failure of (2) leads to undecidability results.

The most natural examples, at this level of generality, are the following four classes $\mathcal{G}$ of profinite groups:

(A) arbitrary;

(B) finite rank;

(C) bounded rank (with a specified bound);

(D) Iwasawa groups.

The definition of this last class may be given in diagrammatic form as follows:

$$\begin{array}{ccccc} & & G & & \\ & & \vdots & \searrow & \\ & & \downarrow & & \\ G & \to & B & \to & A \end{array}$$

Here all maps (including the one sought) are epimorphisms, and A,B are finite. Model theorists may note as a mnemonic device that reversing the arrows and interpreting them as monomorphisms, one obtains the diagrammatic formulation of homogeneity in the sense of Fraisse (quantifier-elimination). This has led to the proposal to call such groups "cohomogeneous".

Of these four classes, only the last two give rise to a decidable theory $T(\mathcal{G})$. I will sketch the Haran-Lubotzky theory whch handles case (D) in the next part. The undecidability results, which are outside the scope of the present lecture, were found independently by Ershov and by van den Dries, Macintyre, and me; Ershov got there first [12,13,14].

This rather sketchy historical outline has brought us more or less up to the present. The main problems connected with p.a.c. fields have been dealt with in a reasonably coherent fashion, and the frontiers of the subject have moved to greener pastures, such as pseudo-real-closed fields (with which Professor Prestel will deal) and beyond.

To recapitulate, then, the issues which we have so far successfully evaded are the following.

I. Axiomatizability.

We naturally want to deal with classes $\mathcal{G}$ of profinite groups for which the class of all fields with absolute Galois group lying in $\mathcal{G}$ is axiomatizable. For this one wants a way to recognize properties of profinite groups which, when considered as properties of absolute Galois groups $G(K)$, can be expressed by sentences in the language of fields. In other words one wants a model theory of profinite groups adapted to the needs of Galois theory. Such a model theory will be sketched in the third part of this lecture.

II. Elementary Invariants.

Here one needs a way to bring the Jarden-Kiehne Embedding Lemma into play, something like the Gaschütz lemma. As I indicated before, the simplest way to get this is to assume that you have it, and we will explore this possibility with Haran and Lubotzky in the next part. One can deal with this issue in other ways, but I think this will keep us sufficiently busy.

II. Realizability.

I refer under this rubric to existence theorems for perfect p.a.c. fields showing that all plausible elementary invariants actually occur--are realized--in connection with some p.a.c. field. What is needed is a theorem of the following type. One considers a field A algebraic over the prime field (in any characteristic), a profinite group G, and a continuous surjection:

$$G \twoheadrightarrow G(A).$$

Call this a "situation". In particular, if K is a p.a.c. field then the natural restriction map:

$$G(K) \twoheadrightarrow G(\text{Abs } K)$$

is a situation; call it a "p.a.c. situation". We need a characterization of the p.a.c. situations.

Van den Dries observed that a complete solution to the realizability problem was implicit in results of Ax, Gruenberg, and Jarden. Call a profinite group <u>projective</u> if it satisfies the usual diagrammatic property:

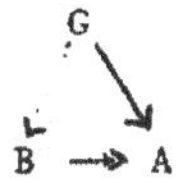

Ax noticed that any p.a.c. field has a projective Galois group (more precisely, van den Dries noticed that Gruenberg had proved that such is the content of Ax's observation). Conversely Jarden's methods provide a rich enough supply of p.a.c. fields to show that any projective profinite group actually occurs as absolute Galois group of a p.a.c. field, more precisely that every situation:

$$G \twoheadrightarrow G(A)$$

with G projective occurs as a p.a.c. situation.

I called this a complete solution to the realizability problem, but this is perhaps a misleading remark. In connection with decision problems, in which one wants to determine the compatibility of certain conditions on our elementary invariants, what is really accomplished at this point is that the question of the existence of a p.a.c. field with certain invariants is traded in for the presumably simpler question, does a projective profinite group exist with certain additional properties? We will see in the next part how this happens in practice.

2. Iwasawa Groups.

Recall the diagrammatic definition of the Iwasawa property given earlier, or the mnemonic device of dualizing homogeneity. In any case, for K perfect, p.a.c. with Iwasawa Galois group, the study of the situation:

$$G(K) \to G(\mathrm{Abs}\ K)$$

reduces to the study of G(K) and of Abs K, separately. In a sense the very formulation of the Iwasawa property ensures this, since we are simply abstracting out the main property used by Jarden and Kiehne in proving the analogous result for $\hat{F}_n$. On the other hand to make this precise it would be convenient to invoke the (rather simple) model-theoretic machinery of the next section. So let us take the following as our starting point (with some of the missing detail to be filled in later): the elementary invariants for a perfect p.a.c. field K with Iwasawa absolute Galois group are of two kinds-

(1) invariants describing Abs K up to isomorphism: a list of all polynomials with integer coefficients having roots in K;

(2) a list of the finite groups which occur as homomorphic images of G(K).

Combining this with the solution to the realizability problem described at the end of the previous section, we see that the decision problem for the theory of all perfect p.a.c. fields with Iwasawa Galois group comes down to the following:

Problem.

Given finite groups $A_1,\ldots,A_k,B_1,\ldots,B_m$-

Does there exist a projective, Iwasawa profinite group having all of the A_i and none of the B_j as a continuous homomorphic image?

Since this is a particularly good example of a nontrivial decision problem concerning profinite groups, I would like to examine briefly the solution found by Haran and Lubotzky [10].

Note at the outset that we may suppose that $k = 1$, and write A for A_1 (the hypothetical map $G \to \Pi A_i$ has as its image one of finitely many possible groups A, which may be considered separately). Now we make five claims.

(A) Any profinite group A has an "Iwasawa cover" $I(A)$, by which we mean an Iwasawa group mapping onto A which is smallest in the sense of the following diagrammatic condition:

$$\begin{array}{ccc} I' & \dashrightarrow & I \\ & \searrow \quad \swarrow & \\ & A & \end{array}$$

(B) If A is finite then $I(A)$ is finite.

(C) Any profinite group A has a "projective cover" $P(A)$, defined in the same spirit (Banaschewski [11]).

(D) If I is a finitely generated Iwasawa group then $P(I)$ is still Iwasawa (and projective).

(E) The group $P(A)$ will have a profinite group B as a homomorphic image just in case there is a map $B \twoheadrightarrow \bar{A}$, with $\bar{A}$ a homomorphic image of A, such that:

no proper closed subgroup of B maps onto $\bar{A}$.

Now if G is a projective Iwasawa group mapping onto A, then by (A,C) this map lifts onto $P(I(A))$, which is another projective Iwasawa group, by (B,D). Thus the original decision problem becomes:

Does $P(I(A))$ have one of the B_j as a homomorphic image?

By (B,E) this question can be answered by inspection. Thus the decision problem has been solved.

Since any decision problem connected with p.a.c. fields will involve the existence of projective profinite groups with various properties, facts (C,E) will be used repeatedly. Haran and Lubotzky contributed the theory summarized in (A,B,D) for the problem at hand.

For the proof of (B), one considers a surjection $F \twoheadrightarrow A$ with F the free profinite group on some finite set of generators, and one takes $I = F/N$ with N the intersection of the kernels of all homomorphisms $F \longrightarrow A$. Then I is a finite Iwasawa group mapping onto A, and granted (A), (B) follows.

The proof of (D) involves some moderately extensive diagram chasing, and I willnot reproduce it here, but it seems appropriate to outline the proof of (A), which is dual to the usual construction of prime models. Call a surjection $B \longrightarrow A$ "I-prime" if B satisfies the diagrammatic condition in (A), without being necessarily Iwasawa itself. Then as any model theorist would expect, the dual of the amalgamation property for I-prime "extensions" plays a major role. That is, given $B_1, B_2 \longrightarrow A$ both I-prime, we wish to complete the following diagram with B I-prime:

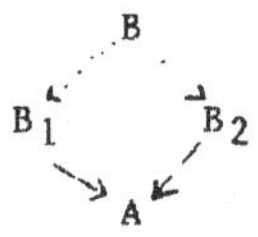

One way to look at this would go as follows. If we "try out" the pullback as our candidate for B, then we have to try to lift surjections $I \longrightarrow A$ with I Iwasawa up to B. We have liftings $I \longrightarrow B_i$ by hypothesis and then automatically $I \longrightarrow B$, but this last may not be surjective. However if the image is $B_0 < B$, this induces a new diagram:

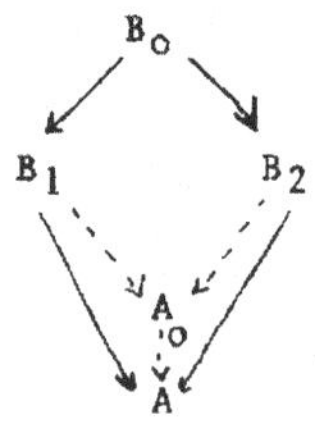

and we can in a sense start over. A suitable compactness argument makes this line of reasoning terminate.

For a precise account of all of this, consult the original paper.

3. Model Theory of Galois Groups.

I want to think of the algebraic closure of a field K as a many-sorted structure, with the n-th sort consisting of the elements of degree at most n over K. This many-sorted structure will be called the <u>algebraic hull</u> of K; from the point of view of first order logic it is not quite the algebraic closure. The absolute Galois group $G = G(K)$ is in some sense dual to the algebraic hull of K, but to analyze this more precisely I will introduce a true duality.

Attach to any profinite group G a dual structure G^* in the following fashion. G^* will be a many-sorted structure whose elements of level n are all the cosets of arbitrary open normal subgroups of index at most n. Equip G^* with the following relations (in the many-sorted manner):

(i) $(Na) \cdot (Ny) = Nz$

(ii) $N_1 x \subseteq N_2 y$.

With a little thought one can see that G^* is essentially just a canonical inverse system of finite groups with inverse limit G, dressed up in a specific model-theoretic garb. One can also check that the category of profinite groups with continuous surjections as morphisms is dual to the category of structures G^* with embeddings as morphisms; one can give the latter category a more intrinsic description, but I for one would not claim that it occurs in nature.

The many-sorted structures G^*, or more precisely the category of all such, come equipped with a model theory. It is this model theory that I will take, by fiat, as the model theory of profinite groups. It is also, for obvious reasons, known as the <u>co</u>model theory of profinite groups, and in any case it seems worthwhile to have a <u>cosatisfaction</u> relation holding between profinite groups G and sentences ϕ about G^*:

$$G \models^* \phi \quad (G \text{ cosatisfies } \phi) \text{ means: } \quad G^* \models \phi .$$

By a natural extension of the terminology, sentences about G^* are called cosentences (about G), and so forth. The point of this terminological exercise lies in the first instance in the following.

Lemma. Any cosentence about Galois groups can be translated into the language of fields.

This means that each cosentence ϕ about profinite groups corresponds (effectively) to a sentence ϕ^* in the language of fields in such a way that:

$$\text{for any field } K, \quad G(K) \models^* \phi \text{ iff } K \models \phi^*.$$

This lemma has been called variously "a complete triviality" and "the central fact"; there is of course no immediate contradiction between these two points of view. In any case one certainly can perform a direct translation from structures $G(K)^*$ (that is, dual Galois groups) into algebraic hulls, so the only question is whether the translation from the algebraic hull back into the original field is trivial or subtle; a point I will cheerfully leave to the reader.

What can one do with this comodel theory? In the first place, one can check directly that the classes of Iwasawa or of projective profinite groups are coaxiomatizable, and hence that the corresponding classes of p.a.c. fields are axiomatizable. This disposes of a point left hanging in part 1.

In the second place, one can apply known results from model theory directly to the structures G^*. Thus, for example, at the beginning of §2 we slid neatly by the following point concerning Iwasawa groups:

Lemma. For G_1, G_2 Iwasawa the following are equivalent:

(i) G_1 and G_2 have the same finite groups as quotients.

(ii) G_1 and G_2 are coelementarily equivalent.

Dualizing, this just becomes an instance of the principle that structures which are homogeneous in the sense of Fraisse and which have the same finite substructures are elementarily equivalent. In precisely the same fashion one can prove:

Lemma. If G_1, G_2 are coelementarily equivalent Iwasawa groups and $\pi_i:G_i \longrightarrow H$ are continuous epimorphisms to a common target group then:

$$(G_1,H) \equiv * (G_2,H).$$

(The point here is that the dual of the pair $G_1 \longrightarrow H$ is the structure G_1^* with a distinguished structure H^*, so it makes sense to speak of coelementary equivalence here.) The dual of this lemma simply involves the passage from Fraisse homogeneity to quantifier elimination.

A more significant application of this formalism running along the same lines is a reformulation of the Jarden-Kiehne Embedding Lemma that should appeal at least to model theorists.

Theorem (Jarden-Kiehne revisited).

For perfect p.a.c. fields K_1, K_2 the following are equivalent.

(i) K_1, K_2 are elementarily equivalent

(ii) There is an isomorphism of pairs

$$f:((\text{Abs } K_2)^-, \text{Abs } K_2) \simeq ((\text{Abs } K_1)^-, \text{Abs } K_1)$$

such that we have:

$$(G(K_1), G(\text{Abs } K_1)) \equiv * \, (G(K_2), G(\text{Abs } K_2))$$

with respect to the identification $\hat{f}$ of the groups $G(\text{Abs } K_i)$ induced by f.

Corollary. In the Iwasawa case the theorem combines with the two lemmas above to yield (immediately) the elementary invariants given at the beginning of §2.

Proof of the theorem:

We may take the fields K_i to be saturated. The groups $G(K_i)$ are then cosaturated (this remark actually goes slightly beyond what has already been said above), and the coelementary equivalence in (ii) becomes an isomorphism. (The precise duality set up above makes this, not an analog of a known result, but a special case of it.) So we have this commutative diagram:

$$\begin{array}{ccc} G(K_1) & \overset{\sim}{\dashrightarrow} & G(K_2) \\ \downarrow & & \downarrow \\ G(\text{Abs } K_1) & \underset{\hat{f}}{\dashrightarrow} & G(\text{Abs } K_2) \end{array}$$

and then the Jarden-Kiehne Embedding Lemma permits us to construct an isomorphism $K_1 \simeq K_2$ by a routine back-and-forth argument.

In summary, then, to a model theorist, or at least to this model theorist, everything seems to revolve around the Jarden-Kiehne embedding criterion (though one should not lose sight of the important role of the realizability result, rather slighted here), and the formalism described here is used to make this more explicit. The model theory of profinite groups has been explored further by Z. Chatzidakis, who has found a nontrivial, and even (more surprisingly) useful, connection with stability theory in Shelah's sense.

I should also mention that the model theory described here is the most expressive one available subject to minimal constraints. More explicitly, if we have any collection Φ of objects called "pseudo sentences" and any relation $\models^{o}$ of "pseudosatisfaction" between profinite groups and pseudoformulas which satisfies two basic conditions:

(1) Isomorphic groups pseudosatisfy the same pseudosentences;

(2) For every pseudosentence ϕ° there is a translation ϕ which is an ordinary sentence in the language of fields, such that for any field K:

$$G(K) \models^{o} \phi^{\circ} \text{ iff } K \models \phi :$$

then we may conclude that in fact every pseudosentence Φ_o has a translation into a cosentence ϕ_* satisfying:

$$G(K) \models^{o} \phi_o \text{ iff } G(K) \models^{*} \phi_*$$

for all fields K. The proof of this is completely straightforward. It suffices to show that for any fields K_1, K_2, if their Galois groups are coelementary equivalent then they are also pseudoelementarily equivalent. But by the Keisler-Shelah theorem, the groups $G(K_i)$ will have isomorphic coultrapowers (a notion which is defined by duality with the many-sorted category), which will in fact be the Galois groups of the corresponding ultrapowers of fields. Applying hypotheses (1,2) in this context, the claim follows.

The original ideological commitment to a workable model theory of Galois groups is due to Macintyre; the idea that this is best expressed in terms of many-sorted structures seems to be due to Chatzidakis and Macintyre independently.

REFERENCES

1. J. Ax: The elementary theory of finite fields. Ann. Math 88 (1968), 239-271.
2. M. Fried and G. Sacerodote: Solving diophantine problems over all residue class fields and all finite fields. Ann. Math 104 (1976), 203-233.
3. M. Jarden: Elementary statements over large algebraic fields. T.A.M.S. 164 (1972), 67-91.
4. M. Jarden: Algebraic extensions of hilbertian fields of finite corank. Israel J. Math 18 (1974), 279-307.
5. M. Jarden, U. Kiehne: The elementary theory of algebraic fields of finite corank. Invent. Math 30 (1975), 275-294.
6. M. Jarden: The elementary theory of ω-free Ax fields. Invent. Math 38 (1976), 187-206.
7. T. Tamagawa, "On regular closed fields," pp. 325-334 in Algebraists' Homage ed. by S. Amitsur et al., AMS, Providence, 1982.
8. W. Gaschütz: Zu einem von B.H. und H. Neumann gestellten Problem. Math. Nachr. 14 (1956), 249-252.
9. K. Gruenberg: Projective profinite groups. J. London Math Soc. 42 (1967), 155-165.
10. D. Haran, A. Lubotzky: Embedding covers and the theory of Frobenius fields. Israel J. Math 41 (1982), 181-201.
11. B. Banaschewski: Projective covers in categories of topological spaces and topological algebras, pp. 63-91 in Proceedings, Kanpur Topology Conferences, 1968, Academic Publishing, Prague, 1971.
12. G. Cherlin, L. van den Dries, A. Macintyre: The elementary theory of regularly closed fields. To appear in Crelle.
13. Yu Ershov: Regularly closed fields. Doklady 251 (1980), 783-785.
14. Yu Ershov, "The undecidability of regularly closed fields (Russian)" Alg. i. Log. 20 (1981), 389-394.

15. M. Fried and M. Jarden: Field Arithmetic. Springer-Verlag, to appear.

G.L.Cherlin

Definability in Power Series Rings of Nonzero Characteristic

Gregory Cherlin
Rutgers University

§1. Introduction

The field F((t)) of formal Laurent series with a pole of finite order over a decidable field F of characteristic zero is itself a decidable field [1,4], and indeed remains decidable even when the language is enriched by predicates representing the valuation ring, the constant subfield, and the set $S = \{t^n : n \in \mathbb{Z}\}$. The first of these new predicates actually turns out to be definable in the pure language of fields, but the other two are not.

In nonzero characteristic the situation appears to be radically different. Let L(F) and L(S) be the extensions of the language L of valued fields by the predicate F or S respectively, denoting the constant subfield or the set S above. The L-theory of F((t)) is not understood in characteristic p, though an example of a nontrivial definable relation is discussed in [2] and repeated in [3] in a more general setting.

Suppose that F is a perfect, infinite field of characteristic $p > 0$. I will discuss the L(F)- and L(S)-theories of F((t)) here, giving a complete account of definability in these two languages. It turns out that these two languages are of essentially the same strength. To be more precise, if we adjoin the parameter t to L(F) to form L(F,t), then L(F,t) and L(S) are equivalent in the sense that they express the same 0-definable relations on F((t)), and indeed these languages coincide with the canonical maximal language L(F,S,Coef) in which Coef(x,s,a) is the relation on F((t)) x S x F defined informally by:

$$x = \Sigma x_i t^i, \quad s = t^n, a = x_n.$$

In particular (for F as above):

I. The L(S)-theory of $F((t))$ is undecidable (and non-analytic).

II. The L(F)-definable relations on F in $F((t))$ are those which are definable in F using quantification over arbitrary countable sequences.

III. The L(F)-theory of $F((t))$ is undecidable.

(Statement I applies also to finite F, c.f. [2], and Statement III was proved already in [3].)

None of this casts any light on the L-theory of $F((t))$, unfortunately. There are also some apparently nontrivial intermediate languages whose expressive capacity is not understood. The most interesting one is L(Init), where Init(x) signifies that the initial coefficient of x is 1. This is easily definable if F is finite, but it is not known for F infinite whether it is definable, or whether the L(Init)-theory of $F((t))$ is undecidable.

It is of course still not known whether the constant subfield F is itself L-definable in $F((t))$. On the other hand, it is also possible that the L(t)-theory of $F((t))$ is model-complete (and hence decidable) when F is algebraically closed or finite.

§2. Definition of F from S

F is assumed to be infinite, perfect, and of characteristic $p > 0$ throughout. We will define F from S in $F((t))$ using the parameter t, which is itself S-definable. We also use the predicate P representing $\{t^{-p^n} : n \in N\}$. It will be clear by the end of this paper that L(P) and L(S) are equivalent languages.

Since the arguments will depend heavily on properties of the set $\tau = \{x^p - x : x \in F((t))\}$, we begin by describing this set more explicitly. If $x = \Sigma x_i t^i$ then $x \in \tau$ iff the following conditions (C_0) and (C_i) ($i < 0$, $i \not\equiv 0 \bmod p$) are satisfied:

(C_0) $\exists a \in F\ x_0 = a^p - a.$

(C_i) $\sum_{n \geq 0} (x_{ip^n})^{1/p^n} = 0.$

In particular, $P = \{s \in S: s-t^{-1} \in \tau\}$ is definable from S.

Let Φ be the set of x satisfying:

$$(\Phi) \qquad \forall s \in P\ \exists y \quad ((xs-y) \in \tau \ \&\ v(y) > -1) \ \&\ v(x) > 0.$$

Then manifestly $F \subseteq \Phi$. The reverse inclusion is proved using the following observation:

$$(F) \qquad F = \bigcap_n F((t))^{p^n}.$$

We show first that $\Phi \subseteq F((t))^p$. If $x \in \Phi$, $x = \sum_{i>0} x_i t^i$, and $s = t^{-p^n} \in P$ then applying criterion (C_j) to $xs-y$ (with $v(y) > -1$) for $j = -p^n + i$, $i < p^n - p^{n-1}$, $i \not\equiv 0 \bmod p$, we find:

$$(i) \qquad x_i = 0 \qquad (i < p^n - p^{n-1},\ i \not\equiv 0 \bmod p)$$

Since n is arbitrary, $x \in F((t))^p$.

On the other hand $\Phi^{1/p} \subseteq \Phi$ by inspection of the definition (Φ) above, so by induction $\Phi \subseteq F((t))^{p^n}$ for all n and hence $\Phi \subseteq F$.

Thus $F = \Phi$ is P-definable without parameters, hence also S-definable.

§3. Definition of P from F

It seems to be considerably less straightforward to get at the results on L(F)-definability, so the argument will be broken up into several stages. In this section I will describe the definition in L(F) of a set $P(x)$ attached to any suitable $x \in F((t))$. $P(x)$ will resemble $\{x^{p^n} : n > 0\}$ fairly closely, and when $x = t^{-1}$ it leads to a simple definition of P.

In order to avoid an annoying case distinction we will adopt a rather unconventional notation. For $x \in F((t))$, we write $v(x) > 0$ with its usual meaning if F satisfies:

$$(\tau) \qquad \forall a \in F\ \exists b \in F \quad b^p - b = a$$

and otherwise "$v(x) > 0$" actually means $v(x) > 0$.

We shall also recall some convenient notation from [3]. If $x = \sum x_i t^i \in F((t))$ then we let

$$x_{[i \bmod p^n]} = \sum_{j \equiv i \bmod p^n} x_j t^j$$

The relation "$y = x_{[i \bmod p^n]}$" is L-definable. We will call $x \in F((t))$ <u>primitive</u> if $v(x_{[0 \bmod p]}) > 0$ and $v(x) < 0$.

<u>Definition</u>

An element y is <u>linked</u> to the element x iff:

$$\forall a \in F \exists b \in F \; ay - bx \in \tau.$$

If x is primitive and y is linked to x then each $a \in F$ determines the corresponding $b \in F$ uniquely. We write $b = y[a]$, if x is fixed.

The <u>ray</u> at x is the set of elements y linked to x which satisfy the condition:

$$y[ab] = y[a]\, y[b] \quad \text{for } a,b \quad F.$$

The ray at x will be denoted $P(x)$.

<u>Lemma</u> If x is primitive then the x-ray $P(x)$ consists of all elements y of the form:

(*) $$y = x^{p^n} + y'$$

with $n \in \mathbb{N}$ and $v(y') > 0$ (in the sense described above).

For the proof, if y is linked to x we have two claims:

(1) $$y = \sum a_i x^{p^i} + y' \quad \text{for some } y' \text{ with } v(y') > 0$$

and then (from (1) alone):

(2) $$y[a] = \sum a_i^{1/p^i} a^{1/p^i}.$$

The lemma then follows from (2), since $y[a]$ is essentially a polynomial function of a and F is supposed infinite.

(1) and (2) are proved simultaneously by a calculation based on the conditions (C_i). This is quite straightforward.

Now we can use the set $P(t^{-1})$ to define P. In fact P coincides with:

(P) $\{x \in P(t^{-1}): \forall y \in P(t^{-1})$ if $v(y) < v(x)$ then $\exists z \in P(t^{-1})\ yx^{-1} \in P(tz)\}$. This follows at once from the lemma. For example, if $x = t^{-p^n} + x'$ with $v(x') \geq 0$ satisfies the stated condition then $x^{-1} = t^{p^n} + x''$ with $v(x'') > p^n$ and applying definition (P) with $y = t^{-p^m}$ for $m > n$ leads to $v(x'') \geq p^m$ for all m, whence $x \notin P$.

§4. Encoding sequences in L(F)

We now develop an apparatus for encoding suitable finite sequences in $F((t))$ by elements of $F((t))$. We will make use of the sets $P(x)$ and P defined in §3.

The first step will be to give a definition for the relation "$ht(x) > s$" on $F((t)) \times P$ whose intended meaning (for $s = t^{-p^n}$) is:

$$\text{"}\exists z\ v(x - z^{p^n}) > 0\text{"}.$$

First consider the auxiliary notion
"$ht'(x) = s$" defined by:

(ht') $\qquad s \in P$ and $\exists z$ primitive $(x \in P(z),\ x + s \notin P(z + t^{-1}))$.

For x in the ray $P(z)$ with $z \neq -t^{-1}$ this means:

$$s = t^{-p^n},\ v(x - z^{-p^n}) \geq 0 \text{ for some } n.$$

The notion "$ht(x) \geq s$" is then defined by:

(ht) $\qquad \exists u,v\ ht'(u) = ht'(v) = s$ and $v(x - (u + s^{-1}v)) > 0$.

To see that this has the intended meaning, suppose $x = z^{p^n}$ and take:

$$u = x - x_{[0 \bmod p^{n+1}]}, \qquad v = t^{-p^n} x_{[0 \bmod p^{n+1}]}.$$

Then u^{1/p^n} and v^{1/p^n} are primitive, so $ht'(u) = ht'(v) = t^{-p^n}$ and hence $ht(x) \geq t^{-p^n}$, as intended.

Our set-theoretical encoding begins as follows.

<u>Definition</u>

For $x \in F((t))$, $s \in P$, let $\alpha(x,s)$ be:

$$\max\{v(x-y): ht(y) \geq s\}$$

if this is negative, and 0 otherwise. Observe that for $s = t^{-p^n}$ if $\alpha(x,s) < 0$ then there is $y \in F((t))$ with $v(x-y^{p^n}) = \alpha(x,s)$, and also $p^n \mid \alpha(x,s)$.

We say that $x \in F((t))$ has <u>content</u> at $s \in P$ if $\alpha(x,s^p) < \alpha(x,s)$, in which case we define the set $Con(x,s)$ of elements y satisfying:

(cn) $\qquad ht'(y) = s$ and $\alpha(x-ys^p) = \alpha(x,s)$.

Notice that for $y \in Con(x,s)$ we have:

(1) $\qquad v(y) = \alpha(x,s^p)$

(2) $\qquad Con(x,s) = \{y': v(y-y') > \alpha(x,s)$ and $ht'(y) = s\}$.

Our encoding has two defects: the degree of ambiguity manifested by (2) above, and fairly substantial limitations on the kinds of sequences which have codes. We will now introduce terminology suggested by the second defect.

<u>Definition.</u>

A sequence $(z_i: 1 \leq i \leq n)$ of primitive elements of $F((t))$ is <u>orderly</u> if there are finite sets I_i of negative integers satisfying:

(or 1) $\qquad pI_1 > p^2 I_2 > \ldots$ (i.e. inf $I_i > p$ sup $I_i + 1$.)

(or 2) $\qquad z_i = \sum_{j \in I_j} z_{ij} t^j$ for some z_{ij} F.

<u>Remark.</u>

If $z_1, \ldots, z_n$ is an orderly sequence of primitive elements then taking $x = \Sigma z_i^{p^i}$ we have:

0. $\qquad \alpha(x, t^{-p}) = 0$ and $\alpha(x, t^{-p^{i+1}}) = p^i v(z_i)$

1. $\qquad x$ has content at t^{-p^i} iff $1 \leq i \leq n$

2. For $1 \leq i \leq n$ $z_i p^i \in Con(x, t^{-p^i})$.

Of course, our final encoding is a modification of the above:

<u>Definition</u>

For $x \in F((t))$, $s \in P$ define $Seq(x,s)$ to be the collection of $z \in F((t))$ satisfying:

$$z \text{ is primitive and } \exists y \in P(z) \cap Con(x,s).$$

Restating the previous remark:

<u>Remark</u>

If $z_1, \ldots, z_n$ is an orderly sequence of primitive elements and $x = \Sigma z_i p^i$ then:

1. $Seq(x, t^{-p^i}) \neq \emptyset$ iff $1 \leq i \leq n$; in which case:

2. $z_i \in Seq(x, t^{-p^i})$, and:

3. $Seq(x, t^{-p^i}) = \{z \text{ primitive: } v(z - z_i) > v(z_{i-1})/p\}$.

§5. Definability of S from F

We exploit the coding of §4. Call an element $x \in F((t))$ <u>special</u> if it satisfies:

(sp1) $\exists s, s' \in P$ $ts' \in Seq(x,s)$ and $Seq(x, s^p) = \emptyset$

(sp2) $\forall s \in P$ If $Seq(x, s^p) \neq 0$ then $\exists y \in Seq(x, s^p)$, $z \in Seq(x,s)$ with $y = t^{-p}z$.

For x special, clearly $v(x) = -(p^n + 1)p^i$ where t^{-p^i}, t^{-p^n} are the elements s, s' referred to in (sp1). Furthermore we see inductively that:

$$\alpha(x, t^{-p^{i-k+1}}) = -(p^n - pk + 1).$$

Hence for $y, y' \in Seq(x, t^{-p^{i-k}})$, we get:

$$v(y - y') \geq -p(p^{n-1} - k - 1/p).$$

Lemma

If x is special and $v(x) = -(p^n+1)p^i$ then for all $k < i$

$t^{-(p^n-pk+1)} \in Seq(x,t^{-p^{i-k}})$.

Proof

Proceeding by induction on k, suppose:

$$y_0 = t^{(p^n-p(k-1)+1)} \in Seq(x,s^p) \quad \text{with} \quad s = t^{-p^{i-k}}$$

By (sp 2) and the remarks above, there is some:

$$y = y_0 + y'$$

with $v(y') \geq -(p^{n-1}-k)/p$ such that $t^p y \in Seq(x,s)$.

Since $v(t^p y') \geq -(p^{n-1}-k)/p + p > -(p^{n-1}-k-1)/p$, therefore $t^p y_0 \in Seq(x,s)$, as claimed.

Remark

A code x for the orderly sequence $(t^{-(pi+1)}: 1 < i < p^n)$ is special, with $v(x) = -(p^{n+1}+1)p^{p^n}$.

Now let S' be the set of y such that for some special x, s P, and z $Seq(x,s)$:

$$v(y-z) \geq (v(z)+1)/p.$$

This is the set of y such that for some $i > 0$:

$$v(y-t^{-(pi+1)}) \geq -i.$$

Let $S'' = \{y: t^{-1}y^p \in S'\}$. Then $y \in S''$ iff:

$$v(y-t^{-i}) \geq (-i+1)/p \qquad \text{for some} \quad i > 0.$$

Finally, it is easy to see that for $v(x) \geq 0$, $x \notin S$ iff

$$xS'' \subseteq S'' \cup \{y: v(y) \geq 0\}$$

Thus S is $L(F)$-definable. Notice also that the relation given by:

$$(*)\qquad \exists i,j,k,\ s_1 = t^i,\ s_2 = t^j,\ s_3 = t^{ij}$$

on S^3 is also definable, since the coding apparatus of §4 is certainly powerful enough to encode arbitrary finite subsets of S, and then multiplication can be defined from addition using quantification over finite sets.

Thus all arithmetically definable relations on S are $L(F)$-definable.

§6. Definability of "Coef" from F

Our final goal is to give a first order definition in $L(F)$ for the relation $\mathrm{Coef}(x,s,a)$ meaning:

$$x = \Sigma x_i t^i,\ s = t^n,\ a = x_n.$$

Let $S^* = \{as\colon a \in F^*, s \in S\}$. If $s^* = at^n \in S^*$, let $c(s^*)$ denote a. For $\alpha \in \mathbb{Z}$ let $\chi(\alpha)$ denote t^α.

We will now define the notion of a <u>regular code</u> γ of depth at least n. Such a code is meant to encrypt a sequence of the following form (reading from $-\infty$ toward 0):

$$a_0 t^{-n_0},\ a_0 t^{-n_1} + a_1 t^{-n_1+k_1},\ a_0 t^{-n_2} + a_1 t^{-n_2+k_1} + a_2 t^{-n_2+k_2},\ldots$$

where the exponents $n_0, n_1, n_2, \ldots$ are chosen to make the reverse of this sequence orderly, and $\Sigma k_i \le n$.

Formally, γ is <u>quasiregular</u> of depth at least n if:

(0) $\exists s \in P\ \exists a \in F\quad a\cdot\chi(v(\gamma)/v(s))\ \mathrm{Seq}(\gamma,s)\ \&\ \alpha(\gamma,s^p) = v(\gamma)$.

(1) $\forall s \in P$ If $\alpha(\gamma,s^p) > v(\gamma)$ then $\exists x \in \mathrm{Seq}(\gamma,s)$

$$\exists s^* \in S^*[\chi(\alpha(\gamma,s^{p^2})/v(s^p) - \alpha(\gamma,s^p/v(s))\ x + \chi(\alpha(\gamma,s^{p^2})/v(s^p))s^*] \in \mathrm{Seq}\ (\gamma,s^p)$$

and $0 < v(s^*) \le n$.

(2) $\forall s \in P$ If $\alpha(\gamma,s) > v(\gamma)$ then $\alpha(\gamma,s) - \alpha(\gamma,s^p) > n.v(s)$.

Condition (2) ensures that the code γ is unambiguous over an interval of length at least n.

Thus conditions (0,1) define a function $s \to s^*$ from $\{s \in P: \alpha(\gamma,S) > v(\gamma)\}$ into S^*. The code γ will be said to be <u>regular</u> if $v(s^*)$ increases with $v(s)$. For γ regular of depth at least n and $s_0 \in S$ with $0 \leq v(s_0) \leq n$ define:

$c(\gamma,s_0) = c(s^*)$ where $s \in P$ is chosen so that $\alpha(\gamma,s) > v(\gamma)$ and $c(s^*) = v(s_0)$;

or $c(\gamma,s_0) = 0$ if no such s exists.

Define $\mathrm{Coef}'(x,s,a)$ as follows:

(Cf') There is a regular code γ of depth at least $v(s)$ such that $x \in \mathrm{Seq}(\gamma,s')$ for some $s' \in P$ and $c(\gamma,s) = a$.

This has the intended meaning if $x_{[i \bmod p]} = 0$ for some i, so that $t^{-n}z$ is primitive for $n \equiv i$ modulo p and suitable codes exist.

In general $\mathrm{Coef}(x, s,a)$ may be defined by:

(Cf) $\exists x_1,x_2,a_1,a_2$ ($x = x_1+x_2$, $\mathrm{Coef}'(x_1,s,a_1)$, $\mathrm{Coef}'(x_2,sa_2)$, and $a = a_1 + a_2$).

REFERENCES

1. J. Ax, S. Kochen, "Diophantine problems over local fields I, II", Amer. J. Math. 187 (1965), 605-630 and 631-648.

2. J. Becker, J. Denef, L. van den Dries, "Further remarks on the elementary theory of formal power series rings," in Model Theory of Algebra and Arithmetic, Pacholski et al. eds., LNM 834, Springer-Verlag NY 1980, pp. 1-9.

3. G. Cherlin, "Undecidability of rational function fields in characteristic p," in Logic Colloquium 1982, Lolli et al. eds., North-Holland Amsterdam 1984, pp. 85-95.

4. Yu. Ershov, "On elementary theories of local fields," Alg. Log. 4 (1965), 5-30.

CONVEXITY PROPERTIES AND ALGEBRAIC CLOSURE OPERATORS

G.L.Cherlin and H.Volger

Dept. of Mathematics
Rutgers university
New Brunswick, N.J.
U.S.A. 08903

Math. Institut der
Universität Tübingen
D 7400 Tübingen
Fed. Rep. of Germany

§.0 Introduction

In dealing with a given class $\underline{A}$ of algebraic systems one frequently requires a good notion of the subalgebra "generated by" a subset S in the sense of the class $\underline{A}$, or the "algebraic closure" of S in the sense of $\underline{A}$. The simplest case is that in which $\underline{A}$ is closed under the formation of arbitrary intersections of subalgebras (in $\underline{A}$) of a given algebra. There is a compactness phenomenon which occurs if the class $\underline{A}$ is (first order) axiomatizable. In this case if $\underline{A}$ is closed under the formation of intersections of pairs of subalgebras, then it in fact satisfies the full closure condition. This applies for example to the class of algebraically closed fields. Examples show that the class of commutative local rings does not have this pleasant property, still this class is closed under the formation of intersections of descending chains. So here we have a strictly weaker closure condition, which we will relate to another notion of algebraic closure.

This paper is devoted to the systematic investigation of such closure conditions, often called "intersection properties" or "convexity properties" (Robinson [63]). Any partial order P determines an intersection property, which asks for closure under intersections of diagrams of substructures of order type P. We shall show that any nontrivial convexity property coincides with one of the two convexity properties mentioned above. One obtains the elementary convexity properties if elementary substructures replace substructures. There is exactly one nontrivial elementary convexity property. In general, there are at most $\omega+2$ nontrivial convexity properties for a given type of generalized embeddings, as we shall show.

The known results on convexity properties can be found in Robinson [51,63], Rabin [62] and Park [64]. Rabin [62] has given a syntactical characterization of those elementary classes which are closed under arbitrary intersections. They can be axiomatized by a set of $\forall\exists$-sentences, which satisfy an additional finiteness condition. In addition, he showed that it suffices to consider binary intersections. All the known results on intersections of countable descending chains can be found in the dissertation of Park [64]. He proved that an elementary class is closed under intersections of countable descending chains iff it can be axiomatized by a set of $\forall\exists$-sentences and it is closed w.r.t. substructures which are closed under a certain algebraic closure operator. In addition, he obtained some results on elementary convexity properties . Park's results indicate that there is a connection between convexity properties of elementary classes and algebraic closure operators on subsets of structures

Park's results can be extended with the help of a new set of *algebraic closure operators* which will be introduced in chapter 1 . The closu operators are algebraic in the following sense . A subset A of a given structure $\underline{B}$ is to be closed under the addition of finite A-definable sub sets of B of a certain type . It is crucial that the algebraic closure can also be described as an intersection of certain substructures of $\underline{B}$ containing A .

This will enable us to show in chapter 4 that arbitrary intersection and intersections of descending chains can be characterized by means of appropriate algebraic closure operators . In chapter 2 we shall give several syntactic characterization results for heredity with respect to algebraic closure operators . This leads to a syntactic characterization of the classes which are closed under intersections of descending chains thus solving an open problem left over from the sixties . Our methods do not yield a syntactic characterization in the case of intersections of elementary substructures , but we obtain Park's characterization of a slightly stronger property . As the syntactic characterizations are quite complicated , the characterization in terms of algebraic closure operator seems to be more useful .

In chapter 3 we shall study an invariant of partial orders which measures the extent to which the partial order is downward directed . Using this invariant we shall classify in chapter 4 for a given type of generalized embeddings . In addition , we shall compare convexity properti for different types of embeddings .

In the following we shall use the following notations and convention L is a fixed first-order language with equality . The set of atomic resp. all formulas is denoted by At resp. Fm . Any set Φ of formulas is assumed to be closed under conjunctions and disjunctions and to contain all quar tifier-free formulas . Φ^* denotes the closure of Φ under the propositiona operations , whereas $\exists\Phi$ resp. $\forall\Phi$ denotes the closure under existential resp. universal quantifications . $\exists(\Phi^*)$ is abbreviated to $\exists\Phi^*$. In additic we define : $\exists_o = At^*$ and $\exists_{n+1} = \exists((\exists_n)^*)$.

If Φ is a set of formulas then $L(\Phi)$ is the extension of L by predicates for the formulas in Φ which are not sentences . If T is a L-theory then $T(\Phi)$ is the $L(\Phi)$-theory determined by T and axioms which state tha each formula in Φ has the same interpretation as the associated new predicate.

An embedding $f : \underline{B}_1 \to \underline{B}_2$ of L-structures is said to be *Φ-elementary* if f preserves and reflects the validity of formulas in Φ with parameters from B_1. We write $\underline{B}_1 \leq_\Phi \underline{B}_2$ if the inclusion of the substructure $\underline{B}_1$ of $\underline{B}_2$ is Φ-elementary. In particular, embeddings resp. elementary embeddings are At*-elementary resp. Fm-elementary embeddings.

§.1 Algebraic Closure Operators

Below we shall introduce several closure operators on the set Sub($\underline{B}$) of substructures of a given structure $\underline{B}$. In the cases considered a subset A of B is to be closed by the addition of certain finite subsets of B which are definable in $\underline{B}$ with parameters from A. Because of the finiteness condition we call these operators algebraic.

Definition 1: Let Φ be a set of L-formulas, T be a L-theory and $\underline{B}$ be a L-structure. For a subset A of B we define the substructures $\mathrm{Alg}_T^\Phi(A,\underline{B})$ and $\mathrm{Alg}^\Phi(A,\underline{B})$ of $\underline{B}$ as follows:

An element b of B belongs to $\mathrm{Alg}_T^\Phi(A,\underline{B})$, if there exist a formula $\varphi(x,\overline{z})$ in Φ, a sequence $\overline{a}$ in A and $k > 0$ such that:

(i) $\underline{B} \models \varphi(b,\overline{a})$, (ii) $T \vee \Delta_{At*}(A) \vdash \exists^{\leq k} x\varphi(x,\overline{a})$.

Since Φ is assumed to be closed under conjunctions and to contain At* the finiteness condition (ii) can be replaced by

(ii') $T \vdash \forall z \exists^{\leq k} x\varphi(x,\overline{z})$.

An element b of B belongs to $\mathrm{Alg}^\Phi(A,\underline{B})$, if there exist a formula $\varphi(x,\overline{z})$ in Φ, a sequence $\overline{a}$ in A and $k > 0$ such that:

(i) $\underline{B} \models \varphi(b,\overline{a})$, (ii) $\underline{B} \models \exists^{\leq k}\varphi(x,\overline{a})$.

The structures $\mathrm{Alg}_T^\Phi(\emptyset,\underline{B})$ and $\mathrm{Alg}^\Phi(\emptyset,\underline{B})$ are defined, whenever the sets defining them are nonempty.

The operator $\mathrm{Alg}_T^\exists(-,\underline{B})$ has been considered by Robinson [63] and Bacsich [73]. The operator $\mathrm{Alg}^{Fm}(-,\underline{B})$ was introduced by Park [64].

Definition 2: Let Φ and Ψ be sets of L-formulas, T be a L-theory and let $\underline{B}$ be a L-structure. For a subset A of B we define the substructures $\Psi\text{-}\mathrm{Alg}_T^\Phi(A,\underline{B})$ and $\Psi\text{-}\overline{\mathrm{Alg}}_T^\Phi(A,\underline{B})$ of $\underline{B}$ as follows:

An element b of B belongs to $\Psi\text{-}\mathrm{Alg}_T^\Phi(A,\underline{B})$, if there exist a formula $\varphi(x,\overline{z})$ in Φ, a sequence $\overline{a}$ in A and a formula $\psi(\overline{y},\overline{z})$ in Ψ capturing the solutions of $\varphi(x,\overline{a})$ in $\underline{B}$ such that:

(i) $\underline{B} \models \varphi(b,\overline{a})$

(ii) $T \vee \Delta_{At*}(A) \vdash \exists\overline{y}\psi(\overline{y},\overline{a})$

$T \vee \Delta_{At*}(A) \vdash \forall x\overline{y}(\varphi(x,\overline{a}) \wedge \psi(\overline{y},\overline{a}) \Rightarrow \bigvee\langle x{\equiv}y_i : i=1,\dots,m\rangle)$.

Since Φ is assumed to be closed under conjunctions and disjunctions and to contain At* the finiteness condition (ii) can be replaced by

(ii') $T \vdash \forall\overline{z}\exists\overline{y}\psi(\overline{y},\overline{z})$, $T \vdash \forall x\overline{y}\overline{z}(\varphi(x,\overline{z}) \wedge \psi(\overline{y},\overline{z}) \Rightarrow \bigvee\langle x{\equiv}y_i : i=1,\dots,m\rangle)$.

An element b of B belongs to $\Psi\text{-}\overline{\mathrm{Alg}}_T^{\Phi}(A,\underline{B})$, if there exist a formula $\varphi(x,\bar{z})$ in Φ , a sequence $\bar{a}$ in A and a formula $\psi(\bar{y},\bar{z})$ in Ψ capturing the solutions of $\varphi(x,\bar{a})$ in $\underline{B}$ such that :

(i) $\underline{B}\models\varphi(b,\bar{a})$

(ii) $\underline{B}\models\exists\bar{y}\psi(\bar{y},\bar{a})$, $T\vee\Delta_{At*}(A)\vdash\forall x\bar{y}(\varphi(x,\bar{a})\wedge\psi(\bar{y},\bar{a})\Rightarrow\bigvee\langle x\equiv y_i:i=1,\dots,m\rangle)$.

Since Φ is assumed to be closed under conjunctions and to contain At* the finiteness condition (ii) can be replaced by

(ii') $\underline{B}\models\exists\bar{y}\psi(\bar{y},\bar{a})$, $T\vdash\forall x\overline{yz}(\varphi(x,\bar{z})\wedge\psi(\bar{y},\bar{z})\Rightarrow\bigvee\langle x\equiv y_i : i=1,\dots,m\rangle)$.

The structures $\Psi\text{-}\mathrm{Alg}_T^{\Phi}(\emptyset,\underline{B})$ and $\Psi\text{-}\overline{\mathrm{Alg}}_T^{\Phi}(\emptyset,\underline{B})$ are defined whenever the sets defining them are nonempty .

The following properties of the operators just defined can easily be verified .

Lemma 3: Let $\Psi\text{-}P^{\Phi}(-,\underline{B})$ be one of the operators defined above.

(1) $\Psi\text{-}P^{\Phi}(-,\underline{B})$ is monotone in Φ and Ψ i.e. $\Psi_1\text{-}P^{\Phi_1}(A,\underline{B})\leq\Psi_2\text{-}P^{\Phi_2}(A,\underline{B})$ whenever $\Phi_1\subseteq\Phi_2$ and $\Psi_1\subseteq\Psi_2$.

(2) $\Psi\text{-}P^{\Phi}(-,\underline{B})$ is monotone and increasing i.e. $\Psi\text{-}P^{\Phi}(A_1,\underline{B})\leq\Psi\text{-}P^{\Phi}(A_2,\underline{B})$ whenever $A_1\subseteq A_2$, and $A\subseteq\Psi\text{-}P^{\Phi}(A,\underline{B})$.

(3) $\Psi\text{-}P^{\Phi}(-,\underline{B})$ is finitary i.e. $\Psi\text{-}P^{\Phi}(A,\underline{B})=\bigcup\langle\Psi\text{-}P^{\Phi}(A',\underline{B}):A'\subseteq A, A' \text{ finite}\rangle$.

The operators are idempotent as well . However , this will follow more easily from a new description of the operators given later .

Lemma 4: Let $\underline{B}$ be a T-model .

(1) $\Psi\text{-}\mathrm{Alg}_T^{\Phi}(A,\underline{B})\leq\Psi\text{-}\overline{\mathrm{Alg}}_T^{\Phi}(A,\underline{B})$

(2) $\Psi^*\text{-}\mathrm{Alg}_T^{\Phi}(A,\underline{B})=\exists\Psi^*\text{-}\mathrm{Alg}_T^{\Phi}(A,\underline{B})$, $\Psi^*\text{-}\mathrm{Alg}_T(A,\underline{B})=\exists\Psi^*\text{-}\mathrm{Alg}_T^{\Phi}(A,\underline{B})$

(3) $\mathrm{Fm}\text{-}\mathrm{Alg}_T^{\Phi}(A,\underline{B})=\mathrm{Alg}_T^{\Phi}(A,\underline{B})$, $\mathrm{Fm}\text{-}\overline{\mathrm{Alg}}_T^{\Phi}(A,\underline{B})=\mathrm{Alg}^{\Phi}(A,\underline{B})$

(4) $\mathrm{Alg}_T^{\mathrm{Fm}}(A,\underline{B})=\mathrm{Alg}^{\mathrm{Fm}}(A,\underline{B})$.

Most of the above inclusions can easily be verified . Making use of the formula $\psi(\bar{y},\bar{z})=\bigwedge\langle\varphi(y_i,\bar{z}) : i=1,\dots,k\rangle$ one can verify easily that $\mathrm{Alg}_T^{\Phi}(A,\underline{B})\leq\mathrm{Fm}\text{-}\mathrm{Alg}_T^{\Phi}(A,\underline{B})$ and $\mathrm{Alg}^{\Phi}(A,\underline{B})\leq\mathrm{Fm}\text{-}\overline{\mathrm{Alg}}_T^{\Phi}(A,\underline{B})$. In order to prove now $\mathrm{Alg}^{\mathrm{Fm}}(A,\underline{B})\leq\mathrm{Alg}_T^{\mathrm{Fm}}(A,\underline{B})$ assume $\underline{B}\models\varphi(b,\bar{a})\wedge\exists^k x\varphi(x,\bar{a})$. Then one has $\underline{B}\models\varphi^*(b,\bar{a})$ and $T\vee\Delta_{At*}(A)\vdash\exists^{\leq k}x\varphi^*(x,\bar{a})$, where $\varphi^*(x,\bar{z})=\varphi(x,\bar{z})\wedge\exists^{\leq k}x\varphi(x,\bar{z})$

Lemma 5: (1) $\mathrm{Alg}^{\exists\Phi^*}(A,\underline{B}_1)=\mathrm{Alg}^{\exists\Phi^*}(A,\underline{B}_2)$, $\Psi\text{-}\mathrm{Alg}_T^{\exists\Phi^*}(A,\underline{B}_1)=\Psi\text{-}\mathrm{Alg}_T^{\exists\Phi^*}(A,\underline{B}_2)$ whenever $\underline{B}_1\leq_{\exists\Phi^*}\underline{B}_2$

(2) $\underline{B}_1\leq_{\Phi}\underline{B}_2\models T$ implies $\Psi\text{-}\mathrm{Alg}_T^{\exists\Phi^*}(A,\underline{B}_1)\leq\Psi\text{-}\mathrm{Alg}_T^{\exists\Phi^*}(A,\underline{B}_2)$

(3) $\underline{A}\leq_{\exists\Phi^*}\underline{B}$ implies $\underline{A}=\mathrm{Alg}^{\exists\Phi^*}(A,\underline{B})$

(4) $\underline{A}\leq_{\exists\Phi^*}\underline{B}\models T$ implies $\underline{A}=\exists\Phi^*\text{-}\overline{\mathrm{Alg}}_T^{\mathrm{Fm}}(A,\underline{B})$.

By our conventions on sets of formulas $\exists^{\geq k}x\varphi(x,\bar{z})$ is in $\exists\Phi^*$ whenever $\varphi(x,\bar{z})$ is in $\exists\Phi^*$. Therefore $\underline{B}_1\leq_{\exists\Phi^*}\underline{B}_2$ ensures that each formula $\varphi(x,\bar{a})$ in $\exists\Phi^*$ has either infinitely many solutions in $\underline{B}_1$ and $\underline{B}_2$ or it has the same finite set of solutions in $\underline{B}_1$ and $\underline{B}_2$. This proves (1) .

The proof of (2) is straightforward and (3) is a consequence of (1). To prove (4) assume $\underline{B} \models \varphi(b,\bar{a}) \wedge \exists\bar{y}\psi(\bar{y},\bar{a})$ and $T \vee \Delta_{At*}(A) \vdash \forall x\bar{y}(\varphi(x,\bar{a}) \wedge (\bar{y},\bar{a}) \rightarrow V\langle x \equiv y_i : i=1,\dots,m\rangle)$, where ψ belongs to $\exists\Phi^*$. Because of $\underline{A} \leq_{\exists\Phi^*} \underline{B}$ there exists $\bar{a}'$ in A such that $\underline{A} \models \psi(\bar{a}',\bar{a})$ and hence $\underline{B} \models \psi(\bar{a}',\bar{a})$. Since $\underline{B}$ is a T-model there exists a_i' in A such that $b = a_i'$, as required.

Quite often the closure of a set can not only be described as a union of smaller sets but also as an intersection of larger sets. The following hull operators will appear naturally in the context of convexity properties which will be discussed later.

Definition 6: Let Φ and Ψ be sets of L-formulas, T be a L-theory and let A be a subset of a L-structure $\underline{B}$.

The structure $C_T^{\Phi}(A,\underline{B})$ is the intersection of all T-models $\underline{C}$ satisfying $A \subseteq \underline{C} \leq_{\Phi} \underline{B}$. Let $C_T^{\Phi}(A,\underline{B})$ be $\underline{B}$ if there is no structure $\underline{C}$ with these properties.

An element b of B belongs to $H_T^{\Phi,\Psi}(A,\underline{B})$, if it belongs to each T-model $\underline{B}'$ containing A for which there exists a T-model $\underline{C}$ satisfying $\underline{B}' \leq_{\Psi} \underline{C} \;{}_{\Phi}\!\geq \underline{B}$. Let $H_T^{\Phi,\Psi}(A,\underline{B})$ be $\underline{B}$ if there is no structure $\underline{B}'$ with these properties.

An element b of B belongs to $\bar{H}_T^{\Phi,\Psi}(A,\underline{B})$, if it belongs to each T-model $\underline{B}'$ A-isomorphic to $\underline{B}$ for which there exists a T-model $\underline{C}$ satisfying $\underline{B}' \leq_{\Psi} \underline{C}$ and $\underline{C} \;{}_{\Phi}\!\geq \underline{B}$. Let $\bar{H}_T^{\Phi,\Psi}(A,\underline{B})$ be $\underline{B}$ if there is no structure $\underline{B}'$ with these properties.

The structures $C_T^{\Phi}(\emptyset,\underline{B})$, $H_T^{\Phi,\Psi}(\emptyset,\underline{B})$ and $\bar{H}_T^{\Phi,\Psi}(\emptyset,\underline{B})$ can be defined as well if the sets defining them are nonempty.

The following properties of the hull operators defined above can easily be verified.

Lemma 7: Let $\underline{B}$ be a T-model.

(1) $C_T^{\Phi}(-,\underline{B})$, $H_T^{\Phi,\Psi}(-,\underline{B})$ and $\bar{H}_T^{\Phi,\Psi}(-,\underline{B})$ are closure operators on $Sub(\underline{B})$

(2) $\Phi_1 \subseteq \Phi_2$ and $\Psi_1 \subseteq \Psi_2$ imply $C_T^{\Phi_1}(A,\underline{B}) \leq C_T^{\Phi_2}(A,\underline{B})$, $H_T^{\Phi_1,\Psi_1}(A,\underline{B}) \leq H_T^{\Phi_2,\Psi_2}(A,\underline{B})$ and $\bar{H}_T^{\Phi_1,\Psi_1}(A,\underline{B}) \leq \bar{H}_T^{\Phi_2,\Psi_2}(A,\underline{B})$

(3) $H_T^{\Phi,Fm}(A,\underline{B}) = H_T^{\Phi,\exists\Phi^*}(A,\underline{B})$, $\bar{H}_T^{\Phi,Fm}(A,\underline{B}) = \bar{H}_T^{\Phi,\exists\Phi^*}(A,\underline{B})$ and $H_T^{Fm,\Phi}(A,\underline{B}) = H_T^{\exists\Phi\ddagger\Phi}(A,\underline{B})$, $\bar{H}_T^{Fm,\Phi}(A,\underline{B}) = \bar{H}_T^{\exists\Phi\ddagger\Phi}(A,\underline{B})$

(4) $\underline{A} = C_T^{\Phi}(A,\underline{B})$ whenever $\underline{A} = \cap\langle \underline{A}_i : i \in I\rangle$ with $\underline{B} \;{}_{\Phi}\!\geq \underline{A}_i \models T$ for i in I

(5) $Alg^{\exists\Phi^*}(A,\underline{B}) \leq H_T^{\exists\Phi\ddagger\exists\Phi^*}(A,\underline{B})$, (6) $H_T^{Fm,\Phi}(A,\underline{B}) \leq C_T^{\Phi}(A,\underline{B})$.

The proofs of (1),(2),(4),(6) are straightforward. To prove (3) one uses the following fact (cf. Simmons [75]): $\underline{A} \leq_{\exists\Phi^*} \underline{B}$ iff there exists $\underline{C}$ such that $\underline{B} \leq_{\Phi} \underline{C} \;{}_{Fm}\!\geq \underline{A}$. To prove (5) let $\underline{B}'$ and $\underline{C}$ be T-models satisfying $A \subseteq \underline{B}' \leq_{\exists\Phi^*} \underline{C} \;{}_{\exists\Phi^*}\!\geq \underline{B}$. By lemma 5(1) we have $Alg^{\exists\Phi^*}(A,\underline{B}) = Alg^{\exists\Phi^*}(A,\underline{C}) = Alg^{\exists\Phi^*}(A,\underline{B}') \leq \underline{B} \wedge \underline{B}'$. This implies $Alg^{\exists\Phi^*}(A,\underline{B}) \leq H_T^{\exists\Phi\ddagger\exists\Phi^*}(A,\underline{B})$, as required.

The proof of the proposition below explains how the definition of the operators $\Psi\text{-}Alg_T^{\Phi}(-,\underline{B})$ and $\Psi\text{-}\overline{Alg}_T^{\Phi}(-,\underline{B})$ was obtained.

Proposition 8: Let $A \subseteq \underline{B} \models T$ be given :
(1) $\exists\Psi^*\text{-}Alg_T^{\exists\Phi^*}(A,\underline{B}) = H_T^{\Phi,\Psi}(A,\underline{B})$, (2) $\exists\Psi^*\text{-}\overline{Alg}_T^{\exists\Phi^*}(A,\underline{B}) = \overline{H}_T^{\Phi,\Psi}(A,\underline{B})$.

To prove (1) it suffices to show that for every b in B − A the following two statements are equivalent :

(*) $b \in H_T^{\Phi,\Psi}(A,\underline{B})$, (**) $b \in \exists\Psi^*\text{-}Alg_T^{\exists\Phi^*}(A,\underline{B})$.

The statement (*) is equivalent to the following condition :

(a) The set $T \vee \Delta_{\Phi^*}(\underline{B}) \vee \Delta_{\exists\ *}(\underline{B}') \vee \{\neg b \equiv b' : b' \in B - A\}$ is inconsistent for every model $\underline{B}'$ of $T \vee \Delta_{At^*}(A)$.

Let $\Sigma(b)$ be the set of those formulas $\psi(\overline{y},\overline{a})$ in $\exists\Psi^*$ which satisfy :

$T \vee \Delta_{\Phi^*}(\underline{B}) \vdash \forall\overline{y}(\psi(\overline{y},\overline{a}) \Rightarrow V\langle b \equiv y_i : i=1,\dots,m\rangle)$.

By means of a compactness argument we are able to replace (a) by

(b) For every model $\underline{B}'$ of $T \vee \Delta_{At^*}(A)$ there exists $\psi(\overline{y},\overline{a})$ in $\Sigma(b)$ such that $\underline{B}' \models \exists\overline{y}\psi(\overline{y},\overline{a})$.

This can be reformulated as follows :

(c) The set $T \vee \Delta_{At^*}(A) \vee \{\neg\exists\overline{y}\psi(\overline{y},\overline{a}) : \psi(\overline{y},\overline{a}) \in \Sigma(b)\}$ is inconsistent.

The set (b) is closed under disjunctions. Using another compactness argument we arrive at :

(d) There exists $\psi(\overline{y},\overline{a})$ in $\Sigma(b)$ such that $T \vee \Delta_{At^*}(A) \vdash \exists\overline{y}\psi(\overline{y},\overline{a})$.

Applying the compactness theorem to the defining condition of (b), we obtain as required :

(**) There exist $\psi(\overline{y},\overline{a})$ in $\exists\Psi^*$ and $\varphi(x,\overline{a})$ in $\exists\Phi^*$ such that $\underline{B} \models \varphi(b,\overline{a})$

$T \vee \Delta_{At^*}(A) \vdash \exists\overline{y}\psi(\overline{y},\overline{a})$ and

$T \vee \Delta_{At^*}(A) \vdash \forall x\overline{y}(\varphi(x,\overline{a}) \wedge \psi(\overline{y},\overline{a}) \Rightarrow V\langle x \equiv y_i : i=1,\dots,m\rangle)$

To prove (2) it suffices to show that for every b in B − A the following two statements are equivalent :

(*) $b \in \overline{H}_T^{\Phi,\Psi}(A,\underline{B})$, (**) $b \in \exists\Psi^*\text{-}\overline{Alg}_T^{\exists\Phi^*}(A,\underline{B})$.

The statement (*) is equivalent to the following condition :

(a) The set $T \vee \Delta_{\Phi^*}(\underline{B}) \vee \Delta'_{\exists\Psi^*}(\underline{B}) \vee \{\neg c_b \equiv c'_{b'} : b' \in B - A\} \vee \{c_a \equiv c'_a : a \in A\}$ is inconsistent.

By means of a compactness argument we are able to replace (a) by

(b) There exist $\varphi(x,\overline{z})$ in $\exists\Phi^*$, $\psi(\overline{y},\overline{z})$ in $\exists\Psi^*$, $\overline{a}$ in A , $\overline{b}$ in B − A such that $\underline{B} \models \varphi(b,\overline{a}) \wedge \psi(\overline{b},\overline{a})$ and

$T \vee \{\varphi(c_b,c_{\overline{a}})\} \vee \{\psi(c'_{\overline{b}},c_{\overline{a}})\} \vdash V\langle c_b \equiv c'_{b_i} : i=1,\dots,m\rangle$.

Obviously, (b) is equivalent to (**) .

As an immediate consequence of proposition 8 we have :

Corollary 9: The operators $Alg^{\exists\Phi^*}(-,\underline{B})$, $Alg_T^{\exists\Phi^*}(-,\underline{B})$, $\exists\Psi^*\text{-}Alg_T^{\exists\Phi^*}(-,\underline{B})$ and $\exists\Psi^*\text{-}\overline{Alg}_T^{\exists\Phi^*}(-,\underline{B})$ are closure operators.

Since we want to study convexity properties, we are interested in a syntactic characterization of those substructures of a given structure $\underline{B}$ which can be represented as the intersection of $\underline{B}$ and another structure $\underline{B}'$ within a larger structure.

Definition 10: Let $\underline{A} \leq \underline{B} \models T$ be given.
$\underline{A}$ is called *(Φ,Ψ)-tight for T in $\underline{B}$*, if there exist T-models $\underline{B}'$ and $\underline{C}$ satisfying $\underline{B}' \leq_\Psi \underline{C} \;{}_\Phi\!\geq \underline{B}$ and $\underline{A} = \underline{B}' \wedge \underline{B}$.
$\underline{A}$ is called *strongly (Φ,Ψ)-tight for T in $\underline{B}$*, if there exist T-models $\underline{B}'$ and $\underline{C}$ satisfying $\underline{B} \simeq_A \underline{B}'$, $\underline{B}' \leq_\Psi \underline{C} \;{}_\Phi\!\geq \underline{B}$ and $\underline{A} = \underline{B}' \wedge \underline{B}$.

Substructures which are strongly (Φ,Fm)-tight for T in $\underline{B}$ were considered in Park [64] and Omarov [67].

The following properties can easily be verified.

Lemma 11: Let $A \leq \underline{B} \models T$ be given.
(1) If $\underline{A}$ is strongly (Φ,Ψ)-tight for T in $\underline{B}$, then $\underline{A}$ is (Φ,Ψ)-tight for T in $\underline{B}$.
(2) If $\underline{A}$ is (Φ,Ψ)-tight resp. strongly (Φ,Ψ)-tight for T in $\underline{B}$ then $\underline{A} = H_T^{\Phi,\Psi}(A,\underline{B})$ resp. $\underline{A} = \overline{H}_T^{\Phi,\Psi}(A,\underline{B})$.
(3) $\underline{A}$ is strongly (Φ,Ψ)-tight for T in $\underline{B}$ iff $\underline{A}$ is strongly (Ψ,Φ)-tight for T in $\underline{B}$.

The proof of proposition 8 can be modified to yield a characterization of tightness and strong tightness.

Proposition 12: Let $\underline{A} < \underline{B} \models T$ be given.
(1) $\underline{A}$ is not (Φ,Ψ)-tight for T in $\underline{B}$ iff there exist $\varphi(\overline{x},\overline{z})$ in Φ^*, $\psi(\overline{y},\overline{z})$ in Ψ^*, $\overline{a}$ in A and $\overline{b}$ in B − A such that
(i) $\underline{B} \models \varphi(\overline{b},\overline{a})$
(ii) $T \vee \Delta_{At*}(A) \vdash \exists\overline{y}\psi(\overline{y},\overline{a})$
$T \vee \Delta_{At*}(A) \vdash \forall\overline{y}\overline{z}(\varphi(\overline{x},\overline{a}) \wedge \psi(\overline{y},\overline{a}) \Rightarrow \bigvee\langle x_i \equiv y_j : i=1,\dots,n; j=1,\dots,m\rangle)$.
(2) $\underline{A}$ is not strongly (Φ,Ψ)-tight for T in $\underline{B}$ iff there exist $\varphi(\overline{x},\overline{z})$ in Φ^*, $\psi(\overline{y},\overline{z})$ in Ψ^*, $\overline{a}$ in A and $\overline{b}$ in B − A such that
(i) $\underline{B} \models \varphi(\overline{b},\overline{a}) \wedge \exists\overline{y}\psi(\overline{y},\overline{a})$
(ii) $T \vee \Delta_{At*}(A) \vdash \forall\overline{y}\overline{z}(\varphi(\overline{x},\overline{a}) \wedge \psi(\overline{y},\overline{a}) \Rightarrow \bigvee\langle x_i \equiv y_j : i=1,\dots,n; j=1,\dots,m\rangle)$.

(1) By definition $\underline{A}$ is not (Φ,Ψ)-tight for T in $\underline{B}$ iff for every model $\underline{B}'$ of $T \vee \Delta_{At*}(A)$ the following set is inconsistent:

$$T \vee \Delta_{\Phi*}(\underline{B}) \vee \Delta_{\Psi*}(\underline{B}') \vee \{\neg b \equiv b' : b \in B - A,\ b' \in B' - A\}.$$

Now one can proceed as in the proof of proposition 8(1).

(2) By definition $\underline{A}$ is not strongly (Φ,Ψ)-tight for T in $\underline{B}$ iff the following set is inconsistent:

$$T \vee \Delta_{\Phi*}(\underline{B}) \vee \Delta'_{\Psi*}(\underline{B}) \vee \{\neg c_b \equiv c'_{b'} : b,b' \in B - A\} \vee \{c_a \equiv c'_a : a \in A\}.$$

Now one can proceed as in the proof of proposition 8(2) .

Lemma 13: $\underline{A}$ is strongly (Φ,Fm)-tight for T in $\underline{B}$ whenever $\underline{A} \leq_{\exists\Phi^*} \underline{B} \models T$.

Otherwise the following set of sentences is inconsistent :

$T \vee \Delta_{\Phi^*}(\underline{B}) \vee \Delta'_{Fm}(\underline{B}) \vee \{\neg c_{b_1} \equiv c'_{b_2} : b_1,b_2 \in B-A\} \vee \{c_a \equiv c'_a : a \in A\}$.

By compactness there exist $\varphi(\overline{x},\overline{z})$ in Φ^* , $\overline{a}$ in A and $\overline{b},\overline{d}$ in B-A such that

$\underline{B} \models \varphi(\overline{b},\overline{a})$, $T \vee \Delta'_{Fm}(\underline{B}) \vdash \forall\overline{y}(\varphi(\overline{y},c_{\overline{a}}) \Rightarrow V\langle y_i \equiv c'_{d_j} : i=1,\dots,n;j=1,\dots,m\rangle)$.

Because of $\underline{A} \leq_{\exists\Phi^*} \underline{B}$ there exists $\overline{a}'$ in A such that $\underline{A} \models \varphi(\overline{a}',\overline{a})$ and hence $\underline{B} \models \varphi(\overline{a}',\overline{a})$. Since $\underline{B}$ is a model of $T \vee \Delta'_{Fm}(\underline{B})$, there exist i,j such that $a'_i = d_j$, which is impossible.

The result in proposition 8(2) can be improved if Φ or Ψ is Fm . In this form it will be very useful later on . - Let $\hat{\overline{Alg}}^{\exists\Phi^*}(A,\underline{B})$ be the following substructure of $\underline{B}$. An element b belongs to $\hat{\overline{Alg}}^{\exists\Phi^*}(A,\underline{B})$ if b belongs to A or there exist $\varphi(\overline{x},\overline{y})$ in Φ^* , $\overline{a}$ in A , $\overline{b}$ in B - A such that :

(i) $\underline{B} \models \varphi(\overline{b},\overline{a})$, $b = b_i$ for some i

(ii) $\underline{B} \models \forall\overline{x}(\varphi(\overline{x},\overline{a}) \Rightarrow V\langle x_i \equiv b_j : i=1,\dots,n;j=1,\dots,m\rangle)$ for the same i .

Proposition 14: The following three statements are equivalent :

(1) $Alg^{\exists\Phi^*}(A,\underline{B}) = \underline{A}$, $\hat{\overline{Alg}}^{\exists\Phi^*}(A,\underline{B}) = \underline{A}$,

(3) $\underline{A}$ is strongly (Φ,Fm)-tight for $Th(\underline{B})$ in $\underline{B}$.

(1) → (2) :

Assume $\hat{\overline{Alg}}^{\exists\Phi^*}(A,\underline{B}) \neq \underline{A}$. Then there exist $\varphi(\overline{x},\overline{y})$ in Φ^* , $\overline{a}$ in A , $\overline{b}$ in B - A such that the following holds :

(*) $\underline{B} \models \varphi(\overline{b},\overline{a})$

(**) There exists a finite , nonempty subset K of B - A such that every solution of $\varphi(\overline{x},\overline{a})$ in $\underline{B}$ meets K .

Consider a formula $\varphi(\overline{x},\overline{y})$ satisfying (*),(**) in which the length of $x = (x_o,\dots,x_{n-1})$ is as small as possible . An ordered solution of $\varphi(\overline{x},\overline{a})$ in $\underline{B}$ is a function $f : n \to B$ such that $\underline{B} \models \varphi(f(o),\dots,f(n-1),\overline{a})$. The defect $d(\varphi,K)$ of $\varphi(\overline{x},\overline{a})$ with respect to K is the minimum of

$\{card\{i : f(i) \in K\} : f$ ordered solution of $\varphi(\overline{x},\overline{a})$ in $\underline{B}\}$.

By our choice we have $d(\varphi,K) \leq n$. Let the defect $d(\varphi)$ be the maximum of

$\{d(\varphi,K) : K$ finite , $\emptyset \neq K \subseteq B-A\}$.

By assumption we have $0 < d(\varphi) \leq n$. Among all formulas $\varphi(\overline{x},\overline{a})$ in Φ^* with x of length n fix a formula with maximal defect d . In addition , fix K such that $d(\varphi,K) = d$. Obviously we have $d \leq k = card(K)$.

Let $I \subseteq n$ with $card(I) = d$ and $f : I \to K$ be fixed . Then the following cases are possible :

(i) There are at most k prolongations f_i of f to solutions of $\varphi(\overline{x},\overline{a}$ which are pairwise disjoint modulo I i.e. $f_i(n-I) \wedge f_j(n-I) = \emptyset$

for $i \neq j$.

(ii) = not (i) .

First we consider case (ii) . Let $\varphi'(x_1,\dots,x_d,\overline{y})$ be the formula which states that (ii) holds i.e. that there are at least k+1 extensions of the function f to ordered solutions $f_1,\dots,f_{k+1}$ of $\varphi(\overline{x},\overline{a})$ which are pairwise disjoint modulo I , where $f(i_j)$ corresponds to x_j . It can be verified that φ' belongs to $\exists\Phi^*$. The formula φ' has the following property :

(iii) $\underline{B} \models \varphi'(b_1,\dots,b_d,a)$ implies $b_1,\dots,b_d \in K$.

Otherwise $d(\varphi,K) = d$ yields for each $j = 1,\dots,k+1$ an element $i_j \in n - I$ such that $f_j(i_j) \in K$, so that K would have to contain at least k+1 elements , contradicting $k = \mathrm{card}(K)$. Thus we can conclude from (ii) that $\varphi'(\overline{x},\overline{a})$ has at most k^d ordered solutions in $\underline{B}$ and it follows easily that they are all contained in $\mathrm{Alg}^{\exists\Phi^*}(A,\underline{B})$. Because of (iii) and $K \subseteq B - A$ they all lie outside of A . Therefore we have $\underline{A} \neq \mathrm{Alg}^{\exists\Phi^*}(A,\underline{B})$ in this case.

Now consider case (i) . Assuming $d = n$ we know that $\varphi(\overline{x},\overline{a})$ has at most k^n solutions in $\underline{B}$, which therefore must lie in $\mathrm{Alg}^{\exists\Phi^*}(A,\underline{B})$. However , (**) implies that each solution lies outside of A . Therefore we have $\underline{A} \neq \mathrm{Alg}^{\exists\Phi^*}(A,\underline{B})$ also in this case .

It remains to be shown that $d < n$ is impossible in the presence of (i) . For any set $I = i_1,\dots,i_d$ and any $f : I \to K$ an application of (i) yields a set K_f such that for any extension g of f to an ordered solution of $\varphi(\overline{x},\overline{a})$ there exists $i \in n - I$ with $g(i) \in K_f$. We obtain $d(\varphi,K') \geq d+1$, thus contradicting the maximality of d . To see this fix an ordered solution g of $\varphi(\overline{x},\overline{a})$. Then there is a set I with d elements such that $g(i) \in K$ for $i \in I$. Hence there exists $j \in n - I$ such that $g(j) \in K_{g|I} \subset K'$. This finishes the proof .

(2) → (3) :

If $\underline{A}$ is not strongly (Φ,Fm)-tight for $\mathrm{Th}(\underline{B})$ in $\underline{B}$ then the following set of sentences is not consistent :

$$\Delta_{\Phi^*}(\underline{B}) \vee \Delta'_{\mathrm{Fm}}(\underline{B}) \vee \{\neg c_{b_1} \equiv c'_{b_2} : b_1,b_2 \in B - A\} \vee \{c_a \equiv c'_a : a \in A\} .$$

By compactness there exist $\varphi(\overline{x},\overline{y})$ in Φ^* , $\overline{a}$ in A , $\overline{b}$ in $B - A$ such that

$\underline{B} \models \varphi(\overline{b},\overline{a})$ and $\underline{B} \models \forall\overline{x}(\varphi(\overline{x},\overline{a}) \Rightarrow \bigvee\langle x_i \equiv b_j : j=1,\dots,m\rangle)$.

The sequence $\overline{b}$ must be nonempty since the set $\Delta_{\Phi^*}(\underline{B}) \vee \Delta'_{\mathrm{Fm}}(\underline{B}) \vee \{c_a \equiv c'_a : a \in A\}$ is consistent . Hence we have $\underline{A} \neq \mathrm{Alg}^{\exists\Phi^*}(A,\underline{B})$, as required .

(3) → (1) :

Let $\underline{B}' \leq_{\mathrm{Fm}} \underline{C} \,{}_{\Phi}\!\geq \underline{B}$ be given such that $\underline{A} = \underline{B} \wedge \underline{B}'$ and $\underline{B} \simeq_A \underline{B}'$. If b belongs to $\mathrm{Alg}^{\exists\Phi^*}(A,\underline{B})$,then there exist $\varphi(x,\overline{y})$ in $\exists\Phi^*$, $\overline{a}$ in A and $k > 0$ such that $\underline{B} \models \varphi(b,\overline{a}) \wedge \exists^k x\varphi(x,\overline{a})$. However , $\underline{B} \simeq_A \underline{B}'$ and $\underline{B}' \leq_{\mathrm{Fm}} \underline{C}$ imply $\underline{B}',\underline{C} \models \exists^k x\varphi(x,\overline{a})$. Because of $\underline{B} \leq_{\Phi} \underline{C}$ we have $\underline{C} \models \varphi(b,\overline{a})$ and hence $b \in \underline{B}' \wedge B = A$.

Corollary 15: $\underline{A} = \mathrm{Alg}^{\exists\Phi^*}(A,\underline{B})$ and $\underline{B} \models T$ implies $\underline{A} = \exists\Phi^*\text{-}\overline{\mathrm{Alg}}_T^{Fm}(A,\underline{B})$.

By proposition 14 there exist T-models $\underline{B}'$ and $\underline{C}$ such that $\underline{B} \simeq_A \underline{B}'$, $\underline{B}' \leq_{Fm} \underline{C} \;_{\Phi}\!\geq \underline{B}$ and $\underline{A} = \underline{B} \wedge \underline{B}'$. It suffices to prove $\exists\Phi^*\text{-}\overline{\mathrm{Alg}}_T^{Fm}(A,\underline{B}') = \underline{A}$. If b' belongs to $\exists\Phi^*\text{-}\overline{\mathrm{Alg}}_T^{Fm}(A,\underline{B}')$ then there exist formulas $\varphi(x,\bar z)$ and $\psi(\bar y,\bar z)$ and $\bar a$ in $\underline{B}'$ such that $\psi \in \exists\Phi^*$ and

$\underline{B} \models \varphi(b',\bar a) \wedge \exists\bar y\psi(\bar y,\bar a)$ and $T \vdash \forall x\bar y(\varphi(x,\bar a)\wedge\psi(\bar y,\bar a) \Rightarrow V\langle x{=}y_i : i=1,\dots,m\rangle)$. Hence there exists $\bar b$ in B satisfying $\underline{B} \models \psi(\bar b,\bar a)$. This implies $\underline{C} \models \varphi(b',\bar a) \wedge \psi(\bar b,\bar a)$ and hence $b' \in B \wedge B' = A$, as required .

§.2 Heredity with respect to Algebraic Closure Operators

We want to study those theories T which have the property that structures which are algebraically closed in a T-model are again T-models .

Definition 1: Let $P(-,\underline{B})$ be a closure operator on $\mathrm{Sub}(\underline{B})$, the set of substructures of a structure $\underline{B}$.
A theory T is said to be *hereditary with respect to* P , if $P(A,\underline{B}) \models T$ whenever $\underline{B} \models T$, $A \subseteq B$ and $P(A,\underline{B}) \neq \emptyset$.
A theory T is said to be Φ-*hereditary with respect to* P , if $P(A,\underline{B}) \models T$ and $P(A,\underline{B}) \leq_\Phi \underline{B}$ whenever $\underline{B} \models T$, $A \subseteq B$ and $P(A,\underline{B}) \neq \emptyset$.
As $P(-,\underline{B})$ is a closure operator it suffices to consider substructures $\underline{A}$ satisfying $\underline{A} = P(A,\underline{B})$.

The following observation can be used to compare different operators with respect to heredity .

Lemma 2: If $P_1(P_2(A,\underline{B}),\underline{B}) = P_2(A,\underline{B})$ holds for all $\underline{A}$ and all T-models $\underline{B}$, then T is hereditary w.r.t. P_2 whenever T is hereditary w.r.t. P_1 . In particular , if $P_1(A,\underline{B}) \leq P_2(A,\underline{B})$ holds for all $\underline{A}$ and all T-models $\underline{B}$, then the above condition is satisfied . - The lemma remains true if 'hereditary' is replaced by 'Φ-hereditary' .

Let us consider the following properties of a theory T :

(1_Φ) T is hereditary w.r.t. $\exists\Phi^*\text{-}\mathrm{Alg}_T^{\exists\Phi^*}$

(2_Φ) T is hereditary w.r.t. $\mathrm{Alg}_T^{\exists\Phi^*}$

(3_Φ) T is hereditary w.r.t. $\mathrm{Alg}^{\exists\Phi^*}$.

By lemma 1.3 and 1.4 we have : $(1_\Phi) \to (2_\Phi) \to (3_\Phi)$. Moreover , because of proposition 1.8 we have : $(3_\Phi) \to (1_{\exists\Phi^*})$.

Proposition 3: (1) If T is hereditary w.r.t. $\mathrm{Alg}^{\exists\Phi^*}$ or $\exists\Phi^*\text{-}\overline{\mathrm{Alg}}_T^{Fm}$ then T is hereditary w.r.t. $\exists\Phi^*$-elementary substructures .
(2) T is hereditary w.r.t. $\exists\Phi^*$-elementary substructures iff T can be axiomatized by a set of $\forall\exists\Phi^*$-sentences iff the class of T-models is close

under unions of chains of Φ-elementary substructures .

(1) follows from (3) resp. (4) in lemma 1.5 . (2) can be found in Keisler [60] .

This shows that theories satisfying (1_Φ) , (2_Φ) or (3_Φ) can be axiomatized by a set of $\forall\exists\Phi^*$-sentences . Below we shall solve the characterization problem in the case $\Phi = At$.

Theorem 4: A theory T is hereditary w.r.t. $Alg^\exists$ iff T can be axiomatized by a set Δ of $\forall\exists$-sentences which satisfies the following condition :

(i) For every sentence $\forall\bar{x}\exists\bar{y}\alpha(\bar{x},\bar{y})$ in Δ with α in At* , there exist r , formulas $\mu_1(\bar{x},\bar{y}),\dots,\mu_r(\bar{x},\bar{y})$ in $\exists$ and $k_1,\dots,k_r > 0$ such that :
$T \vdash \forall\bar{x}\exists\bar{y}(\alpha(\bar{x},\bar{y}) \wedge V\langle\mu_i(x,y) \wedge \exists^{\le k_i}\bar{z}\mu_i(\bar{x},\bar{z}) : i=1,\dots,r\rangle)$.

Theorem 5: A theory T is hereditary w.r.t. $Alg_T^\exists$ iff T can be axiomatized by a set Δ of $\forall\exists$-sentences which satisfies the following condition :

(ii) For every sentence $\forall\bar{x}\exists\bar{y}\alpha(\bar{x},\bar{y})$ in Δ with α in At* , there exist a formula $\mu(\bar{x},\bar{y})$ in $\exists$ and $k > 0$ such that :
$T \vdash \forall\bar{x}\exists\bar{y}(\alpha(\bar{x},\bar{y}) \wedge \mu(\bar{x},\bar{y}) \wedge \exists^{\le k}\bar{z}\mu(\bar{x},\bar{z}))$.

Theorem 6: A theory T is hereditary w.r.t. $\exists\text{-}\overline{Alg}_T^\exists$ iff T can be axiomatized by a set Δ of $\forall\exists$-sentences which satisfies the following condition:

(iii) For every sentence $\forall\bar{x}\exists\bar{y}\alpha(\bar{x},\bar{y})$ in Δ with α in At* , there exist r , formulas $\mu_1(\bar{x},\bar{y}),\dots,\mu_r(\bar{x},\bar{y})$ and $\psi_1(\bar{x},\bar{z}),\dots,\psi_r(\bar{x},\bar{z})$ in $\exists$ such that :
$T \vdash \forall\bar{x}\exists\bar{y}(\alpha(\bar{x},\bar{y}) \wedge V\langle\mu_i(\bar{x},\bar{y}) \wedge \exists\bar{z}\psi_i(\bar{x},\bar{z}) : i=1,\dots,r\rangle)$ and
$T \vdash \forall\overline{xyz}(\mu_i(\bar{x},\bar{y}) \wedge \psi_i(x\ z) \rightarrow V\langle y_j \equiv z_k : k=1,\dots,m\rangle)$
for $i = 1,\dots,r$ and $j = 1,\dots,n$.

Theorem 7: A theory T is hereditary w.r.t. $\exists\text{-}Alg_T^\exists$ iff T can be axiomatized by a set Δ of $\forall\exists$-sentences which satisfies the following condition:

(iv) For every sentence $\forall\bar{x}\exists\bar{y}\alpha(\bar{x},\bar{y})$ in Δ with α in At* , there exist formulas $\mu(\bar{x},\bar{y})$ and $\psi(\bar{x},\bar{z})$ in $\exists$ such that :
$T \vdash \forall\bar{x}\exists\bar{y}(\alpha(\bar{x},\bar{y}) \wedge \mu(\bar{x},\bar{y}) \wedge \exists\bar{z}\psi(\bar{x},\bar{z}))$ and
$T \vdash \forall\bar{x}\exists\bar{y}(\mu(\bar{x},\bar{y}) \wedge \psi(\bar{x},\bar{z}) \rightarrow V\langle y_j \equiv z_k : k=1,\dots,m\rangle)$ for $j = 1,\dots,n$.

The results for $Alg^\exists$, $\exists\text{-}\overline{Alg}_T^\exists$ and $\exists\text{-}Alg_T^\exists$ are new , the result for $Alg_T^\exists$ is contained in Volger [79] . A different syntactic characterization for $\exists\text{-}Alg_T^\exists$ can be extracted from Kueker [75] , as we shall see later on .

The proofs of the four results follow the same pattern . The proof of theorem 4 proceeds as follows . The structure $Alg^\exists(A,\underline{B})$ is the union of the substructures $Alg^\exists(\bar{a},\underline{B})$, where $\bar{a}$ is a finite sequence in A . Moreover, if $\bar{a}$ belongs to $Alg^\exists(A,\underline{B})$ then $Alg^\exists(a,\underline{B}) \le Alg^\exists(A,\underline{B})$. Making use of pro-

position 3 we obtain the following reduction :
T is hereditary w.r.t. $Alg^{\exists}$ iff T can be axiomatized by a set Δ of $\forall\exists$-sentences satisfying : For every sentence $\forall\overline{x}\exists\overline{y}\alpha(\overline{x},\overline{y})$ in Δ with α in At* , for every T-model $\underline{B}$ and every sequence $\overline{a}$ in B we have :

(*) $Alg^{\exists}(\overline{a},\underline{B}) \models \exists\overline{y}\alpha(\overline{a},\overline{y})$.

It is easy to verify that the condition (i) is sufficient to guarantee this . Conversely , if $\exists\overline{y}\alpha(\overline{a},\overline{y})$ holds in $Alg^{\exists}(\overline{a},\underline{B})$ then there exist a formula $\mu(\overline{x},\overline{y})$ in $\exists$ and $l > 0$ such that :

$\underline{B} \models \exists\overline{y}(\alpha(\overline{a},\overline{y}) \wedge \mu(\overline{a},\overline{y})) \wedge \exists^{\leq l}\overline{z}\mu(\overline{a},\overline{z})$.

Now a compactness argument yields the required result .

The proof of theorem 5 proceeds along the same line . However , in this case we have for every T-model $\underline{B}$ and every $\overline{a}$ in B :

(**) $Alg_T^{\exists}(\overline{a},\underline{B}) \models \exists\overline{y}\alpha(\overline{a},\overline{y})$

As before , it is easy to verify that the condition (ii) is sufficient to guarantee this . Conversely , if $\exists\overline{y}\alpha(\overline{a},\overline{y})$ holds in $Alg_T^{\exists}(\overline{a},\underline{B})$ then there exist a formula $\mu(\overline{x},\overline{y})$ in $\exists$ and $l > 0$ such that :

$\underline{B} \models \exists\overline{y}(\alpha(\overline{a},\overline{y}) \wedge \mu(\overline{a},\overline{y}))$ and $T \vdash \forall\overline{x}\exists^{\leq l}\overline{z}\mu(\overline{x},\overline{z})$.

Again a compactness argument yields the desired result , since in this case the disjunction $\bigvee\langle\mu_i : i=1,\dots,r\rangle$ can be replaced by a single formula μ .

The proofs of theorem 6 and 7 can be produced by the same method . The same method could be used to characterize those theories which are hereditary w.r.t. $\exists\text{-}Alg_T^{Fm}$ resp. $\exists\text{-}\overline{Alg}_T^{Fm}$.

The extension $T(\Phi)$ of T by new predicates for the formulas in Φ can be used to transform Φ-heredity into heredity .

<u>Lemma 8</u>: (1) T is Φ-hereditary w.r.t. $Alg^{\exists\Phi^*}$ iff $T(\Phi)$ is hereditary w.r.t. $Alg^{\exists}$.

(2) T is $\exists\Phi^*$-hereditary w.r.t. $Alg^{\exists\Phi^*}$ iff $T(\Phi)$ is $\exists$-hereditary w.r.t. $Alg^{\exists}$.

The lemma remains true if $Alg^{\exists\Phi^*}$ is replaced by $Alg_T^{\exists\Phi^*}$ or $\exists\Phi^*\text{-}\overline{Alg}_T^{\exists\Phi^*}$ or $\exists\Phi^*\text{-}Alg_T^{\exists\Phi^*}$.

Let $\underline{A},\underline{B}$ be $T(\Phi)$-models and let $\underline{A}|L,\underline{B}|L$ be the reducts which are T-models . Then the following holds : (i) $\underline{A} \leq \underline{B}$ iff $\underline{A}|L \leq_{\Phi} \underline{B}|L$

(ii) $\underline{A} \leq_{\exists} \underline{B}$ iff $\underline{A}|L \leq_{\exists\Phi^*} \underline{B}|L$.

Moreover , every T-model $\underline{C}$ has a unique extension to a $T(\Phi)$-model $\underline{A}$ with $\underline{A}|L = \underline{C}$. Hence it suffices to verify for every $T(\Phi)$-model $\underline{B}$:

($\square$) $Alg^{\exists\Phi^*}(\underline{A}|L,\underline{B}|L) = Alg^{\exists}(\underline{A},\underline{B})|L$.

In the other cases one has to verify analogous equations .

Remark 9: Part (1) of the previous lemma can be used to obtain from the results in theorems 4 - 7 syntactic characterizations of those theories which are Φ-hereditary w.r.t. $Alg^{\exists\Phi^*}$, $Alg_T^{\exists\Phi^*}$, $\exists\Phi^*\text{-}\overline{Alg}_T^{\exists\Phi^*}$ and $\exists\Phi^*\text{-}Alg_T^{\exists\Phi^*}$.

Below we shall characterize those theories which are $\exists$-hereditary w.r.t. $Alg^{\exists}$, $Alg_T^{\exists}$, $\exists\text{-}\overline{Alg}_T^{\exists}$ and $\exists\text{-}Alg_T^{\exists}$. Then part (2) of the previous lemma can be used to replace $\exists$ by $\exists\Phi^*$ in these results.

Proposition 10: A theory T is $\exists$-hereditary w.r.t. $Alg^{\exists}$ iff T can be axiomatized by a set of $\forall\exists$-sentences and satisfies the following condition:

(i) For every formula $\alpha(\overline{x},\overline{y})$ in At* there exist r, formulas $\mu_1(\overline{x},\overline{y}),\ldots,\mu_r(\overline{x},\overline{y})$ in $\exists$ and $k_1,\ldots,k_r > 0$ such that:

$T\vdash \forall\overline{x}(\exists\overline{y}\alpha(\overline{x},\overline{y}) \Rightarrow V\langle\exists\overline{y}(\alpha(\overline{x},\overline{y}) \wedge \mu_i(\overline{x},\overline{y})) \wedge \exists^{\le k_i}\overline{z}\mu_i(\overline{x},\overline{z}) : i=1,\ldots,r\rangle)$.

Proposition 11: A theory T is $\exists$-hereditary w.r.t. $Alg_T^{\exists}$ iff T can be axiomatized by a set of $\forall\exists$-sentences and satisfies the following condition:

(ii) For every formula $\alpha(\overline{x},\overline{y})$ in At* there exists a formula $\mu(\overline{x},\overline{y})$ in $\exists$ and $k > 0$ such that:

$T\vdash \forall\overline{x}(\exists\overline{y}\alpha(\overline{x},\overline{y}) \Rightarrow \exists\overline{y}(\alpha(\overline{x},\overline{y}) \wedge \mu(\overline{x},\overline{y})))$ and $T\vdash \forall\overline{x}\exists^{\le k}\overline{y}\mu(\overline{x},\overline{y})$.

Proposition 12: A theory T is $\exists$-hereditary w.r.t. $\exists\text{-}\overline{Alg}_T^{\exists}$ iff T can be axiomatized by a set of $\forall\exists$-sentences and satisfies the following condition:

(iii) For every formula $\alpha(\overline{x},\overline{y})$ in At* there exist r, formulas $\mu_1(\overline{x},\overline{y}),\ldots,\mu_r(\overline{x},\overline{y})$ and $\psi_1(\overline{x},\overline{z}),\ldots,\psi_r(\overline{x},\overline{z})$ such that:

$T\vdash \forall\overline{x}(\exists\overline{y}\alpha(\overline{x},\overline{y}) \Rightarrow V\langle\exists\overline{y}(\alpha(\overline{x},\overline{y}) \wedge \mu_i(\overline{x},\overline{y})) \wedge \exists\overline{z}\psi_i(\overline{x},\overline{z}) : i=1,\ldots,r\rangle)$ and

$T\vdash \forall\overline{xyz}(\mu_i(\overline{x},\overline{y}) \wedge \psi_i(\overline{x},\overline{z}) \Rightarrow V\langle y_j \equiv z_k : k=1,\ldots,m\rangle)$

for $i = 1,\ldots,r$ and $j = 1,\ldots,n$.

Proposition 13: A theory T is $\exists$-hereditary w.r.t. $\exists\text{-}Alg_T^{\exists}$ iff T can be axiomatized by a set of $\forall\exists$-sentences and satisfies the following condition:

(iv) For every formula $\alpha(\overline{x},\overline{y})$ in At* there exist formulas $\mu(\overline{x},\overline{y})$, $\psi(\overline{x},\overline{z})$ in $\exists$ such that:

$T\vdash \forall\overline{x}(\exists\overline{y}\alpha(\overline{x},\overline{y}) \Rightarrow \exists\overline{y}(\alpha(\overline{x},\overline{y}) \wedge \mu(\overline{x},\overline{y})))$, $T\vdash \forall\overline{x}\exists\overline{z}\psi(\overline{x},\overline{z})$ and

$T\vdash \forall\overline{xyz}(\mu(\overline{x},\overline{y})\wedge\psi(\overline{x},\overline{z}) \Rightarrow V\langle y_j \equiv z_k : k=1,\ldots,m\rangle)$ for $j = 1,\ldots,n$.

An application of lemma 8(2) to proposition 13 produces the characterization of those theories which are Fm-hereditary w.r.t. $Alg_T^{Fm} = Alg^{Fm}$ due to Park [64].

As before, the four proofs follow the same pattern. The proof of proposition 10 proceeds as follows. By proposition 3 we can conclude: T is $\exists$-hereditary w.r.t. $Alg^{\exists}$ iff T can be axiomatized by a set of $\forall\exists$-

sentences and one has $Alg^{\exists}(A,\underline{B}) \leq_{\exists} \underline{B}$ for every T-model $\underline{B}$. To prove that $Alg^{\exists}(A,\underline{B}) \leq_{\exists} \underline{B}$ holds it suffices to show for every formula $\alpha(\bar{x},\bar{y})$ in At* and every sequence $\bar{a}$ in B that $\underline{B} \models \exists\bar{y}\alpha(\bar{a},\bar{y})$ implies $Alg^{\exists}(\bar{a},\underline{B}) \models \exists\bar{y}\alpha(\bar{a},\bar{y})$. The latter states that there must exist $\mu(\bar{x},\bar{y})$ in $\exists$ and $k>0$ such that $\underline{B} \models \exists\bar{y}(\alpha(\bar{a},\bar{y}) \wedge \mu(\bar{a},\bar{y})) \wedge \exists^{\leq k}\bar{z}\mu(\bar{a},\bar{z})$. Now a compactness argument yields the desired result.

The same method can be used to prove the three other propositions.

The property described in proposition 13 implies model completeness as the proposition below will show. The syntactic characterization of model completeness (c.f.Chang,Keisler [73],Prop.3.1.7) can be extended as follows:

Lemma 14: The following properties of a theory T are equivalent:

(1) $T(\Phi)$ is model complete

(2) For every $\varphi(\bar{x})$ in Fm there exists $\alpha(\bar{x})$ in $\forall\Phi^*$ such that $T \vdash \forall\bar{x}(\varphi(\bar{x}) \leftrightarrow \alpha(\bar{x}))$

(3) $\underline{A} \leq_{\Phi} \underline{B}$ and $\underline{A}, \underline{B} \models T$ imply $\underline{A} \leq_{Fm} \underline{B}$.

Fm can be replaced by $\exists\Phi^*$ everywhere.

Proposition 15: Let P_T^{Φ} be an operator which satisfies: $\underline{A} \leq_{\Phi} \underline{B}$ and $\underline{A}, \underline{B} \models T$ imply $P_T^{\Phi}(A,\underline{B}) = \underline{A}$. Then the following properties of T are equivalent:

(1) T is $\exists\Phi^*$-hereditary w.r.t. P_T^{Φ}

(2) T is Φ-hereditary w.r.t. P_T^{Φ} and $T(\Phi)$ is model complete

(3) T is Fm-hereditary w.r.t. P_T^{Φ}.

If, in addition, $P_T^{\Phi} = P_T^{Fm}$ holds whenever $T(\Phi)$ is model complete then (3) is equivalent to:

(4) T is Fm-hereditary w.r.t. P_T^{Fm} and $T(\Phi)$ is model complete.

(1) → (2): Let $\underline{A}, \underline{B}$ be T-models with $\underline{A} \leq_{\Phi} \underline{B}$. Our assumptions on T and P_T^{Φ} imply $\underline{A} = P_T(A,\underline{B}) \leq_{\exists\Phi^*} \underline{B}$. Hence $T(\Phi)$ is model complete by lemma 14.

(2) → (3): Consider $P_T^{\Phi}(A,\underline{B})$ for a T-model $\underline{B}$. By assumption we have $\underline{B}\ {}_{\Phi}{\geq}\ P_T^{\Phi}(A,\underline{B}) \models T$. Since $T(\Phi)$ is model complete this implies $\underline{B}\ {}_{Fm}{\geq}\ P_T(A,\underline{B})$ as required.

(3) → (1): This implication is trivial.

We may add in (3) the condition that $T(\Phi)$ is model complete since (2) and (3) are equivalent. The additional assumption on P_T^{Φ} ensures that (3) is equivalent to (4).

Remark 16: It follows from lemma 1.7 that C_T^{Φ} and $H_T^{\Phi,\Phi} = \exists\Phi^*\text{-}Alg_T^{\exists\Phi^*}$ satisfy both conditions. It follows from lemma 1.5 that Alg^{Φ} and Alg_T^{Φ} satisfy the first condition. Thus we obtain the following corollary:

Corollary 17: The following statements are equivalent :

(1) T is $\exists$-hereditary w.r.t. $\exists\text{-Alg}_T^{\exists}$

(2) T is hereditary w.r.t. $\exists\text{-Alg}_T^{\exists}$ and T is model complete

(3) T is Fm-hereditary w.r.t. $\exists\text{-Alg}_T^{\exists}$

(4) T is Fm-hereditary w.r.t. Alg_T^{Fm} and T is model complete .

The results in Kueker [75] can be used to obtain a different syntactic characterization of those theories which are hereditary w.r.t. $\exists\text{-Alg}_T^{\exists}$.

Definition 18: A structure $\underline{C}$ is said to be a *core structure for T* if every T-model contains a unique substructure which is isomorphic to $\underline{C}$.

Let A be a subset of the structure $\underline{B}$. The substructure $!\text{-Alg}_T^{\exists}(A,\underline{B})$ of $\underline{B}$ is defined as follows . An element b of B belongs to $!\text{-Alg}_T^{\exists}(A,\underline{B})$ if there exist $\varphi(x,\bar{y})$ in $\exists$, $\bar{a}$ in A und $k>0$ such that :

(i) $\underline{B}\models\varphi(b,\bar{a})$, (ii) $T\vee\Delta_{At*}(A)\vdash\exists^k x\varphi(x,\bar{a})$.

Condition (ii) can be replaced by (ii') $T\vdash\forall\bar{y}\exists^k x\varphi(x,\bar{y})$.

These two notions were introduced in Kueker [75] .

Lemma 19: Let $\underline{B}$ be a T-model .

(0) $A\subseteq !\text{-Alg}_T(A,\underline{B})\subseteq\exists\text{-Alg}_T^{\exists}(A,\underline{B})$

(1) If $\underline{C}$ is a core structure for T then $\underline{C}$ is isomorphic to a substructure of $!\text{-Alg}_T^{\exists}(\emptyset,\underline{B})$. If , in addition , $\underline{C}$ is a T-model then $\underline{C}\simeq !\text{-Alg}_T^{\exists}(\emptyset,\underline{B})$.

(2) If T has the joint embedding property and if $\exists\text{-Alg}_T^{\exists}(\emptyset,\underline{B})$ is a T-model then $!\text{-Alg}_T^{\exists}(\emptyset,\underline{B})=\exists\text{-Alg}_T^{\exists}(\emptyset,\underline{B})$ and $\exists\text{-Alg}_T^{\exists}(\emptyset,\underline{B})$ is a maximal core structure for T .

(0) Define $\psi(\bar{z},\bar{y})$ by $\bigwedge\langle\varphi(z_i,\bar{y}) : i=1,\dots,k\rangle\wedge\bigwedge\langle z_i\neq z_j : i\neq j\wedge i,j=1,\dots,k\rangle$. $T\vdash\forall\bar{y}\exists^k x\varphi(x,\bar{y})$ implies $T\vdash\forall x\bar{y}\bar{z}(\varphi(x,\bar{y})\wedge\psi(\bar{z},\bar{y})\Rightarrow\bigvee\langle x\equiv z_i : i=1,\dots,k\rangle)$ and $T\vdash\forall\bar{y}\exists\bar{z}\psi(\bar{z},\bar{y})$. This suffices to prove $!\text{-Alg}_T^{\exists}(A,\underline{B})\subseteq\exists\text{-Alg}_T^{\exists}(A,\underline{B})$. To prove $A\subseteq !\text{-Alg}_T(A,\underline{B})$ one uses $T\vdash\forall y\exists^1 x(x\equiv y)$.

(1) We may assume that $\underline{C}$ is a substructure of $\underline{B}$. We use the second half of the proof of theorem 2.1 in Kueker [75] to show $\underline{C}\leq !\text{-Alg}_T^{\exists}(\emptyset,\underline{B})$.

(2) First observe that the structure $\underline{C}=\exists\text{-Alg}_T^{\exists}(\emptyset,\underline{B})$ is independent of $\underline{B}$ since T has the joint embedding property . Here we have to use part (2) of lemma 1.5 . Moreover , every T-model $\underline{A}$ contains the substructure $\exists\text{-Alg}_T^{\exists}(\emptyset,\underline{A})$ which is isomorphic to $\underline{C}$. To prove the uniqueness of $\underline{C}$ let $h:\underline{C}\to\underline{C}'$ be an isomorphism onto another substructure $\underline{C}'$ of $\underline{B}$. By assumption $\underline{C}$ is a T-model and hence $\underline{C}'$ is a T-model as well . As $\underline{C}$ and $\underline{C}'$ are substructures of the T-model $\underline{B}$, we have $\underline{C}=\exists\text{-Alg}_T^{\exists}(\emptyset,\underline{B})=H_T^{At,At}(\emptyset,\underline{B})\leq\underline{C}'$ by proposition 1.8 . This implies $\underline{C}=\underline{C}'$ as required . Thus $\underline{C}$ is a core structure for T . $\underline{C}$ is maximal since $\underline{C}$ itself is a T-model . An application of (0) and (1) yields $!\text{-Alg}_T^{\exists}(\emptyset,\underline{B})=\exists\text{-Alg}_T^{\exists}(\emptyset,\underline{B})$.

Theorem 20: A theory T is hereditary w.r.t. $\exists\text{-Alg}_T^{\exists}$ iff T can be axiomatized by a set Δ of $\forall\exists$-sentences which satisfies the following condition

(iv') For every sentence $\forall\bar{x}\exists\bar{y}\alpha(\bar{x},\bar{y})$ in Δ with α in At* there exist r, formulas $\mu_1(\bar{x},\bar{y}),\dots,\mu_r(\bar{x},\bar{y})$ in , formulas $\gamma_1(x),\dots,\gamma_r(x)$ in $\forall$ and $k_1,\dots,k_r > 0$ such that:

$T \vdash \forall\bar{x}(\bigvee\langle\gamma_i(\bar{x}) : i=1,\dots,r\rangle)$ and

$T \vdash \forall\bar{x}(\gamma_i(\bar{x}) \Rightarrow \exists^{k_i}\bar{y}(\alpha(\bar{x},\bar{y}) \wedge \mu_i(\bar{x},\bar{y})))$ for $i=1,\dots,r$.

In the proof of theorem 7 we have shown that T is hereditary w.r.t. $\exists\text{-Alg}_T^{\exists}$ iff T can be axiomatized by a set Δ of $\forall\exists$-sentences satisfying: For every sentence $\forall\bar{x}\exists\bar{y}\alpha(\bar{x},\bar{y})$ in Δ with α in At*, for every T-model $\underline{B}$ and every $\bar{a}$ in B we have: $\exists\text{-Alg}_T^{\exists}(\bar{a},\underline{B}) \models \exists\bar{y}\alpha(\bar{a},\bar{y})$.

To prove that the condition (iv') is sufficient, we proceed as follows. By proposition 1.8 we have $\exists\text{-Alg}_T^{\exists}(\bar{a},\underline{B}) = H_T^{At,At}(\bar{a},\underline{B})$ for every T-model $\underline{B}$. Let $\underline{B}'$ and $\underline{C}$ be T-models such that $\underline{B}$ and $\underline{B}'$ are substructures of $\underline{C}$ containing $\bar{a}$. Then there exists i such that $\underline{C} \models \gamma_i(\bar{a})$. Since $\gamma_i(\bar{a})$ is in $\forall$ we have also $\underline{B}, \underline{B}' \models \gamma_i(\bar{a})$. Hence $\underline{B}$ and $\underline{B}'$ have the same set of solutions of $\alpha(\bar{a},\bar{y}) \wedge \mu_i(\bar{a},\bar{y})$. Therefore we have $H_T^{At,At}(\bar{a},\underline{B}) \models \exists\bar{y}\alpha(\bar{a},\bar{y})$, as required.

To prove that the condition (iv') is necessary, we introduce the following components of the theory T (cf. Bacsich, Rowlands-Hughes [74]). Let $L(\bar{e})$ be the extension of L by a sequence $\bar{e}$ of constants. For every T-model $\underline{B}$ and every sequence $\bar{a}$ in B of appropriate length let $T(\underline{B},\bar{a})$ be the $L(\bar{e})$-theory determined by the set $T \cup \{\alpha(\bar{e}) : \alpha(\bar{x}) \in \forall, \underline{B} \models \alpha(\bar{a})\}$. A $L(\bar{e})$-structure $(\underline{D},\bar{d})$ is a model of $T(\underline{B},\bar{a})$ iff $\underline{D}$ is a T-model and each sentence in $\forall$ which holds in $(\underline{B},\bar{a})$ holds also in $(\underline{D},\bar{d})$. It is well known that a theory T has the joint embedding property iff for any pair α_1,α_2 of sentences in $\forall$ $T \vdash \alpha_1 \vee \alpha_2$ implies $T \vdash \alpha_1$ or $T \vdash \alpha_2$. This can be used to verify that $T(\underline{B},\bar{a})$ has the joint embedding property. We want to apply part (2) of lemma 19 to the model $(\underline{B},\bar{a})$ of $T(\underline{B},\bar{a})$. Therefore we have to show that $\exists\text{-Alg}_{T(\underline{B},\bar{a})}^{\exists}(\emptyset,(\underline{B},\bar{a}))$ is also a model of $T(\underline{B},\bar{a})$.

This can be verified as follows. By proposition 1.8 we know that the operators $\exists\text{-Alg}_T^{\exists}$ and $H_T^{At,At}$ coincide. Moreover, $(H_T^{At,At}(\bar{a},\underline{B}),\bar{a})$ is a substructure of the structure $H_{T(\underline{B},\bar{a})}^{At,At}(\emptyset,(\underline{B},\bar{a}))$. However, the L-structure determined by the latter structure is an intersection of substructures of $\underline{B}$ which are T-models. The intersection is closed under $H_T^{At,At}$ by part (4) of lemma 1.5, since T is hereditary w.r.t. $H_T^{At,At}$. Therefore it is a T-model which is a substructure of $\underline{B}$. Hence we conclude that $H_{T(\underline{B},\bar{a})}^{At,At}(\emptyset,(\underline{B},\bar{a})) = \exists\text{-Alg}_{T(\underline{B},\bar{a})}^{\exists}(\emptyset,(\underline{B},\bar{a}))$ is a model of $T(\underline{B},\bar{a})$ as required.

From part (2) of lemma 19 we conclude that $\exists\text{-Alg}^{\exists}_{T(\underline{B},\bar{a})}(\emptyset,(\underline{B},\bar{a}))$ and $!\text{-Alg}^{\exists}_{T(\underline{B},\bar{a})}(\emptyset,(\underline{B},\bar{a}))$ coincide. Hence the latter is a model of $T(\underline{B},\bar{a})$ and, in particular, of $\exists\bar{y}\alpha(\bar{a},\bar{y})$. Hence there exist $\mu(\bar{e},\bar{y})$ in $\exists$ and $k>0$ such that: (*) $\underline{B}\models\exists\bar{y}(\alpha(\bar{a},\bar{y})\wedge\mu(\bar{a},\bar{y}))$, (**) $T(\underline{B},\bar{a})\vdash\exists^k\bar{y}\mu(\bar{a},\bar{y})$.
Since $H^{At,At}_{T(\underline{B},\bar{a})}(\emptyset,(\underline{B},\bar{a}))$ is a core structure for $T(\underline{B},\bar{a})$ by lemma 19, this can be improved as follows:

(*) $\underline{B}\models\exists\bar{y}(\alpha(\bar{a},\bar{y})\wedge\mu(\bar{a},\bar{y}))$, (**') $T(\underline{B},\bar{a})\vdash\exists^k\bar{y}(\alpha(\bar{a},\bar{y})\wedge\mu(\bar{a},\bar{y}))$.

Now a compactness argument yields a formula $\gamma(\bar{x})$ in $\forall$ such that:

($\overset{*}{*}$) $\underline{B}\models\gamma(\bar{a})$ and $T\vdash\gamma(\bar{a})\Rightarrow\exists^k\bar{y}(\alpha(\bar{a},\bar{y})\wedge\mu(\bar{a},\bar{y}))$.

The formulas μ and γ and the number k depend on $\underline{B}$ and $\bar{a}$. A final compactness argument yields the condition (iv') as desired.

The syntactic characterizations given in theorem 4 - 7 and proposition 10 - 13 have a common feature. In all cases the theory can be axiomatized by a set Δ of $\forall\exists$-sentences satisfying a condition of the following type. Each sentence in Δ has to satisfy a syntactic condition involving the theory T and the existence of additional formulas. The result below, claimed already by Park [64], shows that the dependence on T cannot be avoided, in general. The counter-example is due to Rabin [62]. A modification of his proof yields the desired generalization. (cf. Volger [79])

<u>Proposition 21</u>: For each structure $\underline{B}$ let $P(-,\underline{B})$ be a closure operator on $\text{Sub}(\underline{B})$ satisfying: $\exists\text{-Alg}^{\exists}_T(A,\underline{B})\leq P(A,\underline{B})\leq \text{Alg}^{Fm}(A,\underline{B})$ for every subset A of B. Then there exists a theory T with the following two properties:
(1) T is hereditary w.r.t. P
(2) T cannot be axiomatized by a set of sentences all of which are hereditary w.r.t. P.

Let T be determined by the following set $\Sigma=\{\varphi_n\wedge\psi_n\wedge\chi_n : n\in\omega\}$ of axioms, where R_n is a (n+1)-ary predicate and

$$\varphi_n=\forall x\bar{y}y(R_{n+1}(x,\bar{y},y)\Rightarrow R_n(x,\bar{y}))$$
$$\psi_n=\forall x\bar{y}y\bar{z}z(R_{n+1}(x,\bar{y},y)\wedge R_{n+1}(x,\bar{z},z)\Rightarrow\bigwedge\langle z_i\equiv y_i : i=1,\dots,n\rangle)$$
$$\chi_n=\forall x\exists\bar{y}R_n(x,\bar{y}).$$

Theorem 7 can be used to show that T is hereditary w.r.t. $\exists\text{-Alg}^{\exists}_T$ and hence w.r.t. P. The verification that T cannot be axiomatized by sentences hereditary w.r.t. P, relies on the following lemma which replaces a similar lemma in Rabin [62].

If $\underline{A}\models\Sigma_o\vee\{\chi_n\}$ then there exist $\underline{B}$ such that $\underline{B}\models\Sigma_o\vee\{\chi_{n+1}\}$ and $\text{Alg}^{Fm}(A,\underline{B})=\underline{A}$, where $\Sigma_o=\{\varphi_n\wedge\psi_n : n\in\omega\}$.

By proposition 1.14 it suffices to find $\underline{B}_1$, $\underline{B}$, $\underline{B}_2$ such that $\underline{A}=$

$\underline{B}_1 \wedge \underline{B}_2$, $\underline{B}_1 \simeq_A \underline{B}_2$, $\underline{B}_1 \leq_{Fm} \underline{B} \ {}_{Fm}\!\geq \underline{B}_2$ and $\underline{B} \models \Sigma_o \vee \{\chi_{n+1}\}$. A compactness argument can be used to show that it suffices to find a structure $\underline{B}$ such that $\underline{A} \leq \underline{B} \models \Sigma_o \vee \{\chi_{n+1}\}$. $\underline{B}$ is an extension of $\underline{A}$ by a disjoint copy of $D \times \omega$, where D is given by $D = \{a \in A : \underline{A} \models \neg \exists \overline{y} R_{n+1}(a, \overline{y})\}$.

Now assume that $T \wedge \Delta$ axiomatizes T , where Δ is the set of sentences hereditary w.r.t. P . Hence there exists δ in Δ such that $T \vdash \delta \vdash \chi_2$. Now consider k_0 , the smallest k such that $\Sigma_o \vee \{\chi_{k+1}\} \vdash \delta$ and $\Sigma_o \vee \{\chi_k\} \not\vdash \delta$. The above lemma implies $\Sigma_o \vee \{\chi_{k_0}\} \vdash \delta$ since every sentence in Δ is hereditary w.r.t. P and hence w.r.t. Alg^{Fm} . This contradicts the minimality of k_0 . Therefore $T \wedge \Delta$ does not axiomatize T .

The following examples show that the properties described in the theorems 4 - 7 are all different :

Examples 22: (1) The theory T_1 defined below is hereditary w.r.t. $Alg^{\exists}$ but not w.r.t. $Alg^{\exists}_{T_1}$ resp. $\exists\text{-}\overline{Alg}^{Fm}_{T_1}$. T_1 has the following axioms :

$$\forall x \exists y (R_1(x,y) \vee R_2(x,y))$$
$$\forall x (\exists^{\geq 2} y R_1(x,y) \Rightarrow \exists^1 y R_2(x,y)) \wedge \forall x (\exists^{\geq 2} y R_2(x,y) \Rightarrow \exists^1 y R_1(x,y))$$

An application of theorem 4 shows that T_1 is hereditary w.r.t. $Alg^{\exists}$. To prove the other claim let $\underline{B}_1$ be the following model of T_1 :
$B_1 = \{a_o, a_1\}$, $\underline{B}_1 \models \delta(a_o, a_1)$, where δ is the conjunction of $\Delta_{At*}(\underline{B}_1)$:

$$\delta(a_o,a_1) = R_1(a_o,a_1) \wedge R_2(a_1,a_o) \wedge \neg\ {}_1(a_1,a_o) \wedge \neg R_2(a_o,a_1) \wedge$$
$$\bigwedge \langle \neg R_1(a_i,a_i) \wedge \neg R_2(a_i,a_i) : i=o,1 \rangle .$$

We claim that $Alg^{\exists}_{T_1}(\{a_o\}, \underline{B}_1) = \{a_o\}$ holds . This produces the desired contradiction since $\{a_o\}$ is not a model of T_1 . It suffices to show that for each formula $\mu(y,x)$ in $\exists$ there does not exist $k > 0$ with $\underline{B}_1 \models \mu(a_1, a_o)$ and $T_1 \vdash \forall x \exists^{\leq k} y \mu(y,x)$. Let $\underline{B}_\omega$ be the following model of T_1 :
$B_\omega = \{a_n : n \in \omega\}$, $\underline{B}_\omega \models \delta(a_o,a_n) \wedge \neg R_1(a_n,a_m) \wedge R_2(a_n,a_m)$ for all n , $m \neq 0$. Then $\underline{B}_1 \models \mu(a_1,a_o)$ implies $\underline{B}_\omega \models \mu(a_n,a_o)$ for all $n > 0$. Hence we have $T \not\vdash \forall x \exists^{\leq k} y \mu(y,x)$ for all k , as required . The structure $\underline{B}_\omega$ can also be used to show that $\exists\text{-}\overline{Alg}^{Fm}_{T_1}(\{a_o\}, \underline{B}_1) = \{a_o\}$.

(2) The theory T_2 defined below is hereditary w.r.t. $Alg^{\exists}_{T_2}$, $\exists\text{-}\overline{Alg}^{\exists}_{T_2}$ and and $\exists\text{-}Alg^{Fm}_{T_2}$ but not w.r.t. $\exists\text{-}Alg^{\exists}_{T_2}$. T_2 is the theory of commutative rings with 1 which are local i.e. which satisfy : $\forall x \exists y (x \cdot y \equiv 1 \vee (1+x) \cdot y \equiv 1)$.

An application of theorem 5 and 6 shows that T_2 is hereditary w.r.t. $Alg^{\exists}_{T_2}$ and $\exists\text{-}\overline{Alg}^{\exists}_{T_2}$. An analogous result can be used to show that T_2 is hereditary w.r.t. $\exists\text{-}Alg^{Fm}_{T_2}$. To prove the other claim let $\underline{Z}_{(p_1)}$, $\underline{Z}_{(p_2)}$ be the localizations of the ring of integers with respect to two different primes p_1 , p_2 . They are subrings of the local ring $\underline{Q}$ of rationals .

However, $Z_{(p_1)} \wedge Z_{(p_2)}$ is not local in general. Because of $H_{T_2}^{At,At} = \exists\text{-}Alg_{T_2}^{\exists}$ and $H_{T_2}^{At,At}(Z_{(p_1)} \wedge Z_{(p_2)}, \underline{Q}) = Z_{(p_1)} \wedge Z_{(p_2)}$ T_2 cannot be hereditary w.r.t. $\exists\text{-}Alg_{T_2}^{\exists}$.

(3) The theory T_3 of sets which have 2 or 3 elements is hereditary w.r.t. $Alg_{T_3}^{\exists}$ and $\exists\text{-}\overline{Alg}_{T_3}^{Fm}$ but not w.r.t. $\exists\text{-}\overline{Alg}_{T_3}^{\exists}$ and $\exists\text{-}Alg_{T_3}^{Fm}$.

An application of theorem 5 and an analogous result for $\exists\text{-}\overline{Alg}_{T_3}^{Fm}$ shows that T_3 is hereditary w.r.t. $Alg_{T_3}^{\exists}$ and $\exists\text{-}\overline{Alg}_{T_3}^{Fm}$. In order to prove that T_3 is not hereditary with respect to $\exists\text{-}\overline{Alg}_{T_3}^{Fm}$ note that $H_{T_3}^{At,At}(A,\underline{B}) = \exists\text{-}\overline{Alg}_{T_3}^{Fm}(A,\underline{B}) = \underline{A}$ if card(A) = 1 and card(B) = 2. In order to prove that T_3 is not hereditary with respect to $\exists\text{-}Alg_{T_3}^{Fm}$ note that $H_{T_3}^{Fm,At}(A,\underline{B}) = \underline{A} = \exists\text{-}Alg_{T_3}^{Fm}(A,\underline{B})$ if card(A) = 1 and card(B) = 3.

§.3 Classification of Partially Ordered Sets

We want to know to what extent a partial order $\underline{P}$ is downward-directed. This question leads to the classification of partial orders considered below. For x in P let xP be the set $\{y : y \leq x\}$. We write $x \sim y$ if x and y are *compatible* elements of P i.e. $xP \wedge yP \neq \emptyset$.

Definition 1: The *width* of a partial order $\underline{P}$ is the largest integer n such that P contains n pairwise incompatible elements or else it is the ordinal ω. If the width is finite $\underline{P}$ is said to be of *finite type*, otherwise it is said to be of *infinite type*.

A *frame* for a partial order $\underline{P}$ of width $n < \omega$ is a set F of n pairwise incompatible elements of P. A frame F *dominates* a frame F' (written $F \geq F'$) if for all x in F there exists x' in F' such that $x \geq x'$.

Lemma 2: Let F_1, F_2 be frames of a partial order $\underline{P}$ of width $n < \omega$.
(1) There exists a frame F such that $F_1, F_2 \geq F$.
(2) If $F_1 \geq F_2$ then the relation $x_1 \geq x_2$ defines a bijection between F_1 and F_2.
(3) The relation $x_1 \sim x_2$ defines a bijection between F_1 and F_2.

(1) Proceed by downward induction on the number $c = card(F_1 \wedge F_2)$. The case $c = n$ is trivial. Assume now $c < n$, and fix x_1 in $F_1 - F_2$. Since F_2 is a frame there is some x_2 in F_2 compatible with x_1. Suppose $a \leq x_1$, x_2 and set $F_i' = (F_i - \{x_i\}) \vee \{a\}$. Then F_i dominates F_i'. F_1' and F_2' are again frames, since if $a \sim x \in F_i$ then $x_i \sim x$ and hence $x = x_i$. Now since $card(F_1' \wedge F_2') = c+1$ we can proceed by induction.

(2) By assumption there is a function $s : F_1 \to F_2$ such that $x \geq s(x)$ for x in F_1. It is clearly injective and hence surjective. If also $x \geq x' \neq s(x)$ with x' in F_2 then we have $s(y) = x'$ for some y in F_1. This implies $x \sim y$ and hence $x = y$, which is impossible. Hence $s(x) = x'$ is uniquely determined.

(3) By (1) there is a frame F with $F \leq F_1, F_2$. By (2) each x in F_1 determines $x^* \in F_2$ by : $x, x^* \geq x' \in F$. The map $* : F_1 \to F_2$ is a bijection by (2). It remains to be shown that $y = x^*$ if $x \in F_1$, $y \in F_2$ and $x \sim y$. Fix $a \leq x, y$. If $a \in F$ then $y = x^*$ as claimed. If $a \notin F$ then $\{a\} \vee F$ is not a frame, so $a \sim z \in F$ for some z. Choose $z_1 \in F_1$, $z_2 \in F_2$ with $z \leq z_1, z_2$. If $b \leq a, z$ then $b \leq x, z_1, y, z_2$, so $x = z_1$ and $y = z_2$. Thus $y = x^*$, as claimed

Lemma 3: Let F and F' be frames of a partial order $\underline{P}$ of finite type.
(1) If $x \in F$ then xP is downward-directed.
(2) If $x \in F$ then xP contains a minimal element of P iff xP has a minimum.
(3) If $x \in F$, $x' \in F'$ and $x \sim x'$ then xP has a minimum iff $x'P$ has a minimum.
(4) The cardinality of $\{x \in F : xP$ has no minimum$\}$ is independent of the choice of F.

(1) If $y_1, y_2 \leq x$ are incompatible then since $\{y_1, y_2\} \vee (F - \{x\})$ is not a frame, we have without loss of generality $y_1 \sim x' \in F - \{x\}$. This implies $x \sim x'$ and hence $x = x'$, which is impossible.

(2) This is an immediate consequence of (1).

(3) If for example $a \in xP$ is minimal in P it suffices to show $a \in x'P$. Choose $b \leq x, x'$. Then $a \leq b$ since $b \in xP$, so $a \leq x'$ as claimed.

(4) This is an immediate consequence of (3) and part (3) of lemma 2.

Definition 4: Let F be a frame of a partial order $\underline{P}$ of finite type. The cardinality of the set $\{x \in F : xP$ has no minimum$\}$ is called the *depth* of $\underline{P}$. The *type* of $\underline{P}$ is the pair (n,d), where n is the width of $\underline{P}$ and d is the depth of P.

If $\underline{P}$ is a partial order of infinite type then the *type* of $\underline{P}$ is the pair $(\omega,1)$. - The class of partial orders of type (n,d) is denoted by $P(n,d)$.

It should be noted that the set of types can be ordered lexicographical.

Lemma 5: Let $\underline{P}$ be a partial order of infinite type. Let P_0 be the set $\{x \in P : xP$ is downward-directed$\}$ and P_1 be the set $\{x \in P : xP \wedge P_0 = \emptyset\}$.

(1) If $P_1 \neq \emptyset$ then P_1 contains an infinite set of pairwise incompatible elements .

(2) If $P_1 = \emptyset$ then P_0 contains an infinite set of pairwise incompatible elements .

Suppose $P_1 \neq \emptyset$. For x in P_1 we can select $x', x'' \leq x$ with $x' \not\sim x''$. Then x', x'' also belong to P_1 . Define inductively $\{x_i : i \geq 0\}$ and $\{y_i : i \geq 0\}$ by : x_0 arbitrary , $y_{i+1} = x_i'$ and $x_{i+1} = x_i''$. It can easily be verified that the elements $\{y_i : i \geq 1\}$ are pairwise incompatible .

Suppose $P_1 = \emptyset$. It is easy to see that the compatibility relation defines an equivalence relation on P_0 . It suffices to show that there are infinitely many equivalence classes . If there are just $n < \omega$ equivalence classes and $x_1, \dots, x_{n+1}$ are pairwise incompatible elements of P , we can choose $x_i' \in (x_i P \wedge P_0)$ for $i = 1, \dots, n+1$. By assumption $x_i' \sim x_j'$ for some $i \neq j$. This implies $x_i \sim x_j$, a contradiction .

Definition 6: For a partial order $\underline{P}$ let $\underline{P}^*$ denote the reverse order on the same underlying set . Let $\underline{P}^{\cdot}$ denote the extension of $\underline{P}$ by a new element greater than all elements of P .

For cardinals $k_1, \dots, k_d$ let $(k_1^*, \dots, k_d^*, 1^{n-d})$ denote the disjoint union of $k_1^*, \dots, k_d^*$ together with n-d additional incomparable elements . An order of the form $(k_1^*, \dots, k_d^*, 1^{n-d})$ or $(k_1^*, \dots, k_d^*, 1^{n-d})^{\cdot}$ with $k_1, \dots, k_d$ regular cardinals is called a *reduced order of type (n,d)* . The order $P(n,d) = (\omega^* \dots, \omega^* 1^{n-d})$ will be called the *canonical order of type (n,d)* .

Lemma 7: Let $\underline{P}$ be a partial order of type (n,d) with $n < \omega$. Then there exists a surjective homomorphism from $\underline{P}$ onto a reduced partial order $\overline{\underline{P}}$ of type (n,d) .

First we prove the result for the type (1,1) . Then $\underline{P}$ is a downward-directed set without first element . Let α be the smallest ordinal for which there is no embedding of $(\alpha+1)^*$ into $\underline{P}$. Then there is an embedding h of α^* into $\underline{P}$ which cannot be extended to $(\alpha+1)^*$. We can replace α by its cofinality . Therefore $\alpha = \mathrm{cof}(\alpha)$ is a regular cardinal . For x in P let $\pi(x)$ be the smallest $\gamma < \alpha$ for which $x < h(\gamma)$ fails . We have $\pi(h(\gamma)) = \gamma$, since h is injective and im(h) is a well-ordering . Therefore π is surjective . To see that $\pi : P \to \alpha^*$ is orderpreserving , suppose $x \leq y$. Then $y < h(\pi(x))$ fails , since otherwise we would have $x < h(\pi(x))$ which is impossible . Hence we have $\pi(y) \leq \pi(x)$, as required .

Now consider a partial order $\underline{P}$ of type (n,d) . Fix a frame F for P and set $F^{\perp} = P - \cup\langle xP : x \in F\rangle$. If $F^{\perp}$ is nonempty , collapse it to a single element p and make p greater than all elements in $\cup\langle xP : x \in F\rangle$. If $x \in F$

and xP contains a smallest element, collapse xP to a single element. If $x \in F$ and xP contains no smallest element, use the first part of the proo to map xP onto a reversed regular cardinal. These three operations are compatible without additional collapsing, and hence yield the desired homomorphism.

Lemma 8: Let $\underline{P}$ be a partial order of type $(\omega,1)$. Then there exists a surjective homomorphism from $\underline{P}$ onto 1^{ω} or $(1^{\omega})^{\cdot}$, where 1^{ω} is a countable set of incomparable elements.

By lemma 5 we know that P contains a countable subset X of pairwise incompatible elements. Then collapse xP onto x if x belongs to X and collapse $P - \bigcup\langle xP : x \in X\rangle$ onto an element which is made greater than all elements in X. This yields the desired homomorphism.

§.4 Convexity Properties

We propose the following general definition of a convexity property, tailored rather closely to the standard examples.

Definition 1: Let $\underline{P}$ be a partial order and let Φ be a set of formulas satisfying the conventions mentioned in the introduction. Let $\underline{B}$ be a model of a theory T. A *T-Φ-representation* R *of* $\underline{P}$ *in* $\underline{B}$ is an orderpreserving function R from $\underline{P}$ to the partially ordered set $\mathrm{Sub}_{\Phi}^{T}(\underline{B})$ of structures $\underline{A}$ such that $\underline{A} \leq_{\Phi} \underline{B}$ and $\underline{A} \models T$.

Let P be a class of partial orders. A theory T is said to be *P-Φ-convex*, if for every T-Φ-representation R of any $\underline{P}$ in P in a T-model $\underline{B}$ the intersection $\bigcap\langle R(p) : p \in P\rangle$ is either empty or again a model of T. T is said to be *P-Φ-superconvex*, if, in addition, $\bigcap\langle R(p) : p \in P\rangle \leq_{\Phi} \underline{B}$.

Examples 2: (1) If P is the class of all partial orders, then P-Φ-convexity is called *Φ-convexity*. The case $\Phi = \mathrm{At}$ was studied in Rabin [62] and Robinson [63].
(2) If $P(n,d)$ is the class of partial orders of type (n,d) then $P(n,d)$-Φ-convexity is called *(n,d)-Φ-convexity*. In particular, (n,0)-Φ-convexity is called *n-Φ-convexity*. The case $\Phi = \mathrm{At}$ and $n = 2$ was studied in Rabin [62] and Robinson [63]. The cases $\Phi = \mathrm{At}$ or Fm and $(n,d) = (1,1)$ were studied in Park [64].
(3) If P is the class of downward-directed partial orders then P-Φ-convexity is called *$\downarrow\Phi$-convexity*.
(4) Another useful example which does not fit into the above set up is *(Φ_1,Φ_2)-convexity*. T is said to be (Φ_1,Φ_2)-convex if $\underline{A}_1 \leq_{\Phi_1} \underline{B} {}_{\Phi_2}\!\geq \underline{A}_2$, $\underline{A}, \underline{B}_1, \underline{B}_2 \models T$ implies $\underline{A}_1 \wedge \underline{A}_2 \models T$ or $A_1 \wedge A_2 = \emptyset$.

The extension $T(\Phi)$ of T by new predicates for the formulas in Φ can be used to transform $\mathcal{P}$-Φ-superconvexity into $\mathcal{P}$-At-convexity. Let $\underline{A}$, $\underline{B}$ be $L(\Phi)$-structures which are $T(\Phi)$-models. $\underline{A}$ is a substructure if $\underline{B}$ iff the reduct $\underline{A}|L$ is a Φ-elementary substructure of the reduct $\underline{B}|L$.

Lemma 3: A theory T is $\mathcal{P}$-Φ-superconvex iff $T(\Phi)$ is $\mathcal{P}$-At-convex.

Lemma 3 can be used to generalize the known result that At-convexity is the same as 2-At-convexity (cf.Chang,Keisler [73],Prop.5.2.14).

Proposition 4: A theory T is Φ-superconvex iff T is 2-Φ-superconvex.

In the following we want to compare the convexity properties for a fixed set of formulas Φ. First we shall show that (n,d)-Φ-convexity is the same as $\mathcal{P}$-Φ-convexity, whenever $\mathcal{P}$ is a class of partial orders of type (n,d). Then we shall show that the ordering of the types can be used to order the convexity properties for Φ.

The following lemma is basic for comparing convexity properties.

Lemma 5: Let $\mathcal{P}$, $\mathcal{Q}$ be classes of partial orders such that for every $\underline{P}$ in $\mathcal{P}$ there exist $\underline{Q}$ in $\mathcal{Q}$ and an orderpreserving map from $\underline{Q}$ to $\underline{P}$ whose image is coinitial in $\underline{P}$. If a theory T is $\mathcal{Q}$-Φ-convex then T is $\mathcal{P}$-Φ-convex. The same result holds for superconvexity.

By our assumption each intersection $\cap\langle R(p) : p \in P\rangle$ can be represented as $\cap\langle R(f(q)) : q \in Q\rangle$, where $f : Q \to P$ is an orderpreserving map whose image is coinitial in P.

The condition of the lemma is satisfied if $\mathcal{P}$ is a subclass of $\mathcal{Q}$. Hence Φ-convexity is the strongest convexity property for Φ. In addition, Φ-convexity is equivalent to $\mathcal{T}$-Φ-convexity, where $\mathcal{T}$ is the class of trivially ordered sets. If $\underline{P}$ is of type (n,d) then (n,d)-Φ-convexity implies $\underline{P}$-Φ-convexity. In particular, (n,d)-Φ-convexity implies P(n,d)-Φ-convexity, where P(n,d) is the canonical order of type (n,d).

To prove the converse we have to improve the results of chapter 1.

Lemma 6: Suppose $\underline{A} \leq \underline{B} \models T$ and let k be an infinite cardinal. Then we have: (1) $\to$ (2) $\to$ (3).

(1) $\underline{A} = \mathrm{Alg}^{\exists\Phi^*}(A,\underline{B})$

(2) There exist T-models $\underline{B}_1$, $\underline{B}'$ such that $\underline{B}_1 \leq_\Phi \underline{B}' \ {}_{Fm}{\geq} \underline{B}$, $\underline{A} = \underline{B}_1 \wedge \underline{B}$ and $\underline{A} = \mathrm{Alg}^{\exists\Phi^*}(A,\underline{B}_1)$

(3) There exist an elementary extension $\overline{\underline{B}}$ of $\underline{B}$ and a T-Φ-representation R of k* in $\overline{\underline{B}}$ such that $\underline{A} = \cap\langle R(i) : i \in k\rangle$.

(1) $\to$ (2): By proposition 1.14 there exist $\underline{B}_1$, $\underline{B}'$ such that $\underline{B}_1 \leq_\Phi \underline{B}' \ {}_{Fm}{\geq} \underline{B}$,

$\underline{A} = \underline{B}_1 \wedge \underline{B}$ and $\underline{B} \simeq_A \underline{B}_1$. Hence $\underline{A} = \mathrm{Alg}^{\exists\Phi^*}(A,\underline{B})$ implies $\underline{A} = \mathrm{Alg}^{\exists\Phi^*}(A,\underline{B}_1)$.

(2) → (3) : Set $\underline{B}^0 = \underline{B}$. We shall define an elementary chain $(\underline{B}^\alpha)_{\alpha<k}$ of structures satisfying the following conditions for every $\alpha < k$:

(iα) $\underline{B}^\alpha$ contains a descending chain $(\underline{B}^\alpha_i)_{i\leq\alpha}$ of Φ-elementary substructures of $\underline{B}^\alpha = \underline{B}^\alpha_0$ each of which is a T-model containing $\underline{A}$ and satisfying $\underline{A} = \mathrm{Alg}^{\exists\Phi^*}(A,\underline{B}^\alpha_i)$

(iiα) For all $\beta < \alpha$: $\langle\underline{B}^\beta, \langle B^\beta_i : i \leq \beta\rangle\rangle <_{Fm} \langle\underline{B}^\alpha, \langle B^\alpha_i : i \leq \alpha\rangle\rangle$

(iiiα) For all $\beta < \alpha$: $\underline{B}^\alpha_\alpha \wedge \underline{B}^\beta_\beta = \underline{A}$.

Once this has been accomplished set $\overline{\underline{B}} = \bigcup\langle\underline{B}^\alpha : \alpha < k\rangle$ and $\overline{\underline{B}}_i = \bigcup\langle\underline{B}_i : i \leq \alpha\rangle$ for $i < k$. We will then have :

(i) $(\underline{B}_i)_{i<k}$ is a descending k-chain of Φ-elementary substructures of $\overline{\underline{B}} = \overline{B}$ each of which is a T-model containing $\underline{A}$

(ii) $\underline{B} \leq_{Fm} \overline{\underline{B}}$

(iii) $\underline{A} = \bigcap\langle\underline{B}_i : i < k\rangle$.

Only the third condition requires verification. We claim first :

(*) $\underline{B}^\gamma_\beta \wedge \underline{B}^\alpha_\alpha = \underline{A}$ for $\alpha < \beta \leq \gamma < k$.

For $\gamma = \beta$ this is just (iiiα). For $\gamma > \beta$, if $b \in \underline{B}^\gamma_\beta \wedge \underline{B}^\alpha_\alpha$ then we have $b \in \underline{A} = \underline{B}^\beta_\beta \wedge \underline{B}^\alpha_\alpha$ as claimed because of $\langle\underline{B}^\beta, B^\beta_\beta\rangle \leq_{Fm} \langle\underline{B}^\gamma, B^\gamma_\beta\rangle$. We claim next :

(**) $\underline{B}^\gamma_\beta \wedge \underline{B}^\alpha = \underline{A}$ for $\alpha < \beta \leq \gamma < k$.

Indeed, if $b \in \underline{B}^\gamma_\beta \wedge \underline{B}^\alpha$ then $b \in \underline{B}^\gamma_\alpha \wedge \underline{B}^\alpha = \underline{B}^\alpha_\alpha$ since $\langle\underline{B}^\alpha, B^\alpha_\alpha\rangle \leq_{Fm} \langle\underline{B}^\gamma, B^\gamma_\alpha\rangle$. Thus $b \in \underline{B}^\gamma_\beta \wedge \underline{B}^\alpha = \underline{A}$, as claimed. Finally (**) yields $\overline{\underline{B}}_\beta \wedge \underline{B}^\alpha = \underline{A}$ for $\alpha < \beta < k$ and (iii) follows if k is infinite.

Since (i) - (iii) are sufficient to prove (3) it remains to carry out the inductive construction. We have to express the conditions (iα) - (iiiα) by a theory T^α. To do this we extend the given language L to a language L^α by adding unary predicates P^α_i for $i \leq \alpha$. Because of (iiα) it suffices to require in (iα) that $\underline{B}^\alpha_\alpha$ is a T-model containing $\underline{A}$ and satisfying $\underline{A} = \mathrm{Alg}^{\exists\Phi^*}(A,\underline{B}^\alpha_\alpha)$. The conditions (iiα) and (iiiα) can easily be expressed in L^α. However, the condition $\underline{A} = \mathrm{Alg}^{\exists\Phi^*}(A,\underline{B}^\alpha_\alpha)$ presents a problem.

<u>Construction at successor ordinals α+1</u> :

We have $\underline{A} = \mathrm{Alg}^{\exists\Phi^*}(A,\underline{B}^\alpha_\alpha)$. The implication (1) → (2) yields T-models $\underline{B}'$ and $\underline{B}^{\alpha+1}_{\alpha+1}$, so that $\underline{B}^\alpha_\alpha \leq_{Fm} \underline{B}' \;{}_\Phi{\geq}\; \underline{B}^{\alpha+1}_{\alpha+1}$, $\underline{B}^{\alpha+1}_{\alpha+1} \wedge \underline{B}^\alpha_\alpha = \underline{A}$ and $\underline{A} = \mathrm{Alg}^{\exists\Phi^*}(A,\underline{B}^{\alpha+1}_{\alpha+1})$. By an easy compactness argument we can find a chain $(\underline{B}^{\alpha+1}_i)_{i\leq\alpha}$ of structures satisfying :

(a) $\langle\underline{B}^\alpha_i : i < \alpha\rangle \leq_{Fm} \langle\underline{B}^{\alpha+1}_i : i < \alpha\rangle$

(b) $\underline{B}' \leq_{Fm} \underline{B}^{\alpha+1}_\alpha$.

Notice that $\underline{B}^{\alpha+1}_{\alpha+1} \leq_\Phi \underline{B}^{\alpha+1} = \underline{B}^{\alpha+1}_0$ since $\underline{B}^{\alpha+1}_{\alpha+1} \leq_\Phi \underline{B}' \leq_{Fm} \underline{B}^{\alpha+1}_\alpha \leq_\Phi \underline{B}^{\alpha+1}$. To prove (iiiα+1) it suffices to have $\underline{B}^{\alpha+1}_{\alpha+1} \wedge \underline{B}^\alpha_\alpha = \underline{A}$ because of (iiα). Thus the construction is complete in this case.

Construction at limit ordinals δ :

Define $\underline{B}_i' = \cup\langle \underline{B}_i^\alpha : i < \alpha < \delta\rangle$ for $i < \delta$. We need an elementary extension $\langle \underline{B}_i^\delta : i < \delta\rangle$ of $\langle \underline{B}_i' : i < \delta\rangle$ together with a structure $\underline{B}_\delta^\delta$ satisfying :

(*) $\underline{B}_\delta^\delta \leq \underline{B}_i^\delta$ and $\underline{B}_\delta^\delta \wedge \underline{B}_i^i = \underline{A}$ for $i < \delta$, $\underline{B}_\delta^\delta \leq_\Phi \underline{B}^\delta = \underline{B}_0^\delta$, $\underline{B}_\delta^\delta \models T$

(**) $\underline{A} = \mathrm{Alg}^{\exists\Phi^*}(A, \underline{B}_\delta^\delta)$.

In order to express (**) in L^δ we have to strengthen it somewhat . For $\varphi(x,\overline{y})$ in $\exists\Phi^*$, $\overline{a}$ in A and $i < \delta$ set :

$V(\varphi,\overline{a},i) = \{b \in B_i : \underline{B}_i \models \varphi(b,\overline{a})\}$ and

$\Phi'(\overline{a}) = \{\varphi \in \exists\Phi^* : V(\varphi,\overline{a},i)$ finite for some $i < \delta\}$.

We have $V(\varphi,\overline{a},i) \subseteq V(\varphi,\overline{a},j)$ for $i \geq j$. Hence for $\varphi \in \Phi'(\overline{a})$ the set $V(\varphi,\overline{a},i)$ is independent of i for i sufficiently large . Set $V(\varphi,\overline{a},*) = V(\varphi,\overline{a},i)$ for i large enough . Note that $V(\varphi,\overline{a},*)$ is a finite subset of A since $\underline{A} = \mathrm{Alg}^{\exists\Phi^*}(A,\underline{B}_i)$ for $i < \delta$. Therefore the condition (**) should be strengthened to :

(a) $\underline{B} \models \forall x(\varphi(x,\overline{a}) \Rightarrow V\langle x \equiv c : c \in V(\varphi,\overline{a},*)\rangle)$ for $\varphi \in \Phi'(\overline{a})$

(b) $\underline{B} \models \exists^{\geq k} x\varphi(x,\overline{a})$ for $k \in \omega$, $\varphi \in \exists\Phi^* - \Phi'(\overline{a})$.

Then (*) and (a) , (b) are expressible by a theory T^δ which is clearly finitely consistent . An application of the compactness theorem completes the construction in this case .

Lemma 7: Let $\underline{P}_1,\dots,\underline{P}_d$ be downward-directed partial orders and let $R_1,\dots, R_d$ be T-Φ-representations of $\underline{P}_1,\dots,\underline{P}_d$ in a T-model $\underline{B}$. For $i = 1,\dots,d$ set $\underline{A}_i = \cap\langle R_i(p) : p \in P_i\rangle$. Then there exists an elementary extension $\underline{B}'$ of $\underline{B}$ containing T-models $\underline{B}_1,\dots,\underline{B}_d$ such that $\underline{B}_i \leq_\Phi \underline{B}'$ and $\underline{A}_i = \mathrm{Alg}^{\exists\Phi^*}(A_i,\underline{B}_i)$ for $i = 1,\dots,d$.

In the proof we shall use again the device introduced in the last proof . For $\varphi(x,\overline{y}) \in \exists\Phi^*$, $\overline{a}$ in A_i , $p \in P_i$ define :

$V_i(\varphi,\overline{a},p) = \{b \in R_i(p) : R_i(p) \models \varphi(b,\overline{a})\}$ and

$\Phi_i'(\overline{a}) = \{\varphi \in \exists\Phi^* : V(\varphi,\overline{a},p)$ finite for some p in $P_i\}$.

As in the previous proof we can define $V_i(\varphi,\overline{a},*) = V_i(\varphi,\overline{a},p)$ for $p \leq p_i$ with p_i in P_i sufficiently small , if $\varphi \in \Phi_i'(\overline{a})$. We will have have $A_i \supseteq V_i(\varphi,\overline{a},*)$ since p_iP_i is a coinitial subset of P_i .

Consider the theory determined by the following conditions for $\underline{B}'$ and its substructures $\underline{B}_1,\dots,\underline{B}_d$:

(a) $\underline{B} \leq_{Fm} \underline{B}'$

(b) $\underline{A}_i \leq_\Phi \underline{B}_i \models T$

(c) $\underline{B}_i \models \forall x(\varphi(x,\overline{a}) \Rightarrow V\langle x \equiv c : c \in V_i(\varphi,\overline{a},*)\rangle)$ for $\varphi \in \Phi'(\overline{a})$

(d) $\underline{B}_i \models \exists^{\geq k} x\varphi(x,\overline{a})$ for $k \in \omega$, $\varphi \in \exists\Phi^* - \Phi'(\overline{a})$.

Clearly , this set is finitely consistent and hence consistent . Any model $\langle \underline{B}', B_1,\dots,B_d\rangle$ will fulfill the requirements of the lemma .

Combining the last two lemmas with proposition 1.14 we obtain the following corollary :

Corollary 8: Let $\underline{A} \leq \underline{B} \models T$ be given . For any infinite cardinal k the following are equivalent :

(1) $\underline{A} = Alg^{\exists\Phi^*}(A,\underline{B})$

(2) There exist T-models $\underline{B}'$, $\underline{B}_1$ containing $\underline{A}$ such that $\underline{B}_1 \leq_\Phi \underline{B}'\ {}_{Fm}\!\geq \underline{B}$, $\underline{A} = \underline{B}_1 \wedge \underline{B}$ and $\underline{B} \simeq_A \underline{B}'$

(3) There exist $\underline{B}'\ {}_{Fm}\!\geq \underline{B}$ and a T-Φ-representation R of k* in $\underline{B}'$ such that $\underline{A} = \bigcap\langle R(i) : i < k\rangle$.

The equivalence of (2) and (3) is contained in theorem I.4.5 of Park [64] .

Lemma 9: Let $\underline{P}$ and $\underline{Q}$ be two reduced partial orders of type (n,d) . Then for any structure $\underline{A}$ the following are equivalent :

(1) There exist a T-model $\underline{B}$ and a T-Φ-representation R of $\underline{P}$ in $\underline{B}$ with intersection $\underline{A}$

(2) There exist a T-model $\underline{B}$ and a T-Φ-representation R of $\underline{Q}$ in $\underline{B}$ with intersection $\underline{A}$.

It suffices to prove that condition (1) is equivalent to the following condition :

(3) There exist a T-model $\underline{B}$ and structures $\underline{B}_1, \dots, \underline{B}_n$ and $\underline{A}_1, \dots, \underline{A}_d$ satisfying : (i) $\underline{B}\ {}_\Phi\!\geq \underline{B}_i \models T$ for $i = 1, \dots, n$
(ii) $\underline{A}_i \leq \underline{B}_i$ and $\underline{A}_i = Alg^{\exists\Phi^*}(A_i, \underline{B}_i)$ for $i = 1, \dots, d$
(iii) $\underline{A} = \underline{A}_1 \wedge \dots \wedge \underline{A}_d \wedge \underline{B}_{d+1} \wedge \dots \wedge \underline{B}_n$.

To show (1) implies (3) fix a T-Φ-representation R of $\underline{P}$ in $\underline{B} \models T$ with $\underline{A} = \bigcap\langle R(p) : p \in P\rangle$. Let $F = \{x_1, \dots, x_n\}$ be a frame for $\underline{P}$ such that x_1P , ... , x_dP have no least element and $x_{d+1}P = \{x_{d+1}\}, \dots, x_nP = \{x_n\}$. Set for $i = 1, \dots, n$ $\underline{A}_i = \bigcap\langle R(p) : p \in x_iP\rangle$. We have $\underline{B}\ {}_\Phi\!\geq \underline{A}_i \models T$ for $i = d+1, \dots, n$ since $\underline{A}_i = R(x_i)$. By lemma 7 there exist T-models $\underline{B}', \underline{B}_1', \dots, \underline{B}_n'$ such that $\underline{B}_i' \leq_\Phi \underline{B}'$ and $\underline{A}_i = Alg^{\exists\Phi^*}(A_i, \underline{B}_i')$ for $i = 1, \dots, n$ since x_1P , ... , x_nP are downward-directed . Clearly , now (3) holds with $\underline{B}_i = \underline{B}_i'$ for $i = 1, \dots, d$ and $\underline{B}_i = \underline{A}_i$ for $i = d+1, \dots, n$ since $\underline{A}_1 \wedge \dots \wedge \underline{A}_n = \underline{A}$.

To show (3) implies (1) we may suppose $\underline{P} = (k_1^*, \dots, k_d^*, 1^{n-d})$. Let $\underline{B}$, $\underline{B}_i$, $\underline{A}_i$, $\underline{A}$ be as specified in condition (3) . It suffices to show that there is $\overline{\underline{B}}\ {}_{Fm}\!\geq \underline{B}$ in which $\underline{A}_i$ is the intersection of a T-Φ-representation of k_i^* in $\underline{B}$. By lemma 6 we can find $\overline{\underline{B}}_i\ {}_{Fm}\!\geq \underline{B}_i$ such that $\underline{A}_i$ is the intersection of a T-Φ-representation of k_i^* in $\overline{\underline{B}}_i$. It then suffices to find $\overline{\underline{B}}\ {}_{Fm}\!\geq \underline{B}$ such that $\overline{\underline{B}}_i \leq_\Phi \overline{\underline{B}}$ for $i = 1, \dots, n$. This follows by an easy compactness argument .

Theorem 10: Let $\mathcal{Q}$ be a nonempty class of partial orders of type (n,d) and let P(n,d) be the canonical partial order of type (n,d), where $n < \omega$. Then the following are equivalent:

(1) T is P(n,d)-Φ-convex , (2) T is (n,d)-Φ-convex ,

(3) T is $\mathcal{Q}$-Φ-convex .

(1) → (2) : Let $\underline{P}$ be a partial order of type (n,d) and let $\mathcal{R}$ be a T-Φ-representation of $\underline{P}$ in a T-model $\underline{B}$ with nonempty intersection $\underline{A}$. We must show that $\underline{A}$ is a T-model. We show first that we may assume that $\underline{P}$ is at most countable.

Form the two-sorted model $<\underline{B},\underline{P};A,R>$, where $(b,p) \in R$ iff $b \in \mathcal{R}(p)$. The fact that $\mathcal{R}$ is a T-Φ-representation of $\underline{P}$ in $\underline{B}$ with intersection $\underline{A}$ can be expressed in the two-sorted language. Hence there exists an elementary substructure $<\underline{B}_0,\underline{P}_0;A_0,R_0>$ with $\underline{P}_0$ at most countable by the Löwenheim-Skolem theorem. R_0 defines a T-Φ-representation of $\underline{P}_0$ in the T-model $\underline{B}_0$ with intersection $\underline{A}_0$ and $\underline{A}_0$ is an elementary substructure of $\underline{A}$. Hence it suffices to show that $\underline{A}_0$ is a T-model.

Fix a frame F for $\underline{P}_0$. Then P(n,d) can be embedded coinitially into $\underline{P}_0$, since $\underline{P}_0$ is again of type (n,d). Then R_0 restricts to a T-Φ-representation of P(n,d) in $\underline{B}_0$ whose intersection is $\underline{A}_0$. Thus $\underline{A}_0$ and hence $\underline{A}$ is a T-model.

(2) → (3) : This is an immediate consequence of lemma 5 because of $\mathcal{Q} \subseteq$ P(n,d).

(3) → (1) : Let $\underline{Q}$ be a partial order in $\mathcal{Q}$. By lemma 5 it suffices to show that $\underline{Q}$-Φ-convexity implies P(n,d)-Φ-convexity. By lemma 3.7 there is a homomorphic image $\overline{\underline{Q}}$ of $\underline{Q}$ which is reduced of type (n,d). By lemma 5 $\underline{Q}$-Φ-convexity implies $\overline{\underline{Q}}$-Φ-convexity. However, $\overline{\underline{Q}}$-Φ-convexity is equivalent to P(n,d)-Φ-convexity by lemma 9, and our claim follows.

We have the following comparison theorem for the convexity properties of finite type.

Theorem 11: Let n be finite.

(1) (n,d_1)-Φ-convexity and $(n,d_1) > (n,d_2)$ implies (n,d_2)-Φ-convexity

(2) (n+1)-Φ-convexity implies (n,n)-Φ-convexity .

(1) This follows from lemma 5 since $P(n,d_2)$ is a homomorphic image of $P(n,d_1)$.

(2) The case $n = 1$ can be verified easily. Let $\mathcal{R}$ be a T-Φ-representation of a downward-directed set $\underline{P}$ in a T-model $\underline{B}$ with intersection $\underline{A}$. It suffices to show that $\underline{A}$ is (Φ,Φ)-tight for T in $\underline{B}$, since then we may use 2-Φ-convexity to show $\underline{A} \models T$. By lemma 7 we know $\underline{A} = \mathrm{Alg}^{\exists\Phi^*}(A,\underline{B})$. Hence

$\underline{A}$ is strongly (Φ,Fm)-tight for T in $\underline{B}$ by proposition 1.14 . This is more than we need .

The proof of the general case is considerably more delicate . It generalizes an argument of Park [64] for the case $n=1$. We shall use the following lemma .

Lemma 12: Let $\underline{P}_1,\dots,\underline{P}_d$ be downward-directed partial orders and let $R_1,\dots R_d$ be T-Φ-representations of $\underline{P}_1,\dots,\underline{P}_d$ in a T-model $\underline{B}$ with intersections $\underline{A}_1,\dots,\underline{A}_d$. Then there exist T-models $\underline{B}',\underline{B}_1,\dots,\underline{B}_d,\underline{C}$ such that :

$\underline{C}\leq_\Phi\underline{B}'$, $\underline{B}\leq_{Fm}\underline{B}'$ and $\underline{B}_i\leq_\Phi\underline{B}'$, $\underline{C}\wedge\underline{B}_i=\underline{A}_i$ for $i=1,\dots,d$.

For $\bar{a}\in\bar{A}=\cup\langle A_i : i=1,\dots,d\rangle$ let $\Phi'(\bar{a})=\{\varphi\in\Phi^* : T\vee\Delta_{At*}(\bar{A})\vdash\exists\bar{x}\varphi(\bar{x},\bar{a})\}$. For $\varphi\in\Phi'(\bar{a})$ choose $\bar{c}=\bar{c}(\varphi)$ in B such that $\underline{B}\models\varphi(\bar{c},\bar{a})$. Consider the theor determined by the following conditions for $\underline{B}',\underline{B}_1,\dots,\underline{B}_d$:

(a) $\underline{B}\leq_{Fm}\underline{B}'$ (b) $\underline{A}_i\leq_\Phi\underline{B}_i\models T$ for $i=1,\dots,d$

(c) $\bar{c}(\varphi)\wedge A_i=\bar{c}(\varphi)\wedge B_i$ if $\bar{a}\in\bar{A}$, $\varphi\in\Phi'(\bar{a})$, $1\leq i\leq d$.

To see that the set is finitely consistent , let $\varphi_1,\dots,\varphi_k$ be formulas in $\cup\langle\Phi'(\bar{a}) : \bar{a}\in\bar{A}\rangle$. Then there exist elements $p_1\in P_1,\dots,p_d\in P_d$ satisfying $\bar{c}(\varphi_j)\wedge R_i(p_i)=\bar{c}(\varphi_j)\wedge A_i$ for all i , j since $\bar{c}(\varphi_1),\dots,\bar{c}(\varphi_d)$ are finite sets and $\underline{A}_1,\dots,\underline{A}_d$ are the intersection of $R_1,\dots,R_d$. Hence $\underline{B}$ together with $R_1(p_1),\dots,R_d(p_d)$ satisfies (a) - (c) for $\varphi_1,\dots,\varphi_k$. Therefore the set is consistent .

Let $\langle\underline{B}',\underline{B}_1,\dots,\underline{B}_d\rangle$ be any model of this theory . Then there cannot exist formulas $\psi,\varphi_1,\dots,\varphi_d$ in Φ^* and sequences $\bar{a}$ in $\bar{A}$, $\bar{b}^1$ in $B_1-A_1,\dots,\bar{b}^d$ in B_d-A_d satisfying :

(1) $T\vee\Delta_{At*}(\bar{A})\vdash\exists\bar{y}\psi(\bar{y},\bar{a})$ (2) $\underline{B}_i\models\varphi_i(\bar{b}^i,\bar{a})$

(3) $T\vee\Delta_{At*}(\bar{A})\vdash\forall\bar{x}\bar{y}(\varphi(\bar{x},\bar{a})\wedge\psi(y,a)\Rightarrow V\langle x_j\equiv y_k : j=1,\dots,n ; k=1,\dots,m\rangle)$,

where $\bar{x}=(\bar{x}^1,\dots,\bar{x}^d)$ and $\varphi(\bar{x},\bar{z})$ is the conjunction of the formulas $\varphi_i(\bar{x}^i,\bar{z})$ together with all equations $x^i_j\equiv x^k_l$ for which $b^i_j=b^k_l$.

For fixed i take $\psi=\varphi_i=(y\equiv a)$ with a in $\bar{A}-A_i$ and $\varphi_j=(z\equiv z)$ for $j\neq i$. The above condition implies $B_i\wedge\bar{A}=A_i$ for $i=1,\dots,d$.

For $\bar{a}$ in $\bar{A}$ let $\Delta'(\bar{a})$ be defined as follows . A formula $\psi(\bar{y},\bar{z})$ in belongs to $\Delta'(\bar{a})$ iff there exist $\varphi_1,\dots,\varphi_d$ in Φ^* , $\bar{b}^1\in B_1-A_1,\dots,b^d\in B_d-A_d$ such that

$\underline{B}_i\models\varphi_i(\bar{b}^i,\bar{a})$ for $1\leq i\leq d$ and

$T\vee\Delta_{At*}(\bar{A})\vdash\forall\bar{x}\bar{y}(\varphi(\bar{x},\bar{a})\wedge\psi(\bar{y},\bar{a})\Rightarrow V\langle x_j\equiv y_k : j=1,\dots,n ; k=1,\dots,m\rangle)$,

where φ is defined as above . The condition mentioned above implies the consistency of the following set of formulas :

$T\vee\Delta_{At*}(\bar{A})\vee\{\neg\exists\bar{y}\psi(\bar{y},\bar{a}) : \psi(\bar{y},\bar{a})\in\Delta'(\bar{a}) , \bar{a}\in\bar{A}\}$.

Let $\underline{C}$ be a model of this set . Now it is easy to see , that the following set is consistent :

$T \vee \Delta_\Phi(\underline{C}) \vee \bigcup\langle \Delta_\Phi(\underline{B}_i) : i=1,\dots,d\rangle \vee \{\neg b{=}c : b \in B_i - A_i,\ c \in C,\ 1 \le i \le d\}$.
Because of $B_i \wedge \bar{A} = A_i$ for $i = 1,\dots,d$ we may use a single set of constants for the elements of $\bar{A}$. Any model $\underline{B}'$ of this set of sentences will have the desired properties.

Now we are able to prove part (2) of theorem 11. By theorem 10 it suffices to show that if, $\underline{A}$ is the intersection of the T-Φ-representation R of $P(n,d)$ in a T-model $\underline{B}$ then $\underline{A}$ is the intersection of a T-Φ-representation R' of $P(n+1,0)$ in some T-model $\underline{B}'$.

In $\underline{B}$ we have $\underline{A} = \underline{A}_1 \wedge \dots \wedge \underline{A}_n$ where each $\underline{A}_i$ is the intersection of a T-Φ-representation of ω^* in $\underline{B}$. By lemma 12 there exist T-models $\underline{B}'$, $\underline{B}_1$, ..., $\underline{B}_{n+1}$ such that $\underline{B} \le_{Fm} \underline{B}'$, $\underline{B}_i \le_\Phi \underline{B}'$ for $i = 1,\dots,n+1$ and $\underline{B}_{n+1} \wedge \underline{B}_i = \underline{A}_i$ for $i = 1,\dots,n$. Hence we have $\underline{A} = \underline{B}_1 \wedge \dots \wedge \underline{B}_n \wedge \underline{B}_{n+1}$ as desired.

The following two results are the counterpart of theorem 10 for infinite types.

Proposition 13: Let $\mathcal{Q}$ be a class of partial orders of type $(\omega,1)$ and let $P(\omega,1)$ be the ordering 1^ω. Then the following are equivalent:

(1) T is $P(\omega,1)$-Φ-convex (2) T is Φ-convex

(3) T is $\mathcal{Q}$-Φ-convex.

(1) → (2): Let $\underline{P}$ be any arbitrary partial order and let R be a T-Φ-representation of $\underline{P}$ in a T-model $\underline{B}$ with nonempty intersection $\underline{A}$. As in the proof of theorem 10 we may assume that $\underline{P}$ is at most countable. Then $\underline{P}$ is a homomorphic image of $1^\omega = P(\omega,1)$. An application of lemma 5 yields the desired conclusion.

(2) → (3): This is obvious.

(3) → (1): Let $\underline{Q}$ be a member of $\mathcal{Q}$. By lemma 3.8 there exists a surjective homomorphism onto 1^ω or $(1^\omega)^\cdot$. However, 1^ω-Φ-convexity is the same as $(1^\omega)^\cdot$-Φ-convexity. Again lemma 5 yields the desired conclusion.

As an immediate consequence of the comparison theorem 11 we obtain the following result:

Proposition 14: Let $\mathcal{Q}$ be a class of partial orders of finite type and assume that there is no finite bound on the types of orders in $\mathcal{Q}$. Then the following statements are equivalent:

(1) T is n-Φ-convex for every n in ω

(2) T is $\mathcal{Q}$-Φ-convex.

Definition 15: The *type* of a class P of partial orders is defined as follows. P is said to be of *finite type* if there exists a finite bound for the type of members of P, otherwise P is said to be of *infinite type* If P is of finite type then the type of P is the supremum of the types of members of P. If P is of infinite type then two cases are possible. Either P contains a partial order of infinite type or P contains only partial orders of finite type. In the first case the type of P is $(\omega,1)$ and in the second case the type of P is $(\omega,0)$.

Now the previous results can be summarized as follows:

Theorem 16: Let P_1 resp. P_2 be a class of partial orders of type (n_1,d_1) resp. (n_2,d_2). If $(n_1,d_1) \geq (n_2,d_2)$ then P_1-Φ-convexity implies P_2-Φ-convexity. In particular, if $(n_1,d_1) = (n_2,d_2)$ then P_1-Φ-convexity is equivalent to P_2-Φ-convexity. Hence there are at most $\omega+2$ convexity properties for a given set Φ of formulas.

The results below show that Φ-convexity and $\downarrow\Phi$-convexity can be characterized with the help of algebraic closure operators.

Proposition 17: A theory T is Φ-convex iff T is hereditary w.r.t. C_T^Φ.

Let T be Φ-convex. $C_T^\Phi(A,\underline{B})$ is the intersection of a T-Φ-representation in the T-model $\underline{B}$. Hence $C_T^\Phi(A,\underline{B})$ must be a T-model as well. Thus T is hereditary w.r.t. C_T^Φ.

Conversely, assume that T is hereditary w.r.t. C_T^Φ. Let $\underline{A}$ be the intersection of a T-Φ-representation R in a T-model $\underline{B}$. From lemma 1.7 we conclude $C_T^\Phi(A,\underline{B}) = \underline{A}$. Hence $\underline{A}$ must be a T-model as desired.

Proposition 18: For every theory T the following are equivalent:

(1) T is 2-Φ-superconvex

(2) T is Φ-superconvex

(3) T is Φ-hereditary w.r.t. C_T^Φ

(4) T is Φ-hereditary w.r.t. $H_T^{\Phi,\Phi}$

(5) T is Φ-hereditary w.r.t. $\exists\Phi^*$-$Alg_T^{\exists\Phi^*}$.

The equivalence of (1) and (2) was proved in proposition 4. The equivalence of (2) and (3) is a variant of proposition 17. (4) implies (3) because of part (6) of lemma 1.7. Since (3) is equivalent to (1), (3) implies $C_T^\Phi(A,\underline{B}) = H_T^{\Phi,\Phi}(A,\underline{B})$. Hence (3) implies (4). The equivalence of (4) and (5) follows from proposition 1.8.

The tight substructures can be used to characterize 2-Φ-convexity and (Φ_1,Φ_2)-convexity.

Lemma 19: (1) T is 2-Φ-convex iff T is hereditary w.r.t. substructures which are (Φ,Φ)-tight for T.

(2) T is (Φ_1,Φ_2)-convex iff T is hereditary w.r.t. substructures which

are (Φ_1,Φ_2)-tight for T .

Proposition 20: For every theory T the following are equivalent :

(1) T is $(1,1)$-Φ-convex

(2) T is $\downarrow\Phi$-convex

(3) T is hereditary w.r.t. $Alg^{\exists\Phi^*}$.

(4) T is hereditary w.r.t. substructures which are strongly (,Fm)-tight for T .

Moreover , the proposition remains true if 'convex' and 'hereditary' are replaced by 'superconvex' and 'Φ-hereditary' .

The equivalence of (1) and (2) follows from theorem 10 . The equivalence of (2) and (3) is a consequence of corollary 8 . The equivalence of (3) and (4) is due to proposition 1.12 .

Making use of theorem I.5.3 in Park [64] we can show that there is another equivalent condition :

(5) T is hereditary w.r.t. Alg^{Fm} and substructures which are $\exists\Phi^*$-elementary .

Remark 21: Proposition 18 and theorem 2.7 yield a syntactic characterization of At-convex theories . Proposition 20 and theorem 2.4 yield a syntactic characterization of $\downarrow$At-convex theories . Making use of lemma 2.8 these results can be extended to theories which are Φ-superconvex resp. $\downarrow\Phi$-superconvex .

The proof of the following fact is obvious .

Lemma 22: If T is P-Φ-convex and $\Phi \subseteq \Psi$ then T is P-Ψ-convex .

The next proposition relates Φ-convexity and $\downarrow\Phi$-convexity for different sets Φ .

Proposition 23: (1) Φ-convexity implies 2-Φ-convexity

(2) 2-Φ-convexity implies (Φ,Fm)-convexity

(3) (Φ,Fm)-convexity implies $\downarrow\Phi$-convexity

(4) $\downarrow\Phi$-convexity implies $\exists\Phi^*$-convexity .

The implications in (1) and (2) are obvious . (3) follows from the equivalence of (2) and (4) in proposition 20 . The implication in (4) is based on the following inclusions which were proved in lemma 1.7 :

$$Alg^{\exists\Phi^*}(A,\underline{B}) \leq H_T^{\exists\Phi^*,\exists\Phi^*}(A,\underline{B}) \leq C_T^{\exists\Phi^*}(A,\underline{B})$$

An application of proposition 17 and 20 yields the desired implication .

By proposition 23 we obtain a chain of weaker and weaker convexity properties by going from $\exists_1$ to $\exists_2$ and so on up to Fm .

The following observation can be useful .

<u>Lemma 24</u>: If T is hereditary w.r.t. $Alg_T^{\exists\Phi^*}$ or $\exists\Phi^*\text{-}Alg_T^{Fm}$ then T is (Φ,Fm)-convex .

Let $\underline{B}_1$, $\underline{B}_2$, $\underline{C}$ be T-models satisfying $\underline{B}_1 \leq_\Phi \underline{C}\ {}_{Fm}{\geq} \underline{B}_2$. Then we have $Alg_T^{\exists\Phi^*}(B_1 \wedge B_2, \underline{B}_1) = \underline{B}_1 \wedge \underline{B}_2$ and $\exists\Phi^*\text{-}Alg_T^{Fm}(B_1 \wedge B_2, \underline{B}_2) = \underline{B}_1 \wedge \underline{B}_2$. Hence $\underline{B}_1 \wedge \underline{B}_2$ has to be a T-model in both cases .

The following result can be found in Park [64] .

<u>Corollary 25</u>: The following properties are equivalent :
(1) Fm-convexity (2) 2-Fm-convexity (3) ↓Fm-convexity
The result remains true if 'convexity' is replaced by 'superconvexity' .

The result for convexity is a consequence of proposition 23 because of Fm = Fm . - To prove the result for superconvexity we proceed as follows . (1) and (2) are equivalent by proposition 4 . The equivalence of (2) and (4) in proposition 20 yields the implication (2) → (3) . If T is ↓Fm-superconvex then T is Fm-hereditary w.r.t. Alg^{Fm} by proposition 20 . However , Alg^{Fm} and Alg_T^{Fm} coincide by lemma 1.4 . An application of lemma 24 shows that T is 2-Fm-superconvex . This yields the implication (3) → (2) .

It should be noted that the results in 2.13 and 2.8 can be used to give a syntactic characterization of those theories which are Fm-superconvex . - However , we do not have a syntactic characterization of those theories which are Fm-convex i.e. which are hereditary w.r.t. Alg .

The following list of examples shows that the implications in proposition 23 cannot be reversed in the case $\Phi = At$.

<u>Examples 26</u>: (1) The theory T_2 defined in 2.22 is hereditary w.r.t. $Alg_{T_2}^{\exists}$ but not hereditary w.r.t. $\exists\text{-}Alg_{T_2}^{\exists}$. By proposition 24 and 14 we conclude that T_2 is (At,Fm)-convex but is not (At,At)-convex .

(2) The theory T_1 defined in 2.22 is hereditary w.r.t. $Alg^{\exists}$. Hence T_1 is ↓At-convex by proposition 20 . The structures $\underline{B}_1$, $\underline{B}_\omega$ introduced in 2.22 can be used to verify that T_1 is not (At,Fm)-convex . It suffices to find a structure $\underline{B}'_\omega$ such that $\underline{B}'_\omega \leq_{Fm} \underline{B}_\omega$ and $B_1 \wedge B'_\omega = \{a_o\}$.

(3) The theory T_4 defined below is ∃-convex but not ↓At-convex . T_4 has the following axiom : $\exists^2 x \forall y R(x,y)$. T_4 is ∃-convex , since $\underline{A}, \underline{B} \models T_4$ and $\underline{A} \leq_\exists \underline{B}$ imply that $\underline{A}$ and $\underline{B}$ have the same number of solutions of $\forall y R(x,y)$. The T_4-model $\underline{B} = (\omega+1, R)$ is defined as follows : $(i,j) \in R$ iff $i \leq j$ or $i = \omega$. For every k in $\omega+1$ let $\underline{B}_k$ be the substructure determined by $\{n \in \omega+1 : n > k\}$. $\underline{B}_k$ is a model of T_4 if $k \in \omega$. However , $\underline{B}_\omega = \cap\langle \underline{B}_n : n \in \omega\rangle$ is not a model of

T_4 . Hence T_4 is not ↓At-convex .

(4) Let T_5 be the theory of fields . T_5 is At-convex and hence Φ-convex for every Φ , since Φ contains At by our conventions . However , T_5 is not ∃Φ*-superconvex as the following example of Park [64] shows . Let F be a finite field and let $F_1 = F(x_{2n+1} : n \in \omega)$, $F_2 = F(x_{2n} : n \in \omega)$, $F_3 = F(x_n : n \in \omega)$ be purely transcendental extensions of F . Then we have $F_1 \wedge F_2 = F$ and $F_1 \leq_{Fm} F_3 \, {}_{Fm}\!\geq F_2$. However , F is not an ∃-elementary substructure of F_3 .

(5) The theory T_6 of dense linear orders without endpoints is not Fm-convex . Making use of the fact that T_6 admits quantifier elimination one can easily verify $Alg(\{q\},\underline{Q}) = \{q\}$, where $\underline{Q}$ is the ordered set of rational numbers .

§.5 Summary

In 1962 M.O.Rabin has given a syntactic characterization of those elementary classes which are closed under *arbitrary intersections* . Extending results of D.M.R.Park in 1964 we are able to give a syntactic characterization of those elementary classes which are closed under *intersections of descending chains* , thus solving a problem left over from the early sixties .

More generally , we consider *convexity (=intersection) properties* for partially ordered sets of models and embeddings for a given partial order $\underline{P}$. Our methods are based on a new set of *algebraic closure operators* , which are algebraic in the following sense : A given substructure $\underline{A}$ of a structure $\underline{B}$ is being closed under the addition of certain finite A-definable subsets . In the case of arbitrary intersections resp. intersections of descending chains we can show that the corresponding convexity property is equivalent to heredity with respect to closed substructures for an appropriate closure operator . This leads to a syntactic characterization in both cases . In the first case we obtain a new syntactic characterization , but Rabin's characterization can be obtained as well . The characterization of the convexity property in terms of the closure operator seems to be more useful . In addition , we are able to show that these are the only convexity properties in this situation .

More generally , we consider Φ-elementary embeddings instead of embeddings , where Φ is a given set of formulas . Using a set of invariants for partially ordered sets we are able to show that there are at most ω+2 convexity properties in the general case . In the case of elementary embeddings there is exactly one convexity property . However , the syntactic characterization of this convexity property remains an open problem .

References :

Bacsich,P.D.[73]: Defining algebraic elements , J.Symb.Logic 38(1973), 93-101

Bacsich,P.D.[75]: The strong amalgamation property , Colloq.Math.33(1975) 13-23

Bacsich,P.D.,Rowlands-Hughes,D.[74]: Syntactic characterisations of amalgamation , convexity and related properties , J.Symb.Logic 39(1974),433-451

Chang,C.C.,Keisler,H.J.[73]: Model theory , North-Holland Publ.Comp., Amsterdam 1973

Cherlin,G.L.,Volger,H.[79]: $\dagger$-convexity and $Alg^{\exists}$, Notices Amer.Math.Soc. 26(1979),A-524

Cherlin,G.L.,Volger,H.[80]: Convexity properties and algebraic closure operators , submitted to Annals Math.Logic in April 1980

Eklof,P.[74]: Algebraic closure operators and strong amalgamation bases, Algebra Universalis 4(1974),89-93

Keisler,H.J.[60]: Theory of models with generalized atomic formulas , J. Symb.Logic 25(1960),1-26

Kueker,D.W.[70]: Generalized interpolation and definability , Annals Math Logic 1(1970),423-468

Kueker,D.W.[75]: Core structures for theories , Fund.Math.89(1975),155-171

Omarov,A.I.[67]: Filtered products of models (russian: O fil'trovannyh proizvedenijah modelei),Algebra i Logika 6,3(1967),77-89

Park,D.M.R.[64]: Set-theoretic constructions in model theory , dissertation , MIT 1964

Pinter,C.C.[77]: Some theorems on omitting types with applications to model completeness , amalgamation and related properties , in: Non-classical logic,model theory and computability,eds.Arruda,A.I.et al.,North-Holland Publ.Comp.,Amsterdam 1977,233-238

Rabin,M.O.[62]: Classes of models and sets of sentences with the intersection property , Ann.Sci.Univ.Clermont 7(1962),39-53

Robinson,A.[51]: On the metamathematics of algebra , North-Holland Publ. Comp.,Amsterdam 1951

Robinson,A.[63]: Introduction to model theory and to the metamathematics of algebra , North-Holland Publ.Comp.,Amsterdam 1963

Simmons,H.[75]: Companion theories (Forcing in model theory),Inst.Math. Pure Appl.,Univ.Cath.de Louvain, 1975

Volger,H.[79]: Preservation theorems for limits of structures and global sections of sheaves of structures , Math.Zeitschr.166(1979),27-53

Zaharov,D.A.[63]: Convex classes of models (russian),Ivanov.Gos.Ped. Inst.Ucen.Zap.31(1963),54-55

REMARKS ON FINITELY BASED LOGICS

Janusz Czelakowski
Polish Academy of Sciences
Institute of Philosophy and Sociology
The Section of Logic
Piotrkowska 179, 90-447 Łódź, Poland

§I. General remarks

There are many nonequivalent definitions of sentential logics. Some logicians prefer to call a logic any invariant (i.e. closed under substitutions) set of formulas that is additionally closed under some explicitly given rules of inference. Thus the term 'logic' is understood here as 'logical theory', i.e. as a set of sentences that tell us what is logically true much the same as, say, laws of physics tell us "truth" about physical phenomena. But logic is also viewed as a tool that serves us to draw valid conclusions from valid premises. According to this standpoint, logic is a set of valid inferences, not just valid formulas. The difference is essential: the notion of a valid formula can be defined in terms of valid inferences but, in general, not vice versa. Thus we face two different methodological perspectives here. The inferential approach was originated by Alfred Tarski [15] in the thirties with his works on deductive systems and consequence operations, and then continued by Łoś, Suszko, Rasiowa, Wójcicki, to mention only a few names. According to this approach, a sentential logic is a pair

$$(\underline{S}, C),$$

where $\underline{S}$ is a sentential language, i.e. an absolutely free algebra freely generated by an infinite set $Var(\underline{S}) = \{p,q,r,\ldots\}$ of sentential variables and endowed with countably many finitary operations (connectives) $\$_1, \$_2, \ldots$, and C is a structural consequence operation on S, the underlying set of the algebra $\underline{S}$, i.e., C satisfies the following conditions:

(a) $X \subseteq C(X)$, for all $X \subseteq S$;
(b) if $X \subseteq Y$ then $C(X) \subseteq C(Y)$, for all $X, Y \subseteq S$;
(c) $C(C(X)) = C(X)$, for all $X \subseteq S$;
(d) $e(C(X)) \subseteq C(e(X))$, for all $X \subseteq S$ and for every endomorphism e of $\underline{S}$.

The members of S are called sentential formulas. The endomorphisms of $\underline{S}$ are customarily called substitutions in $\underline{S}$.

Condition (d) is just the definition of structurality of C.(d) says that for all $\alpha \in S$, $X \subseteq S$ and for every substitution e, $e\alpha \in C(e(X))$ whenever $\alpha \in C(X)$. The structurality of C is the formal counterpart of the fact that valid inferences (arguments) refer only to the form, shape of a sentence which is determined by the occurrences of logical connectives, not to its "content", the latter being represented by the sentential variables the sentence involves. If no confusion is likely, a logic $(\underline{S},C)$ is usually identified with its consequence operation C.

A logic C is standard (or finite) iff

$$C(X) = \bigcup\{C(Y): Y \subseteq X \mathbin{\&} Y \text{ is finite}\},$$

for all $X \subseteq S$.

If X is a finite set of formulas, then instead of C(X) we often write $C(\alpha_1,\dots,\alpha_n)$, where $\alpha_1,\dots,\alpha_n$ is a fixed arrangement of the elements of X. The symbol $C(X,\alpha)$ is also used as an abbreviation for $C(X \cup \{\alpha\})$.

A set $X \subseteq S$ such that $C(X) = X$ is called a theory of a logic C. The set of all theories of C, denoted by

$$Th(C),$$

is a complete lattice under the set-theoretic inclusion. If C is standard then the lattice Th(C) is algebraic.

After Prucnal and Wroński [11], a logic $(\underline{S},C)$ is equivalential iff there exists a set $\Lambda(p,q)$ of formulas of $\underline{S}$ in two variables, hereafter referred to as a C-equivalence, such that for all $\alpha,\beta,\gamma \in S$ the following conditions are satisfied:

(i) $\Lambda(\alpha,\alpha) \subseteq C(\emptyset)$

(ii) $\Lambda(\alpha,\beta) \subseteq C(\Lambda(\beta,\alpha))$

(iii) $\Lambda(\alpha,\gamma) \subseteq C(\Lambda(\alpha,\beta) \cup \Lambda(\beta,\gamma))$

(iv) for each n-ary connective § of $\underline{S}$, $n \geq 0$, and any formulas $\alpha_1,\dots,\alpha_n,\beta_1,\dots,\beta_n$, $\Lambda(\S(\alpha_1\dots\alpha_n),\S(\beta_1\dots\beta_n)) \subseteq C(\Lambda(\alpha_1,\beta_1) \cup \dots \cup \Lambda(\alpha_n,\beta_n))$

(v) $\beta \in C(\Lambda(\alpha,\beta) \cup \{\alpha\})$,

where $\Lambda(\alpha,\beta)$ denotes the set of formulas which result by the simultaneous substitution of α for p and β for q in all formulas in $\Lambda(p,q)$. A logic C is finitely equivalential if it is equivalential and has a finite C-equivalence. The class of finitely equivalential logics includes a great many of the more important sentential logics, among others all implicative logics in the sense of Helena Rasiowa [12]. More information about equivalential logics can be found in [3] and [4].

The notion of a consequence operation has its origin in the notion of a 'provability' relation. The expression '$\alpha \in C(X)$' is usually read 'α is deducible or provable from X by means of some accepted rules of inference'. A rule of inference is usually viewed as a set of instructions of the form:

From X infer α.

In other words, rules of inference are relations between sets of formulas (premises) and formulas (conclusions); thus, formally speaking, a <u>rule on S</u> is a subset of the Cartesian product $P(S) \times S$, where $P(S)$ is the power set of S. A rule r is <u>structural</u> if it is invariant under the substitutions of $\underline{S}$, that is, if for every substitution e in $\underline{S}$ and every $X \subseteq S$, $\alpha \in S$, $(eX, e\alpha) \in r$ whenever $(X, \alpha) \in r$. Given a pair (X, α) from $P(S) \times S$ we define the structural rule

$$X/\alpha$$

to be the set

$$\{(eX, e\alpha):\ e \text{ is a substitution in } \underline{S}\}.$$

The rules of the form X/α are called <u>sequential</u> and the pair (X, α) is called a scheme of X/α. Let us notice that a scheme of X/α is unique up to the choice of variables. A sequential rule X/α is <u>standard</u> when the set X is finite. The standard rule X/α is customarily denoted by $\gamma_1, \ldots, \gamma_k/\alpha$, where $\gamma_1, \ldots, \gamma_k$ is a fixed arrangement of the elements of X. Modus Ponens $p, p \rightarrow q/q$ is an example of a standard rule. Our interest in the notion of a standard rule, apart from its simplicity, lies in the fact that every standard logic can be viewed as a set of standard rules. A standard rule r is <u>axiomatic</u> if it is of the form $\emptyset/\alpha$. In many contexts the axiomatic rule $\emptyset/\alpha$ is identified with the formula α.

Let Q be a set of standard rules in $\underline{S}$ and let $X \cup \{\beta\}$ be a set of formulas of $\underline{S}$. A finite sequence of formulas

$$\beta_1, \ldots, \beta_n$$

is said to be a <u>formal proof</u> of β from X by means of the rules from Q iff $\beta_n = \beta$ and for every i, $1 \leq i \leq n$, either $\beta_i \in X$ or there exist indices $i_1,\ldots,i_k < i$, a rule $\gamma_1,\ldots,\gamma_k/\alpha$ in Q and a substitution e in $\underline{S}$ such that $\beta_{i_1} = e\gamma_1,\ldots,\beta_{i_k} = e\gamma_k$ and $\beta_i = e\alpha$.

By C_Q we denote the logic in $\underline{S}$ determined as follows: $\alpha \in C_Q(X)$ iff there is a formal proof of α from X by means of the rules from Q.

In connection with the considerations which will follow it is worth-while to recall a classical theorem having a considerable theoretical importance:

THEOREM I.1.(Łoś and Suszko [9]). Each standard sentential logic can be formalized by means of a countable set of standard rules which precisely amounts to saying that for any standard logic C there is a countable set Q of standard rules such that $C = C_Q$.

Let C be a logic in a language $\underline{S}$. A sequential rule X/α is called a rule of C if $\alpha \in C(X)$. By a C-proof in $\underline{S}$ we shall mean any formal proof in $\underline{S}$ carried out by means of rules of C. It follows from Theorem I.1 that for every standard logic C and any $X \subseteq S$, $\alpha \in S$:

$$\alpha \in C(X) \text{ iff there is a } C\text{-proof of } \alpha \text{ from } X.$$

Thus, according to Theorem I.1, there is no essential difference between the notion of a logic viewed as a finite structural consequence operation and the one which views logic as a set of standard rules of inference.

A set Q of standard rules is called an inferential base for a given finite logic C iff $C = C_Q$. We can reformulate Theorem I.1 by saying that every standard logic has a countable base. The standard rules of C form the greatest inferential base for C. A logic C is finitely based if it has a finite inferential base, that is, if C can be formalized by means of a finite set of axiomatic rules and a finite set of standard nonaxiomatic ones.

The present paper is motivated by the following question, put forward by Stephen L. Bloom in [1]: when can a logic be described proof-theoretically by a finite number of standard rules of inference? Of course, most of the familiar logics, e.g. those studied in Rasiowa's monograph [12] are defined in this way, i.e., defined by a finite number of rule schema. However, when a logic is defined by means of 'models' of some sort, e.g. matrices, it is not always clear whether an explicit proof-theoretic definition is possible.

We close this section with two definitions. Given a logic C in a language $\underline{S}$, we shall say that a logic C' in $\underline{S}$ is a strengthening of C, in symbols: $C \leq C'$, if $C(X) \subseteq C'(X)$, for all $X \subseteq S$. A strengthening C' of C is called <u>axiomatic</u> if there is a set $A \subseteq S$ such that

$$(1.1) \qquad C'(X) = C(Sb(A) \cup X),$$

for all $X \subseteq S$, where $Sb(A) =_{df} \bigcup\{e(A)\colon e \text{ is a substitution in } \underline{S}\}$. If (1.1) holds for every $X \subseteq S$, then C' is referred to as the strengthening of C by means of the set of axioms A, or, which amounts to the same, the strengthening of C by means of the set of axiomatic rules $\emptyset/\alpha$, where $\alpha \in A$.

§II. Matrices

We shall briefly recall some facts from the theory of logical matrices.

A logical <u>matrix</u> is a pair

$$M = (\underline{A}, D),$$

where $\underline{A}$ is an algebra, referred to as the algebra of M, and D is a subset of A called the set of designated elements of M. (In the sequel, algebras will be denoted by underlined letters $\underline{A}$, $\underline{B}$, $\underline{S}$ etc., while their universes by the corresponding capital characters A,B,S). If the algebra of a matrix M is similar to a sentential language $\underline{S}$, then M is called a <u>matrix for</u> $\underline{S}$.

Since the notion of a matrix falls under the more general concept of an algebraic structure (or a model), we can apply to matrices all the model-theoretic operations which are performed on algebraic structures. Thus $N = (\underline{B}, E)$ is called a submatrix of $M = (\underline{A}, D)$ iff $\underline{B}$ is a subalgebra of $\underline{A}$ and $E = B \cap D$. Similarly we define the direct product $\prod_{i \in I} M_i$ of a non-empty family of matrices $M_i = (\underline{A}_i, D_i)$ to be the matrix $(\prod_{i\in I} \underline{A}_i, \prod_{i \in I} D_i)$. (We admit that the trivial one-element matrix

$$M_e = (\underline{1}, \{1\})$$

is the direct product of the void family. $\underline{1}$ is the trivial algebra with the universe $\{1\}$.) Also the definitions of a reduced product of matrices, of an ultraproduct of matrices etc., are particular instances of the more general definitions from model theory.

Let us recall that a mapping h is a <u>homomorphism</u> of a matrix $(\underline{A}, D)$ into $(\underline{B}, E)$ if h is a homomorphism of the algebra $\underline{A}$ into $\underline{B}$ and $hD \subseteq E$. A homomorphism $h\colon M \longrightarrow N$ is <u>strict</u> if moreover $h(A-D) \subseteq B-E$.

In the sequel we admit the following notation. Given a class K of matrices,

I(K) denotes the class of all isomorphic copies of matrices from K ;

S(K) is the class of all isomorphic copies of submatrices of members of K ;

P(K) is the class of all isomorphic copies of direct products of families of members of K ;

$P_R(K)$ ($P_U(K)$,respectively) is the class of all isomorphic copies of reduced products of non-empty families (of ultraproducts of families) of members of K ;

$H_S(K)$ is the class of all strict homomorphic images of the members of K. Thus $N \in H_S(K)$ iff there is a matrix $M \in K$ and a strict homomorphism h of M onto N ;

$\overleftarrow{H}_S(K)$ is the class of all strict homomorphic pre-images of the members of K. Thus $M \in \overleftarrow{H}_S(K)$ iff there is a matrix $N \in K$ and a strict homomorphism h of M onto N.

Each class K of matrices for a language $\underline{S}$ induces the logic

$$Cn_K$$

in $\underline{S}$: $\alpha \in Cn_K(X)$ iff for every matrix $M = (\underline{A}, D)$ in K and every homomorphism v of $\underline{S}$ into $\underline{A}$, $v\alpha \in D$ whenever $vX \subseteq D$. A class K of matrices is <u>strongly adequate</u> for a logic C iff $C = Cn_K$. Given a logic $(\underline{S}, C)$ we define

$$Matr(C)$$

to be the class of all matrices M which validate C, that is, which have the property: $C \leqslant Cn_M$. (We use here the symbol Cn_M instead of the awkward $Cn_{\{M\}}$). The members of Matr(C) are called C-<u>matrices</u>, for short. Matr(C) is the largest class of matrices strongly adequate for C. If C is standard, then Matr(C) is a quasivariety, i.e., it is closed under the operations S, P and P_U. It is not difficult to check that if M and N are matrices for $\underline{S}$ and h is a strict homomorphism of M onto N, then M and N induce the same logics in $\underline{S}$, i.e., $Cn_M = Cn_N$ (see [19]). It follows that the class Matr(C) is also 'invariant' with respect to strict homomorphisms - this means that $\overleftarrow{H}_S H_S(Matr(C)) \subseteq Matr(C)$.

The following theorem, characterizing the class Matr(C) can be found in [2]:

THEOREM II.1. Let K be a class of matrices for a language $\underline{S}$ and let us assume that the logic Cn_K in $\underline{S}$ is standard. Then

$$\mathrm{Matr}(Cn_K) = \overleftarrow{H}_S H_S S P_R(K) = \overleftarrow{H}_S H_S S P P_U(K).$$

Let $M = (\underline{A}, D)$ be a matrix for a language $\underline{S}$. A congruence Θ of the algebra $\underline{A}$ is called strict if for any $a,b \in A$, $a \equiv b(\Theta)$ entails: $a \in D$ iff $b \in D$. In other words, strict congruences do not paste together designated elements of M with undesignated ones. In every matrix M there exists a greatest strict congruence of M, denoted by Θ_M. If C is equivalential and $M = (\underline{A}, D)$ is in Matr(C), then Θ_M is characterized as follows:

$$a \equiv b(\Theta_M) \quad \text{iff} \quad \Lambda_M(a,b) \subseteq D,$$

where Λ is a C-equivalence and $\Lambda_M(a,b)$ is the set of values in (a,b) of the binary polynomials over $\underline{A}$ corresponding to the formulas of Λ.

A matrix M is factorial (or simple) if Θ_M is the diagonal of M. Thus, if a logic C is equivalential, then a matrix $M = (\underline{A}, D)$ from Matr(C) is factorial if for every a,b,

$$a = b \quad \text{iff} \quad \Lambda_M(a,b) \subseteq D.$$

Given a matrix $M = (\underline{A}, D)$ we let M/Θ_M denote the quotient matrix $(\underline{A}/\Theta_M, D/\Theta_M)$, where $D/\Theta = \{[a]_\Theta : a \in D\}$. Clearly M/Θ_M is factorial.

The importance of factorial matrices lies in the fact that both M and M/Θ_M induce the same logics. The class of all factorial matrices in Matr(C) is denoted by

$$\mathrm{Matr}^*(C).$$

Matr*(C) is also strongly adequate for C. Moreover, if C is standard and finitely equivalential, then Matr*(C) is a quasivariety. More specifically, we have the following theorem:

THEOREM II.2. Let C be a standard finitely equivalential logic and let a class $K \subseteq \mathrm{Matr}^*(C)$ be strongly adequate for C. Then

$$\mathrm{Matr}^*(C) = SP_R(K) = SPP_U(K).$$

For instance, if C is the classical logic, then the class Matr*(C) consists of matrices of the form $(\underline{A}, \{1\})$, where $\underline{A}$ is a Boolean algebra and 1 is the unit of $\underline{A}$. Thus we can identify the members of Matr*(C) with Boolean algebras. Similarly for the intuitionistic logic C - the members of Matr*(C) are of the form $(\underline{A},\{1\})$, where $\underline{A}$ is a Heyting algebra and 1 is the unit of $\underline{A}$.

Let $\underline{S}^*$ be the first-order language with equality describing the class Matr(C). $\underline{S}^*$ has one unary predicate letter, say D. We identify the sentential language $\underline{S}$ in which C is defined, with the algebra of terms of $\underline{S}^*$. For each standard rule of inference $r = \gamma_1,\dots,\gamma_k/\alpha$ let (r) be the universal closure of the following formula of $\underline{S}^*$: $\bigwedge_{1\le i\le k} D(\gamma_i) \Rightarrow D(\alpha)$. (r) is a quasi-identity in $\underline{S}^*$ not involving the equality sign. The following lemma, which gives a connection between the models for a consequence operation and the standard rules which define this operation, is due to Bloom [1]:

LEMMA II.3. Let Q be a set of standard rules in a language $\underline{S}$. Then

$$\mathrm{Matr}(C_Q) = \mathrm{Mod}\{(r):\ r \in Q\}.$$

(Mod(Σ) is the class of all models of the set of sentences Σ.) The above lemma enables us to apply model-theoretic tools in the domain of sentential logics. Bloom's Lemma, Theorem I.1 and the compactness theorem yield:

COROLLARY II.4. A standard logic C is finitely based iff the quasivariety Matr(C) is finitely axiomatizable (in the usual model-theoretic sense).

If C is a standard finitely equivalential logic with a C-equivalence $\Delta(p,q) = \{\gamma_1,\dots,\gamma_n\}$ then adding to a set of quasi--identities which axiomatize Matr(C) the following quasi-identities

$$\forall p\ \forall q\ (\bigwedge_{1\le i\le n} D(\gamma_i) \Rightarrow p = q)$$

$$\forall p\ \forall q\ (p = q \Rightarrow D(\gamma_i)), \qquad \text{for } i = 1,\dots,n,$$

we obtain an axiom system for the quasivariety Matr*(C).

This remark together with Corollary II.4 gives:

COROLLARY II.5. A standard finitely equivalential logic C is finitely based iff the quasivariety Matr*(C) of factorial C-matrices is finitely axiomatizable.

§III. C-prime matrices

Given two logics C_1 and C_2 in a language $\underline{S}$ we write $C_1 < C_2$ whenever C_2 is a strengthening of C_1 and $C_1 \neq C_2$. Generalizing some earlier observations made by Alfred Tarski [15], Bloom has proved in [1] the following syntactical criterion for a logic to be not finitely based:

THEOREM III.1. A logic C in $\underline{S}$ is not finitely based iff there is a strictly increasing sequence $C_1 < C_2 < C_3 < \dots$ of standard logics such that $C = \sup C_n$.

We then face the following question - is the logic induced by a finite matrix always finitely based? Andrzej Wroński answered this question negatively constructing in [20] an appropriate six-element matrix. The next example - a five-element matrix - has been provided by Alasdair Urquhardt [17]. The latter result has been improved by Wroński [21] who showed - by way of skillful adaptation of the techniques worked out by Urquhardt - that the logic induced by the following three-element matrix

$\circ$	0	1	2
0	2	0	2
1	2	2	2
2	2	2	2

is not finitely based. (2 is its only designated value). The Wroński's result is the best possible because of the following theorem due to Wolfgang Rautenberg [13]:

THEOREM III.2. Every logic induced by a two-element matrix with one designated value is finitely based.

We shall give further positive results concerning finitely based logics. To this end we need some definitions. Let $(\underline{S},C)$ be a logic and let $\underline{A}$ be an algebra similar to the language $\underline{S}$. Following Rasiowa [12], we call a subset $\nabla \subseteq A$ a <u>deductive filter</u> relative to C, or a C-<u>filter</u>, for short, iff the matrix $(\underline{A},\nabla)$ validates the logic C, i.e., $(\underline{A},\nabla)$ belongs to Matr(C). The family of all C-filter in $\underline{A}$, denoted by

$$F_C(\underline{A}),$$

is a non-empty closure system on $\underline{A}$. If the logic C is standard, then the lattice $(F_C(\underline{A}), \subseteq)$ is algebraic, for all algebras $\underline{A}$. Clearly, $F_C(\underline{S})$ coincides with Th(C). We call a C-filter $\nabla \in F_C(\underline{A})$ <u>prime in</u> $F_C(\underline{A})$ (or C-<u>prime</u>, for short) if it is finitely meet-irreducible in $F_C(\underline{A})$, that is, for all C-filters $\nabla_1, \nabla_2 \in F_C(\underline{A})$, if $\nabla = \nabla_1 \cap \nabla_2$, then $\nabla = \nabla_1$ or $\nabla = \nabla_2$. For instance, if C is the classical logic and $\underline{A}$ is a Boolean algebra, then $F_C(\underline{A})$ coincides with the lattice of 'usual' filters and a C-filter ∇ is prime in $F_C(\underline{A})$ iff it is an ultrafilter in $\underline{A}$.

A matrix $M = (\underline{A}, D)$ is called C-<u>prime</u> whenever D is a prime filter in $F_C(\underline{A})$, and we denote by

$$\mathrm{Matr}(C)_{\mathrm{prime}}$$

the class of all C-prime matrices. It is easy to show (see Theorem II.2.2 in [3]) that a matrix M is C-prime iff it is 'almost' finitely subdirectly irreducible in the class Matr(C) in the following sense: in any representation of M as a subdirect product of a finite family of matrices from Matr(C), the projection of M onto at least one of the factors is a strict homomorphism.

We denote by

$$\mathrm{Matr}^*(C)_{\mathrm{FSI}}$$

the class of all matrices in $\mathrm{Matr}^*(C)$ that are finitely subdirectly irreducible in the class $\mathrm{Matr}^*(C)$. Thus $M \in \mathrm{Matr}^*(C)_{\mathrm{FSI}}$ iff $M \in \mathrm{Matr}^*(C)$ and in any representation of M as a subdirect product of a finite family of matrices from $\mathrm{Matr}^*(C)$, the projection of M onto at least one of the factors is an isomorphism.

The matrices in the classes $\mathrm{Matr}(C)_{\mathrm{prime}}$ and $\mathrm{Matr}^*(C)_{\mathrm{FSI}}$ respectively, play as if the role of 'building blocks' : it is not difficult to show that if C is standard, then every matrix in Matr(C) is isomorphic with a subdirect product of members of $\mathrm{Matr}(C)_{\mathrm{prime}}$ and accordingly every matrix in $\mathrm{Matr}^*(C)$ is isomorphic with a subdirect product of matrices from $\mathrm{Matr}^*(C)_{\mathrm{FSI}}$. It follows that both the classes $\mathrm{Matr}(C)_{\mathrm{prime}}$ and $\mathrm{Matr}^*(C)_{\mathrm{FSI}}$ are strongly adequate for C.

If C is an equivalential logic, it is not difficult to prove that for every matrix M,

(3.1) $$M \in \mathrm{Matr}(C)_{\mathrm{prime}} \text{ iff } M/\Theta_M \in \mathrm{Matr}^*(C)_{\mathrm{FSI}},$$

where Θ_M is the greatest strict congruence of M (cf. Theorem II.2.4 in [3]). It follows from (3.1) that a factorial matrix M is C-prime if it is finitely subdirectly irreducible in $Matr^*(C)$; in other words we have the equality

$$(3.2) \qquad Matr^*(C)_{FSI} = Matr^*(C) \cap Matr(C)_{prime} .$$

We shall make use of the following auxiliary lemma proved in [4]:

LEMMA III.3. Let C be a standard finitely equivalential logic. Then the class $Matr^*(C)_{FSI}$ is axiomatizable (finitely axiomatizable) iff the class $Matr(C)_{prime}$ is axiomatizable (finitely axiomatizable, respectively).

We shall now focus our attention on logics with disjunction connective. (We shall call them, for brevity, disjunctive logics). Our interest in disjunctive logics lies in the fact that there are very strong positive results concerning finite inferential bases for logics of this kind.

Given a logic C in a language $\underline{S}$, let us recall that a binary connective $\vee$ of $\underline{S}$ is called a disjunction (or an alternative) of C if it satisfies the condition:

$$C(X, \alpha \vee \beta) = C(X, \alpha) \cap C(X, \beta),$$

for all formulas α, $\beta \in S$ and for all $X \subseteq S$.

One can show that a binary connective $\vee$ is a disjunction of a standard logic C iff for all sentential variables p and q, $p \vee p/p$, $p \vee q/q \vee p$, $p/p \vee q$ are rules of C and for every standard nonaxiomatic rule $\gamma_1, \dots, \gamma_k/\alpha$ of C, $\gamma_1 \vee p, \dots, \gamma_k \vee p/\alpha \vee p$ is also a rule of C (see the proof of Theorem III.2 in [4]). Given an algebra $\underline{A}$ similar to $\underline{S}$, and a set $X \cup \{a\} \subseteq A$, denote by $\overline{X,a}$ the least C-filter in $\underline{A}$ including $X \cup \{a\}$. It follows from the above remark that if C is a standard disjunctive logic, then

$$(3.3) \qquad \overline{X,a} \cap \overline{X,b} = \overline{X,a \vee b} \quad ,$$

for all $a,b \in A$. (The operation in $\underline{A}$, corresponding to the connective $\vee$, is denoted by the same symbol $\vee$). Moreover, we also have a very simple description of C-prime matrices:

PROPOSITION III.4. Let C be a standard disjunctive logic. Then a matrix $M = (\underline{A},D)$ is C-prime iff $M \in Matr(C)$ and for every elements $a,b \in A$, $a \vee b \in D$ iff $a \in D$ or $b \in D$.

Proof. Let $M = (\underline{A},D)$ be in $Matr(C)_{prime}$ and let $a \vee b \in D$. We claim that $a \in D$ or $b \in D$. In view of (3.3) we have: $D = \overline{D,a \vee b} = \overline{D,a} \cap \overline{D,b}$. Since, according to the definition of $Matr(C)_{prime}$, D is finitely meet-irreducible in $F_C(\underline{A})$ we then get that $D = \overline{D,a}$ or $D = \overline{D,b}$ whence $a \in D$ or $b \in D$.

Conversely, let $M = (\underline{A},D)$ be in Matr(C) and assume that for all $a,b \in A$, $a \vee b \in D$ iff $a \in D$ or $b \in D$. Let also $D = \nabla_1 \cap \nabla_2$, for some $\nabla_1, \nabla_2 \in F_C(\underline{A})$. We must show that $D = \nabla_1$ or $D = \nabla_2$. Without loss of generality we can assume that D is properly included in both ∇_1 and ∇_2. Hence there are $a_0 \in \nabla_1$, $b_0 \in \nabla_2$ such that

$$a_0 \notin D \text{ and } b_0 \notin D. \tag{3.4}$$

Clearly $\overline{a_0} \cap \overline{b_0} \subseteq \nabla_1 \cap \nabla_2 = D$. But $\overline{a_0} \cap \overline{b_0} = \overline{a_0 \vee b_0}$. Thus $a_0 \vee b_0 \in D$. Then $a_0 \in D$ or $b_0 \in D$. This contradicts (3.4). ∎

It easily follows from formula (3.3) that if C is a standard disjunctive logic, then for every algebra $\underline{A}$, the lattice of C-filters $F_C(\underline{A})$ is distributive.

The class of logics with disjunction is quite large. It comprises e.g. all intermediate logics i.e. the axiomatic strengthening of the intuitionistic logic, in particular - the classical logic, as well as many modal logics.

In the case of disjunctive logics the above-mentioned Corollary II.4 can be reinforced to the following theorem proved in [4]:

THEOREM III.5. Let C be a standard disjunctive logic. Then C is finitely based iff the class $Matr(C)_{prime}$ is finitely axiomatizable.

Theorem III.5 and Lemma III.2 yield:

COROLLARY III.6. Let C be a standard disjunctive finitely equivalential logic. Then C is finitely based iff the class $Matr^*(C)_{FSI}$ is finitely axiomatizable.

Every intermediate logic C is assigned in a one-to-one manner the variety H(C) of Heyting algebras which validate C. It follows from Corollary III.6 that an intermediate logic C is finitely based iff the class $H(C)_{FSI}$ of finitely subdirectly irreducible members of H(C) is finitely axiomatizable. It is easy to characterize the members of $H(C)_{FSI}$: $\underline{A} \in H(C)_{FSI}$ iff $\underline{A} \in H(C)$ and for all $a,b \in A$, $a \vee b = 1$ iff $a = 1$ or $b = 1$, where 1 is the unit of $\underline{A}$.

In order to estimate how large the class $Matr(C)_{prime}$ is, we shall prove a theorem which 'localizes' C-prime matrices in the class Matr(C). But first we state the following simple lemma:

LEMMA III.7. For every logic C,

(3.5) $\qquad H_S(Matr(C)_{prime}) \subseteq Matr(C)_{prime}$

Moreover, if C is disjunctive or equivalential, then also

(3.6) $\qquad \overleftarrow{H}_S(Matr(C)_{prime}) \subseteq Matr(C)_{prime}$.

Proof. Let $M = (\underline{A},D)$, $N = (\underline{B},E)$ and let h be a strict homomorphism of M onto N. Let us assume that $M \in Matr(C)_{prime}$. We must show that E is finitely meet-irreducible in $F_C(\underline{B})$. Let us suppose that $E = E_1 \cap E_2$ for some C-filters E_1 and E_2 in $\underline{B}$. Set $D_i =_{df} \overleftarrow{h}(E_i)$ for $i = 1,2$. Since h is a strict homomorphism of $(\underline{A},D_i)$ onto $(\underline{B},E_i)$ and E_i is a C-filter we get that D_i is a C-filter in $\underline{A}$, $i = 1,2$. We claim that $D = D_1 \cap D_2$. Indeed, if $a \in D$ then $ha \in E = E_1 \cap E_2$ whence $a \in D_i = \overleftarrow{h}(E_i)$, $i = 1,2$. Hence $a \in D_1 \cap D_2$. Conversely, if $a \in D_1 \cap D_2$, then $ha \in E_1 \cap E_2 = E$. Since h is a strict homomorphism of $(\underline{A},D)$ onto $(\underline{B},E)$ we get that $a \in D$. Thus $D = D_1 \cap D_2$. Since M is prime in Matr(C) we then have: $D = D_1$ or $D = D_2$. Let us assume for simplicity that $D = D_1$. It follows that $E = E_1$. Indeed, let $b \in E_1$. Then there is $a \in D_1 = D$ such that $ha = b$. But $E = \overleftarrow{h}(D)$ whence $b \in E$. Thus $E_1 \subseteq E$. The opposite inclusion $E \subseteq E_1$ is obvious. Thus formula (3.5) has been proved.

Now, let us assume that C has a disjunction connective. Let $N = (\underline{B},E) \in Matr(C)_{prime}$. Then, according to Proposition III.4, for every $c,d \in B$, if $c \vee d \in E$ then $c \in E$ or $d \in E$. We must show that $M = (\underline{A},D) \in Matr(C)_{prime}$. So let $a \vee b \in D$. Then $ha \vee hb \in E$ whence $ha \in E$ or $hb \in E$. Since h is a strict homomorphism we thus get that $a \in D$ or $b \in D$. Thus M is C-prime.

Let C be equivalential and let us assume that h is a strict homomorphism of M onto N and that $N \in Matr(C)_{prime}$. Then, by formula

(3.1), $N/\Theta_N \in \mathrm{Matr}^*(C)_{FSI}$, where Θ_N is the greatest strict congruence of N. But N/Θ_N is isomorphic with M/Θ_M whence $M/\Theta_M \in \mathrm{Matr}^*(C)_{FSI}$. Applying again (3.1) we then have: $M \in \mathrm{Matr}(C)_{prime}$. This completes the proof of the lemma. ∎

Given a class K of matrices for a sentential language $\underline{S}$, we denote by

$$K_e$$

the class $K \cup \{M_e\}$, i.e. the class K augmented with the trivial one-element matrix $M_e = (\underline{1}, \{1\})$.

THEOREM III.8. Let C be a standard logic such that $\overleftarrow{H_S}(\mathrm{Matr}(C)_{prime}) \subseteq \mathrm{Matr}(C)_{prime}$. (In particular, let C be a standard equivalential or disjunctive logic). Let a class K of matrices be strongly adequate for C. Then

(3.7) $$\mathrm{Matr}(C)_{prime} \subseteq \overleftarrow{H_S} H_S SP_U(K_e).$$

Proof. In view of the first statement of Lemma III.7, it suffices to show that every factorial C-prime matrix belongs to $H_S SP_U(K_e)$, i.e.,

(3.8) $$\mathrm{Matr}^*(C)_{FSI} \subseteq H_S SP_U(K_e).$$

Clearly, $M_e \in \mathrm{Matr}^*(C)_{FSI}$. (Let us notice that if $C(\emptyset)$ is empty, then the trivial one-element matrix $(\underline{1}, \emptyset)$ with the empty set of designated values also belongs to $\mathrm{Matr}^*(C)_{FSI}$). Thus to prove (3.8) we must show that every factorial C-prime matrix, nonisomorphic with M_e, belongs to $H_S SP_U(K)$.

According to Theorem II.1, $\mathrm{Matr}(C) = \overleftarrow{H_S} H_S SPP_U(K)$, whence $\mathrm{Matr}^*(C) \subseteq H_S SPP_U(K)$. So let $M = (\underline{A}, D) \in \mathrm{Matr}^*(C)_{FSI}$ be non-isomorphic with M_e. Since $M \in H_S SPP_U(K)$, there is a family $M_i = (\underline{A}_i, D_i)$, $i \in I$, of members of $P_U(K)$, a submatrix $N = (\underline{B}, E)$ of $\prod_{i \in I} M_i$ and a strict homomorphism h of N onto M. In turn, since M is C-prime and $\overleftarrow{H_S}(\mathrm{Matr}(C)_{prime}) \subseteq \mathrm{Matr}(C)_{prime}$ we obtain $N \in \mathrm{Matr}(C)_{prime}$.

For $J \subseteq I$ let ∇_J be a subset of $\prod_{i \in I} A_i$ defined as follows:

$$f \in \nabla_J \quad \text{iff} \quad J \subseteq \{i \in I : f(i) \in D_i\}.$$

∇_J is a C-filter in the algebra $\prod_{i\in I}\underline{A}_i$ because the matrix $(\prod_{i\in I}, \nabla_J)$ is isomorphic with the direct product of the following C-matrices: $(\underline{A}_i, D_i)$, for all $i \in J$, and $(\underline{A}_i, A_i)$, for all $i \in I-J$. Thus $(\prod_{i\in I}\underline{A}_i, \nabla_J)$ is also in Matr(C).

Let $\xi = \{J \subseteq I: \nabla_J \cap B = E\}$. Then, as is easy to check, ξ has the following properties:

(*) $I \in \xi$, also $J \in \xi$, $J \subseteq L$ imply $L \in \xi$

(**) $K \cup L \in \xi$ implies that $K \in \xi$ or $L \in \xi$.

(**) follows directly from the fact that N is C-prime. Finally, $\emptyset \notin \xi$, since $B \neq E$. (If B were equal to E, then A = D. But M, being a factorial matrix, would be then isomorphic with the trivial matrix M_e).

Let F be a filter of subsets of I, maximal with respect to the inclusion $F \subseteq \xi$. (*) ensures that such a filter F exists. Moreover a straightforward argument, which makes use of (**) shows that F is actually an ultrafilter over I.

Let $\nabla_F =_{df} \bigcup\{\nabla_J: J \in F\}$. The set ∇_F is a C-filter in $\prod_{i\in I}\underline{A}_i$ since the lattice $F_C(\prod_{i\in I}\underline{A}_i)$ is algebraic and the family $\{\nabla_J: J \in F\}$ is upward directed. Let us also notice that $\nabla_F \cap B = E$.

We claim that the mapping ϕ,

$$\phi(f) = (f(i): i \in I)_F, \quad f \in B,$$

is a strict homomorphism of the matrix N into the ultraproduct $\prod_F M_i$. It is clear that ϕ is a homomorphism of the algebra $\underline{B}$ into the algebra $\prod_F \underline{A}_i$. Let $f \in B$. Then $f \in E$ iff $f \in \nabla_F$ iff there is a $J \in F$ such that $f \in \nabla_J$ iff $J \subseteq \{i \in I: f(i) \in D_i\}$ for some $J \in F$ iff $\{i \in I: f(i) \in D_i\} \in F$ iff $(f(i): i \in I)_F \in \prod_F D_i$. This proves that ϕ is a strict homomorphism.

Since $M_i \in P_U(K)$, for all $i \in I$, we thus get $N \in \overleftarrow{H}_S SP_U(P_U(K)) = \overleftarrow{H}_S SP_U(K)$. Consequently $M \in H_S \overleftarrow{H}_S SP_U(K)$. It is easy to show that for every class L of matrices $H_S \overleftarrow{H}_S(L) \subseteq \overleftarrow{H}_S H_S(L)$. It follows that $M \in \overleftarrow{H}_S H_S SP_U(K)$. Since M is factorial, we thus get that $M \in H_S SP_U(K)$.

This completes the proof of Theorem III.8. ∎

COROLLARY III.9. Let C be a standard logic such that the class $Matr(C)_{prime}$ is axiomatizable by means of a set of universal sentences of $\underline{S}^*$ which do not involve the equality sign. If a class $K \subseteq Matr(C)_{prime}$ is strongly adequate for C, then

$$(3.9) \qquad Matr(C)_{prime} = \overleftarrow{H_S}H_S SP_U(K_e).$$

Proof. By the assumption, $K_e \subseteq Mod(\Sigma) = Matr(C)_{prime}$ whence $P_U(K_e) \subseteq Mod(\Sigma)$. Since the sentences of Σ are universal, we get that $SP_U(K_e) \subseteq Mod(\Sigma)$.

Let us note that if $h: M \longrightarrow N$ is an onto strict homomorphism then $M \models \sigma$ iff $N \models \sigma$, for every formula σ of $\underline{S}^*$ which does not contain the equality sign. (The proof of this equivalence is by induction on the complexity of formulas of $\underline{S}^*$). Consequently, the fact that each sentence of Σ does not involve = entails that $\overleftarrow{H_S}H_S SP_U(K_e) \subseteq Mod(\Sigma)$ and that $\overleftarrow{H_S}(Matr(C)_{prime}) \subseteq Matr(C)_{prime}$. Thus $\overleftarrow{H_S}H_S SP_U(K_e) \subseteq Matr(C)_{prime}$. The last inclusion together with formula (3.7) yield (3.9). ∎

COROLLARY III.10. Let C be a standard disjuctive logic. If a class $K \subseteq Matr(C)_{prime}$ is strongly adequate for C, then

$$Matr(C)_{prime} = \overleftarrow{H_S}H_S SP_U(K_e).$$

Proof. In view of Proposition III.4 and Lemma II.3, the class $Matr(C)_{prime}$ is universally axiomatizable by a set of sentences not involving the equality sign. Then apply Corollary III.9. ∎

COROLLARY III.11. Let C be a standard disjunctive and finitely equivalential logic. If a class $K \subseteq Matr^*(C)_{FSI}$ is strongly adequate for C, then

$$Matr^*(C)_{FSI} = SP_U(K_e).$$

Proof. According to Theorem II.2, $Matr^*(C) = SPP_U(K) = SPP_U(K_e)$. Then formula (3.2) and Corollary III.10 yield $Matr^*(C)_{FSI} = Matr^*(C) \cap Matr(C)_{prime} = SPP_U(K_e) \cap \overleftarrow{H_S}H_S SP_U(K_e)$. Since every matrix in $SPP_U(K_e)$ and in $SP_U(K_e)$ is factorial, it follows that $SPP_U(K_e) \cap \overleftarrow{H_S}H_S SP_U(K_e) = SPP_U(K_e) \cap SP_U(K_e) = SP_U(K_e)$. ∎

We shall now prove a theorem, being a slight generalisation of the result obtained by Shoesmith and Smiley (cf. Theorem 19.19 in [14]):

THEOREM III.12. Let C be a standard disjunctive logic in a language $\underline{S}$ with a finite stock of connectives. Let us also assume that there is a finite family K_0 of finite matrices strongly adequate for C. Then C is finitely based.

Proof. In view of Theorem III.5, we must show that the class $Matr(C)_{prime}$ is finitely axiomatizable.

Given a matrix $M = (\underline{A}, D)$ in K_0, we define

$$K_M = \{(\underline{A}, \nabla): \nabla \text{ is a C-prime filter in } \underline{A} \text{ and } \nabla \supseteq D\}.$$

Let $K =_{df} S(\bigcup_{M \in K_0} K_M)$. Since the class $Matr(C)_{prime}$ is closed under the formation of submatrices we see that K is a finite family of finite C-prime matrices strongly adequate for C. Clearly C and K satisfy the hypothesis of Corollary III.10 and $SP_U(K) = I(K)$. Thus

$$(3.10) \qquad Matr(C)_{prime} = \overleftarrow{H_S} H_S(K_e).$$

Since $Matr(C)_{prime}$ is axiomatizable by a set of sentences which do not involve = , we easily get that for every family M_i, $i \in I$, of matrices for $\underline{S}$ and for every ultrafilter F over I,

$$(3.11) \qquad M_i \in Matr(C)_{prime} \quad \text{iff} \quad M_i/\theta_i \in Matr(C)_{prime} ,$$

where θ_i is the greatest congruence of M_i, and

$$(3.12) \qquad \prod_F M_i \in Matr(C)_{prime} \quad \text{iff} \quad \prod_F M_i/\theta_i \in Matr(C)_{prime} .$$

We shall make use of the following lemma due to Zygmunt [22]:

LEMMA III.13. Let N_i be factorial matrices for a language $\underline{S}$ with a finite stock of connectives and let $\prod_F N_i$ be their ultraproduct. If there is a strict homomorphism of $\prod_F N_i$ onto a finite matrix, then $\prod_F N_i$ is factorial (and thereby it is also finite).

We shall prove that for every family M_i, $i \in I$, of matrices and for every ultrafilter F over I, if $\prod_F M_i \in Matr(C)_{prime}$, then

$M_i \in Matr(C)_{prime}$ for some $i \in I$. This, in view of the well-known Keisler-Shelah's Theorem, will yield finite axiomatizability of the class $Matr(C)_{prime}$.

So let $M = \prod_F M_i$ be an ultraproduct of matrices and let us assume that $M \in Matr(C)_{prime}$. By (3.12),

$$(3.13) \qquad \prod_F M_i/\Theta_i \in Matr(C)_{prime} .$$

According to formula (3.10), there is then a matrix N in $H_S(K_e)$ and a strict homomorphism h of $\prod_F M_i/\Theta_i$ onto N. Clearly N is finite. Thus, by Lemma III.13, the matrix $\prod_F M_i/\Theta_i$ is factorial and finite. Then a straightforward model-theoretic argument shows $\prod_F M_i/\Theta_i$ must be isomorphic with some matrix M_i/Θ_i. Thus $M_i/\Theta_i \in Matr(C)_{prime}$ whence, by (3.11), $M_i \in Matr(C)_{prime}$.

This completes the proof of Theorem III.12. ∎

REMARK. Theorem III.12 has been also independently obtained by Zygmunt [2] who proved it applying some results from the theory of structural consequence relations.

Theorem III.12 can be strengthened in the case of disjunctive finitely equivalential logics. Let us at first recall that a class K of matrices is <u>weakly adequate</u> for a logic C if K determines the set of theses of C, i.e., $Cn_K(\emptyset) = C(\emptyset)$. One can prove (see [3]) that a class $K \subseteq Matr(C)$ is weakly adequate for C iff $Matr(C) \subseteq HSP(K)$, where H(K) is the class of all matrices homomorphic to the members of K. A logic C is called <u>tabular</u> if there is a finite matrix in Matr(C) weakly adequate for C.

THEOREM III.14. Let C be a standard disjunctive and finitely equivalential logic in a language $\underline{S}$ with a finite stock of connectives. If C is tabular, then it is finitely based.

Proof. We shall apply the following auxiliary lemma, being the matrix counterpart of the well-known Jónsson's lemma from universal algebra (see Theorem 39.6 in [7] and Theorem II.2.13 in [3]):

LEMMA III.15. Let C be a standard finitely equivalential logic in a language $\underline{S}$. Let us assume that the lattice Th(C) of theories of C is distributive and that

$$K \subseteq \mathrm{Matr}^*(C) \subseteq \mathrm{HSP}(K)$$

for some class K. Then

$$\mathrm{Matr}^*(C)_{FSI} \subseteq \mathrm{HSP}_U(K).$$

As we already remarked, for every standard disjunctive logic C, the lattice Th(C) is distributive. Moreover, it follows from tabularity of C that there is a finite matrix M in $\mathrm{Matr}^*(C)$ weakly adequate for C. Hence, in view of Lemma III.15, $\mathrm{Matr}^*(C)_{FSI} \subseteq \mathrm{HSP}_U(M) = \mathrm{HS}(M)$. Thus the class $\mathrm{Matr}^*(C)_{FSI}$ consists, up to isomorphism, of finitely many finite matrices. Since $\underline{S}$ has finitely many connectives, it follows that $\mathrm{Matr}^*(C)_{FSI}$ is described by a single sentence of $\underline{S}^*$. Then, according to Lemma III.3, the class $\mathrm{Matr}(C)_{prime}$ is finitely axiomatizable. This together with Theorem III.5 yield the thesis of the theorem. ∎

§IV. Final remarks

Some particular instances of Theorem III.14 were earlier known. For instance, Theorem III.14 is a direct generalisation of Theorem 3.4 in [18]. On the other hand, in his [16] Troelstra gives information that de Jongh has achieved an (unpublished) result that every intermediate logic with a weakly adequate finite Heyting algebra is finitely based (or, equivalently, each such logic is formalized by means of a finite set of axioms and Modus Ponens). The very elegant proof of the Jongh's result is published by McKay in [10].

A Heyting algebra $\underline{A} = (A; \cup, \cap, \Rightarrow, 1)$ is <u>strongly compact</u> iff there exists a greatest element in the partially ordered set $(A - \{1\}, \leqslant)$, where $\leqslant$ is the 'natural' ordering of $\underline{A}$, that is, $a \leqslant b$ iff $a \Rightarrow b = 1$, for all a,b. One easily verifies that if a logic C in the language of intuitionism has a strongly adequate class of strongly compact Heyting algebras, then it is disjunctive. Clearly each such a logic is a strengthening (not necessarily an axiomatic one) of the intuitionistic logic. It follows from Theorem III.12 that every logic with a strongly adequate finite family of finite strongly compact Heyting algebras is finitely based. This observation also yields the de Jongh's result. (One should make use of Lemma III.15 and the fact that for each intermediate logic C, the subdirectly irreducible matrices of $\mathrm{Matr}^*(C)$ are precisely the strongly compact Heyting algebras in $\mathrm{Matr}^*(C)$.)

But the de Jongh's result also easily follows from the celebrated Baker's theorem stating that for every congruently distributive variety V of algebras of a finite similarity type, if the class V_{FSI} of finitely subdirectly irreducible members of V is finitely axiomatizable, then V is finitely based, i.e. V is defined by means of a finite set of identities (see [8], Theorem 62.2). Baker's Theorem is also a convenient tool in investigations into finite axiomatizability of many modal logics.

Baker's Theorem directly appeals to distributivity of congruence lattices. But Theorem III.5 also, but indirectly, makes use of distributivity of the lattices $F_C(A)$ - as we already mentioned, if C is a standard disjunctive logic C in $\underline{S}$, then for every algebra $\underline{A}$ similar to $\underline{S}$ the lattice $F_C(\underline{A})$ is distributive. One may then ask whether there is a matrix counterpart of Baker's Theorem that more exhibits the lattice-theoretic structure of the $F_C(\underline{A})$'s. This question gives rise to the following definition and the following conjecture. Call a logic C in $\underline{S}$ <u>filter distributive</u> (see [5]) if for every algebra $\underline{A}$ similar to $\underline{S}$ the lattice $F_C(\underline{A})$ is distributive.

CONJECTURE IV.1. Let C be a standard filter distributive logic in a language $\underline{S}$ with a finite stock of connectives. If the class $Matr^*(C)_{prime}$ is finitely axiomatizable, then C is finitely based.

We must however remember that for every standard logic C, the class Matr(C) is merely a quasivariety. Moreover, in contradistinction to the congruence distributive varieties of algebras, the fact that C is filter distributive, cannot be, in general, expressed by means of a finite set of identities of the language $\underline{S}^*$ (see Theorem II.5 in [5], cf. also Theorem 60.2 in [8]).

The situation seems to be more promising in the case of finitely equivalential logics.

CONJECTURE IV.2. Let C be as in Conjecture IV.1 with the proviso that it is finitely equivalential. If the class $Matr^*(C)_{FSI}$ is finitely axiomatizable, then C is finitely based.

In view of Lemma III.3, the above conjecture is a particular instance of IV.1.

The class of filter distributive logics includes of course all standard disjunctive logics. But it also includes the broad class of logics with so called Uniform Deduction Theorem Schemas (UDT schemas, for short). By a UDT scheme for a logic C in $\underline{S}$ we simply understand any finite non-empty set P of sentential formulas of $\underline{S}$ in two variables such that

$$\beta \in C(X,\alpha) \quad \text{iff} \quad P(\alpha,\beta) \subseteq C(X),$$

for all $X \subseteq S$ and for all α, $\beta \in S$, One can show that every standard logic with a UDT scheme is filter distributive.

Finally, let us make mention of some positive results. The theorem below contains an analysis of axiomatic strengthenings of filter distributive logics (see Theorem V.2 in [5]):

THEOREM IV.3. Let C be a standard filter distributive logic such that the class $\mathrm{Matr}(C)_{\mathrm{prime}}$ is axiomatizable. Then every axiomatic strengthening C' of C for which the class $\mathrm{Matr}(C')_{\mathrm{prime}}$ is finitely axiomatizable results from C by enlarging it with a finite set of axioms.

The proof is technically complicated so I shall omit it. The proof employs the following observation concerning distributive closure systems, made by Dzik and Suszko in [6]:

PROPOSITION IV.4. Let C be a standard logic in a language $\underline{S}$ and let $\underline{A}$ be an algebra similar to $\underline{S}$. The lattice $F_C(\underline{A})$ is distributive iff $\overline{a} \cap \overline{b} \subseteq \nabla$ implies $a \in \nabla$ or $b \in \nabla$, for any C-prime filter ∇ in $\underline{A}$ and any $a,b \in A$.

Theorem IV.3 enables us to derive the following corollary (cf. Theorem III.14):

COROLLARY IV.5. Let C_0 be a standard equivalential logic in $\underline{S}$ satisfying the conditions:

(i) C_0 is filter distributive

(ii) the class $\mathrm{Matr}^*(C_0)_{\mathrm{FSI}}$ is axiomatizable.

If C is a tabular axiomatic strengthening of C_0, then C is obtained from C_0 by enlarging it with a finite set of axioms.

References

[1] S.L.Bloom, Some theorems on structural consequence operations, Studia Logica 34(1975),1-9

[2] J.Czelakowski, Reduced products of logical matrices, Studia Logica 39(1980), 19-43

[3] ——— , Equivalential logics, Part I, Studia Logica, Vol.40, No.3(1981), 227-236; Part II, ibidem, Vol.40, No.4(1981), 355-372

[4] ——— , Matrices, primitive satisfaction and finitely based logics, Studia Logica, Vol.42, No.1(1983), 89-104

[5] ——— , Filter distributive logics, to appear in Studia Logica, Vol.43, No.4(1984)

[6] W.Dzik and R.Suszko, On distributivity of closure systems, Bulletin of the Section of Logic (IFiS PAN), Vol.6(1977), 64-66

[7] G.Grätzer, Universal Algebra (second edition), Van Nostrand, Princeton, New Jersey, 1978

[8] B.Jónsson, Appendix 3 to [7] - Congruence varieties

[9] J.Łoś and R.Suszko, Remarks on sentential logics, Indagationes Mathematicae 20(1958), 177-183

[10] C.G.McKay, On finite logics, Indagationes Mathematicae 29(1967), 363-365

[11] T.Prucnal and A.Wroński, An algebraic characterization of the notion of structural completeness, Bulletin of the Section of Logic (IFiS PAN), Vol.3(1974), 30-33

[12] H.Rasiowa, An Algebraic Approach to Non-Classical Logics, PWN - North-Holland,Warszawa-Amsterdam 1974

[13] W.Rautenberg, 2-element matrices, Studia Logica, Vol.40, No.4 (1981), 315-353

[14] D.J.Shoesmith and T.J.Smiley, Multiple-Conclusion Logic, Cambridge University Press, Cambridge 1978

[15] A.Tarski, Fundamental concepts of the methodology of the deductive sciences, in: Logic, Semantics and Metamathematics, Oxford 1956

[16] A.S.Troelstra, On intermediate propositional logics, Indagationes Mathematicae 27(1965), 141-152

[17] A.Urquhardt, A finite matrix whose consequence relation is not finitely axiomatizable, Reports on Mathematical Logic 9(1977), 71-73

[18] P.Wojtylak, Strongly finite logics: finite axiomatizability and the problem of supremum, Bulletin of the Section of Logic (IFiS PAN), vol.8(1979), 99-111

[19] R.Wójcicki, Matrix approach in sentential calculi, Studia Logica 32(1973), 7-37

[20] A.Wroński, On finitely based congruence operations, Studia Logica 35(1976), 453-458

[21] A.Wroński, A three-element matrix whose consequence operation is not finitely based, Bulletin of the Section of Logic (IFiS PAN), Vol.8(1978), 68-71

[22] J.Zygmunt, On structural entailment relation, to appear in Acta Universitatis Wratislaviensis

MONADICITY IN TOPOLOGICAL PSEUDO-BOOLEAN ALGEBRAS

Josep M. Font
Faculty of Mathematics, University of Barcelona
Gran Via 585, Barcelona 7, Spain

0. INTRODUCTION

It is generally accepted that in intuitionistic modal logic the two modal operators cannot be completely dual in the sense of the two classical laws: $L \leftrightarrow \neg M \neg$ and $M \leftrightarrow \neg L \neg$. Most authors have chosen to work considering both operators as primitive and independent, linking them with other weaker relations; this is the case of Prior, Bull, Ono, Fischer-Servi, Sotirov, and others. In some cases, such as in part of [Bul] and [O], they avoid having M. In this paper we begin an algebraic study of the application of Gödel's proposal (that is, to have a primitive L and define $M \leftrightarrow \neg L \neg$) to an intuitionistic base. It is worth noting that the remaining alternative, having M primitive and defining $L \leftrightarrow \neg M \neg$, is not viable, as example 5.10 shows.

The work has been done and written in the algebraic side of the subject, and in order to avoid repetitions we will not refer, outside of this section, to the equivalent logical formulations of several results; some of them will be dealt with in another paper. Here we start from a system of (propositional) intuitionistic modal logic analogous to S4, whose algebraic models are topological pseudo-Boolean algebras (tpBa); these are defined as in [O] using only the interior operator I, in the Gödelian style. A deductive and implicational study of tpBa and of the logical system, whithout a mention to M, has been published in [Fo].

In addition to the interior operator I there are two closure operators in tpBas: the well known operator of double negation $\neg\neg$ and the topological closure $\delta = \neg I \neg$ (dual to I and thus corresponding to M). In section 1 we give their general properties and define three types of "regular" elements which naturally appear in this algebraic and topological context. One of the central topics of the paper is a comparative study of those operators and of the corresponding sets of fixed points (open, H-regular and closed elements, resp.: B, Reg_H,T) and their relations to "dense" and the remaining "regular" elements. In some cases we find the algebraic structure of some of these subsets, taking into account new operations with additional logical significance.

The other central topic of the paper is the analysis of several conditions (axioms or inference rules) which can be added to the S4 system to obtain S5, that is, conditions which turn a topological Boolean algebra into a monadic one. In such structures they are all equivalent, but it is not so in topological pseudo-Boolean algebras, due to the peculiar features of intuitionistic negation; hence the interest of its study (which was begun by H. Ono concerning some conditions where M does not appear).

We examine fourteen different conditions involving L and/or M, that is, I and/or δ. Some of them come from classical modal logic, such as the law of reduction of modalities, Becker's rule and axioms, or the M" axiom of von Wright. Some others originate in mainly algebraic works, as are those of Halmos, Davis, Monteiro, Bull and Beth-Nieland mentioned here. We determine all the equivalences and implications that hold between these conditions and give the adequate counterexemples. Among them we find four main groups of equivalent conditions with some additional interest, and we use them to define four subclasses of tpBas, which turn out to be characterized also by natural algebraic conditions: weakly monadic (T closed under $\neg$), monadic (B closed under $\neg$), strongly monadic (B closed under $\rightarrow$) and semisimple ($B \subseteq T$). These special tpBas share with monadic Boolean algebras several of their algebraic properties, and we can compare the relative strength of some classical properties with a "monadic" character. All this analysis is contained in sections 2,3 and 4 of the paper. In section 5 we have gathered all the counterexamples we use in many different places throughout the paper.

The four logical systems that would correspond to the subclasses of tpBa here defined are all of type S5 in the sense of [Bu2], although the last one is not intuitionistically plausible. We have made no attempt to single out one of them as a "true" analogue of S5, but we rather study the properties that establish differences between them, thus finding they have an increasing "degree of monadicity". On the other hand, H. Ono has proven in [O] that there is an infinite number of systems of intuitionistic modal logics analogous to S5, and G. Fisher-Servi has worked on this subject with remarkable deepness in [Fi1], [Fi2] and [Fi3], among other papers.

We recall some of the definitions and notational conventions introduced in [Fo] and that will be used here. A <u>topological pseudo-Boolean algebra (tpBa)</u> $\langle A, I, \neg, \wedge, \vee, \rightarrow \rangle$ is a pseudo-Boolean algebra $\langle A, \neg, \wedge, \vee, \rightarrow \rangle$ where 1 denotes the maximum and 0 the minimum, together with a lattice -(or multiplicative)- interior operator I on A. The <u>open</u> elements are those of $B = \{a \in A: Ia = a\}$, and the <u>deductive systems</u> of A are the sets in $\mathcal{D} = \{D \subseteq A: 1 \in D, \text{ if } a, a \rightarrow b \in D \text{ then } Ia, b \in D\}$. If for each $D \in \mathcal{D}$ we define $\equiv_D$ by $a \equiv_D b$ iff $a \rightarrow b \in D$ and $b \rightarrow a \in D$ then the correspondence $D \mapsto \equiv_D$ is a lattice isomorphism between $\mathcal{D}$ and the congruence lattice of A. We denote by $\underline{D}$ the consequence operator associated with $\mathcal{D}$ (where $\underline{D}(X) = \bigcap \{D \in \mathcal{D}: D \supseteq X\}$ for all $X \subseteq A$) and $\underline{L} = \langle A, \underline{D} \rangle$ is the associated <u>abstract logic</u>. This abstract logic has the following properties: The <u>Adjunction Principle</u> $\underline{D}(a,b) = \underline{D}(a \wedge b)$, the <u>Strong Disjunction Principle</u> $\underline{D}(X,a) \cap \underline{D}(X,b) = \underline{D}(X, a \overset{*}{\vee} b)$ where $a \overset{*}{\vee} b =$

$= Ia \vee Ib$, the Deduction Principle $b \in \underline{D}(X,a)$ iff $a*b \in \underline{D}(X)$ and the Pseudo-Reductio ad Absurdum Principle $\neg *a \in \underline{D}(X)$ iff $\underline{D}(X,a) = A$, where $\neg *a = a*0$, and $*$ is any of the following implication operations: The weak implication $a \twoheadrightarrow b = Ia \rightarrow b$, the intuitionist implication $a \Rightarrow b = I(Ia \rightarrow Ib)$, and the strange implication $a \rightsquigarrow b = Ia \rightarrow Ib$. Hence the logic $\underline{L}$ has an intuitionistic character when we take into account the preceding connectives, and this is reinforced by the following construction: with each $D \in \mathcal{D}$ we associate the relation $\sim_D$ defined as $a \sim_D b$ iff $\underline{D}(D,a) = \underline{D}(D,b)$ or iff $a*b \in D$ and $b*a \in D$. Then $\sim_D$ is a logical congruence of $\underline{L}$ in the sense of [B-S] and the quotient $A/\sim_D$ is a pseudo-Boolean algebra with respect to the operations there induced by $\neg *, \wedge, \overset{*}{\vee}$ and $*$.

The purely intuitionistic character of these structures is shown in results as the following: for all $D \subseteq A$, $D \in \mathcal{D}$ iff $1 \in D$ and D is closed by Modus Ponens with respect to $*$ (that is, if $a, a*b \in D$ then $b \in D$). We can extend the analogy with intuitionistic structures by introducing two kinds of elements related to implication and negation as in pseudo-Boolean algebra: the $*$-dense elements $D_* = \{a \in A: \neg *a = 0\}$ and the $*$-Peircean elements $P_* = \{a \in A: a = ((b*c)*b)*b$ for some $b,c \in A\}$. We have the usual characterization of the (maximal) radical of A: $R(A) = \underline{D}(P_*)$ and the following relations: $P_{\Rightarrow} \subsetneq P_{\rightsquigarrow} \subsetneq P_{\twoheadrightarrow} = R(A) = D_{\Rightarrow} \supsetneq D_{\rightsquigarrow} = D_{\twoheadrightarrow}$, and $D_{\twoheadrightarrow} \cap B \subseteq P_{\Rightarrow} = I(P_{\twoheadrightarrow}) = P_{\rightsquigarrow} \cap B = R(A) \cap B$. Moreover for each $a \in A$ we have that $a \in P_{\Rightarrow}$ iff $a = ((a \Rightarrow 0) \Rightarrow a) \Rightarrow a$. In the present paper these results will be strengthened for some of the subclasses of tpBas studied and more properties of these special elements and of other related concepts will be obtained.

1. NEGATION AND POSSIBILITY IN TPBAS

In all pseudo-Boolean algebras we have the closure operator $x \longmapsto \neg\neg x$ whose properties are well-known; see for instance [M-T] or [R-S]. In tpBas there is another closure operator which is defined as follows:

1.1. Definition. In every tpBa A the closure operator associated with the interior operator I is $\delta a = \neg I \neg a$ for all $a \in A$. The closed elements of A are those of $T = \{a \in A: a = \delta a\}$.

1.2. Proposition. In all tpBa A the following hold:

(1) $Ia \leq a \leq \neg\neg a \leq \delta a$ for all $a \in A$,

(2) $\delta 0 = 0$, $a \leq \delta a = \delta^2 a$ for all $a \in A$, and if $a \leq b$ then $\delta a \leq \delta b$ for all $a, b \in A$, and

(3) T is $\wedge$-closed, contains 0 and 1, and for all $a \in A$, $\delta a = \min\{t \in T: a \leq t\}$.

Proofs: All trivial. ■

We observe that (2) states that δ is an order-closure satisfying $\delta 0=0$ and, from (3), $\delta(\delta a \wedge \delta b) = \delta a \wedge \delta b$ for all $a,b \in A$. While I is a lattice-interior operator, δ is not a lattice-closure operator because it does not necessarily satisfy $\delta(a \vee b) = \delta a \vee \delta b$ nor even $\delta(\delta a \vee \delta b) = \delta a \vee \delta b$, as example 5.7 shows. We now give some relations and properties satisfied by I,δ and $\neg$ which we shall use from now on without mentioning them.

1.3. Proposition. In every tpBa A and for all $a \in A$:

(1) $\delta a = \delta \neg\neg a = \neg\neg \delta a$,

(2) if $a \in T$ then $a = \neg\neg a$,

(3) if $\neg a = 0$ then $\delta a = 1$,

(4) $\delta I \neg a = \neg I \delta a$,

(5) $I \neg a \leq \neg \delta a \leq \neg a \leq \delta \neg a \leq \neg I a$, and

(6) $\neg I a \in T$.

Proof: All reduce to easy computations dealing only with the definitions of δ and T and with elementary properties of negation in pseudo-Boolean algebras.■

Now we ask whether the negation induces any relation between open and closed elements. As is well-known, in topological Boolean algebras, as in ordinary topological spaces, there is a perfect duality between them, and there are four valid implications: (1) if a is open then $\neg a$ is closed; (2) if $\neg a$ is closed then a is open; (3) if a is closed then $\neg a$ is open; and (4) if $\neg a$ is open then a is closed. In our case the results are more limited:

1.4. Theorem. If a is an open element of a tpBa then $\neg a$ is closed.

Proof: From $a \leq \neg\neg a$ we deduce $a = Ia \leq I \neg\neg a$ and then $\delta \neg a = \neg I \neg\neg a \leq \neg Ia = \neg a$, so $\delta \neg a = \neg a$.■

We have proven that implication (1) is always true. The remaining three are not true in general, as we see in example 5.3, although (2) and (3) can hold in some cases, as in 5.2. This is not the case of (4), which turns out to be characteristic for topological Boolean algebras:

1.5. Theorem. A tpBa A is a topological Boolean algebra iff for all $a \in A$, if $\neg a$ is open then a is closed.

Proof: If A is Boolean there is nothing to prove. If it is not, then there is an $a \in A$, $a \neq 1$, such that $\neg a = 0$. This implies that $\neg a$ is open and that a is not closed since by 1.3(3), $\delta a = 1 \neq a$.■

Other most direct relations between open and closed elements, namely $B \subseteq T$ and $T \subseteq B$, will be shown to be equivalent to the definitions of some subclases of tpBa in sections 3 and 4. Another kind of relations between B and T are given by the so-called "connecting conditions" of [Fi3]; here we have one of them:

1.6. Proposition. In every tpBa A the following hold:

(1) $\delta(a \rightarrow b) \leq Ia \rightarrow \delta b$ for all $a, b \in A$, and

(2) $B \rightarrow T \subseteq T$.

Proof: We have $a \rightarrow b \leq \neg b \rightarrow \neg a = (b \rightarrow 0) \rightarrow (a \rightarrow 0)$ and therefore $b \rightarrow 0 \leq (a \rightarrow b) \rightarrow (a \rightarrow 0) = a \rightarrow ((a \rightarrow b) \rightarrow 0)$ and so $a \leq (b \rightarrow 0) \rightarrow ((a \rightarrow b) \rightarrow 0)$, that is, $a \leq \neg b \rightarrow \neg(a \rightarrow b)$; now $Ia \leq I(\neg b \rightarrow \neg(a \rightarrow b)) \leq I\neg b \rightarrow I\neg(a \rightarrow b) \leq \neg I \neg(a \rightarrow b) \rightarrow \neg I \neg b = \delta(a \rightarrow b) \rightarrow \delta b$ which is equivalent to (1). Now (2) is a trivial consequence of (1). ■

The remaining condition $\delta a \rightarrow Ib \leq I(a \rightarrow b)$, which is equivalent to $T \rightarrow B \subseteq B$, will be dealt with at the end of section 3.

The rest of this section is devoted ot the introduction of the concept of regularity in tpBas. The concept of regular element comes from topology and was introduced in [M-T] by using the negation of pseudo-Boolean algebras. In a tpBa A, we denote by $Reg_H(A) = \{a \in A : a = \neg\neg a\}$ the set of H-regular elements, that is, the set of regular elements of the underlying pseudo-Boolean algebra (or Heyting algebra). Now 1.3 (2) is read $T \subseteq Reg_H(A)$; it is not possible to strengthen this relation: in 5.7 the inclusion is proper, and there we see that even in monadic Boolean algebras it can be so. In topological pseudo-Boolean algebras there are three operations $\neg *$ which have the behaviour of logical negations of the intuitionistic type; but we also have a topological interior operator and an associated closure operator, and so it can have some interest trying to write down the original topological ideas. We think that finding some coincidences between the two formulations is not merely casual.

1.7. Proposition. In every tpBa A and for all $a \in A$ the following conditions are equivalent:

(1) $a = \delta Ia$,

(2) $a = \neg_{\rightarrowtail} \neg_{\rightarrowtail} a$, and

(3) $a = \neg_{\leadsto} \neg_{\leadsto} a$.

Proof: From the definition of $\neg_{\rightarrowtail}$ and $\neg_{\leadsto}$ we see that $\neg_{\rightarrowtail} a = \neg_{\leadsto} a = \neg Ia$ and then $\neg_{\rightarrowtail} \neg_{\rightarrowtail} a = \neg_{\leadsto} \neg_{\leadsto} a = \neg I \neg Ia = \delta Ia$. ■

1.8. Proposition. In every tpBa A and for all $a \in A$ the following conditions are equivalent:

(1) $a = I\delta a$, and

(2) $a = \neg_{\Rightarrow} \neg_{\Rightarrow} a$.

Proof: From the definition of $\neg_{\Rightarrow}$ we see that $\neg_{\Rightarrow} a = I\neg Ia$ and then $\neg_{\Rightarrow}\neg_{\Rightarrow} a = I\neg I\neg Ia = I\delta Ia$, but $a = I\delta a$ implies a is open and so $a = I\delta Ia = \neg_{\Rightarrow}\neg_{\Rightarrow} a$.

Conversely $a = \neg\Rightarrow\neg\Rightarrow a = I\delta Ia$ also implies a is open and therefore $a = I\delta a$. ■

In this situation we are nearly forced to give the two following definitions.

<u>1.9. Definitions.</u> In a tpBa A the elements $a \in A$ such that $a = \delta Ia$ will be called <u>regular</u>, and those satisfying $a = I\delta a$ will be called <u>I-regular</u>. We denote by $Reg(A)$ and $Reg_I(A)$ the sets of regular and I-regular elements, respectively.

The most immediate properties of regular elements are the following:

<u>1.10. Proposition.</u> In every tpBa A we have:

(1) $\{0,1\} \subseteq Reg(A) \subseteq T \subseteq Reg_H(A)$,

(2) $\{0,1\} \subseteq T \cap B = Reg(A) \cap B \subseteq Reg_I(A) \subseteq B$, and

(3) $Reg(A) \cap R(A) = Reg(A) \cap D_{\leftrightarrow} = \{1\}$.

(4) If $a \in Reg(A)$ then $Ia \in Reg_I(A)$; if $a \in Reg_I(A)$ then $\delta a \in Reg(A)$; and these two correspondences are inverse.

<u>Proof.</u> (1), (2) and (4) are direct consequences of the definitions. For (3), we have $Reg(A) \cap D_{\leftrightarrow} \subseteq Reg(A) \cap R(A)$ and if $a \in Reg(A) \cap R(A)$ then, taking into account that $R(A) = D_{\Rightarrow}$, we have $a = \neg\leftrightarrow\neg\leftrightarrow a = (a \leftrightarrow 0) \leftrightarrow 0 = (a \Rightarrow 0) \leftrightarrow 0 = 0 \leftrightarrow 0 = 1$. ■

Examples 5.2 and 5.3 show us that the inclusions of 1.10 are not equalities in general. Later we will complete the analysis of the concept(s) of regularity and some of this properties will be partially improved.

<u>2. WEAKLY MONADIC TPBAS</u>

The concept of monadic Boolean algebras was invented by P.R. Halmos to set out an algebraic description of the monadic predicate calculus. He defined them in [H] with the specific axiom $\delta(a \wedge \delta b) = \delta a \wedge \delta b$, and showed that this was equivalent to the addition of the condition $\delta\neg\delta = \neg\delta$ to a topological Boolean algebra. Independently and at nearly the same time, C. Davis defined in [D] the class of "S5 operators" on a Boolean algebra with the condition "if $a \wedge b = 0$ then $\delta a \wedge \delta b = 0$", and showed that it was also equivalent to $\delta\neg\delta = \neg\delta$. The motivation for Davis' work was modal logic, and in fact this last condition corresponds to the specific axiom of von Wright's system M" presented in [vW]; as it is well-known, B. Sobociński showed the equivalence between M" and S5. Now we see the equivalence of these conditions on a topological pseudo-Boolean algebra.

<u>2.1. Theorem.</u> In every tpBa A the following conditions are equivalent:

(1) $\delta\neg\delta a = \neg\delta a$ for all $a \in A$ (that is, T is closed under negation),

(2) if $a \wedge \delta b = 0$ then $\delta a \wedge \delta b = 0$ for all $a,b \in A$, and

(3) $\delta(a \wedge \delta b) = \delta a \wedge \delta b$ for all $a,b \in A$.

Proof. (1)$\Rightarrow$(2): $a \wedge \delta b = 0$ is equivalent to $a \leq \neg\delta b$ and from this and (1) we have $\delta a \leq \delta\neg\delta b = \neg\delta b$, which is equivalent to $\delta a \wedge \delta b = 0$.

(2)$\Rightarrow$(1) because we always have $\neg\delta a \leq \delta\neg\delta a$, and since $\neg\delta a \wedge \delta a = 0$, by (2) we have $\delta\neg\delta a \wedge \delta a = 0$, thus establishing $\delta\neg\delta a \leq \neg\delta a$.

(2)$\Rightarrow$(3): $a \leq \delta a$ and so $a \wedge \delta b \leq \delta a \wedge \delta b \in T$ as T is $\wedge$-closed. By 1.2(3), to show (3) it suffices to show that for any $t \in T$, if $a \wedge \delta b \leq t$ then $\delta a \wedge \delta b \leq t$. But $a \wedge \delta b \leq t$ is equivalent to $\delta b \leq a \rightarrow t \leq \neg t \rightarrow \neg a$ which in turn is equivalent to $\delta b \wedge \neg t \leq \neg a$ and hence to $\delta b \wedge \neg t \wedge a = 0$. Now $\delta b \wedge \neg t \in T$ and so we have $\delta(\delta b \wedge \neg t) \wedge \delta a = 0$, and applying (2) we obtain $\delta b \wedge \neg t \wedge \delta a = \delta(\delta b \wedge \neg t) \wedge \delta a = 0$ from where we can infer $\delta a \wedge \delta b \leq \neg\neg t = t$ by 1.3(2). Thus we have proved that $\delta(a \wedge \delta b) = \delta a \wedge \delta b$.

(3)$\Rightarrow$(2) is trivial. ■

The preceding result justifies the following

2.2. Definition. A tpBa is called weakly monadic iff it satisfies any of the conditions in Theorem 2.1.

We give a list of several useful rules for weakly monadic tpBas and some elementary properties.

2.3. Proposition. In every weakly monadic tpBa A the following hold:

(1) $\delta I \neg a = \neg\delta a = \neg I\delta a$ for all $a \in A$,

(2) $\neg Ia = \neg\delta Ia$ for all $a \in A$,

(3) $\neg\neg Ia = \delta Ia$ for all $a \in A$,

(4) if $a \in B$ then $\delta a = \neg\neg a$, and

(5) if $a \in T$ then $\neg a = \neg Ia$ and $a = \neg\neg Ia$.

Proofs: They all are straightforward computations making use of 2.1(1); for (5) recall that $T \subseteq Reg_H(A)$. ■

2.4. Theorem. In every tpBa the following conditions are equivalent:

(1) A is weakly monadic,

(2) $\neg\neg Ia \in T$ for all $a \in A$, and

(3) $a = \neg\neg Ia$ for all $a \in T$.

Proof. In 2.3(3) we have seen that (1) implies (2), and in 2.3(5) we show that (1) implies (3). If we assume (2) and apply it to $\neg a$ we have $\neg\delta a = \neg\neg I\neg a \in T$ and we see that T is closed under negation, which is equivalent to (1). Similarly if we assume (3) and apply it to δa we find $\delta a = \neg\neg I\delta a = \neg\neg I\neg I\neg a =$

$= \neg\neg I \neg\neg\neg I \neg a = \neg\delta\neg\delta a$ and by negation $\neg\delta a = \neg\neg\delta\neg\delta a = \delta\neg\delta a$ because $T \subsetneq Reg_H(A)$. Therefore we have (1) again. ■

<u>2.5. Proposition</u>. In every weakly monadic tpBa A and for every $a \in A$ we have:

(1) $\neg^{\Rightarrow} a=0$ iff $\neg^{\leftrightarrow} a=0$, that is, $I\neg Ia = 0$ iff $\neg Ia = 0$, and

(2) if $a \in T$ and $Ia = 0$ then $a=0$.

<u>Proof</u>. For (1) there is nothing to prove in one direction; and if $I\neg Ia = 0$ then using 2.3(5) for $\neg Ia \in T$ we obtain $\neg Ia = \neg\neg I\neg Ia = \neg\neg 0 = 0$. (2) is also a direct consequence of 2.3 (5). ■

<u>2.6. Corollary</u>. In every weakly monadic tpBa A, $R(A) = D_{\Rightarrow} = D_{\leftrightarrow} = D_{\rightsquigarrow}$ and if we put $R_H(A) = \{a \in A: \neg a=0\}$ (the maximal radical of the pseudo-Boolean algebra under A) then $R(A) \subseteq R_H(A)$. ■

It is worth noting that, according to 2.6, in weakly monadic tpBas there is only one kind of <u>dense</u> elements, $D_* = \{a \in A: \neg^* a = 0\}$ because of the coincidence of $D_{\Rightarrow}$, $D_{\leftrightarrow}$ and $D_{\rightsquigarrow}$ (example 5.1 shows that this is not true in general). Therefore, there is a unified characterization of the radical in terms of dense elements: $R(A) = D_*$. This can be logically interpreted as stating the equivalence of the two kinds of "almost true" sentences: those represented by the dense elements D_*, as their logical negation $\neg^*$ is the false 0, and those represented by the elements in the radical $R(A)$, as they belong to every complete consistent theory (the maximal deductive systems).

There is no coincidence among the three types of Peircean elements, as exemple 5.3, where the inclusions $P_{\Rightarrow} \subsetneq P_{\rightsquigarrow} \subsetneq P_{\leftrightarrow}$ are proper, shows. In the same example we see that it is not possible to improve, for weakly monadic tpBas, the results obtained in section 1 concerning the relations between open and closed elements via negation. Concerning 2.5(1) we announce that it is not true in every weakly monadic tpBa A that $\neg^{\Rightarrow} a = \neg^{\leftrightarrow} a = \neg^{\rightsquigarrow} a$ for all $a \in A$: in 3.5 we will show that this fact characterizes a proper subclass of those algebras.

It is well known that the set of H-regular elements in a pseudo-Boolean algebra A is a Boolean algebra with respect to $\neg, \wedge, \overset{+}{\vee}$ and $\rightarrow$, where the join is $a \overset{+}{\vee} b = \neg\neg(a \vee b)$, and also that it is isomorphic to the ordinary quotient of A by its radical. In tpBas we have two kinds of regular elements and a radical linked with the dense elements. We shall obtain for these concepts, in weakly monadic tpBas, several results partially similar to the classical ones.

<u>2.7. Theorem</u>. In every weakly monadic tpBa A, we have that $T = Reg(A)$ and this is a Boolean algebra with respect to $\neg, \wedge, \overset{+}{\vee}, \rightarrow$.

<u>Proof</u>. According to 1.10(1) we always have $Reg(A) \subseteq T$; and if $a \in T$ then using

2.3(1) twice we obtain $\delta Ia = \delta I\delta a = \delta I\neg I\neg a = \neg\delta I\neg a = \neg\neg I\delta a = \neg\neg\delta a = \neg\neg a =$ $= a$, and so $a \in Reg(A)$. On the other hand $T \subseteq Reg_H(A)$ and we know that $Reg_H(A)$ is a Boolean algebra with respect to the desired operations; hence we only need to prove that T is closed with respect to them. T is always closed under $\wedge$, and if A is weakly monadic then T is closed under $\neg$. Moreover we have $a \overset{+}{\vee} b =$ $= \neg\neg(a \vee b) = \neg(\neg a \wedge \neg b)$, therefore T is closed under $\overset{+}{\vee}$, and since in $Reg_H(A)$ $a \to b = \neg a \overset{+}{\vee} b$ then T is also closed under $\to$. As a result $T = Reg(A)$ is a Boolean algebra with $\neg, \wedge, \overset{+}{\vee}$ and $\to$. ∎

Note that in weakly monadic tpBa the set T is closed under $\neg$ and $\to$. However it is not a subalgebra of A, since it does not need to be closed under $\vee$ (see 5.7); this condition will later on play a role. The point whether B is or is not closed under $\neg$ and $\to$ will also play an important role in the next section, but we can say nothing about this now. On the other hand, in 2.7 we have turned one of the inclusions of 1.10 into an equality; we cannot do the same for the reminaning ones, for all weakly monadic tpBas, as the examples 5.3 (for the first inclusion of 1.10(1) and the ones of (2)) and 5.2 (for the rest) show. Therefore $B \cap Reg(A) \subseteq$ $\subseteq Reg_I(A)$ and the inclusion can be proper; thus we still have two different kinds of regular elements. This is not an impediment to show that one of them has all the properties cited above. We first need a general result:

<u>2.8. Lemma</u>. In every tpBa A we have that $a \sim_{R(A)} \neg\neg Ia$ for all $a \in A$.

<u>Proof</u>: On one side we have $Ia \leq I\neg\neg a$ and from this $a \leftrightarrow \neg\neg Ia = Ia \leftrightarrow \neg\neg Ia =$ $= 1 \in R(A)$. On the other side, from $0 \leq a$ we deduce $(a \leftrightarrow 0) \to 0 \leq (a \leftrightarrow 0) \leftrightarrow 0 \leq$ $\leq (a \leftrightarrow 0) \leftrightarrow a$ and so $\neg\neg Ia \leftrightarrow a = I\neg\neg Ia \to a = ((a \leftrightarrow 0) \to 0) \leftrightarrow a \geq$ $\geq ((a \leftrightarrow 0) \leftrightarrow a) \leftrightarrow a \in P_{\leftrightarrow} = R(A)$ and, as $R(A)$ is a deductive system and hence an order filter, we find that $\neg\neg Ia \to a \in R(A)$. Thus $a \sim_{R(A)} \neg\neg Ia$. ∎

It is easy to show (see [Fo]) that in every tpBa A and for every $D \in \mathcal{D}$, the quotient $A/\sim_D$ is a semisimple pseudo-Boolean algebra (i.e., a Boolean algebra) if and only if $D \supseteq R(A)$; this is the case for $R(A)$ itself, therefore $A/\sim_{R(A)}$ is a Boolean algebra with respect to the operations $\bar{\neg}, \bar{\wedge}, \bar{\vee}, \bar{\to}$ induced in the quotient by $\neg, \wedge, \overset{*}{\vee}, {}^*$ respectively. Then we have:

<u>2.9. Theorem</u>. In every weakly monadic tpBa A the Boolean algebras T and $A/\sim_{R(A)}$ are isomorphic.

<u>Proof</u>: We define the following mapping from A to T: $h(a) = \delta Ia$ for all $a \in A$. Recall that here $\delta Ia = \neg\leftrightarrow\neg\leftrightarrow a = \neg\neg Ia$. This mapping is onto, because $T = Reg(A)$, and $Shell(h) = h^{-1}(\{1\}) = \{a \in A: \delta Ia = 1\} = \{a \in A: \neg\leftrightarrow\neg\leftrightarrow a = 1\} =$ $= \{a \in A: \neg\leftrightarrow a = 0\} = D_{\leftrightarrow} = R(A)$.

We first show that for all $a, b \in A$, $h(a \leftrightarrow b) \leq h(a) \leftrightarrow h(b)$: using 1.6(1) we have $h(a \leftrightarrow b) = \delta I(Ia \to b) \leq \delta(Ia \to Ib) \leq Ia \to \delta Ib$, and so $\delta I(a \leftrightarrow b) \wedge Ia \leq \delta Ib$ from where, applying 2.1(3), we obtain $\delta I(a \leftrightarrow b) \wedge I\delta Ia \leq \delta I(a \leftrightarrow b) \wedge \delta Ia = \delta(\delta I(a \leftrightarrow b) \wedge Ia) \leq \delta Ib$ and therefore $\delta I(a \leftrightarrow b) \leq I\delta Ia \to \delta Ib = \delta Ia \leftrightarrow Ib$, that is, $h(a \leftrightarrow b) \leq h(a) \leftrightarrow h(b)$ as we desired.

We can now show that for all $a, b \in A$, $h(a) = h(b)$ iff $a \sim_{R(A)} b$: if $h(a) = h(b)$ we have $\delta Ia = \delta Ib$ and by 2.8 we have $a \sim_{R(A)} \delta Ia$ and $b \sim_{R(A)} \delta Ib$, so $a \sim_{R(A)} b$. Conversely if $a \sim_{R(A)} b$ we have that $a \leftrightarrow b \in R(A) = \text{Shell } h$ and $b \leftrightarrow a \in R(A) = \text{Shell } h$, and then $1 = h(a \leftrightarrow b) \leq h(a) \leftrightarrow h(b)$ and $1 \leq h(b) \leftrightarrow h(a)$; therefore $h(a) \leftrightarrow h(b) = h(b) \leftrightarrow h(a) = 1$ which implies $Ih(a) = Ih(b)$. But $h(a), h(b) \in T = \text{Reg}(A)$, so $h(a) = \delta Ih(a)$ and $h(b) = \delta Ih(b)$, and finally $h(a) = h(b)$.

It follows from what has been done until now that the induced mapping $\bar{h}$ maps $A/\sim_{R(A)}$ onto T and is indeed a bijection between them. We must only prove that h is a morphism, and then it will be the required isomorphism. It suffices to show it for the negation $\neg$ and the meet $\wedge$: $\bar{h}(\neg \bar{a}) = \bar{h}(\overline{\neg \leftrightarrow a}) = \bar{h}(\overline{\neg Ia}) = \delta I \neg Ia = \neg \delta Ia = \neg \bar{h}(\bar{a})$ using 2.3(1); if we recall that $\delta Ia \in T$ for all $a \in A$ and that $T = \text{Reg}(A)$ is $\wedge$-closed, then we have that $\bar{h}(\bar{a} \wedge \bar{b}) = \bar{h}(\overline{\delta Ia} \wedge \overline{\delta Ib}) = \bar{h}(\overline{\delta Ia \wedge \delta Ib}) = \delta I(\delta Ia \wedge \delta Ib) = \delta Ia \wedge \delta Ib = \bar{h}(\bar{a}) \wedge \bar{h}(\bar{b})$. ∎

3. MONADIC AND STRONGLY MONADIC ALGEBRAS

In this section we present two subclasses of the topological pseudo-Boolean algebras related to five classical conditions belonging to S5, that is, to monadic Boolean algebras. First, the condition $I \neg I = \neg I$, dual to another one studied in the previous section, has been used by A. Monteiro in [M], and its logical form appears in the 1933 axiomatics given by M. Wajsberg in [W]. We must say that A.N. Prior calls it insistently "the Gödelian axiom for S5" (see for instance [P1] page 20 and [P2] page 312) although the reference given [G] does not seem to provide reasons for this. Second, the law of reduction of modalities $I\delta = \delta$ already used in its logical and strict form by C.I. Lewis to define S5 over S1. Third, the axiom $a \leq I\delta a$ which can be used to produce S5 from S4 and characterizes a system called "Brouwerian" because of a comment of O. Becker in [B] about the intuitionistic character of a property of the strict negation or impossibility in some modal systems. Fourth, the rule "if $\delta a \leq b$ then $a \leq Ib$" which has been classically regarded as equivalent to the last axiom. And finally ~~an interesting~~ axiom involving the most elementary operators, namely implication and interior: $I(Ia \to b) = Ia \to Ib$; it was used by E.W. Beth and J.F.F. Nieland to give an axiomatization of S5 based on S4 in [B-N]. We begin our study by showing that the first four preceding conditions are equivalent in our case.

3.1. Theorem. In every tpBa A the following conditions are equivalent:

(1) $I\neg Ia = \neg Ia$ for all $a \in A$, that is, B is closed under negation,

(2) $I\delta a = \delta a$ for all $a \in A$, that is, $T \subseteq B$,

(3) $a \leq I\delta a$ for all $a \in A$, and

(4) if $\delta a \leq b$ then $a \leq Ib$ for all $a, b \in A$.

<u>Proof</u>: (1) $\Rightarrow$ (2) $\Rightarrow$ (3) are trivial. If $\delta a \leq b$ then $I\delta a \leq Ib$ and by (3) $a \leq Ib$ and so we have (4). Finally, as for every $a \in A$ we have $\delta \neg Ia = \neg Ia$ by 1.3(6), in particular $\delta \neg Ia \leq \neg Ia$, and if we apply (4) then we have $\neg Ia \leq I\neg Ia$ and we obtain (1). ■

<u>3.2. Definition</u>. A tpBa A is called <u>monadic</u> iff it satisfies any of the conditions in Theorem 3.1.

<u>3.3. Proposition</u>. Every monadic tpBa is weakly monadic.

<u>Proof</u>: From 3.1(1) we have $\delta\neg\delta a = \neg I\neg\neg\neg I\neg a = \neg I\neg I\neg a = \neg\neg I\neg a = \neg\delta a$. So we obtain 2.1(1). ■

The converse of 3.3 is not true, as example 5.3 shows. Thus we have a proper subclass of all weakly monadic tpBa. We now give some properties of the several operations of negation we have at hand; we begin by improving the relations between open and closed elements via negation.

<u>3.4. Proposition</u>. In every monadic tpBa if $a \in A$ is closed then $\neg a$ is open.

<u>Proof</u>: $a \in T$ implies $\neg a \in T$ by 3.3, and $T \subseteq B$ by 3.1(2). ■

We already know that the converse does not hold unless A is Boolean (Prop. 1.5). Only one implication remains (if $\neg a$ is closed then a is open) and 5.6 shows that it is not true for all monadic tpBas, although it can be true for particular cases, as in 5.7.

<u>3.5. Proposition</u>. A tpBa is monadic iff the three logical negations $\neg^{\Rightarrow}$, $\neg^{\leftrightarrow}$ and $\neg^{\rightsquigarrow}$ coincide.

<u>Proof</u>: It is enough to observe that $\neg^{\rightsquigarrow} a = \neg^{\leftrightarrow} a = \neg Ia$ and $\neg^{\Rightarrow} a = I\neg Ia$ and use 3.1(1). ■

<u>3.6. Corollary</u>. In every monadic tpBa A, we have $T = \mathrm{Reg}(A) = \mathrm{Reg}_I(A) = B \cap \mathrm{Reg}_H(A)$.

<u>Proof</u>: In weakly monadic tpBas $T = \mathrm{Reg}(A)$ and by 3.5 $\mathrm{Reg}(A) = \mathrm{Reg}_I(A)$. On the other hand by 2.3(4) we have $B \cap \mathrm{Reg}_H(A) \subseteq T$ and always $T \subseteq \mathrm{Reg}_H(A)$; but if A is monadic we have in addition that $T \subseteq B$ and so the last equality is proved. ■

Therefore we see that in monadic tpBa there is only one kind of regular elements, thus emphasizing the result in 2.9. On the other hand, we see in 5.4 and 5.5 that we still have three types of Peircean elements.

3.7. Theorem. In every monadic tpBa A, if $a \in A$ is open then $a \vee \neg a$ and $\neg\neg a \to a$ are dense.

Proof: If a is open so will be $\neg a$ and $a \vee \neg a$, and then $\neg^{\leftrightarrow}(a \vee \neg a) = \neg I(a \vee \neg a) = \neg(a \vee \neg a) = \neg a \wedge \neg\neg a = 0$, that is, $a \vee \neg a$ is dense. In every pseudo-Boolean algebra $\neg a \vee b \leq a \to b$ for all $a,b \in A$, so $\neg a \vee a = \neg\neg\neg a \vee a \leq \neg\neg a \to a$, and then $\neg\neg a \to a$ is dense as $R(A)$ is an order filter. ■

The logical interpretation of this result is as follows: If we read dense as "almost true" and open as "necessary" then 3.7 describes a partially classical behaviour of necessary sentences in the sense that two strictly classical laws concerning negation are almost true when refereed to necessary sentences. This classical character of the set of open elements will be total in the semisimple tpBas. But before let us say something about an intermediate class of tpBa given by the condition of Beth and Nieland.

3.8. Definition: A tpBa A will be called strongly monadic iff it satisfies $I(Ia \to b) = Ia \to Ib$ for all $a,b \in A$.

3.9. Proposition. In every tpBa the following conditions are equivalent:

(1) A is strongly monadic,

(2) $I(Ia \to Ib) = Ia \to Ib$ for all $a,b \in A$, that is, B is closed under $\to$,

(3) B is a subalgebra of A, and

(4) $a \Rightarrow b = a \rightsquigarrow b$ for all $a,b \in A$.

Proof: (2) and (4) say exactly the same thing, and they are equivalent to (3) because B is always closed under $\wedge, \vee$ and $0 \in B$; to be a subalgebra of A it only needs to be closed under $\neg$ and $\to$, but $\neg a = a \to 0$, so we see that (2) and (3) are equivalent. Putting Ib for b in the definition we see that (1) implies (2), and if we assume (2) then $Ia \to Ib = I(Ia \to Ib) \leq I(Ia \to b) \leq Ia \to Ib$, so we have (1). ■

3.10. Corollary. Every strongly monadic tpBa is monadic. ■

Example 5.4 shows that the converse is not true. Now we shall see two kinds of topological pseudo-Boolean algebras which are always strongly monadic. One of them is any tpBa defined on a linearly ordered set:

3.11. Proposition. If A is a tpBa whose underlying ordering relation is linear, then A is strongly monadic.

Proof: In a linearly ordered set there is only one binary operation $\to$ which can give it the structure of a Hilbert algebra (and hence of a pseudo-Boolean algebra), namely $a \to b = 1$ iff $a \leq b$ and $a \to b = b$ otherwise. It is then trivial that any interior operator on this set will produce an open set B closed by $\to$. ∎

Another kind of strongly monadic tpBa are functional algebras as defined in [H] for the Boolean case. We call functional every tpBa of the form $A = H^X$ where H is a complete (or at least inf-complete) pseudo-Boolean algebra and $X \neq \emptyset$ is any set, with the pointwise defined pseudo-Boolean structure and the interior operator is $(If)(y) = \inf\{f(x): x \in X\}$ for all $f \in A$, $y \in X$. We then have:

3.12. Proposition. Every functional tpBa is strongly monadic.

Proof: From the definition it follows that the open elements of a functional tpBa are the constant functions, and these have the structure of H, thus forming a subalgebra of A. ∎

In the following result we see that 3.9(4) is the only general coincidence among the impliction operations that can hold in any kind of (non trivial) tpBa.

3.13. Proposition. In every tpBa A the following conditions are equivalent:

(1) $a \Rightarrow b = a \twoheadrightarrow b$ for all $a,b \in A$,

(2) $a \rightsquigarrow b = a \twoheadrightarrow b$ for all $a,b \in A$, and

(3) $Ia = a$ for all $a \in A$.

Proof: (3) trivially implies (1) and (2); and putting a=1 in (1) or in (2) we obtain (3). ∎

A trivial although worth mentionning consequence of 3.9(4) is:

3.14. Proposition. In every strongly monadic tpBa, $P_{\Rightarrow} = P_{\rightsquigarrow}$. ∎

However, this set of Peircean elements must not be equal to $P_{\twoheadrightarrow}$, as example 5.6 shows. We also observe that examples 5.6 and 5.7 are strongly monadic algebras and so we cannot improve the relations between open and closed elements we have found for monadic tpBas. Finally we find here the second "connecting condition" of Fischer-Servi:

3.15. Proposition: In every strongly monadic tpBa A we have

(1) $T \to B \subseteq B$, and

(2) $\delta a \to Ib \leq I(a \to b)$ for all $a,b \in A$.

Proof: By 3.10, 3.1(2) and 3.9(2) we have $T \to B \subseteq B \to B \subseteq B$ and so (1) holds. In every tpBa $a \leq \delta a$ and $Ib \leq b$, so $\delta a \to Ib \leq a \to b$; by (1) $\delta a \to Ib \in B$ and therefore $\delta a \to Ib \leq I(a \to b)$. ■

Example 5.5 shows us that condition (2) does not hold in every monadic tpBa. Moreover, it cannot be characteristic of strongly monadic tpBas because it can hold even in non-weakly-monadic tpBas, as 5.2.

4. SEMISIMPLICITY IN TPBAS

Semisimplicity is a property of every monadic Boolean algebra, as Halmos showed, and it is a sufficient condition for a topological Boolean algebra to be monadic, as Monteiro (easily) showed. Actually, semisimple topological Boolean algebras are exactly monadic Boolean algebras. We shall examine in our case the logical significance of this algebraic concept and some interesting consequences. The concepts of simple and semisimple tpBa are of course the usual ones of universal algebra: an algebra is simple iff its only congruences are the two trivial ones, and these are different; in tpBa this is equivalent to say that $B = \{0,1\}$ and $0 \neq 1$. An algebra is semisimple iff it is (isomorphic to) a subdirect product of simple algebras. We then have:

4.1. Proposition. Every semisimple tpBa is strongly monadic.

Proof: $\{0,1\}$ is always a subalgebra, so every simple tpBa is strongly monadic. The condition in 3.8 is an equation, so it is preserved under the formation of direct products and subalgebras; thus every semisimple tpBa will satisfy it. ■

The converse is obviously false, as for instance in 5.6, 5.7 and 5.8. Although we have not shown any example, it is easy to see that there are semisimple tpBa which are not simple. We now give several characterizations of semisimplicity, the first ones with more algebraic contents and the others having a more logical significance.

4.2. Theorem. In every tpBa A the following conditions are equivalent:

(1) A is semisimple,

(2) $R(A) = \{1\}$,

(3) $((a*b)*a)*a = 1$ for all $a,b \in A$,

(4) $((a*0)*a)*a = 1$ for all $a \in A$, and

(5) $Ia = (a \Rightarrow 0) \Rightarrow a$ for all $a \in A$.

where $*$ stands for any of $\Rightarrow$, $\rightsquigarrow$ and $\leftrightarrow$.

Proof: It is easy to see that the simple quotients of any tpBa are in correspondence with its maximal deductive systems. Then by known results of universal algebra (1) becomes equivalent to (2). The equivalence between (2) and (3) results from

$R(A) = \underline{D}(P_*)$. (3) trivially implies (4), and from $0 \leq b$ follows that $((a*0)*a)*a \leq$ $\leq ((a*b)*a)*a$, so (3) follows from (4). Finally, $Ia \leq (a \Rightarrow 0) \rightarrow Ia$, so $Ia \leq (a \Rightarrow 0) \Rightarrow a$ in general; replacing * by $\Rightarrow$, (4) says that $(a \Rightarrow 0) \Rightarrow a \leq Ia$, so we see that (4) for $\Rightarrow$ is equivalent to (5). But the three versions of (4) are mutually equivalent, as it is clear from (1) and (2), so (4) and (5) are completely equivalent. ∎

4.3. Proposition. A strongly monadic tpBa A is semisimple iff $Ia = (a \rightsquigarrow 0) \rightsquigarrow a$ for all $a \in A$.

Proof: From 3.9(4) and 4.2(5). ∎

4.4. Proposition. A weakly monadic tpBa is semisimple iff 1 is the only dense element of the algebra.

Proof: Trivial by 4.2(2) and 2.6. ∎

Note that the preceding result could have been stated for general tpBa by refering only to $\Rightarrow$-dense elements. This result reminds us that a pseudo-Boolean algebra is Boolean (semisimple) iff 1 is its only (H-)dense element.

4.5. Theorem. A tpBa A is semisimple iff its set of open elements B is a Boolean subalgebra of A.

Proof: We already know that if A is semisimple then B is a subalgebra of A and thus a pseudo-Boolean algebra where $\rightarrow$ and the three * coincide. Now 3.7 and 4.4 tell us that $a \vee \neg a = 1$ for all $a \in B$, so B is Boolean. Conversely, if B is a Boolean subalgebra of A, we must have $((Ia \rightarrow 0) \rightarrow Ia) \rightarrow Ia = 1$ for all $a \in A$, and this implies $Ia = (Ia \rightarrow 0) \rightarrow Ia = (a \rightsquigarrow 0) \rightsquigarrow a$ because B is closed by $\rightarrow$. But we also have that A is strongly monadic, so 4.3 completes the proof. ∎

We observe that 4.5 says that a tpBa is a semisimple algebra iff its open set is a semisimple subalgebra of it. In connection with this setting we note that a tpBa is a simple algebra iff its open set is a simple subalgebra of it.

The third important characterization of semisimplicity involves three conditions already known, namely the law of reduction of modalities $\delta I = I$, the law $\delta Ia \leq a$ which is dual to Becker's 3.1(3), and the axiom $I \neg Ia \vee Ia = 1$ which appears in [Bul].

4.6. Theorem. In every tpBa A the following conditions are equivalent:

(1) A is semisimple,

(2) $\delta a = \min\{b \in B : b \geq a\}$ for all $a \in A$,

(3) $\delta Ia = Ia$ for all $a \in A$, that is, $B \subseteq T$,

(4) $\delta Ia \le a$ for all $a \in A$, and

(5) $I \neg Ia \vee Ia = 1$ for all $a \in A$,

Proof: (1) $\Rightarrow$ (2): If A is semisimple then it is also monadic, and so $T \subseteq B$; then
$\delta a \in B$ and $a \le \delta a$, and if $b \in B$ is such that $a \le b$, it follows that $\neg b \in B$
and $\neg b \le \neg a$, so $\neg b \le I \neg a$ and then $\delta a \le \neg \neg b = b$ as B is Boolean. Thus
we have (2).

(2) $\Rightarrow$ (3) $\Rightarrow$ (4) are trivial.

(4) $\Rightarrow$ (5): If we assume (4) we have $\neg I \neg Ia \to Ia = 1$, but $\neg a + b \le \neg\neg (a \vee b)$
for all $a,b \in A$, so we also have $1 = \neg\neg (I \neg Ia \vee Ia) = \neg\neg I(I \neg Ia \vee Ia) \le$
$\le \delta I(I \neg Ia \vee Ia) \le I \neg Ia \vee Ia$ applying (4) once more. Now we have obtained (5).

(5) $\Rightarrow$ (1): If $a \ne 1$ then $Ia \ne 1$ and the assumption of (5) forces us to
accept that $I \neg Ia \ne 0$. From this it follows that there is a maximal deductive
system $D \in \mathscr{D}$ such that $\neg Ia \in D$ and therefore $a \notin D$ (see theorem 3.7 of [Fo]);
then $a \notin R(A)$ and this establishes (1) via 4.2(2). ■

As a consequence of the two preceding theorems we observe that in a semisimple
tpBa the open elements and the closed elements are the same and form a subalgebra
which is Boolean, that is, these elements have a completely classical behaviour.
Hence we have, among other properties, that for all $a \in A$, $\neg\neg Ia = Ia$. The validity of such formula is considered by Bull as an "intuitionistically implausible
thesis" in [Bul], and consequently all systems containing it are rejected as genui-
ne intuitionistic analogues of S5 according to the criteria of [Bu2]. We must say
that the logical system that would correspond to semisimple tpBa is weaker than
the one initially considered by Bull, because this one has the interdefinability
of the two modal operators, which is not true in every semisimple tpBa as 5.8 shows.

We next examine the semisimplicity of the two special kinds of tpBas dealt
with in 3.11 and 3.12; we find that there is no proper semisimple tpBa among them:

4.7. Proposition. A linearly ordered tpBa is semisimple iff it is simple.

Proof: If A is a semisimple tpBa, then B will be a linearly ordered Boolean algebra, and this implies $B = \{0,1\}$, so A is simple. The converse is general. ■

4.8. Proposition. A functional tpBa $A = H^X$ is semisimple iff H is a semisimple
            pseudo-Boolean algebra, that is, iff A is a monadic Boolean algebra.

Proof: We only need to consider that the set of open elements of A has the same
structure of H, as we alreday said in 3.12. ■

We now come to the last two conditions of the fourteen ones mentioned at the
beginning of the paper.

4.9. Proposition. In every semisimple tpBa A:

(1) T is a subalgebra of A,

(2) $I(a \vee Ib) = Ia \vee Ib$ for all $a,b \in A$, and

(3) $\delta(a \vee b) = \delta a \vee \delta b$ for all $a,b \in A$.

Proof: (1) is a consequence of 4.5 and 4.6, as we have already observed. (2) and (3) are true in every simple tpBa, as it is easily checked, and therefore they are true in every semisimple tpBa. ■

Condition (1) is equivalent to saying that T is a Boolean subalgebra of A, and in this form it appears in Halmos' definition of monadic Boolean algebras as equivalent to 2.1(1). Condition (2) is dual to 2.1(3) and it has been explicitly used by Monteiro. Condition (3) is of a different character: it is a property of all topological Boolean algebras. This three conditions have in common that they are not equivalent to the semisimplicity of tpBas, as 5.6 shows. We are going to find all the relations between these conditions and between the four classes of tpBas we have introduced in sections 2 and 3. In the first place, taking 2.1(1) into account we see that

4.10. Proposition. If in a tpBa A the set T is a subalgebra of A then A is weakly monadic. ■

Example 5.7 shows us that the convese is not true even in monadic or strongly monadic tpBas. However in 5.3 we see that condition (1) can hold in any kind of weakly monadic tpBa. Condition (2) is even more independent from monadicity than (1), beacuse it can hold in very general tpBa (as the one of 5.1) and it can fail in strongly monadic tpBa (such as in 5.8). The same is true for condition (3), as it is shown in examples 5.1 and 5.7. The only relation among the three conditions is the following:

4.11. Proposition. If in a tpBa A the set T is a subalgebra of A then $\delta(a \vee b) = \delta a \vee \delta b$ for all a,b A.

Proof: We always have $\delta(a \vee b) \leq \delta a \vee \delta b \leq t$ for all $t \in T$ such that $a \leq t$, $b \leq t$. If T is a subalgebra then $\delta a \vee \delta b \in T$ and by 1.2(3) this gives us $\delta a \vee \delta b = \delta(a \vee b)$. ■

There are no other implications, because the converse of 4.11 is not true as we see in 5.1. Condition (1) does not imply (2) as 5.4 shows, so (3) does not either, and conversely (2) does not imply (3) because of 5.7, and then it does not imply (1) either. The following scheme summarizes all implications we have found between the classically equivalent conditions; note that we have shown that there are no other implications between them.

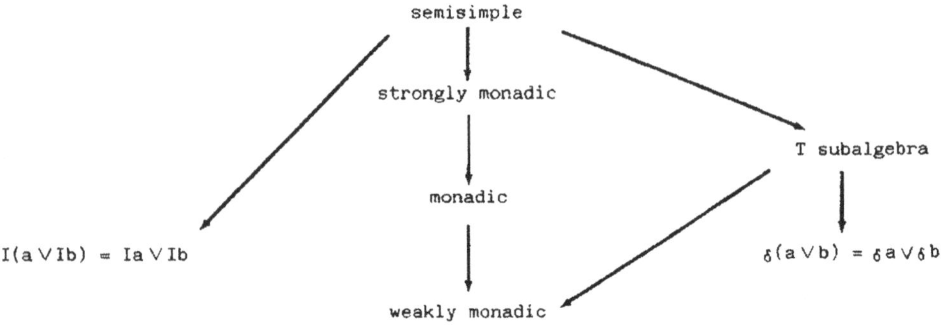

We will close this section by giving several characterizations of semisimplicity in terms of the lattice $\mathscr{D}$ of all deductive systems of A and in terms of the abstract logic $\underline{L}$. Particularizing what we have summarized in section 0, we can associate an equivalence relation to the least deductive system $|1|$: $a \sim b$ iff $\underline{D}(a) = \underline{D}(b)$ iff $Ia = Ib$, which is a logical congruence with respect to $\neg^*, \wedge, \overset{*}{\vee}$ and $*$, and by Theorem IX-6 of [B-S] it is the maximum of $\Theta_{\underline{L}}$. Then we have:

<u>4.12. Theorem</u>. In every tpBa A the following conditions are equivalent:

(1) A is semisimple,

(2) $A/\sim$ is semisimple (that is, a Boolean algebra), and

(3) $\underline{L}$ satisfies the <u>Reductio ad Absurdum Principle</u> with respect to $\neg^*$: $\underline{D}(X, \neg^* a) = A$ iff $a \in \underline{D}(X)$ for all $a \in A$, $X \subseteq A$.

<u>Proof</u>: By 4.2(3) A is semisimple iff $((a*b)*a)*a = 1$ for all $a,b \in A$, and this is equivalent to $((\bar a \,\bar\to\, \bar b)\,\bar\to\, \bar a)\,\bar\to\, \bar a = \bar 1$ because $\bar 1 = \{1\}$, and this is equivalent to the semisimplicity of $A/\sim$ as a pseudo-Boolean algebra. Thus (1) and (2) are equivalent. To see that (2) and (3) are equivalent we consider the canonical projection from A onto $A/\sim$: it is a bilogical morphism between $\underline{L}$ and the logic $\tilde{\underline{L}}$ associated with all filters of $A/\sim$. Then $\tilde{\underline{L}}$ satisfies the four principles mentioned in section 0, and by Theorem 13 of [V], in such situation a necessary and sufficient condition for $A/\sim$ to be Boolean (that is, semisimple) is that $\tilde{\underline{L}}$ satisfies the Reductio ad Absurdum Principle with respect to $\bar\neg$. Again by the already mentioned bilogical morphism this is equivalent to the fact that $\underline{L}$ satisfies it. ■

The Strong Disjunction Principle $\underline{D}(X,a) \cap \underline{D}(X,b) = \underline{D}(X, a\overset{*}{\vee}b)$ gives the operation $\overset{*}{\vee}$ a certain character of a "logical disjunction". This can be extended if we call <u>prime</u> every deductive system $D \in \mathscr{D}$ such that for all $a,b \in A$, if $a \overset{*}{\vee} b \in D$ then $a \in D$ or $b \in D$. The prime deductive systems turn out to be exactly the (finitely) irreducible deductive systems, and we also find the following characterization of semisimplicity, of a clear Boolean flavour:

4.13. Theorem. A tpBa is semisimple iff the prime deductive systems and the maximal deductive systems of $\mathcal{D}$ coincide.

Proof: The semisimplicity of A is equivalent by 4.12 to the semisimplicity of $A/\sim$, and this one is equivalent to the coincidence of its irreducible and maximal filters. But the bilogical morphism between A and $A/\sim$ induces a lattice isomorphism between $\mathcal{D}$ and the lattice of all filters of $A/\sim$. Therefore the last coincidence is equivalent to the one stated in the Theorem, which is proved. ∎

More details about the most basic concepts that are involved in the last part of this section can be seen in [Fo]. Some of them were introduced in [R] for topological Boolean algebras.

5. EXAMPLES AND COUNTEREXAMPLES

We gather here all distinct examples of tpBa that have been used in several places throughout the paper. They are in all cases finite algebras. For brevity's sake we do not give the table of any operation; they all can be produced from the Hasse diagram of the algebra (recall that there is one and only one operation on a finite distributive lattice, namely $a \to b = \max\{c \in A: a \wedge c \leq b\}$, which gives it the structure of a pseudo-Boolean algebra). The open elements are indicated by a circle in the diagrams; then the interior operator is $Ia = \max\{b \in B: b \leq a\}$. We do not show any explicit computation but simply state the properties that each algebra has or has not and which have been mentioned in the paper.

5.1. Example

Here $B = \{0,a,1\}$ and $T = \{0,b,1\}$. This tpBa is not weakly monadic ($c \wedge \delta b = 0$ but $\delta c \wedge \delta b = b \neq 0$) and therefore T is not a subalgebra ($b \in T$ but $\neg b = c \notin T$). It satisfies $I(x \vee Iy) = Ix \vee Iy$ and $\delta(x \vee y) = \delta x \vee \delta y$ for all $x, y \in A$. Here we have $D_{\leftrightarrow} = \{1\}$ and $D_{\Rightarrow} = \{1,a,c,d\}$, so $D_{\leftrightarrow} \subsetneq D_{\Rightarrow}$.

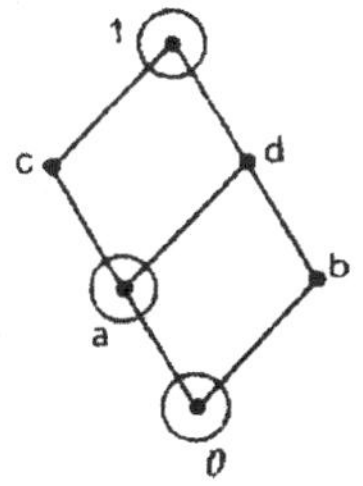

5.2. Example

Here $B = \{0,b,d,1\}$ and $T = \{0,b,1\}$. This is also a non weakly monadic tpBa ($\delta\neg\delta c = 1 \neq b = \neg\delta c$). It satisfies: if $x \in T$ then $\neg x \in B$ and if $\neg x \in T$ then $x \in B$ for all $x \in A$. It satisfies $T \to B \subseteq B$. Here $Reg(A) = \{0,1\}$ and $Reg_H(A) = \{0,b,c,1\}$, therefore $Reg(A) \subsetneq T \subsetneq Reg_H(A)$.

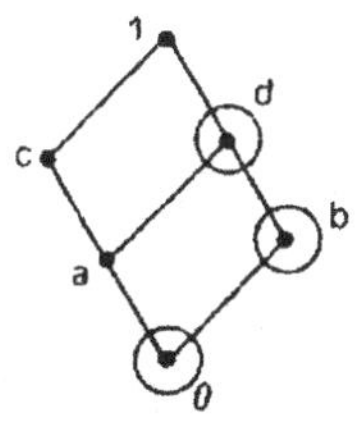

5.3. Example

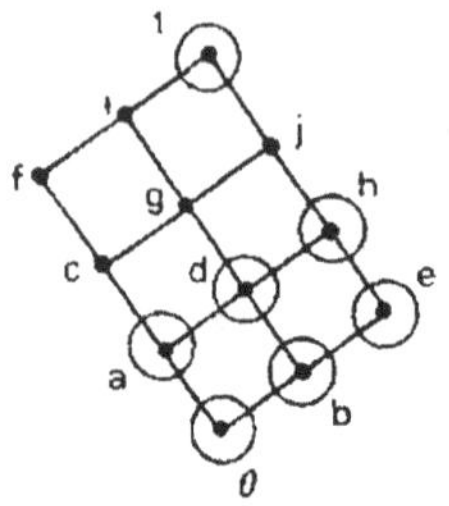

Here $B = \{0,a,b,d,e,h,1\}$ and $T = \{0,e,f,1\}$. This is a weakly monadic tpBa which is not monadic ($e \in B$ but $\neg e = f \notin B$). Here $e \in T$ but $\neg e = f \notin B$, $\neg c = e \notin T$ but $c \in B$ and $\neg c = e \notin B$ but $c \notin T$. Here $Reg(A) = T = \{0,e,f,1\}$ is a subalgebra of A (indeed, a Boolean one), $Reg_I(A) = \{0,a,e,1\}$, $P_{\Rightarrow} = \{1,d,h\}$, $P_{\sim\rightarrow} = \{1,d,h,i\}$ and $P_{\leftrightarrow} = \{1,d,g,h,i,j\}$. We see that $P_{\Rightarrow} \subsetneq P_{\sim\rightarrow} \subsetneq P_{\leftrightarrow}$, that $\{0,1\} \subsetneq T \cap B \subsetneq$ $\subsetneq Reg_I(A) \subsetneq B$ and that $\{0,1\} \subsetneq Reg(A)$.

5.4. Example

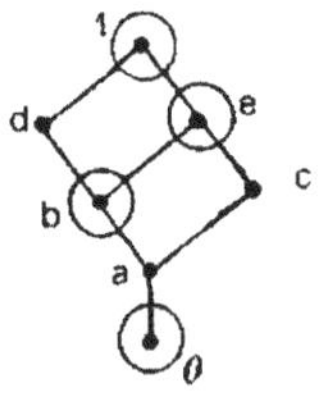

Here $B = \{0,b,e,1\}$ and $T = \{0,1\}$. This constitutes a monadic tpBa which is not strongly monadic ($e,b \in B$ but $e \rightarrow b = d \notin B$). T is a subalgebra of A. $I(c \vee Ib) = e \neq b = Ic \vee Ib$. $P_{\sim\rightarrow} = \{1,b,e\}$ and $P_{\leftrightarrow} = \{1,b,d,e\}$, so $P_{\sim\rightarrow} \subsetneq P_{\leftrightarrow}$.

5.5. Example

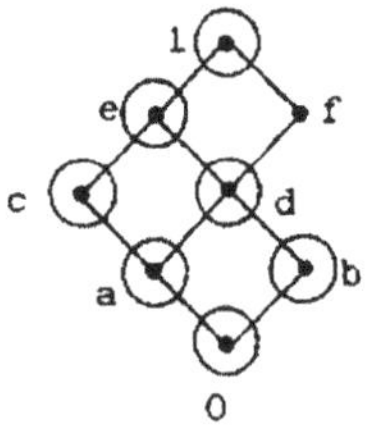

Here $B = \{0,a,b,c,d,e,1\}$ and $T = \{0,b,c,1\}$. This is monadic tpBa not strongly monadic ($c,d \in B$ but $c \rightarrow d = f \notin B$). It does not satisfy the condition $T \rightarrow B \subseteq B$ (because $c \in T$, $d \in B$ but $c \rightarrow d = f \notin B$). $P_{\Rightarrow} = \{1,d,e\}$ and $P_{\sim\rightarrow} = P_{\leftrightarrow} =$ $= \{1,d,e,f\}$, so $P_{\Rightarrow} \subsetneq P_{\sim\rightarrow}$.

5.6. Example

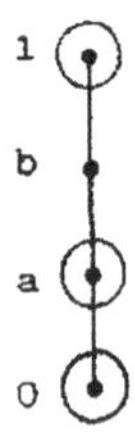

Here $B = \{0,a,1\}$ and $T = \{0,1\}$. This tpBa is strongly monadic (see 3.11) but is not semisimple, as it is not simple (see also 4.7). $\neg b = 0 \in T$ but $b \notin B$. This algebra satisfies that T is a subalgebra of A, $\delta(x \vee y) = \delta x \vee \delta y$, and $I(x \vee Iy) =$ $= Ix \vee Iy$ for all $x,y \in A$. $P_{\Rightarrow} = P_{\sim\rightarrow} = \{a,1\}$ and $P_{\leftrightarrow} = \{a,b,1\}$, so $P_{\sim\rightarrow} \subsetneq P_{\leftrightarrow}$.

5.7. Example

$B = \{0,c,e,h,1\}$ and $T = \{0,c,e,1\}$. This is a strongly monadic tpBa which is not semisimple: $D = \{1,h\} \neq \{1\}$. T is not a subalgebra of A, as it is not closed by $\vee$. We have that $c, e \in T$ and $\delta(c \vee e) = 1 \neq h = \delta c \vee \delta e$. This algebra satisfies that if $\neg x \in T$ then $x \in B$ for all $x \in A$, and $I(x \vee Iy) = Ix \vee Iy$ for all $x, y \in A$.

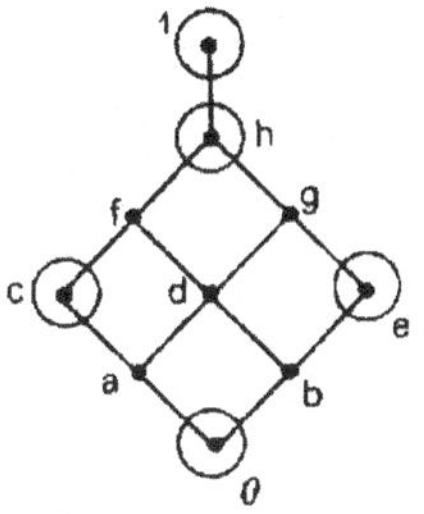

5.8. Example

$B = \{0, e, 1\}$ and $T = \{0, 1\}$, which is a subalgebra of A. The tpBa is strongly monadic and not semisimple. $I(d \vee Ie) = 1 \neq e = Id \vee Ie$. $a = \neg\neg a$, $b = \neg\neg b$ although $a, b \notin T$, and also $\delta a \neq \neg\neg a$ for $a \notin B$.

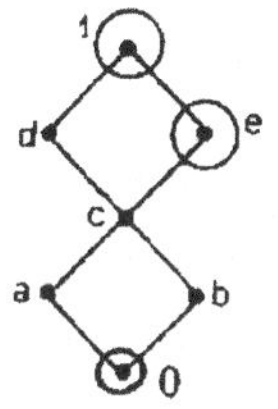

5.9. Example

$B = T = \{0, 1\}$, that is, we have a semisimple tpBa (actually a simple one!) where $I \neq \neg\delta\neg$ because $Ia = 0 \neq 1 = \neg\delta\neg a$.

5.10. Example

This is an example of a totally different kind from the preceding ones. Take the pseudo-Boolean algebra A of four linearly ordered elements of the diagram. Define the following δ operator: $\delta 0 = 0$, $\delta a = \delta b = b$, $\delta 1 = 1$. This is a closure operator on A, and indeed a lattice closure, as it satisfies $\delta(x \vee y) = \delta x \vee \delta y$ for all $x, y \in A$. Moreover it satisfies $\delta(x \wedge \delta y) = \delta x \wedge \delta y$ and $\delta\neg\delta x = \neg\delta x$ for all $x, y \in A$. The operator I defined on A as $Ix = \neg\delta\neg x$ for all $x \in A$ is $I0 = 0$, $Ia = Ib = I1 = 1$. It satisfies $I1 = 1$, $I^2x = Ix$, and $I(x \wedge y) = Ix \wedge Iy$ for all $x, y \in A$. Moreover it satisfies $x \leq I\delta x$ and if $\delta x \leq y$ then $x \leq Iy$ for all $x, y \in A$. But it is not an interior operator of any kind, because $b < Ib$, $a < Ia$.

This example shows that it is not a good choice to have a primitive operator of possibility M and then define necessity as $L \leftrightarrow \neg M \neg$, even if we put stronger conditions on M. It is easy to prove that $\delta = \neg I \neg$ always produces a closure operator from an interior operator, while the dual formula $I = \neg \delta \neg$ does not.

REFERENCES

[B] O. Becker, Zur Logik der Modalitäten, Jahrbuch für Philosophie und Phänomenologische Forschung 11 (1930) 497-548.

[B-N] E. Beth and J.F.F. Nieland, Semantic Construction of Lewis Systems S4 and S5, in: Addison, Henkin, Tarski, eds., The Theory of Models (North-Holland, Amsterdam, 1965) 17-24.

[B-S] D.J. Brown and R. Suzsko, Abstract Logics, Diss. Math. CII (1973) 9-40.

[Bu1] R.A. Bull, Some modal calculi based on IC, in: Crossley and Dummett, eds., Formal Systems and Recursive Functions (North-Holland, Amsterdam, 1965) 3-7.

[Bu2] ----------, A modal extension of intuitionist logic, Notre Dame J. Formal Logic 6 (1965) 142-145.

[D] C. Davis, Modal operators, equivalence relations and projective algebras, Am. J. Math. 76 (1954) 747-762.

[Fi1] G. Fischer-Servi, On modal logic with an intuitionistic base, Studia Logica 36 (1977) 141-149.

[Fi2] ----------, Semantics for a class of intuitionistic modal calculi, in: Dalla Chiara, ed., Italian Studies in the Philosophy of Science (Reidel, Dordrecht, 1980) 59-72.

[Fi3] ----------, An intuitionistic analogue of S4 as a logical modeling for science, J. Philos. Logic, to appear.

[Fo] J. M. Font, Implication and deduction in some intuitionistic modal logics, Rep. Math. Logic 17 (1983) to appear.

[G] K. Gödel, Eine Interpretation des Intuitionistischen Aussagenkalküls, Ergebnisse eines Mathematischen Kolloquiums 4 (1933) 39-40.

[H] P.R. Halmos, Algebraic Logic I: Monadic Boolean algebras, Compos. Math. 12 (1955) 217-249.

[M] A. Monteiro, La semisimplicité des algèbres de Boole topologiques et les systèmes déductifs, Rev. Union Mat. Argent. 25 (1971) 417-448.

[M-T] J.C.C. McKinsey and A. Tarski, On closed elements in closure algebras, Ann. Math. 47 (1946) 122-162.

[O] H. Ono, On some intuitionistic modal logics, Publ. Res. Inst. Math. Sci. Kyoto Univ. 13 (1977) 687-722.

[P1] A.N. Prior, Time and Modality (Oxford Univ. Press, 1957).

[P2] ----------, Formal Logic (Oxford Univ. Press, 1962).

[R] H. Rasiowa, An algebraic approach to non-classical logics (North-Holland, Amsterdam, 1974).

[R-S] H. Rasiowa and R. Sikorski, The mathematics of the metamathematics (P.W.N., Warszawa, 1970).

[V] V. Verdú, Distributive and Boolean logics. Stochastica 3 (1979) 97-108.

[vW] G.H. von Wright, An essay in modal logic (North-Holland, Amsterdam, 1951).

[W] M. Wajsberg, Ein erweiterter Klassenkalkül, Monatsh. Math. Phys. 40 (1933) 113-126.

FINITE EXTENSIONS OF FINITE GROUPS

Wilfrid Hodges
Bedford College
Regent's Park
London NW1 4NS

We examine some questions in the model theory of finite and locally finite groups.

In §1 we show that for every finite group G, every finite number n and every function $\beta: G^n \to G$, there are a finite group $H \supseteq G$ and a polynomial $t(\bar{x},\bar{h})$ such that in H, $\beta(\bar{g}) = t(\bar{g},\bar{h})$ for all $\bar{g}$ in G^n; moreover $t(\bar{x},\bar{y})$ can be chosen depending only on n. This answers a question raised by Belegradek [3].

In §2 we show that if G is Philip Hall's countable universal locally finite group, then the first-order theory of G is recursively equivalent to complete first-order arithmetic.

In §3 we raise and discuss the problem: for a given family $E(\bar{x},\bar{y})$ of equations, what are necessary and sufficient conditions for a finite group G with elements $\bar{g}$ to have a finite group extension H with elements $\bar{h}$ such that $E(\bar{g},\bar{h})$ holds? Nearly all the interesting parts of this problem seem to be open.

I thank O. V. Belegradek for sending me his [3], and Hans Kaiser, Simon Thomas and Martin Ziegler for some helpful comments.

1. Representing a function

Finite groups will be written G, H etc., and their elements a, b, g, h etc. We write $\bar{a}$ for a finite sequence $(a_1,\ldots,a_n)$. Group words in variables $\bar{x}$, $\bar{y}$ etc. are written $w(\bar{x})$, $t(\bar{x},\bar{y})$ etc., and $w_H(\bar{x},\bar{h})$ is the function defined in the group H by the word $w(\bar{x},\bar{y})$ when the variables $\bar{y}$ are replaced by the elements $\bar{h}$ of H. We use Greek letters ϕ, ψ, χ for formulas and α, β, γ, δ for maps.

THEOREM 1. For each positive integer n there is a word $t(x_1,\ldots,x_n,\bar{y})$ such that if G is any finite group and $\beta: G^n \to G$ any map, then there are a finite group $H \supseteq G$ and elements $\bar{h}$ of H such that β is the restriction of $t_H(\bar{x},\bar{h})$ to G.

The proof will occupy most of §1.

LEMMA 2 (Strong amalgamation for finite groups). Let K_1 and K_2 be finite extensions of the group G. Then there exist a finite group $H \supseteq K_1$ and an embedding $\gamma: K_2 \to H$ such that $\gamma|G$ is the identity and $K_1 \cap \gamma(K_2) = G$.

Proof. Cf. B. H. Neumann [10] section 3. []

Lemma 2 will simplify our workings as follows. Suppose we want to show that G has a finite extension H in which both the systems E_1 and E_2 of equations and inequations have solutions. Then it will suffice to construct one extension K_1 with a solution of E_1, and another extension K_2 with a solution of E_2. Lemma 2 says that K_1 and K_2 can be amalgamated into a single finite group H which must then contain solutions of both E_1 and E_2. By induction the same holds for any finite collection of systems $E_1, \ldots, E_n$.

LEMMA 3 (P. Hall [4] Lemma 1). Let G be a finite group and $\gamma: K_1 \to K_2$ an isomorphism between subgroups of G. Then there is a finite group $H \supseteq G$ with an element h such that $h^{-1}kh = \gamma(k)$ for all $k \in K_1$.

Proof. Use the regular representation. Kegel and Wehrfritz [7] p. 180 report the details. []

LEMMA 4. Let G be a finite group, K a subgroup of G and $\gamma: K \to G$ a homomorphism. Then there is a finite group $H \supseteq G$ with elements h, j such that $h^{-1}khj^{-1}k^{-1}j = \gamma(k)$ for all $k \in K$.

Proof. Identify G with the first factor in G×K. Let $\alpha: K \to G\times K$ and $\beta: K \to G\times K$ be defined by:

$$\alpha(k) = \langle\gamma(k),k\rangle \quad \text{and} \quad \beta(k) = \langle e,k\rangle \qquad \text{for each } k \in K.$$

Then α and β are isomorphisms. So by Lemma 3 (and Lemma 2) there is a finite group $H \supseteq G\times K$ with elements h, j such that for all $k \in K$, $h^{-1}kh = \alpha(k)$ and $j^{-1}kj = \beta(k)$. We finish by observing that $\gamma(k) = \alpha(k)\beta(k)^{-1}$. []

Now we need some universal algebra. Let V be a variety of groups and G a group which is generated by elements $g_1, \ldots, g_n$; G is not necessarily in V. We shall say that $g_1, \ldots, g_n$ are V-<u>independent</u> (in G, or in any group extending G) iff either of the following two equivalent conditions hold:

(a) For every group word $w(x_1,\ldots,x_n)$, if $w_G(g_1,\ldots,g_n) = e$ then $w(x_1,\ldots,x_n)$ is a law in V.

(b) For every group K in V and every map α: $\{g_1,\ldots,g_n\} \to K$ there is a homomorphism γ: $G \to K$ which extends α.

Every group G generates a variety V(G); the laws of V(G) are those words $w(\bar{x})$ such that $w_G(\bar{g}) = e$ for every sequence $\bar{g}$ of elements of G. If G is finite then in V(G) the free groups on a finite number of generators are also finite (cf. H. Neumann [11] p. 18).

LEMMA 5. Let G be a finite group. Then there is a finite group $H \supseteq G$ with an element c such that the family of elements $g^{-1}cg$ $(g \in G)$ is V(G)-independent.

Proof. List the elements of G without repetition as $g_1, \ldots, g_n$. Let F be the V(G)-free group with basis $a_1, \ldots, a_n$. Let H be the symmetric group Sym(F×G). For each $g \in G$ we define an element α_g of H by:

$$\alpha_g(f,g') = (f,gg') \qquad \text{for each } f \in F \text{ and } g' \in G.$$

Clearly the map $g \mapsto \alpha_g$ embeds G into H. We define c to be the following element of H:

$$c(f,g_i) = (a_i f,g_i) \qquad \text{for each } f \in F \text{ and } g_i \in G.$$

Then $\alpha_{g_i}^{-1} c \alpha_{g_i}(f,e) = \alpha_{g_i}^{-1} c(f,g_i) = \alpha_{g_i}^{-1}(a_i f,g_i) = (a_i f,e)$. Let h_i be the restriction of $\alpha_{g_i}^{-1} c \alpha_{g_i}$ to the set F×{e}; then the map $a_i \mapsto h_i$ is the regular representation of F on F×{e}. It follows that the elements h_i $(1 \le i \le n)$ are V(G)-independent, and hence so are the elements $\alpha_g^{-1} c \alpha_g$ $(g \in G)$. Identifying g with α_g, we have the lemma.

[]

Now we prove the theorem. Put n = 1 and let β: $G \to G$ be any map. Use Lemma 5 to extend G to a finite H_1 containing an element c such that the elements $g^{-1}cg$ $(g \in G)$ are V(G)-independent. Let K be the subgroup of H_1 generated by the elements $g^{-1}cg$ $(g \in G)$. By V(G)-independence there exists a homomorphism γ: $K \to G$ such that $\gamma(g^{-1}cg) = \beta(g)$ for each $g \in G$. So by Lemma 4 there is a finite group $H \supseteq H_1$ with elements h, j such that $h^{-1}khj^{-1}k^{-1}j = \gamma(k)$ for all $k \in K$. In particular if $g \in G$ then

$$\beta(g) = t_H(g,c,h,j) \quad \text{where } t(x,y,z,v) \text{ is } z^{-1}x^{-1}yxzv^{-1}x^{-1}y^{-1}xv.$$

(By Lemma 2 we can choose the same finite group H to serve all the finitely many maps $\beta: G \to G$, with the same term t as above.) This proves the theorem when $n = 1$.

For the general case, let $\tilde{G}$ be an isomorphic copy of G. Identify G with the first factor of $G \times \tilde{G}^n$, and let $\gamma: G^n \to \tilde{G}^n$ be the natural isomorphism from the n-th power of the first factor to the second factor. Let $\beta: G^n \to G$ be any map. Then β can be factored as $\beta = \beta'\gamma$, where β' is now a 1-ary map from $\tilde{G}^n$ to G in $G \times \tilde{G}^n$. So by the case $n = 1$ above, there is a finite $H \supseteq G \times \tilde{G}^n$ containing elements c, h, j such that for each $\bar{g}$ in G^n,

$$\beta(\bar{g}) = \beta'\gamma(\bar{g}) = t_H(\gamma(\bar{g}),c,h,j)$$

where t is as before. Now by Lemma 3 we can suppose that H contains elements $b_1, \ldots, b_n$ such that for each i, the map $g \mapsto b_i^{-1} g b_i$ is the natural isomorphism of G onto the i-th factor of $\tilde{G}^n$. Clearly then $\gamma(g_1,\ldots,g_n)$ is $b_1^{-1} g_1 b_1 \ldots b_n^{-1} g_n b_n$. Hence finally

$$\beta(\bar{g}) = t_H^n(\bar{g},\bar{b},c,h,j) \quad \text{where } t^n(\bar{x},\bar{u},y,z,v) \text{ is } t(\Pi_i u_i^{-1} x_i u_i, y,z,v).$$

Theorem 1 is proved. []

COROLLARY 6. There is a group word $t(x,y,z,v)$ with the following property. Let G be a finite group and $g_1, \ldots, g_n$ any elements of G. Then there is a finite group $H \supseteq G$ with elements b, c, h, j such that for each i, $g_i = t_H(b^i,c,h,j)$. In particular each finite group G can be embedded into a four-generator finite group as a subgroup of form $\{t_H(b^i,c,h,j) : i < \omega\}$.

<u>Proof</u>. First embed G in $G \times C$ where C is generated by an element b of order n. Then define $\beta: C \to G$ by $\beta(b^i) = g_i$, and apply the proof of the theorem. []

O. V. Belegradek [3], quoting Belegradek [2] and M. Trofimov [13], remarks that Theorem 1 holds when the groups are allowed to be infinite. He asks ([3] p. 10) whether for every finite group G and map $\beta: G^n \to G$ there is a finite group $H \supseteq G$ in which β is represented by a term with parameters; this is Theorem 1 without the uniform choice of t, so that Theorem 1 answers his question affirmatively But Hans Kaiser kindly pointed out to me that Belegradek's question (without uniform t) is already answered by the theorem of Maurer and Rhodes [8] that every finite simple non-abelian group is functionally complete. H. K. Kaiser and R. Lidl [6] contains more on this theme.

2. Universal locally finite groups

In [4] Philip Hall considered groups G which have the following three properties: (i) G is locally finite, (ii) every finite group can be embedded in G, and (iii) any isomorphism between finite subgroups of G is induced by some inner automorphism of G. Such groups are said to be universal locally finite, ULF for short. Hall showed that there is a unique countable ULF group G.

We write T for the complete first-order theory of Hall's group G. Let me remark without proof or explanation that any two ULF groups are back-and-forth equivalent and therefore $L_{\infty\omega}$-equivalent; hence T is also the first-order theory of any ULF group, not necessarily countable. This comment is irrelevant to everything below.

THEOREM 7. (a) Complete first-order arithmetic is 1-1 reducible to T. (b) T is 1-1 reducible to complete first-order arithmetic. [See Postscript.]

To prove (a), we shall define a 1-1 recursive map which takes each first-order formula $\phi(\bar{x})$ in the language of arithmetic to a first-order formula "ϕ"$(\bar{x})$ in the language of groups. Let G be the countable ULF group and let N be the structure of natural numbers. For each group element g let P(g) be the set of all powers of g except e; thus if g has order n+1 then $|P(g)| = n$. The mapping of formulas will satisfy the condition

$$(*) \qquad G \models \text{"}\phi\text{"}(g_1,\ldots,g_n) \qquad \text{iff} \qquad N \models \phi(|Pg_1|,\ldots,|Pg_n|),$$

for all formulas $\phi(\bar{x})$ and sequences $\bar{g}$ of group elements. Restricted to sentences ϕ, condition (*) reduces complete first-order arithmetic to T.

The main coding device we shall use is the term t(x,y,z,v) in Corollary 6 above. This term allows us to encode arbitrary non-empty finite sequences of elements of G as ordered 4-tuples, in the following way. Given a sequence $(g_1,\ldots,g_n)$, let K be the subgroup of G generated by $g_1, \ldots, g_n$. From Corollary 6 we know that there is a finite group $H \supseteq K$ containing elements b, c, h, j such that each g_i is $t_H(b^i,c,h,j)$ and b has order n. By (ii) and (iii) in the definition of a ULF group, we can choose H to be a subgroup of G. Then each g_i is $t_G(b^i,c,h,j)$. This would allow a first-order description of $(g_1,\ldots,g_n)$, at least up to permutation of the terms, if we could express 'x is a power of b' by a first-order formula. But in G we can do just that:

LEMMA 8. Let K_1 be a finite group with a subgroup J and let g be an element in K_1 but not in J. Then there exists a finite group $H \supseteq K_1$ containing an element a which commutes with every element in J but not with g.

Proof. Let K_2 be an isomorphic copy of K_1 such that $K_1 \cap K_2 = J$. By Lemma 2 we can choose K_2 so that both K_1 and K_2 are subgroups of a finite group H_1. Let $\alpha: K_1 \to K_2$ be an isomorphism which is the identity on J. By Lemma 3 there is a finite group $H \supseteq H_1$ with an element a such that $a^{-1}ka = \alpha(k)$ for each $k \in K_1$. Since α is the identity on J, a commutes with each element of J. But $g \notin J$ and hence $\alpha(g) \notin K_1$; so $a^{-1}ga = \alpha(g) \neq g$ and therefore a doesn't commute with g.

[]

The converse of Lemma 8 is obviously true: if there is an element which commutes with everything in J but not with g, then g is not in J. Also if K_1 is a subgroup of the ULF group G, then (ii) and (iii) in the definition of ULF groups again allow us to choose H inside G. Putting all this together, we infer that if b and g are elements of G, then g is a power of b iff $G \models \neg\exists x([x,b] = e \neq [x,g])$.

Now we can define the map $\phi \mapsto$ "ϕ", by induction on the complexity of ϕ. As a technical convenience I shall take the atomic formulas of the language of arithmetic to be the formulas of form x=y, x=0, x=1, x+y=z or x + y + xy = z. Recall that P(g) is the set of all powers of g except e, and that g encodes the number $|P(g)|$.

1. "x=y" is $\exists z\ z^{-1}xz = y$.
2. "x=0" is x=e.
3. "x=1" is $x \neq e \wedge x^2 = e$.
4. "x+y=z" is the first-order formula which says: There is an element a such that $P(a^{-1}xa) \cap P(y) = \emptyset$ and there are elements b, c, h, j such that the map $g \mapsto t(g,c,h,j)$ sets up a bijection from $P(b^{-1}zb)$ to $P(a^{-1}xa) \cup P(y)$.
5. "x + y + xy = z" is the first-order formula which says: There is an element a such that $P(a^{-1}xa) \cap P(y) = \emptyset$ and $a^{-1}xa$ commutes with y, and there are elements b, c, h, j such that the map $g \mapsto t(g,c,h,j)$ sets up a bijection from $P(b^{-1}zb)$ to the set of elements $\neq$ e which are generated by $a^{-1}xa$ and y.
6. "$\neg\phi$" is $\neg$"ϕ", "$\phi\wedge\psi$" is "ϕ"$\wedge$"ψ", etc.
7. "$\forall x\phi$" is $\forall x$"ϕ" and "$\exists x\phi$" is $\exists x$"ϕ".

I leave it to the reader to verify that everything in 4 and 5 can be written as a first-order formula, using the devices of Corollary 6 and Lemma 8. Definition 1 is justified by (iii) in the definition of a ULF group: elements have the same order iff they are conjugate. The condition in definition 5 says that there is a subgroup of G of form C×D where C is generated by an element of the same order as x and D is generated by y, and z has the same order as C×D. Thus if x and y have orders m+1 and n+1 respectively, then C×D and z have order $(m+1)(n+1) = m + n + mn + 1$.

This completes the proof of Theorem 7(a).

For part (b) of the theorem, let T_0 be the universal theory of finite groups, i.e. the set of universal first-order sentences which are true in every finite group. Since we can effectively list the existential sentences which are true in some finite group, T_0 is a Π_1 set. A. M. Slobodskoĭ [12] showed that T_0 is not recursive. Clearly the ULF group G is a model of T_0.

We shall consider finite-generic models of T_0. Finite-generic models were introduced by Abraham Robinson (cf. Barwise and Robinson [1]). Robinson's definition was rather ad hoc. The following description seems more natural and will suit us better (cf. M. Ziegler [14]). For simplicity I ignore relation symbols - there are none in groups. A basic sentence is an equation or inequation which has no variables in it.

Suppose we have a countable set S of universal first-order sentences. Two players I and II play a game in ω steps as follows. To the language of S we add countably many new individual constants, forming a language L. The players I and II alternately choose the stages of a countable increasing tower $X_0 \subseteq X_1 \subseteq X_2 \subseteq \ldots$ of finite sets of basic sentences of L; each set X_n must be consistent with S. Let ϕ be a sentence of L and p a finite set of basic sentences. We say that p forces ϕ iff whenever player I starts by playing a set X_0 which contains p, player II can continue the game so that regardless of what player I does at later stages, the final set $\bigcup_{n<\omega} X_n$ will be a description (more precisely, the diagram) of a structure which is a model of $S \cup \{\phi\}$. A countable model M of S is said to be finite-generic iff when we add constants to name all the elements of M, and take X to be the set of all basic sentences which are true in M, we find that every first-order sentence of L which is true in M is forced by some finite subset of X. We write S^f for the set of those first-order sentences in the language of S which are true in every finite-generic model of S; these are exactly the sentences which are forced by the empty set. The set S^f is known as the finite-forcing companion of S.

We shall need three facts. In these facts, S is assumed to be a countable set of universal first-order sentences.

FACT A. Suppose S has a countable model. Then in the game described above, regardless of how player I moves, player II can play so that the model M described by the set $\bigcup_{n<\omega} X_n$ is a finite-generic model of S. (Ziegler [14] Lemma V.2.4.)

FACT B. Suppose that for any two models of S there is a third model of S in which the first two can be embedded. Then S^f is a complete first-order theory. (Barwise and Robinson [1] Theorem 4.6.)

FACT C. If S is an arithmetical set then S^f is 1-1 reducible to complete first-order arithmetic. (Hirschfeld and Wheeler [5] Theorem 7.7.)

Returning to the theory T_0, let us bring Fact B into play:

LEMMA 9. For any two models of T_0 there is a model of T_0 in which both can be embedded.

Proof. Add constants for the elements of the two models M_1 and M_2. The compactness theorem shows that if the conclusion of the lemma fails, then there are finite sets X_1, X_2 of equations and inequations true in M_1, M_2 respectively, such that $X_1 \cup X_2$ is not true in any model of T_0 (in the language with the added constants). But by choice of T_0 there exist finite groups G_1, G_2 in which X_1, X_2 respectively are true. Then $G_1 \times G_2$ is a finite group, hence a model of T_0, in which $X_1 \cup X_2$ is satisfiable. []

LEMMA 10. The countable ULF group is a finite-generic model of T_0.

Proof. Let two players play the game described above, with $T_0 = S$. By Fact A, player II can play so that the resulting structure M is a countable finite-generic model of T_0. Let him do so. Then M is a group.

We assert that M obeys (ii) and (iii) in the definition of a ULF group. Consider (ii). If it fails, then there is a first-order sentence ϕ which is true in M and which states that a certain finite group K is not embeddable in M. Since M is finite-generic, some finite set p of equations and inequations which are true in M forces ϕ. But from the definition of 'forces', this is impossible: there is no way that player II can play so as to prevent player I from slipping in a copy of K. The proof of (iii) is similar, using Lemma 3.

The group M is not necessarily locally finite - but we have not yet said what player I is to do. Assume the finite sets of individual constants are listed in some order. At stage 2i, player I shall consider the i-th such set, let it be C. Player II has just played a set X_{2i-1} (= $\emptyset$ if i = 0). Since the rules of the game say that X_{2i-1} must be consistent with T_0, some finite group H (with added constants) is a model of X_{2i-1}. If any constants in C don't occur in X_{2i-1}, player I shall read them as names of e in H. Then he shall choose X_{2i} to be a set which contains X_{2i-1} and completely specifies the multiplication table of H. So X_{2i} implies that the elements named by constants in C generate a finite group.

When player I plays as described, the resulting group M will be locally finite. Hence it will be a ULF group, and so it must be the unique countable ULF group. []

Now we prove Theorem 7(b). By Lemma 9 and Fact B, T_0^f is a complete first-order theory. By Lemma 10, the countable ULF group is a model of T_0^f. Hence T, which is the complete first-order theory of this group, is exactly T_0^f. By construction, T_0 is a Π_1 set and hence is arithmetical. Therefore by Fact C, T = T_0^f is 1-1 reducible to first-order arithmetic. This completes the proof of Theorem 7.

[]

3. Conditions for extensions

In [9] B. H. Neumann raised the general problem: Given a group G and a finite set of equations with parameters in G, when does there exist a group $H \supseteq G$ such that the equations are simultaneously satisfied by some elements of H? The fullest model-theoretical treatment of Neumann's problem is in Ziegler [14]; it incorporates important results of G. Higman and A. Macintyre. Since Ziegler's work is only available in German at the time of writing, let me summarise what is known.

Suppose $E(\bar{x},\bar{y})$ is a finite set of equations. Then there is a recursive set Φ of quantifier-free strict Horn formulas $\varphi(\bar{x})$ such that for every group G and every sequence $\bar{g}$ of elements of G the following are equivalent:

(a) $\bar{g}$ satisfies each formula $\phi(\bar{x})$ from Φ.

(b) There is a group $H \supseteq G$ with a sequence $\bar{h}$ of elements which satisfy the equations $E(\bar{g},\bar{y})$.

(A quantifier-free strict Horn formula is a formula $\theta_1 \wedge \ldots \wedge \theta_n \rightarrow \psi$ where $\theta_1, \ldots, \theta_n$, ψ are equations.) Conversely for each recursive set Φ of quantifier-free strict Horn formulas $\phi(\bar{x})$, there is a finite system $E(\bar{x},\bar{y})$ of equations such that for every group G and every $\bar{g}$ in G, (a) is equivalent to (b). Also for every finite system $F(\bar{x},\bar{y})$ of equations and inequations there is a finite system $E(\bar{x},z,\bar{y}')$ of equations such that for every group G with elements $\bar{g}$ and element $d \neq e$, the following are equivalent:

(c) There is a group $H \supseteq G$ with a sequence $\bar{h}$ of elements which satisfy $F(\bar{g},\bar{y})$.

(d) There is a group $H' \supseteq G$ with a sequence $\bar{h}'$ of elements which satisfy $E(\bar{g},d,\bar{y}')$.

Thus in a sense all inequations can be eliminated in favour of a single inequation "$d \neq e$".

We can pose Neumann's problem for <u>finite</u> groups. There is no reason <u>a priori</u> to expect an answer like that for groups. For groups the existence of the set Φ above rests on the fact that the class of groups is a variety; the class of finite groups is not even first-order axiomatisable. Nevertheless I wager that there are some interesting facts to be found here. The facts proved below are all very preliminary, which is why I label them Observations and not Theorems.

It will be helpful to introduce some model-theoretic language. A <u>primitive</u> formula is one of form $\exists\bar{y}(\theta_1 \wedge \ldots \wedge \theta_n)$ where each θ_i is either an equation or an inequation; if each θ_i is an equation, the formula is said to be <u>positive primitive</u>. We shall say that a sentence ϕ is <u>potentially true</u> in the finite group G, in symbols $G \models \Diamond\phi$, iff there is a finite group $H \supseteq G$ such that $H \models \phi$.

The elimination of inequations goes very much as it did for all groups:

OBSERVATION 11. Let $\phi(\bar{x})$ be a primitive formula. Then there is a positive primitive formula $\psi(\bar{x},z)$ such that for every finite group G, all elements $\bar{g}$ in G and each element $d \neq e$ in G the following are equivalent:

(a) $G \models \Diamond\phi(\bar{g})$.
(b) $G \models \Diamond\psi(\bar{g},d)$.

<u>Proof</u>. I show only how to eliminate one inequation. To say "$a \neq b$", given an element $d \neq e$, it suffices to take the term t in Corollary 6 above and say:

"In some finite $H \supseteq G$ there are elements c, h, j such that $t(a,c,h,j) = e$ and $t(b,c,h,j) = d$." By Lemma 2, this is true in G iff it is potentially true in G. []

Next we ask whether the existence of an extension in which given equations are satisfied can be expressed by a set of quantifier-free strict Horn formulas. It can, but I failed to make the set recursive:

OBSERVATION 12. Let $\psi(\bar{x})$ be a positive primitive formula. Then there exists a Π_1 set Φ of quantifier-free strict Horn formulas $\phi(\bar{x})$, such that for every finite group G with elements $\bar{g}$ the following are equivalent:

(a) $G \models \phi(\bar{g})$ for every $\phi(\bar{x}) \in \Phi$.
(b) $G \models \Diamond\psi(\bar{g})$.

Proof. If G is a finite group with elements $\bar{g}$, write $gp(\bar{g})$ for the subgroup of G generated by $\bar{g}$. Clearly if $\theta(\bar{x})$ is a quantifier-free formula then $G \models \theta(\bar{g})$ iff $gp(\bar{g}) \models \theta(\bar{g})$. We claim something stronger: If $\theta(\bar{x})$ is a primitive formula then $G \models \Diamond\theta(\bar{g})$ iff $gp(\bar{g}) \models \Diamond\theta(\bar{g})$. This follows from Lemma 2 and the fact that embeddings preserve primitive formulas. It follows that when we verify the equivalence of (a) and (b), we can confine ourselves to finite groups G which are generated by their elements $\bar{g}$.

Let $(G_i,\bar{g}_i)$ $(i < \omega)$ be a listing, up to isomorphism and without repetitions, of all the pairs $(G,\bar{g})$ such that G is a finite group which is generated by the sequence $\bar{g}$ of the same length as $\bar{x}$. For each $i < \omega$ let $\chi_i(\bar{x})$ be a conjunction of equations and inequations which describes $(G_i,\bar{g}_i)$ up to isomorphism. Let I be the set of all i such that $G_i \models \Diamond\psi(\bar{g}_i)$, and let Ξ be the set $\{\neg\chi_i : i \notin I\}$. Then for every finite group G with generators $\bar{g}$, we have

(**) $\quad G \models \Diamond\psi(\bar{g}) \quad$ iff $\quad G \models \xi(\bar{g})$ for every $\xi(\bar{x}) \in \Xi$.

This is nearly home, but the formulas in Ξ are not quantifier-free strict Horn.

Each formula $\xi(\bar{x})$ in Ξ is the negation of a conjunction of equations and inequations, so that it is logically equivalent to a disjunction of form $\neg\eta_1 \vee \ldots \vee \neg\eta_m \vee \theta_1 \vee \ldots \vee \theta_n$, where the η_j's and the θ_j's are equations. If $n = 1$, this disjunction is in turn logically equivalent to the quantifier-free strict Horn formula $\eta_1\wedge\ldots\wedge\eta_m \to \theta_1$. Now we do know that n must be <u>at least</u> 1. For consider the trivial group $\{e\}$. Every finite system of equations with parameters in this group is already satisfiable in the group. Hence if $i \notin I$ then the group G_i must have at least two elements, and χ_i must contain at least one inequation. Thus $n \geq 1$. We must show next that n can always be reduced to 1.

Let us call a pair $(G,\bar{g})$ <u>good</u> iff G is a finite group with elements $\bar{g}$ and $G \models \Diamond\psi(\bar{g})$. Consider a formula $\xi(\bar{x})$ in Ξ, which is logically equivalent to the disjunction written above. We claim that there is some j $(1 \leq j \leq n)$ such that in every good pair $(G,\bar{g})$, $\bar{g}$ satisfies $\neg\eta_1 \vee \ldots \vee \neg\eta_m \vee \theta_j$.

For suppose not. Then for each j $(1 \leq j \leq n)$ there is a good pair $(K_j,\bar{k}_j)$ such that $\bar{k}_j$ satisfies $\eta_1 \wedge \ldots \wedge \eta_m \wedge \neg\theta_j$. Since $(K_j,\bar{k}_j)$ is good, there is a finite group $H_j \supseteq K_j$ such that $H_j \models \psi(\bar{k}_j)$. Necessarily also $H_j \models (\eta_1 \wedge \ldots \wedge \eta_m \wedge \neg\theta_j)(\bar{k}_j)$ since the formulas involved are quantifier-free. Form the product $(H,\bar{k})$ of the $(H_j,\bar{k}_j)$ $(1 \leq j \leq n)$. Since $H = H_1 \times \ldots \times H_n$ and the H_j are finite groups, H is a finite group. Also by familiar properties of products we have

$$H \models \psi(\bar{k}), \qquad H \models (\eta_1 \wedge \ldots \wedge \eta_m)(\bar{k}) \qquad \text{and for each } j,\ H \models \neg\theta_j(\bar{k}).$$

It follows that $(gp(\bar{k}),\bar{k})$ is a good pair but $gp(\bar{k}) \models \neg\xi(\bar{k})$. This contradicts (**) above.

Thus the claim is proved. It follows that for every formula $\xi(\bar{x})$ in Ξ there is a quantifier-free strict Horn formula $\xi'(\bar{x})$ which logically implies $\xi(\bar{x})$ (it comes from a formula logically equivalent to $\xi(\bar{x})$ by removing some disjuncts), such that the implication from left to right in (**) still holds if we replace Ξ by the set $\Xi' = \{\xi' : \xi \in \Xi\}$. Since Ξ' is logically stronger than Ξ, the implication from right to left must still hold too.

Unfortunately the set Ξ' is not necessarily Π_1. However, if (**) holds with Ξ replaced by some set of quantifier-free strict Horn formulas, then it must certainly hold when Ξ is replaced by the set Φ, where Φ is the set of <u>all</u> quantifier-free strict Horn formulas $\phi(\bar{x})$ such that for every good pair $(G,\bar{g})$ where $\bar{g}$ generates G, $G \models \phi(\bar{g})$. This set Φ is indeed Π_1. The formula ϕ is in Φ iff [ϕ is quantifier-free strict Horn] $\wedge$ $\forall m \forall n$[m encodes a pair $(G,\bar{g})$ where G is a finite group generated by $\bar{g}$, and n encodes a finite group $H \supseteq G$ such that $H \models \psi(\bar{g})$ $\Rightarrow G \models \phi(\bar{g})$]. Here everything inside the square brackets is recursive. Hence Φ proves the Observation. []

QUESTION 13. Can the set Φ in Observation 12 always be chosen to be recursive? (Conjecture: No.)

Question 13 is only one small part of a larger question. Let us consider conditions on pairs $(G,\bar{g})$ such that G is a finite group with elements $\bar{g}$. Let us say that such a condition C is a PP condition (resp. P condition) iff there is some positive primitive formula (resp. primitive formula) $\psi(\bar{x})$ such that $(G,\bar{g})$ meets condition C iff $G \models \Diamond \psi(\bar{g})$. Thanks to Ziegler's work quoted above, we have a nice answer to the following question when the groups are allowed to be infinite:

QUESTION 14. Characterise the PP conditions and the P conditions.

I can say a very little about Question 14. First, by Observation 11 the characterisation of P conditions will not be essentially harder than the characterisation of PP conditions. Next, the following two conditions are respectively P and PP:

(1) "g_1 is not in the subgroup generated by $g_2,\ldots,g_n$"; take $\psi(\bar{x})$ to be $\exists y([x_1,y] \neq e \wedge [x_2,y] = e \wedge \ldots \wedge [x_n,y] = e)$, and refer to Lemma 8 above.

(2) "the order of g_1 divides the order of g_2"; take $\psi(\bar{x})$ to be $\exists y \exists z \exists u(y^{-1}z^{-1}x_2 z u^{-1}x_2^{-1}uy = x_1 \wedge [z^{-1}x_2 z, u^{-1}x_2 u] = e)$, and refer to Lemmas 3 and 4 above.

The only other P conditions known to me can be got from (1), (2) and equations and inequations by conjunction and existential quantification. I cannot believe that the world is really so boring.

Next a question of a more model-theoretic cast. Let us say that a formula $\psi(\bar{x})$ is preserved upwards iff for every pair of finite groups H, G and every sequence of elements $\bar{g}$ of G, if $H \supseteq G$ and $G \models \psi(\bar{g})$ then $H \models \psi(\bar{g})$. Let us say that formulas $\psi(\bar{x})$ and $\phi(\bar{x})$ are finite-equivalent iff for every finite group G with elements $\bar{g}$, $G \models \psi(\bar{g})$ iff $G \models \phi(\bar{g})$.

QUESTION 15. Suppose a first-order formula $\psi(\bar{x})$ is preserved upwards. Then is it finite-equivalent to a finite disjunction of primitive formulas? (Conjecture: No.)

Remark: it is certainly finite-equivalent to a countable disjunction of primitive formulas, viz. a disjunction of complete descriptions of all finite pairs $(G,\bar{g})$ which satisfy it.

Finally a question which may have occurred to the reader in section 2 above:

QUESTION 16. Let T_0 be the universal theory of finite groups. Does T_0 have the amalgamation property? (Conjecture: No.)

REFERENCES

[1] Jon Barwise and Abraham Robinson, Completing theories by forcing, Ann. Math. Logic 2 (1970) 119-142.

[2] O. V. Belegradek, Elementary properties of algebraically closed groups (Russian), Fund. Math. 98 (1978) 83-101.

[3] O. V. Belegradek, Classes of algebras with internal mappings (Russian), in Investigations in theoretical programming (Russian), State University of Alma Ata 1981, pp. 3-10.

[4] P. Hall, Some constructions for locally finite groups, J. London Math. Soc. 34 (1959) 305-319.

[5] Joram Hirschfeld and William H. Wheeler, Forcing, arithmetic, division rings, Lecture Notes in Mathematics 454, Springer, Berlin 1975.

[6] H. K. Kaiser and R. Lidl, Erweiterungs- und Rédeipolynomvollständigkeit universaler Algebren, Acta Math. Acad. Sci. Hungar. 26 (1975) 251-257.

[7] Otto Kegel and Bertram A. F. Wehrfritz, Locally finite groups, North-Holland, Amsterdam 1973.

[8] W. D. Maurer and John L. Rhodes, A property of finite simple non-abelian groups, Proc. Amer. Math. Soc. 16 (1965) 552-554.

[9] B. H. Neumann, Adjunction of elements to groups, J. London Math. Soc. 18 (1943) 4-11.

[10] B. H. Neumann, Permutational products of groups, J. Austral. Math. Soc. Ser. A 1 (1959/60) 299-310.

[11] Hanna Neumann, Varieties of groups, Springer, New York 1967.

[12] A. M. Slobodskoĭ, The undecidability of the universal theory of finite groups (Russian), Algebra i Logika 20 (1981) 207-230.

[13] M. Yu. Trofimov, On definability in algebraically closed systems (Russian), Algebra i Logika 14 (1975) 320-327; Eng. tr. Algebra and Logic 14 (1975) 198-202.

[14] Martin Ziegler, Algebraisch abgeschlossene Gruppen, in Word problems II, the Oxford book, ed. S. I. Adian, W. W. Boone and G. Higman, North-Holland, Amsterdam 1980, pp. 449-576.

Postscript

At the talk some people (I think Scott and Jockusch) kindly pointed out what I had stupidly overlooked, viz. the proper way to prove Theorem 7(b). Hall's construction of the countable ULF group G in [4] clearly makes G a recursive object. So the theory of G is interpretable in that of N. It would have been more sensible to state Theorem 7 in the stronger form which this proof allows:

STRONGER THEOREM 7. T and complete first-order arithmetic are each interpretable in the other.

CONSTRUCTING CHOICE SEQUENCES FROM LAWLESS SEQUENCES OF NEIGHBOURHOOD FUNCTIONS

G. F. van der Hoeven
(Twente University of Technology)

I. Moerdijk
(University of Amsterdam)

1. INTRODUCTION

The aim of this paper is to illustrate how various notions of choice sequence can be derived from, or reduced to, the notion of a lawless sequence. More accurately, we will construct a sequence of models, starting with a model for lawless sequences of neighbourhood functions, and arriving by subsequent modifications at a model for the theory CS of Kreisel & Troelstra(1970).

Such a process of gradually transforming a model for the theory of lawless sequences into a model for the theory of CS provides an answer to the question posed in Kreisel(1968), p.243, "How fundamental are lawless sequences", in the sense that it shows that many concepts of choice sequence can be derived from a notion of lawlessness.

The first model to be discussed in section 4 will be a model for lawless sequences of neighbourhood functions, which is completely analogous to a model for the theory LS of lawless sequences of natural numbers (for LS, see Kreisel(1968), Troelstra(1977)). This model for LS will be presented in section 3, after a short introduction to forcing over sites given in section 2.

With a lawless sequence of neighbourhood functions ξ one can associate a "potential" sequence of natural numbers α: given an initial segment $(f_0,\ldots,f_n)$ of ξ, the information we have about α is that it lies in the image of $f_0 \circ \ldots \circ f_n$ (where the f_i's are regarded as lawlike continuous operations $\mathbb{N}^{\mathbb{N}} \to \mathbb{N}^{\mathbb{N}}$, so composing them makes sense).

A first modification of the site serves to eliminate an intensional aspect of the information we have about such a potential sequence α: two initial segments $(f_0,f_1,\ldots,f_n)$ and $(f_0 \circ f_1, f_2,\ldots,f_n)$ represent the same information about α, and should therefore be identified, given that α is the sequence we are interested in, rather than the lawless sequence of neighbourhood functions that α is constructed from.

A next modification turns these potential sequences of natural numbers into actual ones, simply by refining the Grothendieck topology of the underlying site. We will see that the universe of choice sequences obtained at this stage is of little interest.

This situation changes radically if we modify the site once more, this time in order to obtain closure properties of the universe of choice sequences, and, in a next step, eliminate the intensional aspects introduced with these closure properties. We then have a model in which the universe of choice sequences satisfies the CS-axioms of *analytic data* and $\forall\alpha\exists n$ - *continuity*, i.e.

$$\forall\alpha(A(\alpha) \to \exists F(\alpha \in im(F) \wedge \forall\beta \in im(F)A(\beta)))$$

$$\forall\alpha\exists nA(\alpha,n) \to \exists F\forall\alpha A(\alpha,F\alpha),$$

where F ranges over lawlike continuous operations $\mathbb{N}^{\mathbb{N}} \to \mathbb{N}^{\mathbb{N}}$, and $\mathbb{N}^{\mathbb{N}} \to \mathbb{N}$ respectively.

Moreover, this model has a natural notion of independence, which is decidable (i.e. $\alpha \# \beta \vee \neg\alpha \# \beta$ is valid, where we write $\alpha \# \beta$ for " α is independent from β "). Using this notion, we can formulate several variants of $\forall\alpha\exists\beta$ - continuity which are valid in this model, such as

$$\forall\alpha\exists\beta(\neg\alpha\#\beta \wedge A(\alpha,\beta)) \to \exists F\forall\alpha A(\alpha,F\alpha)$$

$$\forall\alpha\exists\beta(\alpha\#\beta \wedge A(\alpha,\beta)) \to \exists e \in K\forall u(e(u) \neq 0 \to \exists\beta\forall\alpha \in u(\alpha\#\beta \to A(\alpha,\beta))).$$

A multiple parameter version of analytic data also holds:

$$\forall\alpha_1,\ldots\alpha_n(\#(\alpha_1,\ldots,\alpha_n) \wedge A(\alpha_1,\ldots,\alpha_n) \to \exists F_1,\ldots,F_n(\bigwedge_{i\leq n} \alpha_i \in im(F_i) \wedge$$
$$\wedge \forall\beta_1,\ldots,\beta_n(\#(\beta_1,\ldots,\beta_n) \to A(F\beta_1,\ldots,F\beta_n))),$$

where $\#(\alpha_1,\ldots,\alpha_n)$ abbreviates $\bigwedge\{\alpha_i\#\alpha_j \mid 1\leq i<j\leq n\}$.

However, the usual version of $\forall\alpha\exists\beta$ - continuity (which is an axiom of CS),

$$\forall\alpha\exists\beta A(\alpha,\beta) \rightarrow \exists F\forall\alpha A(\alpha,F\alpha),$$

does not hold.

Finally, we modify the site by introducing the possible creation of certain dependencies between sequences. This is done in two steps. After the first step, we obtain validity of the usual CS-version of $\forall\alpha\exists\beta$ - continuity. So the only thing that is missing for a CS-model is the axiom of *pairing*,

$$\forall\alpha,\beta\exists\gamma\exists F,G(\alpha = F\gamma \wedge \beta = G\gamma).$$

A second step will accomplish the validity of this axiom, and we have arrived at a model for CS.

The constructions of these models and the proofs of their properties can be performed in an intuitionistic system like IDB (see Kreisel & Troelstra(1970)). This means that the theories of choice sequences that we provide models for are all consistent with Church's thesis ("all *lawlike* sequences are recursive") and lawlike countable choice.

As we said above, this sequence of models illustrates how various concepts of choice sequence can be reduced to the concepts of lawlessness. There is an interesting parallel here between the material of this paper and the program of "imitating" notions of choice sequence by means of "projections of lawless sequences" (cf. van Dalen & Troelstra(1970), van der Hoeven & Troelstra(1980), van der Hoeven (1982)), which has a similar purpose of reducing arbitrary choice sequences to lawless ones.

For example, in van der Hoeven(1982) a restricted version of CS is modelled by sequences constructed from a lawless sequence of neigbourhood functions and two lawless sequences of natural numbers, of which the latter two serve to make potential sequences into actual ones and to create dependencies between members of the universe of choice sequences.

There are some important differences between these two approaches, however, the main one being that here we obtain new notions of choice sequence by modifying the underlying site, that is, by modifying the notion of truth, whereas on the projections approach new universes of choice sequences are constructed by applying more complex continuous operations to lawless sequences.

Our present approach is technically simpler, because the changes in the forcing definition really make the intensional differences between sequences invisible. Using projections, the forcing definition remains the same, but long formula inductions are needed to show that for formulas in the language of analysis the property of being forced is independent of intensional differences in the parameters.

On the other hand, choice sequences projected from lawless sequences give a clearer picture of a construction process. In the sites we discuss here there are obvious representatives of steps in such a process ("going back along the arrows"), but the process as a whole is not explicitly presented.

Summarizing the results of this paper, then, we find models which have properties similar to the models for the CS-like systems constructed by projections. In particular, our last model but one, in which all of CS except pairing holds, is closely related to the models of van der Hoeven & Troelstra(1980). Technically, however, the projections approach is much more involved than the present one. The full generality of the models we obtain here has (so far) not been achieved along the projections approach: the projection models are all models of restricted variants of the theories we model here. Moreover, we obtain some new models for - so it seems to us - interesting systems with a primitive relation of independence.

In our paper van der Hoeven & Moerdijk(to appear) we constructed two models for the system CS by using forcing over sites, as we do here. Especially the first model (section 2.2 of that paper) is in some sense much simpler than the present one, but its construction is not motivated by a "reduction to lawless sequences" and, contrary to the present approach, we do not meet interesting (sub-)systems on the way of the construction of that model. The second CS-model in that paper (section 4) bears a relation to lawless sequences, but since it is constructed from the first one simply by considering what would be needed to prove it first order equivalent, this relation is less natural as a reduction. (See the remarks in Troelstra(1983), pp. 245-6.)

Thus, the constructions of these three CS-models are motivated rather differently, and the relation between these models needs closer investigation. This is a problem, however, that we do not touch upon in this paper.

2. FORCING OVER SITES

To make this paper accessible to readers who are less familiar with forcing over sites, we will review some of the basic notions of this theory, otherwise known as sheaf semantics or Beth-Kripke-Joyal semantics.

Let $\mathbb{C}$ be a category. If C is an object of $\mathbb{C}$, a *sieve* on C is a collection of morphisms S with codomain C which is closed under right composition, i.e. if $D \xrightarrow{f} C \in S$ and $E \xrightarrow{g} D$ is any morphism of $\mathbb{C}$ then $f \circ g \in S$.

A *Grothendieck topology* on $\mathbb{C}$ is a function which associates to every object C of $\mathbb{C}$ a family $J(C)$ of sieves on C, called *covering sieves*, such that

(i) (trivial cover) For each C, the maximal sieve
$\{f \mid \text{codomain}(f) = C\} \in J(C)$.

(ii) (stability) If $S \in J(C)$ and $D \xrightarrow{f} C$ is a morphism of $\mathbb{C}$ then
$f^*(S) = \{E \xrightarrow{g} D \mid f \circ g \in S\} \in J(D)$.

(iii) (transitivity) If $R \in J(C)$ and S is a sieve on C such that for each $D \xrightarrow{f} C \in R$, $f^*(S) \in J(D)$, then $S \in J(C)$.

A *site* is a category equipped with a Grothendieck topology. A site is called *consistent* if $\phi \notin J(C)$ for some $C \in \mathbb{C}$, i.e. at least one object is not covered by the empty family.

(i), (ii), (iii) are closure conditions, so the intersection of a family of Grothendieck topologies is again a Grothendieck topology. Consequently, if for *some* objects $C \in \mathbb{C}$ we specify a couple of families (not necessarily sieves) $\{C_i \xrightarrow{f_i} C\}_i$ with codomain C ("basic covering families"), then there exists a smallest Grotendieck topology J with the property that for each of these selected objects C, and for each sieve S on C, $S \in J(C)$ whenever S contains one of these basic covering families. This smallest topology J is called the topology *generated* by the basic covering families.

In general, it is rather hard to keep track of what a collection of basic covers generates, in particular, it is hard to see whether the generated Grothendieck topology is consistent. For this reason, it is more convenient to work with basic covers of the following form: for *each* object C we specify a collection $K(C)$ of families $\{C_i \xrightarrow{f_i} C\}_i$ such that

(i') (trivial cover) The one-element family $\{C \xrightarrow{id} C\} \in K(C)$.

(ii') (stability) If $C_i \xrightarrow{f_i} C\}_i \in K(C)$ and $D \xrightarrow{g} C$ is a morphism of $\mathbb{C}$, then there is a family $\{D_j \xrightarrow{h_j} D\}_j \in K(D)$ such that for each j there is an i and a morphism k with $f_i \circ k = g \circ h_j$.

(iii') (transitivity) If $\{C_i \xrightarrow{f_i} C\}_i \in K(C)$ and for each i we have a family $\{C_{ij} \xrightarrow{g_{ij}} C_i\}_j \in K(C_i)$, then $\{C_{ij} \xrightarrow{f_i \circ g_{ij}} C\}_{i,j} \in K(C)$.

If we have a family of *basic covers* $K(C)$ for each $C \in \mathbb{C}$ satisfying (i')-(iii'), then the Grothendieck topology J generated by K is defined by

$$R \in J(C) \iff \exists S \in K(C) S \subseteq R.$$

In particular, J is consistent iff $\phi \notin K(C)$ for some object C (we say that K is consistent).

In section 4, we will define (models over) sites by some basic covers which in general do not satisfy (i')-(iii'). So the way to show that our models are consistent is to find a bigger collection K of basic covers which does satisfy (i')-(iii'), and is consistent. This is rather straightforward in all cases, and will in general not be shown in detail.

A *domain* X on a site $(\mathbb{C},J)$ is a functor $\mathbb{C}^{op} \to \text{Sets}$, i.e. a collection of sets $\{X(C) \mid C \in \mathbb{C}\}$ together with *restriction maps*

$$X(D) \to X(C), \qquad x \mapsto x|f,$$

for every morphism $C \xrightarrow{f} D$ of $\mathbb{C}$, such that $(x|f)|g = x|(f \circ g)$, and $x|id = x$. The elements of $X(C)$ are to be thought of as *partially constructed* members of the domain X, C is the "stage" of construction, and by the restriction along $D \xrightarrow{f} C$ we gain more information about such a partially constructed member of X, i.e. we perform a construction step.

A *lawlike domain* (more precisely, a domain of lawlike objects) is a domain which consists of *complete* objects: there is nothing to be constructed. So X is lawlike if $X(C)$ = a fixed set X, and all restriction maps are identities. Thus, for each "external" set X there is a corresponding lawlike domain, also denoted by X, with $X(C) = X$. The main examples that occur in section 4 are the lawlike domain of natural numbers ($\mathbb{N}(C) = \mathbb{N}$ for all C), and the domain of lawlike neighbourhood functions K ($K(C) = K$, the set of inductively defined neighbourhood functions).

Given a collection of domains on $(\mathbb{C}, J)$ we define forcing for a many sorted language L. Each sort of L is identified with a certain domain. And each constant c of L, of sort X say, is identified with a family of elements $c(D) \in X(D)$, D an object of $\mathbb{C}$, coherent in the sense that $c(D) \mid f = c(E)$ for any morphism $E \xrightarrow{f} D$.

Moreover, we assume to be given an interpretation of each relation symbol R (taking n arguments of sorts $X_1, \ldots, X_n$ say). The interpretation of R is an assignment of a subset $R(C) \subseteq X_1(C) \times \ldots \times X_n(C)$ to each object C, such that for $D \xrightarrow{f} C$,

$$(x_1, \ldots, x_n) \in R(C) \Rightarrow (x_1|f, \ldots, x_n|f) \in R(D).$$

The *forcing relation*

$$C \Vdash \varphi(x_1, \ldots, x_n),$$

where φ has free variables among $v_1, \ldots v_n$, v_i of sort X_i, and $x_i \in X_i(C)$, is now defined by induction. For atomic formulas we have

$$C \Vdash x = y \Leftrightarrow \text{there is an } S \in J(C) \text{ such that } x|f = y|f \text{ for all } f \in S$$

$$C \Vdash R(x_1, \ldots, x_n) \Leftrightarrow \text{there is an } S \in J(C) \text{ such that } (x_1|f, \ldots, x_n|f) \in R(D) \text{ for all } D \xrightarrow{f} C \in S.$$

Furthermore,

$$C \Vdash \bot \Leftrightarrow \emptyset \in J(C)$$

$$C \Vdash \varphi \wedge \psi(x_1, \ldots, x_n) \Leftrightarrow C \Vdash \varphi(x_1, \ldots, x_n) \text{ and } C \Vdash \psi(x_1, \ldots, x_n)$$

$$C \Vdash \varphi \wedge \psi(x_1,\dots,x_n) \iff C \Vdash \varphi(x_1,\dots,x_n) \text{ and } C \Vdash \psi(x_1,\dots,x_n)$$

$$C \Vdash \varphi \vee \psi(x_1,\dots,x_n) \iff \{D \xrightarrow{f} C \mid D \Vdash \varphi(x_1|f,\dots,x_n|f) \text{ or } D \Vdash \psi(x_1|f,\dots,x_n|f)\} \in J(C)$$

$$C \Vdash \varphi \to \psi(x_1,\dots,x_n) \iff \text{for all morphisms } D \xrightarrow{f} C, \text{ if } D \Vdash \varphi(x_1|f,\dots,x_n|f) \text{ then } D \Vdash \psi(x_1|f,\dots,x_n|f).$$

and for variables v of sort Y we have

$$C \Vdash \exists v\varphi(v,x_1,\dots,x_n) \iff \{D \xrightarrow{f} C \mid \exists y \in Y(D)\ D \Vdash \varphi(y,x_1|f,\dots,x_n|f)\} \in J(D)$$

$$C \Vdash \forall v\varphi(v,x_1,\dots,x_n) \iff \text{for all } D \xrightarrow{f} C \text{ and all } y \in Y(D),\ D \Vdash \varphi(y,x_1|f,\dots,x_n|f).$$

By induction, one can show that the forcing relation has the important properties of being *monotone* and *local*:

(monotone) If $D \xrightarrow{f} C$ and $C \Vdash \varphi(x_1,\dots,x_n)$ then $D \Vdash \varphi(x_1|f,\dots,x_n|f)$.

(local) If $S \in J(C)$ and $D \Vdash \varphi(x_1|f,\dots,x_n|f)$ for every $D \xrightarrow{f} C$ in S, then $C \Vdash \varphi(x_1,\dots,x_n)$.

A formula $\varphi(v_1,\dots,v_n)$ is called *valid* (notation: $\models \varphi(v_1,\dots,v_n)$) if for each object C and each n-tuple $(x_1,\dots,x_n) \in X_1(C) \times \dots \times X_n(C)$,

$$C \Vdash \varphi(x_1,\dots,x_n).$$

Function symbols F of L, taking n arguments of sorts $X_1, \dots, X_n$ to a value of sort Y, are treated as $n+1$-place relation symbols such that $\forall x_1 \dots x_n \exists! y$ $F(x_1,\dots,x_n) = y$ is valid.

This interpretation makes all of intuitionistic predicate calculus valid, and when higher order sorts (exponentials and powersets) are properly defined it provides a model for intuitionistic type theory with full comprehension. When the sort $\mathbb{N}$ of natural numbers is interpreted by the corresponding lawlike domain, we obtain a model for (higher order) intuitionistic arithmetic (These are well-known facts, but they are not needed for the understanding of the rest of this paper.) The first order part of arithmetic is classical if we work in a classical metatheory, since (as is easily shown by induction) we have

$$C \Vdash \varphi(x_1,\dots,x_n) \leftrightarrow \varphi(x_1,\dots,x_n) \quad \text{is true}$$

if the sorts X_i are lawlike, and all quantifiers in φ range over lawlike sorts.

3. A MODEL FOR LS

As a preparation to the next section, which is the core of this paper, we will now describe a model for the theory LS of lawless sequences of natural numbers. This model is not new, and was first described in Fourman(1982).

The underlying site of the model has as *objects* finite products of basic open subspaces of Baire space. We write such objects as

$$V_{u_1} \times \dots \times V_{u_n},$$

where $u_i \in \mathbb{N}^{<\mathbb{N}}$ and $V_{u_i} = \{x \in \mathbb{N}^{\mathbb{N}} \mid x \text{ has initial segment } u_i\}$. The empty product, which is the one point space, is denoted by 1. *Morphisms* from one such object to another

$$\Phi\colon V_{u_1}\times\dots\times V_{u_n} \to V_{v_1}\times\dots\times V_{v_m},$$

are continuous maps induced by injections $\varphi\colon \{1,\dots,m\} \rightarrowtail \{1,\dots,n\}$ such that $u_{\varphi(i)}$ extends v_i, via $\Phi(x_1,\dots,x_n) = (x_{\varphi(1)},\dots,x_{\varphi(m)})$. The *Grothendieck topology* is generated by basic covers of two sort ($*$ denotes concatenation):

(i) (open covers) $\{V_{u*n} \hookrightarrow V_u\}_{n\in\mathbb{N}}$ is a cover.

(ii) (projections) the singleton $\{V_u \times V_v \to V_u\}$ is a cover.

Classically, the generated Grothendieck topology can be described as: a family $\{\Phi_i\colon U_i \to U\}$ covers iff the images $\Phi_i(U_i) \subseteq U$ form an (open) cover of U. In an intuitionistic metatheory like IDB, we do not get all open covers, but only the inductively defined ones (cf. the remarks at the end of this section).

The relevant domains over this site are the following: we have the lawlike domains $\mathbb{N}$ of natural numbers and K of neighbourhood functions, the lawlike domain of continuous operations $\mathbb{N}^{\mathbb{N}} \to \mathbb{N}^{\mathbb{N}}$ corresponding to neighbourhood functions, and the lawlike domain $\mathbb{N}^{\mathbb{N}}$ of lawlike sequences (so all the "external" sequences appear in the model as lawlike sequences). The domain L of lawless sequences is the domain of projections,

$$L(V_{u_1}\times\ldots\times V_{u_n}) = \{\pi_i : V_{u_i} \to \mathbb{N}^{\mathbb{N}} \mid i=1,\ldots,n\},$$

with restrictions defined by composition: If $\Phi: V_{u_1}\times\ldots\times V_{v_n} \to V_{v_1}\times\ldots\times V_{v_m}$ is a morphism induced by φ as above, then $\pi_i|\Phi = \pi_i\circ\Phi = \pi_{\varphi(i)}$ $(i=1,\ldots,m)$. If U is an object of the site and $\alpha \in L(U)$, then α is interpreted as a sequence of natural numbers by

(1) $U \Vdash \alpha(n) = m \Longleftrightarrow \forall x \in U\ \alpha(x)(n) = m,$

in other words, if $U = V_{u_1}\times\ldots\times V_{u_n}$ and $\alpha = \pi_i$ then $U \Vdash \alpha \in v$ iff u_i extends v, for any finite sequence v (as usual, $\alpha \in v$ stands for $\forall i < \ell th(v)$ $\alpha(i) = v(i)$). Note that definition (1) is monotone and local, i.e. if $\Phi: W \to U$ and $U \Vdash \alpha(n) = m$ then $W \Vdash (\alpha|\Phi)(n) = m$, and if $\{\Phi_i: W_i \to U\}_i$ covers and each $W_i \Vdash (\alpha|\Phi_i)(n) = m$ then $U \Vdash \alpha(n) = m$.

This completes the description of the model.

The validity of the two simpler LS-axioms, *density*: $\forall v\exists\alpha(\alpha\in v)$ and *decidable equality*: $\forall\alpha,\beta(\alpha=\beta \vee \neg\alpha=\beta)$ is easily verified. For density, take a $v \in \mathbb{N}^{<\mathbb{N}}$ and an object $V_{u_1}\times\ldots\times V_{u_n}$. Then the projection $V_{u_1}\times\ldots\times V_{u_n}\times V_v \to V_{u_1}\times\ldots\times V_{u_n}$ covers, and $V_{u_1}\times\ldots\times V_{u_n}\times V_v \Vdash \pi_{n+1} \in v$. So $V_{u_1}\times\ldots\times V_{u_n} \Vdash \exists\alpha(\alpha\in v)$. Furthermore, it is easily seen that

$$V_{u_1}\times\ldots\times V_{u_n} \Vdash \pi_i \neq \pi_j \quad \text{iff} \quad i \neq j,$$

from which decidable equality follows immediately.

Before we prove the validity of open data and continuity in the model we state three simple observations about the forcing relation.

Observation 1. *If* $A(\alpha_1,\ldots,\alpha_n)$ *is a formula which has all its non-lawlike parameters among* $\alpha_1, \ldots, \alpha_n$, *and* $i_1, \ldots, i_n$ *are distinct numbers in* $\{1,\ldots,k\}$, *then*

$$V_{u_1}\times\ldots\times V_{u_k} \Vdash A(\pi_{i_1},\ldots,\pi_{i_n}) \quad \textit{iff} \quad V_{u_{i_1}}\times\ldots\times V_{u_{i_n}} \Vdash A(\pi_{i_1},\ldots,\pi_{i_n})$$

(proof: $\Vdash$ is local and monotone (section 2), and the projection $\phi: V_{u_1}\times\ldots\times V_{u_k} \to V_{u_{i_1}}\times\ldots\times V_{u_{i_n}}$, $\phi(x_1,\ldots,x_k) = (x_{i_1},\ldots,x_{i_k})$, is a cover.)

Observation 2. *If* A *has only lawlike parameters, then* $\models A$ *iff for some object* U, $U \Vdash A$.

(proof: if $U \Vdash A$ then by observation 1, $1 \Vdash A$, so by monotonicity, $V \Vdash A$ for any object V since 1 is terminal, i.e. there is a unique morphism $V \to 1$ in the site.)

Observation 3. *Let* U *be any object in the site. Then*

$U \Vdash \underline{\forall}\alpha_1\ldots\underline{\forall}\alpha_n(A(\alpha_1,\ldots,\alpha_n) \to B(\alpha_1,\ldots,\alpha_n))$ *iff for all* $u_1,\ldots,u_n \in \mathbb{N}^{<\mathbb{N}}$, $V_{u_1}\times\ldots\times V_{u_n} \Vdash A(\pi_1,\ldots,\pi_n)$ *implies* $V_{u_1}\times\ldots\times V_{u_n} \Vdash B(\pi_1,\ldots,\pi_n)$.

Here $\underline{\forall}\alpha_1 \ldots \underline{\forall}\alpha_n$ (..) abbreviates $\forall\alpha_1 \ldots \alpha_n(\bigwedge_{i<j} \alpha_i\neq\alpha_j \to (..))$ and $A(\alpha_1,\ldots,\alpha_n)$, $B(\alpha_1,\ldots,\alpha_n)$ have all their non-lawlike parameters among $\alpha_1, \ldots, \alpha_n$.

(proof: by observation 2, we may assume $U = 1$. But if $W = V_{w_1}\times\ldots\times V_{w_k}$ is any object, and $i_1, \ldots, i_n$ are indices such that $W \Vdash \bigwedge_{\ell<\ell'} \pi_{i_\ell}\neq\pi_{i_{\ell'}} \wedge A(\alpha_{i_1},\ldots,\alpha_{i_n})$, then $i_1, \ldots i_n$ are all distinct, and by observation 1, $V_{v_1}\times\ldots\times V_{v_n} \Vdash A(\pi_1,\ldots,\pi_n)$ where $v_j = w_{i_j}$. So we may restrict ourselves to the case $W = V_{v_1}\times\ldots\times V_{v_n}$, as was to be shown.)

Note that as a consequence of observation 3 we have:

Genericity lemma. $\models \underline{\forall}\alpha_1 \ldots \underline{\forall}\alpha_n A(\alpha_1,\ldots,\alpha_n)$ *iff* $V_{()}\times\ldots\times V_{()} \Vdash A(\pi_1,\ldots,\pi_n)$ (n-fold product).

Using these observations, validity of the open data axiom and the axiom of continuity is easily established.

Open data reads

$$\underline{\forall}\alpha_1 \ldots \underline{\forall}\alpha_n(A(\alpha_1,\ldots,\alpha_n) \to \exists u_1, \ldots, u_n(\alpha_1\in u_1 \wedge\ldots\wedge \alpha_n\in u_n \wedge$$
$$\wedge\, \underline{\forall}\beta_1\in u_1 \ldots \underline{\forall}\beta_n\in u_n A(\beta_1,\ldots,\beta_n))).$$

By observation 3, it suffices to show that for any n-tuple $u_1, \ldots, u_n$, if

(1) $V_{u_1}\times\ldots\times V_{u_n} \Vdash A(\pi_1,\ldots,\pi_n)$,

then also

(2) $V_{u_1}\times\ldots\times V_{u_n} \Vdash \underline{\forall}\beta_1\in u_1 \ldots \underline{\forall}\beta_n\in u_n\, A(\beta_1,\ldots,\beta_n)$.

So assume (1). To prove (2), it suffices by observation 3 again to show that if

(3) $V_{w_1}\times\ldots\times V_{w_n} \Vdash \pi_1\in u_1 \wedge \ldots \wedge \pi_n\in u_n$

then also

(4) $V_{w_1}\times\ldots\times V_{w_n} \Vdash A(\pi_1,\ldots,\pi_n)$.

But if (3) holds, then w_i extends u_i so we have an inclusion morphism $V_{w_1}\times\ldots\times V_{w_n} \hookrightarrow V_{u_1}\times\ldots\times V_{u_n}$, restriction along which shows that (4) now follows from (1).

The axiom of continuity is

$$\underline{\forall}\alpha_1 \ldots \underline{\forall}\alpha_n \exists m\, A(\alpha_1,\ldots,\alpha_n, m) \rightarrow \exists F\underline{\forall}\alpha_1 \ldots \underline{\forall}\alpha_n\, A(\alpha_1,\ldots,\alpha_n, F(\alpha_1,\ldots,\alpha_n)),$$

where F ranges over lawlike continuous operations $\mathbb{N}^{\mathbb{N}}\times\ldots\times\mathbb{N}^{\mathbb{N}}\rightarrow\mathbb{N}$ (induced by neighbourhood functions $\mathbb{N}^{<\mathbb{N}}\times\ldots\times\mathbb{N}^{<\mathbb{N}}\rightarrow\mathbb{N}$), and $A(\alpha_1,\ldots,\alpha_n, m)$ has no non-lawlike parameters other than $\alpha_1, \ldots, \alpha_n$. By observation 2, proving continuity is equivalent to showing that if

(1) $\models \underline{\forall}\alpha_1 \ldots \underline{\forall}\alpha_n \exists m\, A(\alpha_1,\ldots,\alpha_n, m)$

then also

(2) $\models \exists F\underline{\forall}\alpha_1 \ldots \underline{\forall}\alpha_n\, A(\alpha_1,\ldots,\alpha_n, F(\alpha_1,\ldots,\alpha_n))$.

So assume (1). Then in particular for the n-fold product of $V_{()}$,

$$V_{()}\times\ldots\times V_{()} \Vdash \exists m\, A(\pi_1,\ldots,\pi_n, m),$$

so we find a cover $\{\phi_i: W_i \rightarrow V_{()}\times\ldots\times V_{()}\}$ and natural numbers m_i such that for each i,

$$W_i \Vdash A(\pi_1 \mid \phi_i,\ldots,\pi_n \mid \phi_i, m_i).$$

By observation 1, we may assume Φ_i to be a canonical inclusion $W_i = V_{w_{1,i}} \times \ldots \times V_{w_{n,i}} \hookrightarrow V_{()} \times \ldots \times V_{()}$, and by passing to a disjoint refinement of the inductive open cover $\{W_i\}_i$ of $V_{()} \times \ldots \times V_{()}$, we can define F by

$$F(x_1,\ldots,x_n) = m_i \quad \text{iff} \quad (x_1,\ldots,x_n) \in W_i.$$

Then $W_i \Vdash F(\pi_1,\ldots,\pi_n) = m_i$, so $W_i \Vdash A(\pi_1,\ldots,\pi_n, F(\pi_1,\ldots,\pi_n))$. Since the W_i cover, it follows that $V_{()} \times \ldots \times V_{()} \Vdash A(\pi_1,\ldots,\pi_n, F(\pi_1,\ldots,\pi_n))$. So by the genericity lemma, $\models \underline{\forall}\alpha_1 \ldots \underline{\forall}\alpha_n A(\alpha_1,\ldots,\alpha_n, F(\alpha_1,\ldots,\alpha_n))$, hence (2) holds.

The treatment of the model above is completely constructive, i.e. can be performed in an intuitionistic metatheory like IDB. (By definition of the Grothendieck topology, every cover has a corresponding characteristic neighbourhood function in K, so the map F defined in the proof of the continuity axiom can indeed be defined in IDB.) It is a corollary of the elimination theorem (Troelstra(1977)) that any interpretation of LS in IDB is equivalent to the elimination translation. In particular, for first order sentences A in the language of LS,

$$IDB \vdash (\Vdash A) \leftrightarrow \tau(A),$$

where $\Vdash$ is forcing over the model described above (formalized in IDB) and τ is the elimination translation. In fact, a simple formula induction shows that $V_{u_1} \times \ldots \times V_{u_n} \Vdash A(\pi_1,\ldots,\pi_n)$ and $\tau(\underline{\forall}\alpha_1 \in u_1 \ldots \underline{\forall}\alpha_n \in u_n A(\alpha_1,\ldots,\alpha_n))$ are literally the same (A not containing lawless parameters other than $\alpha_1,\ldots,\alpha_n$), provided one includes observation 2 in the forcing definition to get rid of vacuous quantifiers.

4. CHOICE SEQUENCES CONSTRUCTED FROM LAWLESS SEQUENCES OF NEIGHBOURHOOD FUNCTIONS

4.1. Lawless sequences of neighbourhood functions.

As usual, K denotes the (inductively defined) class of neighbourhood functions $\mathbb{N}^{<\mathbb{N}} \to \mathbb{N}$. Neighbourhood functions induce lawlike continuous operations $F: \mathbb{N}^{\mathbb{N}} \to \mathbb{N}$ and $F: \mathbb{N}^{\mathbb{N}} \to \mathbb{N}^{\mathbb{N}}$ (cf. Troelstra(1977)), and we will often identify F and f, writing things like $F \in K$, or $f \circ g$ for the composition of the corresponding continuous operations $\mathbb{N}^{\mathbb{N}} \to \mathbb{N}^{\mathbb{N}}$, etc.

Completely analogous to the model for lawless sequences described in section 3, we may construct a model for lawless sequences of neighbourhood functions from K (or lawless sequences of lawlike continuous operations $\mathbb{N}^{\mathbb{N}} \to \mathbb{N}^{\mathbb{N}}$). *Objects* of the site are now finite products of basic open subsets of $K^{\mathbb{N}}$,

$$V_{\xi_1} \times \ldots \times V_{\xi_n} \qquad (n \geq 0),$$

where $\xi_i \in K^{<\mathbb{N}}$, $\xi_i = (f_{i1}, \ldots f_{ik_i})$ (and each f_{ij} is identified with the corresponding continuous operation $\mathbb{N}^{\mathbb{N}} \to \mathbb{N}^{\mathbb{N}}$). *Morphisms* of the site are functions

$$\Phi\colon V_{\xi_1} \times \ldots \times V_{\xi_n} \to V_{\zeta_1} \times \ldots \times V_{\zeta_m}$$

which are induced by injections $\varphi\colon \{1,\ldots,m\} \rightarrowtail \{1,\ldots,n\}$ such that $\xi_{\varphi(i)}$ extends ζ_i $(i=1,\ldots,m)$, and $\Phi(x_1,\ldots,x_n) = (x_{\varphi(1)},\ldots,x_{\varphi(m)})$. The *Grothendieck topology* is generated by open covers and projections, just as in section 3. And as in this preceding section, lawless sequences are interpreted as projections. The logical properties of the model are of course exactly the same as the properties of the model of section 3. For the record:

<u>Theorem</u>. *The site described above gives a model for the theory of lawless sequences of neighbourhood functions, i.e. it satisfies the axioms of density, decidable equality, open data, and continuity* (the continuity axiom has to be rephrased using inductively defined neighbourhood functions on the tree $K^{<\mathbb{N}}$).

4.2. Potential choice sequences of natural numbers.

With each of the lawless sequences π_i $(i=1,\ldots,n)$ at an object $V_{\xi_1} \times \ldots \times V_{\xi_n}$ of the site of 4.1 we can associate a potential sequence α_i of natural numbers, by setting

$$V_{\xi_1} \times \ldots \times V_{\xi_n} \Vdash \alpha_i(n) = m \quad \text{iff} \quad \forall x \in \mathbb{N}^{\mathbb{N}} \;\; f_{i1} \circ \ldots \circ f_{ik_i}(x)(n) = m.$$

(Here $\xi_i = (f_{i1},\ldots,f_{ik_i})$, and the f_{ij} are regarded as operations $\mathbb{N}^{\mathbb{N}} \to \mathbb{N}^{\mathbb{N}}$, so composing them makes sense.) α_i is not an actual sequence, i.e.

$$V_{\xi_1} \times \ldots \times V_{\xi_n} \not\Vdash \forall n \exists m \; \alpha_i(n) = m,$$

since the extensions $(f_{i1},\dots,f_{ik_i}, g_1,\dots,g_\ell)$ of ξ_i such that $f_{i1}\circ\dots\circ g_\ell(x)(n)$ is constant in x do not form a cover of V_{ξ_i}. However, we can always extend a sequence in $K^{<\mathbb{N}}$ by a constant function (constant when regarded as an operation $\mathbb{N}^{\mathbb{N}} \to \mathbb{N}^{\mathbb{N}}$), and therefore

$$V_{\xi_1}\times\dots\times V_{\xi_n} \Vdash \neg\neg\forall n\exists m\, \alpha_i(n) = m\,.$$

The information we have about such a potential sequence α at an object $V_{(f_1,\dots f_k)}$ is that (in case α ever turns out to be a real sequence) $\alpha \in \mathrm{im}(f_1\circ\dots\circ f_k)$. Thus, since α is the only non-lawlike element at $V_{(f_1,\dots,f_n)}$ that we are interested in here, the two objects $V_{(f_1,f_2,f_3,\dots,f_n)}$ and $V_{(f_1\circ f_2,f_3,\dots,f_n)}$ represent equivalent information. Therefore we will identify two such objects by passing to a quotient space of $K^{\mathbb{N}}$:

<u>Definition</u>. *Let* $\sim$ *be the equivalence relation on* $K^{\mathbb{N}}$ *generated by*

$$(f_1,f_2,f_3,\dots) \sim (f_1\circ f_2,f_3,\dots).$$

The space X *is the quotient space* $K^{\mathbb{N}}/\sim$, *with the quotient topology.*

<u>Lemma</u>. *The canonical projection* $p\colon K^{\mathbb{N}} \to X$ *is open.*

<u>Proof</u>. The equivalence relation $\sim$ can also be described by

$$(f_n)_n \sim (g_n)_n \quad \text{iff} \quad \exists k,\ell(f_1\circ\dots\circ f_k = g_1\circ\dots\circ g_\ell \;\&\; \forall n\geq 1\; f_{k+n} = g_{\ell+n}).$$

So

$$p^{-1}p(V_{(f_1,\dots,f_n)}) = \{(g_n)_n \mid \exists\ell\exists h\in K\; g_1\circ\dots\circ g_\ell = f_1\circ\dots\circ f_n\circ h\},$$

which is open in $K^{\mathbb{N}}$. □

We will now modify the definition of our site, by replacing each basic open $V_{(f_1,\dots,f_k)}$ of $K^{\mathbb{N}}$ by the corresponding open $p(V_{(f_1,\dots,f_k)})$ of X. Note that $p(V_{(f_1,\dots,f_k)}) = p(V_{(f_1\circ\dots\circ f_k)})$, so we only need to work with sequences of length 1. We will write

$$V_f = p(V_{(f)}),$$

and sometimes by abuse of notation for sequences of length other than 1,

$V_{(f_1,\ldots,f_k)} = p(V_{(f_1,\ldots,f_k)})$, so $V_{(\,)}$ denotes $V_{id} = X$.

As our site we now take the site obtained by applying p to everything in the site of 4.1. So *objects* are now finite products

$$V_{f_1} \times \ldots \times V_{f_n},$$

and *morphisms* $\Phi: V_{f_1} \times \ldots \times V_{f_n} \to V_{g_1} \times \ldots \times V_{g_m}$ come from injections $\varphi: \{1,\ldots,m\} \to \{1,\ldots,n\}$ such that for each $i = 1,\ldots,m$ there exists an $h_i \in K$ with $g_i \circ h_i = f_{\varphi(i)}$, and $\Phi(x_1,\ldots,x_n) = (x_{\varphi(1)},\ldots,x_{\varphi(n)})$ as before, except that Φ is now a function on equivalence classes (note that this is well defined). The *Grothendieck topology* is generated by basic covers of two kinds:

(i) (open inclusions) $\{V_{f\circ g} \hookrightarrow V_f\}_{g\in K}$ is a cover

(ii) (projections) $V_f \times V_g \to V_f$ is a cover.

4.3. Actual choice sequences of natural numbers.

Our next modification will be to force the potential sequences α of 4.2 to become real sequences by allowing to pass to a bar in $\mathbb{N}^{<\mathbb{N}}$ to find the nth value of α. Each finite sequence u induces a lawlike continuous operation $\bar{u}: \mathbb{N}^{\mathbb{N}} \to \mathbb{N}^{\mathbb{N}}$ defined by

$$\bar{u}(x) = u \mid x \quad (\text{"overwrite } u\text{"})$$
$$= (u(0),\ldots,u(\ell th(u)-1),\ x(\ell th(u)),\ x(\ell th(u)+1),\ldots),$$

and we now add to the site of 4.2 as new basic covers the families of inclusions

$$\{V_{f\circ\bar{u}_i} \hookrightarrow V_f\}_i$$

for each inductive bar $\{u_i\}_i$ for $\mathbb{N}^{\mathbb{N}}$. (Note that this makes the covers of type (i) in 4.2 redundant.) By stability and transitivity of the induced Grothendieck topology, this means that for each inductive bar $\{u_i^1 \times \ldots \times u_i^k\}_i$ for $\mathbb{N}^{\mathbb{N}} \times \ldots \times \mathbb{N}^{\mathbb{N}}$, we have a corresponding family

$$\{V_{f_1 \circ \bar{u}_i^1} \times \ldots \times V_{f_k \circ \bar{u}_i^k} \hookrightarrow V_{f_1} \times \ldots \times V_{f_k}\}_i$$

in the site.

Observe that covers of this form are stable. For example, if $\{V_{f \circ \bar{u}_i} \to V_f\}_i$ is a cover and $V_g \to V_f$ is an inclusion in the site (so $g = f \circ h$ for some h) then by continuity of h there is an inductive cover $\{v_j\}_j$ of $\mathbb{N}^{\mathbb{N}}$ such that h maps each v_j into some u_i (i.e. if $x \in \mathbb{N}^{\mathbb{N}}$ extends v_j then $h(x)$ extends u_i, we write this as $h(v_j) \subseteq u_i$), so $\bar{u}_i \circ h \circ \bar{v}_j = h \circ \bar{v}_j$, $g \circ \bar{v}_j = f \circ h \circ \bar{v}_j = f \circ \bar{u}_i \circ h \circ \bar{v}_j$, and hence there is a commutative diagram

$$\begin{array}{ccc} V_{f \circ \bar{u}_i} & \longrightarrow & V_f \\ \uparrow & & \uparrow \\ V_{g \circ \bar{v}_j} & \longrightarrow & V_g \end{array}$$

We now have obtained actual sequences:

<u>Proposition</u>. *In this model* $\models \forall\alpha\forall n\exists m\, \alpha(n) = m$, *i.e. at each object* $V_{f_1} \times \ldots \times V_{f_k}$, *for each choice sequence* $\alpha_i (i=1,\ldots,k)$, $V_{f_1} \times \ldots \times V_{f_k} \Vdash \forall n \exists m\, \alpha_i(n) = m$

<u>proof</u>. It suffices to take $k = 1$ (cf. observation 1, section 3), i.e. to show that for all n, $V_f \Vdash \exists m\, \alpha(n) = m$. But f is continuous (as a map $\mathbb{N}^{\mathbb{N}} \to \mathbb{N}^{\mathbb{N}}$) so there is an inductive bar $\{v_j\}_j$ for $\mathbb{N}^{\mathbb{N}}$ such that for all j there exists an m_j with $f(x)(n) = m_j$ whenever $x \in v_j$. But then $V_{f \circ \bar{v}_j} \Vdash \alpha(n) = m_j$ for each j, so $V_f \Vdash \exists m\, \alpha(n) = m$. □

The universe of choice sequences of natural numbers we have now obtained models a rather poor theory. Most importantly (since we are on our way to a model for CS) the universe is not closed under application of lawlike continuous operations. We do not have analytic data, or $\forall\alpha\exists\beta$ − continuity. Also, of the LS-like properties not much is left. For example, the model does not satisfy decidable equality for choice sequences.

On the positive side, we have

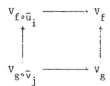

Proposition. *In the model under consideration, the following are valid:*

(i) *(density)* $\forall\alpha \neg\neg \exists a \alpha = a$

(α ranges over choice sequences, a over lawlike sequences)

(ii) *($\forall\alpha\exists n$ - continuity)* $\forall\alpha\exists n A(\alpha,n) \rightarrow \exists F \forall\alpha A(\alpha, F\alpha)$

(here, as usual, A does not contain non-lawlike parameters other than α; and F ranges over lawlike continuous operations $\mathbb{N}^{\mathbb{N}} \rightarrow \mathbb{N}$).

In the proof we use

Lemma. *(genericity of the choice sequence α at $V_{()}$) Let $A(\alpha)$ be a formula with α as its only non-lawlike parameter. If $V_{()} \Vdash A(\alpha)$ then $\models \forall\alpha A(\alpha)$.*

proof. Take a choice sequence α_i at $V_{f_1} \times \ldots \times V_{f_n}$ ($i \leq n$). Since all other parameters in $A(\alpha)$ are (interpreted by) constant (elements), it suffices to show that α_i is a restriction of the sequence α at $V_{()}$ (by monotonicity of $\Vdash$). But restricting α along

$$V_{f_1} \times \ldots \times V_{f_n} \xrightarrow{\pi_i} V_{f_i} \hookrightarrow V_{()}$$

yields α_i. □

proof of proposition. (i) We show $V_{()} \Vdash \neg\neg\exists a \alpha = a$. By observation 1 of section 2 it suffices to show that for each f, $V_f \not\Vdash \neg \exists a \alpha = a$. But this is indeed the case, since if $g: \mathbb{N}^{\mathbb{N}} \rightarrow \mathbb{N}^{\mathbb{N}}$ is constant with value b, then $V_{f \circ g} \Vdash \alpha_i = a$ when we let $a = f(b)$.

(ii) By observation 2 of section 2, we have to show that if $\models \forall\alpha\exists n A(\alpha,n)$ then $\models \exists F \forall\alpha A(\alpha, F\alpha)$. But if $\models \forall\alpha\exists n A(\alpha,n)$ then $V_{()} \models \exists n A(\alpha,n)$, so (cf. observation 1) there are an inductive cover $\{u_i\}_i$ and numbers n_i such that $V_{\bar{u}_i} \Vdash A(\alpha, n_i)$. We may assume the u_i to be disjoint (incompatible), so we can define a lawlike continuous operation $F: \mathbb{N}^{\mathbb{N}} \rightarrow \mathbb{N}$ with value n_i on u_i, i.e. $F(x) = n_i$ if x extends u_i. Then $V_{\bar{u}_i} \Vdash A(\alpha, F\alpha)$, hence since $\{V_{\bar{u}_i} \rightarrow V_{()}\}_i$ is a cover, $V_{()} \Vdash A(\alpha, F\alpha)$. By the genericity lemma, $\models \forall\alpha A(\alpha, F\alpha)$. □

4.4. Closure under lawlike continuous operations.

We will now enlarge our universe of choice sequences, by "projecting" from the lawless sequence $(f_1, f_2, \ldots)$ of neighbourhood functions not only the single sequence α defined by $\alpha(n) = m$ iff $\exists k \forall x \in \mathbb{N}^{\mathbb{N}}\ f_1 \circ \ldots \circ f_k(x)(n) = m$ as in 4.2, but using one lawless sequence to generate an indefinite number of sequences of the form $b(\alpha)$, where b is a lawlike continuous operation $\mathbb{N}^{\mathbb{N}} \to \mathbb{N}^{\mathbb{N}}$.

At the level of the site this means that instead of having finite products of spaces V_f as objects, we now take finite products of objects of the form

$$V_f^{a_1, \ldots, a_k},$$

where $a_1, \ldots, a_k \in K$. If $\{b_1, \ldots, b_m\} \subset \{a_1, \ldots, a_k\}$ we add a morphism

$$V_f^{a_1, \ldots, a_k} \to V_f^{b_1, \ldots, b_m}$$

to our site. On the underlying spaces, this is just the identity map. Going back along this morphism corresponds to the "step in the construction" by which we decide to consider some more choice sequences projected from the single lawless sequence about which we know that it starts with f.

Since we should be able to consider an arbitrary (finite) number of such choice sequences without narrowing our information about the sequences we already had, we should declare this morphism a *cover* in the site. (Stability of this new type of basic cover is trivial, since the underlying function of topological spaces is the identity.)

The model we obtain in this way indeed satisfies closure, that is

$$\forall \alpha \forall F \exists \beta\ F(\alpha) = \beta$$

is valid, where α, β range over the new domain of choice sequences, and F over lawlike continuous operations, but the extra logical properties that we obtain are rather uninteresting. The reason is that the information we have about the single choice sequence $\beta = b(\alpha)$ at the object $V^b_{f \circ g}$ say, is too intensional. This information expresses that $\beta \in \mathrm{im}(b \circ f \circ g)$, and that f and g are the first two elements in our lawless sequence of neighbourhood functions, while b is the lawlike operation that we apply to extract the sequence $\beta = b(\alpha)$. We want to abstract from the different rôles played by b and f in the construction of β, i.e. to pass to a stage of information where f is regarded as an operation used for closing off. This means adding a morphism

$$V_g^{b\circ f} \to V_{f\circ g}^{b}$$

to the site, which on the level of underlying spaces is defined by concatenation, $x \longmapsto f{*}x$ (this is obviously well-defined on equivalence classes $x \in X = K^{\mathbb{N}}/\sim$, and it does not depend on the choice of f, i.e. if f' would be another function such that $b\circ f = b\circ f'$ and $f\circ g = f'\circ g$, the same morphism is defined). Since we wish to ignore the "intensional difference" between the information at $V_{f\circ g}^{b}$ and the information at $V_g^{b\circ f}$ completely, we should moreover declare this morphism $V_g^{b\circ f} \to V_{f\circ g}^{b}$ to cover.

(Digression: A similar abstraction is made in the theory of lawless sequences. A lawless sequence α is usually conceived of as constructed by fixing a finite initial segment u, and then starting to make free choices (throwing a die). At each stage of the construction, the information we have about α is an initial segment $u{*}v$, but we abstract from the extra "intensional" information that u is the initial "deliberate" placings of the die, whereas v comes from making free choices. See Troelstra(1977).)

When compared to the earlier models of this section 4, the properties that the universe of choice sequences has in this model are much richer and much more interesting.

Before we investigate some of these properties, however, we give an explicit description of the site that we have obtained at this stage. For easy reference, we call this site $\mathbb{K}$. *Objects* of the site $\mathbb{K}$ are finite products

$$V_{f_1}^{\bar{a}_1} \times \dots \times V_{f_n}^{\bar{a}_n}$$

where $f_i \in K$ and $\bar{a}_i \in K^{<\mathbb{N}}$. *Morphisms* of $\mathbb{K}$ are best described by: all compositions of morphisms of the various types mentioned above. A more explicit but rather tiresome description is as follows: a morphism

$$\phi: V_{f_1}^{\bar{a}_1}\times\ldots\times V_{f_n}^{\bar{a}_n} \to V_{g_1}^{\bar{b}_1}\times\ldots\times V_{g_m}^{\bar{b}_m}$$

with $\bar{a}_i = (a_{i1},\ldots,a_{ik_i})$ and $\bar{b}_j = (b_{j1},\ldots,b_{j\ell_j})$ is induced by an injection $\varphi: \{1,\ldots,m\} \rightarrowtail \{1,\ldots,n\}$ together with injections $\rho_j: \{1,\ldots,\ell_j\} \rightarrowtail \{1,\ldots,k_{\varphi(j)}\}$ for each $j = 1,\ldots,m$, such that there are maps h_j and k_j $(j=1,\ldots,m)$ with

$$h_j\circ f_{\varphi(j)} = g_j\circ k_j$$

and for each $p = 1,\ldots,\ell_j$,

$$b_{jp}\circ h_j = a_{\varphi(j)\rho_j(p)}.$$

On the underlying spaces we have

$$\phi(x_1,\ldots,x_n) = (h_1 * x_{\varphi(1)},\ldots,h_m * x_{\varphi(m)})$$

($*$ for concatenation; this is well-defined on equivalence classes and does not depend on the choice of h_j, k_j). The *Grothendieck topology* of $\mathbb{K}$ is generated by basic covers of four kinds:

(i) ("open covers") $\{V_{f\circ\bar{u}_i}^{\bar{a}} \to V_f^{\bar{a}}\}_i$ is a cover, for each inductive bar $\{u_i\}_i$ for $\mathbb{N}^{\mathbb{N}}$.

(ii) (projections) $V_f^{\bar{a}}\times V_g^{\bar{b}} \to V_f^{\bar{a}}$ is a cover.

(iii) (adding choice sequences) $V_f^{\bar{a}} \to V_f^{\bar{b}}$ is a cover for the map which is the identity on the level of spaces, and is induced by an inclusion of the sequence $\bar{b}$ as a subsequence of $\bar{a}$.

(iv) (abstraction) $V_g^{\bar{a}\circ f} \to V_{f\circ g}^{\bar{a}}$ is a cover, where $\bar{a}\circ f = (a_1\circ f,\ldots,a_k\circ f)$ if $\bar{a} = (a_1,\ldots,a_k)$.

The *universe of choice sequences* at an object

$$V_{g_1}^{\bar{b}_1}\times\ldots\times V_{g_m}^{\bar{b}_m}$$

consists of all sequences of the form $\beta = b_{jp} \circ \alpha_j$ $(p=1,\dots,\ell)$, where $\bar{b}_j = (b_{j1},\dots,b_{jp})$, and β is a sequence by

$$V_{g_1}^{\bar{b}_1} \times \dots \times V_{g_m}^{\bar{b}_m} \Vdash \beta(n) = m \quad \text{iff} \quad \forall x \in \mathbb{N}^{\mathbb{N}}\ b_{jp} \circ g_j(x)(n) = m.$$

Restriction of choice sequences along morphisms is defined in the obvious way. If Φ is a morphism as just described and $\beta = b_{jp} \circ \alpha_j$ as above, then the restriction of β along Φ is the sequence

$$\beta \mid \Phi = b_{\varphi(j)\rho_j(p)} \circ \alpha_{\varphi(j)}$$

at the object $V_{f_1}^{\bar{a}_1} \times \dots \times V_{f_n}^{\bar{a}_n}$. This definition of restrictions is compatible with the definition of $\Vdash \beta(n) = m$ in the sense that for a morphism Φ with codomain W and domain U, $W \Vdash \beta(n) = m$ implies $U \Vdash (\beta|\Phi)(n) = m$, and if $\{\Phi_i : U_i \to W\}_i$ is a cover in $\mathbb{K}$ such that $U_i \Vdash (\beta|\Phi)(n) = m$ for each i, then also $W \Vdash \beta(n) = m$.

In this model, all one parameter axioms of the theory CS are valid:

Theorem 1. *In the model over* $\mathbb{K}$ *just described, the following are valid:*

(i) (*closure*) $\forall\alpha\forall F\exists\beta\ \beta = F(\alpha)$

(ii) (*analytic data*) $\forall\alpha(A(\alpha) \to \exists F(\alpha \in \mathrm{im}(F) \wedge \forall\beta \in \mathrm{im}(F)A(\alpha)))$

(iii) ($\forall\alpha\exists n$ - *continuity*) $\forall\alpha\exists n A(\alpha,n) \to \exists F\forall\alpha A(\alpha,F\alpha)$.

Here, as usual, F ranges over lawlike continuous operations (into $\mathbb{N}$ or $\mathbb{N}^{\mathbb{N}}$), α,β over choice sequences, and all non-lawlike parameters in A are shown.

Before we prove the theorem, let us reformulate the genericity lemma of 4.3:

Genericity lemma. *Let* $A(\alpha)$ *be a formula with* α *as its only non-lawlike parameter. Then in the model over* $\mathbb{K}$, $V_{()}^{id} \Vdash A(\alpha)$ *implies* $\models \forall\alpha A(\alpha)$.

proof. As before, we have to show that every choice sequence at an object $V_{f_1}^{\bar{a}_1} \times \dots \times V_{f_n}^{\bar{a}_n}$ is the restriction of the (single) choice sequence α at $V_{()}^{id}$. But if $\beta = a_{ij} \circ \alpha_i$ is a choice sequence at $V_{f_1}^{\bar{a}_1} \times \dots \times V_{f_n}^{\bar{a}_n}$, then $\beta = \alpha \mid \Phi$ where Φ is the composite

$$V_{f_1}^{\bar{a}_1}\times\ldots\times V_{f_n}^{\bar{a}_n} \xrightarrow{\pi_i} V_{f_i}^{\bar{a}_i} \longrightarrow V_{f_i}^{a_{ij}} \hookleftarrow V_{()}^{a_{ij}} \longrightarrow V_{a_{ij}}^{id} \hookleftarrow V_{()}^{id}. \qquad \square$$

proof of theorem 1.

(i) By the lemma, it suffices to show $V_{()}^{id} \Vdash \forall F\exists\beta\ \alpha = F(\beta)$. But if f is a law-like continuous operation, then $V_{()}^{id,f} \to V_{()}^{id}$ covers, and at $V_{()}^{id,f}$ we have two sequences α and β with $V_{()}^{id,f} \Vdash \alpha = f(\beta)$.

(ii) Suppose $V_{f_1}^{\bar{a}_1}\times\ldots\times V_{f_n}^{\bar{a}_n} \Vdash A(\beta)$ for some sequence $\beta = a_{ij}\circ\alpha_i$. Since $V_{f_1}^{\bar{a}_1}\times\ldots\times V_{f_n}^{\bar{a}_n} \to V_{f_i}^{a_{ij}}$ covers, we find $V_{f_i}^{a_{ij}} \Vdash A(\beta)$. In other words, we may assume that $n = 1$ and $\bar{a}_1$ has length 1, and we can write $V_f^a \Vdash A(\beta)$. Then also $V_{()}^{a\circ f} \Vdash A(\beta)$ by restricting along the morphism $V_{()}^{a\circ f} \to V_f^a$, so if we add a choice sequence α corresponding to id and put $F = a\circ f$ we find $V_{()}^{af,id} \Vdash A(F\alpha)$. Hence since $V_{()}^{af,id} \to V_{()}^{id}$ covers, $V_{()}^{id} \Vdash A(F\alpha)$. By the genericity lemma, $\models \forall\alpha A(F\alpha)$, so a fortiori $V_f^a \Vdash \forall\alpha A(F\alpha)$, while $V_{()}^{af,id} \Vdash \beta = F\alpha$, so $V_f^a \Vdash \beta\in \mathrm{im}(F)$ since $V_{()}^{af,id} \to V_f^a$ is a cover.

(iii) The proof of $\forall\alpha\exists n$ - continuity is analogous to the one we gave in 4.3. $\square$

Corollary. *Let* A *be a sentence of the language of* CS *containing only one choice variable. Then* A *is valid in this model over* $\mathbb{K}$.

proof. From the proof of the elimination theorem for CS (Kreisel & Troelstra(1970)) we conclude that if $CS \vdash A$ then $CS^1 \vdash A$, where CS^1 is the theory axiomatized by the axioms of CS which contain only one choice variable. But the theorem above states that these axioms are valid over $\mathbb{K}$. $\square$

The model does not satisfy the axioms for CS in more choice variables, notably the *pairing axiom* $\forall\alpha,\beta\exists F,G\exists\gamma(\alpha=F\gamma\wedge\beta=G\gamma)$, and $\forall\alpha\exists\beta$ - continuity.

The properties of the universe of choice sequence in this model can be more closely analysed if we introduce a primitive predicate of *independence*

$\alpha \# \beta$ "α and β are independent choice sequences"

into the language. The interpretation of # in the model is given by

$$V_{f_1}^{\bar{a}_1}\times\ldots\times V_{f_n}^{\bar{a}_n} \Vdash a_{ij}\circ\alpha_i \# a_{k\ell}\circ\alpha_\ell \quad \text{iff} \quad i \neq \ell,$$

i.e. two choice sequences are independent if they come from different factors of the product $V_{f_1}^{\bar{a}_1} \times \ldots \times V_{f_n}^{\bar{a}_n}$, meaning that they are extracted from distinct processes for choosing a lawless sequence of neighbourhood functions. Obviously, # is *decidable*, i.e.

$$\alpha \# \beta \vee \neg\alpha \# \beta$$

is valid in the model (just as decidable equality in section 3).

With this primitive # added to the language, the model can be shown to satisfy a set of axioms that allows elimination of choice sequences. For example, we have multiple parameter versions of analytic data and $\forall\alpha\exists n$ - continuity:

Theorem 2. *The model over* $\mathbb{K}$ *satisfies the following multiple parameter versions of analytic data and continuity, where* $\#(\alpha_1,\ldots,\alpha_n)$ *abbreviates* $\bigwedge_{i<j} \alpha_i \# \alpha_j$.

$$\forall\alpha_1\ldots\alpha_n(\#(\alpha_1,\ldots,\alpha_n) \wedge A(\alpha_1,\ldots,\alpha_n) \rightarrow \exists F_1\ldots F_n$$
$$(\bigwedge_{i=1}^{n} \alpha_i \in im(F_i) \wedge \forall\beta_1\ldots\beta_n(\#(\beta_1,\ldots,\beta_n) \rightarrow A(F\beta_1,\ldots F\beta_n))))$$
$$\forall\alpha_1\ldots\alpha_n(\#(\alpha_1,\ldots,\alpha_n) \rightarrow \exists m\, A(\alpha_1,\ldots,\alpha_n, m)) \rightarrow$$
$$\rightarrow \exists F\forall\alpha_1\ldots\alpha_n(\#(\alpha_1,\ldots,\alpha_n) \rightarrow A(\alpha_1,\ldots,\alpha_n, F(\alpha_1,\ldots,\alpha_n))).$$

the proofs are easy modifications of the proofs given for the one-parameter case, using a "genericity lemma" for independent n-tuples, saying that the independent n-tuple $(\alpha_1,\ldots,\alpha_n)$ at $V_{()}^{id} \times \ldots \times V_{()}^{id}$ (n-fold product) is the generic such.

We do have *dependent* versions of pairing and $\forall\alpha\exists\beta$ - continuity:

Theorem 3. *The model over* $\mathbb{K}$ *satisfies*

(i) $\forall\alpha,\beta(\neg\alpha\#\beta \rightarrow \exists F,G\exists\gamma(\alpha=F\gamma \wedge \beta=G\gamma))$

(ii) $\forall\alpha\exists\beta(\neg\alpha\#\beta \wedge A(\alpha,\beta)) \rightarrow \exists F\forall\alpha A(\alpha,F\alpha)$.

proof. (i) If α,β are given at $V_{f_1}^{\bar{a}_1} \times \ldots \times V_{f_n}^{\bar{a}_n}$ and are not dependent, then $\alpha = a_{ij}\circ\alpha_i$, $\beta = a_{ik}\circ\alpha_i$ for some i and some neighbourhood functions a_{ij}, a_{ik} occurring in $\bar{a}_i$. Take $F = a_{ij}$, $G = a_{ik}$, $\gamma = id\circ\alpha_i$, which exists at the object

$V_{f_1}^{\bar{a}_1} \times \ldots \times V_{f_i}^{\bar{a}_i,id} \times \ldots \times V_{f_n}^{\bar{a}_n}$ covering $V_{f_1}^{\bar{a}_1} \times \ldots \times V_{f_n}^{\bar{a}_n}$.

(ii) Suppose $\forall\alpha\exists\beta(\neg\alpha\#\beta \wedge A(\alpha,\beta))$ is forced somewhere, or equivalently (since there are no non-lawlike parameters) everywhere. Then $V_{()}^{id} \Vdash \exists\beta(\neg\alpha\#\beta \wedge A(\alpha,\beta))$, and from this it follows that there is a disjoint inductive bar $\{u_i\}_i$ and elements $f_i \in K$ such that $V_{\bar{u}_i}^{id,f_i} \Vdash A(\alpha,\beta)$ (where α still corresponds to id, and β to f_i). Let $F: \mathbb{N}^{\mathbb{N}} \to \mathbb{N}^{\mathbb{N}}$ be the lawlike continuous operation with $F|u_i = f_i$, i.e. $F(x) = f_i(x)$ if x extends u_i. Then $V_{\bar{u}_i}^{id,f_i} \Vdash \beta = F(\alpha)$, so $V_{\bar{u}_i}^{id,f_i} \Vdash A(\alpha,F\alpha)$, and hence since the objects $V_{\bar{u}_i}^{id,f_i}$ cover $V_{()}^{id}$, also $V_{()}^{id} \Vdash A(\alpha,F\alpha)$. By genericity, $\Vdash \forall\alpha A(\alpha,F\alpha)$. □

The continuity axiom for the quantifier combination $\forall\alpha_1 \ldots \forall\alpha_n\exists\beta$ in fact splits into several variants. Without proof we state some for $n = 1$.

Theorem 4. *The following versions of* $\forall\alpha\exists\beta$ *- continuity are valid in the model over* $\mathbb{K}$.

(i) (*uniformity*) $\forall\alpha\exists\beta(\alpha\#\beta \wedge A(\alpha,\beta)) \to \exists e \in K\forall u(e(u) \neq 0 \to \exists\beta\forall\alpha \in u(\alpha\#\beta \to A(\alpha,\beta)))$

(ii) $\forall\alpha\exists\beta A(\alpha,\beta) \to \exists e \in K\forall u(e(u) \neq 0 \to (\exists F\forall\alpha \in uA(\alpha,F\alpha) \vee \exists\beta\forall\alpha \in u(\alpha\#\beta \to A(\alpha,\beta))))$.

(By analytic data and continuity for the quantifier combination $\forall\alpha\exists f \in K$, (i) may equivalently be formulated as

(i') $\forall\alpha\exists\beta(\alpha\#\beta \wedge A(\alpha,\beta)) \to \forall\alpha\exists F\forall\beta(\alpha\#\beta \to A(\alpha,F\beta))$.)

4.5. Identifying independent processes.

The obstruction to having the usual form of $\forall\alpha\exists\beta$ - continuity at the end of 4.4 lies in the fact that there are independent "parallel" processes for constructing lawless sequences of neighbourhood functions. As a further abstraction, we will now allow identification of independent processes which "until now" have yielded the same result. This abstraction is formalized by adding more morphisms to the site $\mathbb{K}$ of 4.4: we add morphisms

$$V_f^{\bar{a},\bar{b}} \to V_f^{\bar{a}} \times V_f^{\bar{b}}$$

which, at the level of the underlying spaces, are just the diagonal maps. We thus obtain a new site which has as morphisms all compositions of morphisms of the type described in 4.4 and these new diagonal maps, and with the same basic covers (i)-(iv) above generating the Grothendieck topology. We call this new site $\mathbb{L}$. The model over $\mathbb{L}$, with the universe of choice sequences defined as for the model over $\mathbb{K}$, now comes very close to validating the CS-axioms:

Theorem 1. *In the model over the site* $\mathbb{L}$ *just described, the following are valid:*

(i) (*closure*) $\forall\alpha\forall F\exists\beta(\beta=F\alpha)$

(ii) (*analytic data*) $\forall\alpha(A\alpha\to\exists F(\alpha\in\mathrm{im}(F)\wedge\forall\beta\in\mathrm{im}(F)A\beta))$

(iii) ($\forall\alpha\exists n$ - *continuity*) $\forall\alpha\exists nA(\alpha,n)\to\exists F\forall\alpha A(\alpha,F\alpha)$

(iv) ($\forall\alpha\exists\beta$ - *continuity*) $\forall\alpha\exists\beta A(\alpha,\beta)\to\exists F\forall\alpha A(\alpha,F\alpha)$

proof. (i)-(iii) are proved just as in 4.4 (cf. theorem 1 of 4.4), since the genericity lemma remains valid over $\mathbb{L}$. For $\forall\alpha\exists\beta$ - continuity, suppose $\Vdash\forall\alpha\exists\beta A(\alpha,\beta)$ (at every object, if at any). Then in particular $V^{id}_{()}\Vdash\exists\beta A(\alpha,\beta)$. From this it follows that there is a cover $\{\phi_i: U_i\to V^{id}_{()}\}_i$ where either $U_i=V^{id}_{\bar{u}_i}\times V^{f_i}_{g_i}$ and $U_i\Vdash A(\alpha,\beta_i)$ (with α coming from $V^{id}_{\bar{u}_i}$ and β_i from $V^{f_i}_{g_i}$), or $U_i=V^{id,h_i}_{\bar{u}_i}$ and $U_i\Vdash A(\alpha,\beta_i)$ (with α corresponding to id, β_i to h_i), all this for some inductive bar $\{u_i\}_i$. But if U_i is of the first type, we can restrict along

$$V^{id,f_i\circ g_i}_{\bar{u}_i}\longrightarrow V^{id}_{\bar{u}_i}\times V^{f_i\circ g_i}_{\bar{u}_i}\longrightarrow V^{id}_{\bar{u}_i}\times V^{f_i\circ g_i}_{()}\longrightarrow V^{id}_{\bar{u}_i}\times V^{f_i}_{g_i},$$

and we conclude that $V^{id,f_i\circ g_i}_{\bar{u}_i}\Vdash A(\alpha,\beta_i)$. In other words, without loss all U_i are of the second type. Now let $F:\mathbb{N}^{\mathbb{N}}\to\mathbb{N}^{\mathbb{N}}$ be the lawlike continuous operation such that $F\mid u_i=h_i$. Then $V^{id,h_i}_{\bar{u}_i}\Vdash F(\alpha)=\beta_i$, so $V^{id,h_i}_{\bar{u}_i}\Vdash A(\alpha,F\alpha)$. Since $\{V^{id,h_i}_{\bar{u}_i}\to V^{id}_{()}\}_i$ is a cover, it follows that $V^{id}_{()}\Vdash A(\alpha,F\alpha)$, so by the genericity lemma, $\models\forall\alpha A(\alpha,F\alpha)$. □

The model over $\mathbb{L}$ is not a model for CS, since the pairing axiom is not satisfied. With one small modification, however, we obtain pairing. Let $\mathbb{M}$ be the site with the same objects and morphisms as $\mathbb{L}$, and with the Grothendieck topology generated by the basic covers (i)-(iv) of 4.4, and in addition for each object $V_{f_1}^{\bar{a}_1} \times \ldots \times V_{f_n}^{\bar{a}_n}$ of the site,

(v) (diagonals) the collection of morphisms

$$\{V_{()}^{\bar{a}_1 f_1 h_1} \times \ldots \times V_{()}^{\bar{a}_n f_n h_n} \longrightarrow V_{f_1}^{\bar{a}_1} \times \ldots \times V_{f_n}^{\bar{a}_n} \mid \text{all} \quad (h_1,\ldots,h_n) \in K^n\}$$

is a cover, where on the underlying spaces the morphism for an n-tuple $h_1, \ldots, h_n$ is given by

$$(x_1,\ldots,x_n) \longmapsto (f_1 \circ h_1 * x_1,\ldots,f_n \circ h_n * x_n).$$

Over the site $\mathbb{M}$, we define a universe of choice sequences exactly as in 4.4. We then obtain

Theorem 2. *The model over* $\mathbb{M}$ *is a model for* CS.

proof. All the axioms are verified exactly as for theorem 1 above, except for pairing. But pairing is trivially forced to hold by the new covers of type (v). □

Footnotes.

1. The second author acknowledges financial support by the Netherlands Organization for the Advancement of Pure Research (ZWO).

2. The contents of this paper differs from the lecture given by the second author at the Logic Colloquium in Aachen, where some results from van der Hoeven & Moerdijk (to appear), (to appear 2) were presented.

References.

D. van Dalen & A.S. Troelstra(1970), Projections of lawless sequences, in Kino, Myhill and Vesley(eds.), *Intuitionism and Proof Theory* (North Holland, Amsterdam).

M. Fourman(1982), Notions of choice sequence, in van Dalen and Troelstra(eds.) *Proceedings of the Brouwer Centenary Conference* (North Holland, Amsterdam).

G.F. van der Hoeven(1982), *Projections of Lawless Sequences*, Mathematical Centre Tracts 152 (Mathematisch Centrum, Amsterdam).

G.F. van der Hoeven & I. Moerdijk(to appear), Sheaf models for choice sequences, to appear in Ann.Math.Logic.

G.F. van der Hoeven & I. Moerdijk(to appear 2), On choice sequences determined by spreads, to appear in Journ.Symb.Logic.

G.F. van der Hoeven & A.S. Troelstra(1980), Projections of lawless sequences II, in Boffa, van Dalen, Mc Aloon(eds.), *Logic Colloquium* 78 (North Holland, Amsterdam).

G. Kreisel(1968), Lawless sequences of natural numbers, Comp.Math. 20.

G. Kreisel & A.S. Troelstra(1970), Formal systems for some branches of intuitionistic analysis, Ann.Math.Logic 1.

A.S. Troelstra(1977), *Choice Sequences* (Clarendon Press, Oxford).

A.S. Troelstra(1983), Analysing choice sequences, Journ.Phil.Logic 12.

PARTITIONS AND HOMOGENEOUS SETS FOR ADMISSIBLE ORDINALS

Evangelos Kranakis[(1)]
Mathematische Institut
der Universität Heidelberg

and

Iain Phillips[(2)]
Mathematical Institute
Oxford University

ABSTRACT

In the present paper we explore partition properties of admissible ordinals. In §2 the connection between partition properties, collection, separation and reflection principles is studied. In §3 we give characterizations of partition properties which are satisfied by certain definable subsets of κ. Covering properties, which are studied in §4, are a convenient generalization of partition properties and are used to investigate the strength of certain partition properties of κ. Finally in §5 we study the definability of the homogeneous set in partitions of exponent greater than one.

(1) The author is grateful to the Minna-James-Heineman Stiftung, Hannover, for its financial support during the preparation of the present paper at the Universität Heidelberg.

(2) Some of the results of the present paper have also appeared in the author's 1983 Oxford doctoral thesis. He was assisted by an award from the Science Research Council and by the advice and encouragement of his supervisor Dr. Robin Gandy.

(1) PRELIMINARIES AND NOTATION

Lower case Greek letters $\alpha,\beta,\ldots,\kappa,\lambda,\ldots$ represent ordinals; sequences of ordinals are denoted by $\bar{\alpha},\bar{\beta},\ldots$. The symbol cf will be used throughout as an abbreviation of the word cofinal. The symbol $f:X\to Y$ (resp. $f:X\xrightarrow{onto}Y$, $f:X\xrightarrow{cf}Y$) means that f is a function with domain X and range a subset of Y (resp. onto Y, the range of f is cf in Y), where X,Y are sets of ordinals. If $f:X\to Y$ and $A\subseteq X$ (resp. $B\subseteq Y$) then $f[A]=\{f(x):x\in A\}$ (resp. $f^{-1}[B]=\{x:f(x)\in B\}$). If I is a set of ordinals, then $[I]^n$ denotes the set of all increasing n-tuples of ordinals from I. $<.,.>$ denotes the standard coding function on ordinals; $(\cdot)_i$ (i=1,2) denote the standard decoding functions such that for all $\alpha,\beta,(<\alpha,\beta>)_1=\alpha$ and $(<\alpha,\beta>)_2=\beta$. Lower case Latin letters $n,m,\ldots$ will always denote nonnegative integers.

Remark 1.1.

To avoid unnecessary complications we will assume from now on that κ denotes an ordinal such that $\omega\kappa=\kappa>0$ and $n\geq 1$ an integer.

The Jensen hierachy $<J_\alpha:\alpha\in Ord>$ of constructible sets is defined in [D1]. The constructible universe is $L=\bigcup\{J_\alpha:\alpha\in Ord\}$. We shall consider structures of the form $<J_\alpha,\in>$; to simplify our notation we shall write J_α instead of $<J_\alpha,\in>$ but it will be clear from the context what we mean.

The Levy hierarchy Σ_m,Π_m of $\in$-formulas (i.e., in the language $\{\in\}$) is defined as usual; $\Sigma_\omega=\bigcup\{\Sigma_m:m\geq 0\}$; C_m = the set of formulas of the form $\varphi\wedge\psi$, where φ is in Σ_m and ψ is in Π_m; B_m = the set of Boolean combinations of formulas in Σ_m. A relation R on J_α is in $\Phi(J_\alpha)$, where Φ is a class of $\in$-formulas, if there exists a formula $\varphi(u_1,\ldots,u_s,x_1,\ldots,x_r)$ in Φ and elements $a_1,\ldots,a_r\in J_\alpha$ such that $R=\{(b_1,\ldots,b_s)\in J_\alpha^s : J_\alpha\models\varphi(b_1,\ldots,b_s,a_1,\ldots,a_r)\}$ (J_α^s denotes the Cartesian product of J_α, s times). We also define $\Delta_m(J_\alpha)=\Sigma_m(J_\alpha)\cap\Pi_m(J_\alpha)$. Moreover a relation R belongs to $\forall^b\Sigma_n(J_\alpha)$ if there exists a relation $S\in\Sigma_n(J_\alpha)$ and a constant $a\in J_\alpha$ such that $R=\{b\in J_\alpha:\forall x\in a\ S(x,b)\}$. A formula is Π_1^1 if it is of the form $\forall X_1\ldots\forall X_s\ \varphi(X_1,\ldots,X_s,x_1,\ldots,x_r)$ where $X_1,\ldots,X_s$ are second-order variables and φ is a first-order $\in$-formula.

Recall that (for $n\geq 1$) the satisfaction predicate $\models_{J_\alpha}^{\Sigma_n}$ (resp. $\models_{J_\alpha}^{\Pi_n}$) is $\Sigma_n(J_\alpha)$ (resp. $\Pi_n(J_\alpha)$) uniformly for all $\alpha\geq 1$ (see [D1]). The following is a consequence of Jensen's uniformization theorem.

Theorem 1.2. (Σ_n selection theorem).

For any relation $R\in\Sigma_n(J_\kappa)$, there exists a $\Sigma_n(J_\kappa)$ function r such that

$$\text{(i)}\quad dom(r)=dom(R)$$

$$\text{(ii)}\quad \forall x\in J_\kappa(\exists y\in J_\kappa\ R(x,y)\longleftrightarrow R(x,r(x)))$$

Σ_n-collection (resp. Π_n-collection) is the axiom schema $\forall x\in u\exists\bar{y}\varphi\to\exists w\forall x\in u\,\exists\bar{y}\in w\varphi$

where φ is a Σ_n (resp. Π_n) formula. We say that J_α satisfies the schema of Σ_n-collection (abbreviated $J_\alpha \models \Sigma_n$-coll.) if all instances of Σ_n-collection are valid in J_α. We say that J_α satisfies the schema of Σ_n-separation (abbreviated $J_\alpha \models \Sigma_n$-sep.) if for any $R \in \Sigma_n(J_\alpha)$ and any $a \in J_\alpha$, $R \cap a \in J_\alpha$. $J_\alpha \models \Delta_n$-sep. is defined similarly. A limit ordinal $\alpha \geq 1$ is called Σ_n admissible if $J_\alpha \models \Sigma_n$-coll.

We will need the following concepts which are defined in the fine structure.

ρ^n_α = the least ρ such that there is a $\Sigma_n(J_\alpha)$ map of a subset of $\omega\rho$ onto J_α;
η^n_α = the least η such that there is a $\Sigma_n(J_\alpha)$ map from $\omega\eta$ onto J_α;
$cf^n\alpha$ = the least σ such that there exists a $\Sigma_n(J_\alpha)$ map from σ cofinally into $\omega\alpha$.

The following result is proved using standard results in the fine structure (parts (i), (ii) require the uniformization theorem).

Theorem 1.3.

(i) $\rho^n_\kappa = \kappa \Longleftrightarrow J_\kappa \models \Sigma_n$-sep.

(ii) $\eta^n_\kappa = \kappa \Longleftrightarrow J_\kappa \models \Delta_n$-sep.

(iii) $cf^n\kappa = \kappa \Longleftrightarrow J_\kappa \models \Sigma_n$-coll.

The symbol $J_\alpha \prec_m J_\beta$ means that $\alpha \leq \beta$ and for any Σ_m formula $\varphi(x_1,\ldots,x_r)$ and any parameters $a_1,\ldots,a_r \in J_\alpha$, $J_\alpha \models \varphi(a_1,\ldots,a_r)$ if and only if $J_\beta \models \varphi(a_1,\ldots,a_r)$. We also define the class $S^m_\kappa = \{\alpha < \kappa : J_\alpha \prec_m J_\kappa\}$. $J_\alpha \prec J_\beta$ means that for all $m < \omega$, $J_\alpha \prec_m J_\beta$.

If Φ is a class of $\in$-formulas, then we say that the ordinal κ is Φ-reflecting on X (where X is a nonempty subset of κ) if for any formula $\varphi(x_1,\ldots,x_r)$ from Φ and any parameters $a_1,\ldots,a_r \in J_\kappa$ if $J_\kappa \models \varphi(a_1,\ldots,a_r)$, then there exists $\alpha \in X$ such that $J_\alpha \models \varphi(a_1,\ldots,a_r)$. If $X = \kappa$, then "κ is Φ-reflecting on κ" is abbreviated by "κ is Φ-reflecting".

Finally PSA denotes the power set axiom, i.e., $\forall a\, \exists b\, \forall x\, (x \in b \longleftrightarrow x \subseteq a)$.

(2) RESULTS OF REFLECTION

The present section begins by giving a different proof of a theorem of Simpson on reflection (see [S1]); the proof which is along the lines of Simpson's proof is based on a theorem of Kranakis (see Theorem 2.3 in [K1]). Next the relationship between separation and collection axioms on the one hand and separation axioms and reflection principles on the other hand is studied. This leads to an improvement of a theorem of Kranakis (see Theorem 2.2 in [K1]) which in turn is a generalization of a theorem of Kripke and Platek (see Theorem 7.11 in [B]).

Theorem 2.1. [Simpson]

$J_\kappa \models \Delta_n$-sep. $\Rightarrow \kappa$ is Π_{n+1}-reflecting.

Proof. Assume that $n \geq 1$ and $J_\kappa \models \Delta_n$-sep. It was noticed by Sy Friedman that one can prove, using results of R. Jensen, that $\omega\eta^n_\kappa = \max\{\omega\rho^n_\kappa, cf^n\kappa\}$. If $J_\kappa \models \Delta_n$-sep., then $\omega\eta^n_\kappa = \kappa$ and hence either $\omega\rho^n_\kappa = \kappa$ or else $cf^n_\kappa = \kappa$. However if $cf^n_\kappa = \kappa$ then κ is Σ_n admissible and hence Π_{n+1}-reflecting on S^{n-1}_κ (see Theorem 2.3 in [K1]); in particular κ must be Π_{n+1}-reflecting. Hence without loss of generality it can be assumed that $\omega\rho^n = \kappa$, which in turn is equivalent to the statement $J_\kappa \models \Sigma_n$-sep. Now the rest of the proof can be completed as in the proof of Theorem 6.7. in [S1].

The following result is well known; its proof is similar to that of Lemma 15 in [D2].

Lemma 2.2.

$$J_\kappa \models \Sigma_n\text{-coll} \Longrightarrow J_\kappa \models \Delta_n\text{-sep}.$$

The next two results are interesting because they show what is the effect of the power set axiom on the axiom of separation.

Theorem 2.3. Assume $J_\kappa \models \neg$PSA

The following are equivalent

(i) $J_\kappa \models \Delta_n$-sep.

(ii) $J_\kappa \models \Sigma_n$-coll.

Proof. (ii)$\Longrightarrow$(i)

This is just Lemma 2.2.

(i)$\Longrightarrow$(ii)

Assume (ii) is false; this means that $J_\kappa \not\models \Sigma_n$-coll. Let $\lambda = cf^n\kappa < \kappa$ and consider a function $f:\lambda \to \kappa$ which is $\Sigma_n(J_\kappa)$ definable and such that the range of f is cofinal in κ. Since by assumption J_κ does not satisfy the power set axiom, there exists a largest cardinal in J_κ, say $\delta < \kappa$. This means that

$$\forall\gamma < \kappa \quad \exists h \in J_\kappa \quad h:\delta \xrightarrow{\text{onto}} \gamma$$

By the Σ_1 selection theorem there exists a $\Sigma_1(J_\kappa)$ definable function $H:\kappa \to J_\kappa$ such that for all $\gamma < \kappa$, $H(\gamma):\delta \xrightarrow{\text{onto}} \gamma$. Define a function $g:\lambda \times \delta \xrightarrow{\text{onto}} \kappa$ by

$$g(\xi,\eta) = H(f(\xi))(\eta).$$

It is clear that g is $\Sigma_n(J_\kappa)$ definable. Let β be the J_κ cardinality of $\lambda \times \delta$ (notice that $\lambda \times \delta \in J_\kappa$) and let $\ell:\beta \xrightarrow{\text{onto}} \lambda \times \delta$, where ℓ is some map in J_κ. Then the composition $g \circ \ell$ gives rise to a $\Sigma_n(J_\kappa)$ definable function with domain β and range κ. However $\beta < \kappa$; this is enough to yield $J_\kappa \not\models \Delta_n$-sep. This completes the proof of the theorem.

Theorem 2.4. Assume $J_\kappa \models$ PSA.

The following are equivalent

(i) $J_\kappa \models \Sigma_n$-sep.

(ii) $J_\kappa \models \Delta_n$-sep.

(iii) κ is Π_{n+1}-reflecting.

Proof. (i) $\Longrightarrow$ (ii)

This is trivial.

(ii) $\Longrightarrow$ (iii)

This is Simpson's theorem, Theorem 2.1.

(iii) $\Longrightarrow$ (i)

Assume (iii) $\wedge\ \neg$(i) is true and let $g: \subseteq \gamma \xrightarrow{\text{onto}} \kappa$ be a $\Sigma_n(J_\kappa)$ definable function, where $\gamma < \kappa$. Since J_κ satisfies the power set axiom, there is a J_κ-cardinal $\delta > \gamma$. Consider the Σ_n formula $\psi(x,y)$ which has parameters in J_κ and satisfies

$$J_\kappa \models \psi(\xi,\eta) \Longleftrightarrow g(\xi) = \eta \wedge \eta < \delta .$$

The following formula φ is true in J_κ:

$$\forall\eta < \delta\ \exists\xi < \gamma\ \psi(\xi,\eta) \wedge \forall\xi < \gamma\quad \forall\eta,\eta' < \delta(\psi(\xi,\eta) \wedge \psi(\xi,\eta') \rightarrow \eta = \eta')$$

φ is clearly a Π_{n+1} formula, and so by hypothesis there exists $\beta < \kappa$ such that $J_\beta \models \varphi$. Let f be the $\Sigma_n(J_\beta)$ function defined by

$$f(\xi) = \eta \Longleftrightarrow J_\beta \models \psi(\xi,\eta).$$

It is clear that $f: \subseteq \gamma \xrightarrow{\text{onto}} \delta$ and $f \in J_\kappa$. It follows that the J_κ cardinality of δ (which is equal to δ, since δ is a J_κ cardinal) must be $\leq\gamma$, something which contradicts the choice of δ. This completes the proof of the theorem.

Remark 2.5.

Notice that the formula φ in the proof of (iii) $\Longrightarrow$ (i) is in fact the conjunction of a $\forall^b\ \Sigma_n$ and a Π_n formula. Hence if $\forall^b\ \Sigma_n \wedge \Pi_n$ denotes the class of formulas which are the conjunction of a $\forall^b\Sigma_n$ and a Π_n formula, then it follows that under the hypothesis of Theorem 2.4 each of the statements (i) - (iii) is equivalent to the following statement

(iv) κ is $\forall^b\ \Sigma_n \wedge \Pi_n$-reflecting.

This remark will be used later in the proof of Corollary 4.6.

The following is an immediate corollary of the above Theorems 2.2. - 2.4.

Corollary 2.6.

(i) $J_\kappa \models \neg PSA \wedge J_\kappa \models \Sigma_n$-sep. $\Longrightarrow J_\kappa \models \Sigma_n$-coll.

(ii) $J_\kappa \models PSA \wedge J_\kappa \models \Sigma_n$-coll. $\Longrightarrow J_\kappa \models \Sigma_n$-sep.

Also an immediate corollary of Theorem 2.2 in [K1] and Theorem 2.3 above is the following result which gives a generalization of a theorem of Kripke-Platek (see Theorem 7.11. in [B]).

Corollary 2.7. If $J_\kappa \models \neg PSA$, then

$$S^n_\kappa \text{ is cofinal in } \kappa \Longleftrightarrow J_\kappa = \Sigma_n\text{-sep.}$$

Remarks 2.8.

(1) Notice that for the case $n = 1$, the conclusion of Corollary 2.7 is true without the assumption $J_\kappa \models \neg PSA$. But this hypothesis is necessary for $n \geq 2$. For example, let κ = the ωth L-cardinal. Since κ is an L-cardinal, it is clear that $J_\kappa \models \Sigma_2$-sep. However S^2_κ cannot be cofinal in κ; in fact $S^2_\kappa = \emptyset$.

(2) The same κ as in (1) shows that the hypothesis $J_\kappa \models \neg PSA$ is necessary in Theorem 2.3, if $n \geq 2$. In case $n = 1$ the theorem is true without $J_\kappa \models \neg PSA$; this result due to R. Jensen can be found in Theorems 42, 43 in [D2].

(3) Theorem 2.4 is false without the assumption $J_\kappa \models PSA$ for all $n \geq 1$. If $n = 1$, then (ii), (iii) are equivalent (because both are equivalent to Σ_1 admissibility, see Theorem 42 in [D] and Theorem 1.8 in [R-A]). But Σ_1-sep., which is equivalent to non-projectibility, is stronger than admissibility. The following example however will work even for $n \geq 2$. It is known that for countable κ, κ is Π^1_1-reflecting if and only if $J_\kappa \prec_1 J_{\kappa^+}$, where κ^+ = next admissible ordinal $>\kappa$. (See Theorem 6.4 in [R-A]). If κ is the least Π^1_1-reflecting ordinal, then J_κ cannot even satisfy the schema of Σ_1-separation.

(3) PARTITIONS WITH EXPONENT ONE

In this section partitions of exponent one (i.e., partition properties of an ordinal κ) are considered. The section begins with a new proof of a partition theoretic characterization of Σ_{n+1}-admissibility (see Theorem 4.3. in [K1]) which does not use Jensen's uniformization theorem. Next we consider further partition theoretic properties which are satisfied by every subset of an ordinal κ which belongs to a fixed definable class of subsets of κ. This in turn gives partition theoretic characterizations of all ordinals κ which are either Σ_{n+2} admissible or which satisfy the schemas of Σ_{n+1}-collection and Σ_{n+1}-separation.

Definition 3.1. If Φ is any class of $\in$-formulas, the partition symbol $\kappa \xrightarrow{\Phi} (\alpha)^n_\beta$ means that for any $f:[\kappa]^n \to \beta$ which is $\Phi(J_\kappa)$ definable there exists $I \subseteq \kappa$ of order type α such that f is constant on $[I]^n$.

Theorem 3.2. The following are equivalent

(i) κ is Σ_{n+1} admissible

(ii) $\kappa \to {}^{\Sigma_n}(cf)^1_{<\kappa}$.

(iii) $\kappa \to {}^{\Sigma_n}(2)^1_{<\kappa}$.

Proof. The proof of (iii) $\Longrightarrow$(ii) is trivial and the proof of (i) $\Longrightarrow$(iii) is standard (see for example Theorem 4.3 in [Kl]). The proof of (ii) $\Longrightarrow$(i) is based on the following.

Lemma 3.3. Let Φ be a class of $\in$-formulas which is closed under bounded existential quantification. If J_κ satisfies the schema of $\neg\Phi$-selection and $\kappa \to {}^{\Phi}(cf)^1_{<\kappa}$, then κ satisfies the schema of Φ-collection.

Proof. Let R be a relation in $\Phi(J_\kappa)$ such that the following holds:

$$(*) \qquad \forall\xi < \alpha\ \exists\eta\ R(\xi,\eta)$$

We want to prove that the quantifier $\exists\eta$ in (*) above can be founded by an ordinal $\nu < \kappa$. Assume on the contrary that this is not true. This means that

$$\forall\nu\ \exists\xi < \alpha\ \forall\eta < \nu \neg R(\xi,\eta)$$

Put $S(\nu,\xi)\ \forall\eta < \nu \neg R(\xi,\eta)$. Clearly the relation S is in $\neg\Phi(J_\kappa)$. Moreover

$$\forall\nu\ \exists\xi < \alpha\ S(\nu,\xi).$$

Using the fact that κ satisfies $\neg\Phi$-selection there exists a function $f:\kappa \to \alpha$ which is $\neg\Phi(J_\kappa)$ definable and satisfies

$$\forall\nu\ S(\nu,f(\nu)).$$

Let I be cofinal in κ and homogeneous for f. If $\xi_0 < \alpha$ is the constant value of f when restricted to I, then it is clear that

$$\forall\nu \in I\ \forall\eta < \nu \neg R(\xi_0,\eta).$$

Since I is cofinal in κ it is immediate that $\forall\eta \neg R(\xi_0,\eta)$ which contradicts (*) and completes the proof of the lemma.

Now we can complete the proof of the theorem.

Proof of (ii) $\Longrightarrow$(i).

This is proved by induction on $n \geq 1$. If $n = 1$, then apply the lemma to the class PR of primitive recursive relations to conclude that κ satisfies the schema of PR-collection. It follows from Theorem 3.7 in [R-A] that κ is Σ_1 admissible. Next apply the lemma to the class $\Phi = \Pi_1$ (note that $\Sigma_1(J_\kappa)$ is closed under bounded quantification because κ is Σ_1 admissible) to conclude that κ satisfies the schema of Π_1-collection and hence must be Σ_2 admissible. In both of the above cases it is not hard to see that κ satisfies the schema of PR-(respectively Σ_1 -)

selection without using Jensen's uniformization theorem. Assuming the result to be true for $n-1 \geq 1$, we want to prove it for n. By induction hypothesis κ is Σ_n admissible and hence it satisfies the schema of Σ_n-selection; moreover the class $\Phi = \Pi_n$ is closed under bounded quantification. It follows from the lemma that κ satisfies the schema of Π_n-collection and hence it is Σ_{n+1} admissible. This completes the proof of the theorem.

Definition 3.4. i) If Φ is any class of $\in$-formulas and X is any subset of κ which is cofinal in κ, the symbol $\subseteq\kappa \to^{\Phi}(\alpha\text{-}X)^1_\beta$ means that for any $f: \subseteq\kappa \to \beta$ which is $\Phi(J_\kappa)$ definable with $\mathrm{dom} f \cap X$ cofinal in κ, there exists $I \subseteq \mathrm{dom} f$ such that f is constant on I and $I \cap X$ is of order type α.

ii) The symbol $\kappa \to^{\Phi}(\alpha\text{-}X)^1_\beta$ is defined similarly.

iii) If in i) $X = \kappa$, then we denote the corresponding partition by $\subseteq\kappa \to^{\Phi}(\alpha)^1_\beta$.

iv) The sumbol $\subseteq\kappa \to^{\Phi}(\alpha)^1_{<\kappa}$ means that $\forall\beta<\kappa \subseteq\kappa \to^{\Phi}(\alpha)^1_\beta$. The symbols $\subseteq\kappa \to^{\Phi}(\alpha\text{-}X)^1_{<\kappa}$ and $\subseteq\kappa \to^{\Phi}(<\kappa)^1_{<\kappa}$ are defined similarly.

v) If Ψ is another class of $\in$-formulas, then the symbol $\kappa \to^{\Phi}(\alpha\text{-}\Psi)^m_\alpha$ is defined as before, only now the homogeneous set is assumed to be definable in J_κ, with parameters in J_κ, by a formula in Ψ.

Recall the following theorem from [Kl]:

Theorem 3.5. If κ is Σ_n admissible the following are equivalent

(i) J_κ satisfies Σ_n-sep.

(ii) $\kappa \to^{\Sigma_n}(<\kappa)^1_{<\kappa}$

(iii) $\kappa \to^{\Sigma_n}(2)^1_{<\kappa}$

As an application one can prove the following theorems which give an idea of the "strength" of the partition relation defined in 3.4.

Theorem 3.6. The following are equivalent

(i) J_κ satisfies Σ_n-coll. and Σ_n-sep.

(ii) $\subseteq\kappa \to^{\Sigma_n}(<\kappa)^1_{<\kappa}$

(iii) $\subseteq\kappa \to^{\Sigma_n}(2)^1_{<\kappa}$

Proof. (i) $\Longrightarrow$ (ii). By 3.5 $\kappa \to^{\Sigma_n}(<\kappa)^1_{<\kappa}$.

Let $f: \subseteq\kappa \to \alpha$ be $\Sigma_n(J_\kappa)$ such that $\mathrm{dom}\, f$ is cofinal in κ. Clearly $\mathrm{dom}\, f$ is $\Sigma_n(J_\kappa)$ and cofinal in κ; hence there exists a $\Sigma_n(J_\kappa)$ strictly increasing function $g:\kappa \to \mathrm{dom}\, f$. Let $h = f \circ g : \kappa \to \alpha$. Then for any $\alpha < \kappa$ there exists $\xi < \alpha$ such that the set $h^{-1}[\{\xi\}]$ has order type $> \sigma$. Since g is strictly increasing, for any $\sigma < \kappa$ there exists $\xi < \alpha$ such that $f^{-1}[\{\xi\}]$ has order type $> \sigma$.

(ii) $\Longrightarrow$ (iii) is trivial.

(iii) $\Longrightarrow$ (i). Suppose $f: \subseteq \alpha \to \kappa$ cofinally. Let $R(\xi,\eta) \longleftrightarrow f(\eta) = \xi$. Let g be $\Sigma_n(J_\kappa)$ such that dom g = dom R and for all ξ, $\exists\eta\, R(\xi,\eta) \longleftrightarrow R(\xi,g(\xi))$. Then g must be 1-1 and $g: \subseteq \kappa \to \alpha$. But this contradicts the hypothesis unless $\alpha = \kappa$. It follows that J_κ satisfies Σ_n-coll. and Σ_n-sep.

Definition 3.7. Let Φ be a class of $\in$-formulas

(i) Φ cf(κ) = least σ such that there exists a function $f:\sigma \xrightarrow{cf} \kappa$ which is $\Phi(J_\kappa)$ definable.

(ii) Φ pcf(κ) = least σ such that there exists a function $f: \subseteq \sigma \xrightarrow{cf} \kappa$ which is $\Phi(J_\kappa)$ definable.

According to a theorem of Simpson (see [S3]) it is true that

Theorem 3.8.

$$\Pi_n \mathrm{cf}(\kappa) = \kappa \Rightarrow \Sigma_{n+1}\, \mathrm{cf}(\kappa) = \kappa$$

The idea of the proof of the above theorem is also used in the proof of (ii) $\Longrightarrow$ (i) in the theorem below.

Theorem 3.9. The following are equivalent

(i) $\Sigma_{n+1}\mathrm{pcf}(\kappa) = \kappa$

(ii) $\Pi_n\mathrm{pcf}(\kappa) = \kappa$

(iii) $\subseteq\kappa \xrightarrow{\Sigma_{n+1}} (<\kappa)^1_{<\kappa}$

(iv) $\subseteq\kappa \xrightarrow{\Pi_n} (2)^1_{<\kappa}$

Proof (ii) $\Rightarrow$ (i). Let $f: \subseteq \alpha \xrightarrow{cf} \kappa$ be $\Sigma_{n+1}(J_\kappa)$. Let R be a $\Sigma_{n-1}(J_\kappa)$ relation (if $n = 1$, then R is $\Delta_1(J_\kappa)$) such that

$$f(\xi) = \eta \longleftrightarrow \exists\sigma\forall\tau\, R(\xi,\eta,\sigma,\tau) .$$

Define a function $g: \subseteq \alpha \to \kappa$ by

$$g(\xi) = \langle\eta,\sigma,\rho\rangle \longleftrightarrow \forall\tau R(\eta,\eta,\sigma,\tau) \wedge \forall\zeta < \sigma\, \exists\tau < \rho\, \neg R(\xi,\eta,\zeta,\tau)$$
$$\wedge\ \forall\nu < \rho\ \exists\zeta < \sigma\ \forall\tau < \nu\ R(\xi,\eta,\zeta,\tau)$$

By assumption the Π_n partial cofinality of κ is equal to κ and hence κ must be Σ_n admissible. It follows that g is Π_n; moreover it can be checked easily that dom g = dom f and if $f(\xi) = \eta$ then $g(\xi) = \langle\eta,\sigma,\rho\rangle$, for some $\sigma,\rho < \kappa$. Since the range of f is cofinal in κ, so is the range of g. But this contradicts the hypothesis $\Pi_n\mathrm{pcf}(\kappa) = \kappa$.

(i) $\Rightarrow$ (iii) is immediate in view of Theorem 3.6.

(iii) $\Rightarrow$ (iv) is trivial.

(iv) $\Rightarrow$ (ii) If $n = 1$, then J_κ satisfies Σ_1-sep. (see Theorem 3.5.) and hence κ is Σ_1 admissible (see Theorem 42 in [D2]). If $n \geq 2$, then since $\underline{\subset}\kappa \to^{\Sigma_{n-1}} (2)^1_{<\kappa}$, κ is Σ_{n-1} admissible. Let $f: \underline{\subset}\, \alpha \xrightarrow{cf} \kappa$ be $\Sigma_n(J_\kappa)$ definable ($\alpha < \kappa$). Let R be $\Sigma_{n-1}(J_\kappa)$ definable (if $n = 1$, then R is $\Delta_1(J_\kappa)$) such that

$$f(\xi) = \eta \longleftrightarrow \forall\theta\, R(\xi,\eta,\theta).$$

Define a $\Pi_n(J_\kappa)$ function $g: \underline{\subset}\, \alpha \xrightarrow{1-1} \kappa$ by

$$\begin{aligned} g(\xi) = \langle\eta\ \rho\rangle \longleftrightarrow\ & \forall\theta\, R(\xi,\eta,\theta)\ \wedge \\ & (\forall\nu < \xi)(\exists\tau < \rho)\ \neg R(\nu,\eta,\tau)\ \wedge \\ & (\forall\zeta < \rho)(\exists\nu < \xi)(\forall\tau < \zeta) R(\nu,\eta,\tau). \end{aligned}$$

Assume for the moment that κ is Σ_n admissible; then it is easy to show that

$$\eta \in \text{range}(f) \to \exists\rho < \kappa(\langle\eta,\rho\rangle \in \text{range}(g)).$$

Then range(g) is cofinal in κ and the function $g^{-1}: \underline{\subset}\, \kappa \xrightarrow{1-1} \alpha$ gives a contradiction. It remains to give the proof that κ is Σ_n-admissible. If $n = 1$, then this is true. If $n \geq 2$, then it is enough to prove that $\Pi_{n-1}cf(\kappa) = \kappa$ (see Theorem Theorem 3.8.) So let $f:\alpha \xrightarrow{cf} \kappa$ be $\Pi_{n-1}(J_\kappa)$ definable $(\alpha < \kappa)$ and proceed as before to get $g: \underline{\subset}\, \alpha^{cf} \to \kappa$ which is 1-1 and $\Pi_{n-1}(J_\kappa)$ definable (using the Σ_{n-1} admissibility of κ). As before the function $g^{-1}: \underline{\subset}\kappa \xrightarrow{1-1} \alpha$ gives a contradiction. This completes the proof of the claim and hence of the theorem.

It is an easy consequence of Theorem 3.2. that

Theorem 3.10. If X is a $\Sigma_n(J_\kappa)$ cofinal subset of κ then the following are equivalent

(i) κ is Σ_{n+1} admissible

(ii) $\underline{\subset}\kappa \to^{\Sigma_n} (\kappa - X)^1_{<\kappa}$

On the other hand it is interesting that the partition relation $\underline{\subset}\kappa \to^{\Pi_n} (\kappa)^1_{<\kappa}$ is stronger than $\underline{\subset}\kappa \to^{\Sigma_n} (\kappa)^1_{<\kappa}$. This follows immediately from the characterization below:

Theorem 3.11. The following are equivalent

(i) κ is Σ_{n+2} admissible

(ii) $\underline{\subset}\, \kappa \to^{\Pi_n} (\kappa)^1_{<\kappa}$

(iii) $\underline{\subset}\, \kappa \to^{\Delta_1} (\kappa - S^n_\kappa)^1_{<\kappa}$.

<u>Proof</u> <u>(ii) $\Longrightarrow$ (i).</u> In view of Theorem 3.2. it is enough to show that $\kappa \to {}^{\Sigma_{n+1}}(\kappa)^1_{<\kappa}$. So let $\alpha < \kappa$ and $f:\kappa \to \alpha$ be a $\Sigma_{n+1}(J_\kappa)$ definable function f. Let R be a $\Sigma_{n-1}(J_\kappa)$ definable relation (if n = 1, then R is assumed to be $\Delta_1(J_\kappa)$ definable) such that

$$f(\xi) = \eta \longleftrightarrow \exists\sigma\forall\tau \quad R(\xi,\eta,\sigma,\tau)$$

Define a mapping $g:\kappa \to \kappa$ by

$$\begin{aligned} g(\xi) = \langle\eta,\sigma,\rho\rangle \longleftrightarrow \forall\tau \quad & R(\xi,\eta,\sigma,\tau) \quad \wedge \\ & \neg\exists\zeta < \sigma \ \forall\tau < \rho \quad R(\xi,\eta,\zeta,\tau) \quad \wedge \\ & \forall\nu < \rho \ \exists\zeta < \sigma \ \forall\tau < \nu \quad R(\xi,\eta,\zeta,\tau). \end{aligned}$$

It is easy to show that g is well-defined and $\Pi_n(J_\kappa)$ definable. Moreover

$$f(\xi) = \eta \longleftrightarrow \exists\sigma,\rho \ (g(\xi) = \langle\eta \ \sigma \ \rho\rangle)$$

Next define a mapping $h: \subseteq \kappa \to \alpha$ by

$$h(\xi) = \eta \longleftrightarrow g((\xi)_1) = \langle\eta,(\xi)_2,(\xi)_3\rangle$$

The mapping h is $\Pi_n(J_\kappa)$ definable and dom(h) is cofinal in κ. It follows from the hypothesis that there exists $\beta < \alpha$ such that $h^{-1}[\{\beta\}]$ is cofinal in κ. The proof will be complete if we show that

<u>Claim.</u> $f^{-1}[\{\beta\}]$ is cofinal in κ.

<u>Proof of claim.</u> Assume otherwise and let $\gamma_0 < \kappa$ such that $f^{-1}[\{\beta\}] \subseteq \gamma_0$. From the definition of g we have

$$\forall\xi < \gamma_0 \quad \exists\eta,\sigma,\rho(g(\xi) = \langle\eta,\sigma,\rho\rangle).$$

By Σ_{n+1}-collection there exists $\delta_0 < \kappa$ such that

$$\forall\zeta < \gamma_0 \quad \exists\eta,\sigma,\rho < \delta_0(g(\xi) = \langle\eta,\sigma,\rho\rangle).$$

But then it is easy to see that

$$\begin{aligned} h(\xi) = \beta & \Longrightarrow g((\xi)_1) = \langle\beta,(\xi)_2,(\xi)_3\rangle \\ & \Longrightarrow f((\xi)_1) = \beta \wedge (\xi)_2,(\xi)_3 < \delta_0 \\ & \Longrightarrow (\xi)_1 < \gamma_0 \wedge (\xi)_2, \ (\xi)_3 < \delta_0 \\ & \Longrightarrow \xi < \eta_0, \end{aligned}$$

For some $\eta_0 < \kappa$. Therefore $h^{-1}[\{\beta\}] \subseteq \eta_0$, which is a contradiction.

<u>(i) $\Longrightarrow$ (iii).</u> This is similar to the proof of (i) $\Longrightarrow$ (iii) in Theorem 3.2.

<u>(iii) $\Longrightarrow$ (ii).</u> Let $\alpha < \kappa$ and $f: \subseteq \kappa \to \alpha$ be a $\Pi_n(J_\kappa)$ definable mapping, defined by a Π_n formula $\varphi(x,y,z)$ such that

$$f(\xi) = \eta \longleftrightarrow J_\kappa \models \varphi(\xi,\eta,p),$$

where p is a parameter and $\mathrm{dom}\, f$ is cofinal in κ. Define a function $g: \subseteq \kappa \to \alpha$ by

$g(\xi) \simeq$ least $\eta < \alpha$ such that $p \in J_\xi$ and S^n_ξ has a greatest element σ and $J_\xi \models \exists\rho \geq \sigma\, \varphi(\rho,\eta,p)$.

It is clear that g is $\Delta_1(J_\kappa)$ (moreover $\mathrm{dom}\, g$ is $\Delta_1(J_\kappa)$). If $\xi \in S^n_\kappa$ and $g(\xi) = \eta$ then $p \in J_\xi$; if σ = the greatest element of S^n_ξ, then $J_\xi \models \exists\rho \geq \sigma\ \varphi(\rho,\eta,p)$. But the last formula is Σ_{n+1} and $J_\xi \prec_n J_\kappa$. Hence this last formula is also true in J_κ. This means that there exists ρ such that $\xi > \rho \geq \sigma$ and $f(\rho) = \eta$.

Claim. $\mathrm{dom}\, g \cap S^n_\kappa$ is cofinal in κ.

Proof of Claim. Let $\beta < \kappa$ and let $\rho > \beta$ such that $\rho \in \mathrm{dom}\, f$. Let ξ be the least such that $\xi > \rho$, $p \in J_\xi$ and $\xi \in S^n_\kappa$. Then ξ is not a limit point of S^n_κ. Let σ = the greatest element of S^n_ξ. Then $\sigma \leq \rho$. It follows that $\xi \in \mathrm{dom}(g) \cap S^n_\kappa$, and hence the proof of the claim is complete.

By the hypothesis $\subseteq \kappa \to^{\Delta_1} (\kappa - S^n_\kappa)^1_{<\kappa}$ there exists $\gamma < \alpha$ such that $g^{-1}[\{\gamma\}]$ is cofinal in κ. It follows immediately that $f^{-1}[\{\gamma\}]$ is cofinal in κ. The proof of the theorem is complete.

For an application of the preceding theorem to the theory of "Σ_n end extensions" see the second author's doctoral thesis ([P]).

(4) COVERING PROPERTIES

If Φ is any class of $\in$-formulas then the property $\kappa \to^{\Phi} (\kappa)^1_{<\kappa}$ can be viewed as a partition theoretic property of κ, i.e., for any $\alpha < \kappa$ and any partition $\langle A_\xi : \xi < \alpha \rangle$ of κ which is $\Phi(J_\kappa)$ definable, there exists $\xi < \alpha$ such that A_ξ is cofinal in κ. The Φ covering property is a slight generalization of the above partition property; namely now we assume that $\kappa = \cup \{A_\xi : \xi < \alpha\}$ but the sets A_ξ do not have to be disjoint. In this section the above mentioned Φ covering property is studied for different classes of $\in$-formulas Φ; in particular we study the relative strength of those κ which satisfy the Φ covering property. We also consider the relation between covering and reflection properties of κ. The results of this section have consequences for the theory of "definable ultrafilters"; see the second author's doctoral thesis ([P]).

Definition 4.1. Let X be a cofinal subset of κ and Φ a class of $\in$-formulas.

(i) κ has the Φ covering property on X (abbreivated Φ c.p. on X) if and only if for any $\alpha < \kappa$ and any $\Phi(J_\kappa)$ definable family $\langle A_\xi : \xi < \alpha \rangle$ such that $X \subseteq \cup\{A_\xi : \xi < \alpha\}$, there exists $\xi < \alpha$ such that $A_\xi \cap X$ is cofinal in κ.

(ii) If in the above $X = \kappa$, then "κ has the Φ c.p. on κ" is abbreviated by "κ has the Φ c.p."

(iii) The notion of Δ_n c.p. is defined similarly.

The first result is an easy consequence of Theorem 3.2.

Theorem 4.1. Let X be a $\Sigma_n(J_\kappa)$ cofinal subset of κ. Then the following are equivalent

(i) κ is Σ_{n+1} admissible.

(ii) κ has the Δ_n c.p.

(iii) κ has the Σ_n c.p. on X.

In connection with Theorem 4.1. Simpson showed in the terminology of this paper that "κ is Σ_{n+1} admissible" is equivalent to "κ is Σ_n admissible and has the Σ_n c.p." in the unpublished [S3], although this proof, which does not require any selection, was not known to the first author when he proved Theorem 3.2 and when he formulated the notion of a covering property.

Proposition 4.2.

κ has the C_n c.p. $\Longleftrightarrow$ κ has the B_n c.p.

Proof. ($\Longleftarrow$) is trivial.

($\Longrightarrow$) Let $\alpha < \kappa$ and let $\langle A_\xi : \xi < \alpha\rangle$ be a $B_n(J_\kappa)$ covering of κ. There exists Σ_n formulas $\varphi_0,\ldots,\varphi_{m-1}$ and Π_n formulas $\psi_0,\ldots,\psi_{m-1}$ and a parameter $p \in J_\kappa$ such that

$$\gamma \in A_\xi \longleftrightarrow J_\kappa \models \bigvee_{i<m} [\varphi_i(\gamma,\xi,p) \wedge \psi_i(\gamma,\xi,p)].$$

Define a new covering $\langle A'_\eta : \eta < \alpha m\rangle$ by

$$\gamma \in A'_{\alpha i+\xi} \longleftrightarrow J_\kappa \models \varphi_i(\gamma,\xi,p) \wedge \psi_i(\gamma,\xi,p).$$

Let θ be a universal Σ_n formula and let $f,g : m \to \kappa$ be elements of J_κ such that for $i < m$

$$\varphi_i(\gamma,\xi) \longleftrightarrow \theta(f(i),\gamma,\xi)$$

$$\psi_i(\gamma,\xi) \longleftrightarrow \neg\theta(f(i),\gamma,\xi)$$

Then it is clear that

$$\gamma \in A'_\eta \longleftrightarrow \exists\xi < \alpha\ \exists i < m\ \exists\rho\ [\eta = \alpha i+\xi \wedge f(i) = \rho \wedge \theta(\rho,\gamma,\xi)]$$
$$\wedge\ \forall\xi < \alpha\ \forall i < m \forall\rho\ [\eta = \alpha i+\xi \wedge g(i) = \rho \rightarrow \neg\theta(\rho,\gamma,\xi)]$$

and hence $\langle A'_\eta : \eta < \alpha m\rangle$ is $C_n(J_\kappa)$. Since $A'_{\alpha i+\xi} \subseteq A_\xi$, the result follows immediately.

Though it is not known whether the Π_n c.p. is stronger than the Σ_n c.p. (Theorem 4.1 shows that it is at least as strong), one can show that the B_n c.p.

(which is equivalent to the C_n c.p.) is stronger than the Σ_n c.p. This is done in the result below.

Theorem 4.3. If κ has the B_n c.p., then κ is Σ_{n+1} admissible and a limit of Σ_{n+1} admissibles.

Proof. Assume that κ has the B_n c.p. By Theorem 4.1. κ is Σ_{n+1} admissible. Assume on the contrary that κ is not the limit of Σ_{n+1} admissibles. As in Lemma 3.21 of [K-K] one can prove that there exists $\alpha < \kappa$ such that for all $\xi < \kappa$, the singleton $\{\xi\}$ is Π_n definable in J_κ with parameters in $\alpha \cup \{\alpha\}$. Let $\beta < \kappa$ and $<\theta_\gamma : \gamma < \beta>$ be an effective enumeration of all the Π_n formulas which have parameters in $\alpha \cup \{\alpha\}$. For each $\gamma < \beta$ define the set

$$X_\gamma = \{\xi < \kappa : \models^{\Pi_n}_{J_\kappa} \theta_\gamma(\xi) \wedge \forall\eta < \xi \models^{\Sigma_n}_{J_\kappa} \neg\theta_\gamma(\eta)\} .$$

It is clear that the family $<X_\gamma : \gamma < \beta>$ covers κ; moreover each X_γ can have at most one element. This is a contradiction.

The following results relate covering and reflecting properties.

Lemma 4.4. If X is a $\Pi_n(J_\kappa)$ cofinal subset of κ and for any $\Delta_1(J_\kappa)$ cover $<X_\xi : \xi < \alpha>$ $(\alpha < \kappa)$ of X there exists a $\xi_0 < \alpha$ such that $X_{\xi_0} \cap S^n_\kappa$ is cofinal in κ, then κ is $\forall^b\Sigma_{n+2}$-reflecting on X.

Proof. Let $\alpha < \kappa$ and $\varphi(x,y)$ be a Π_{n+1} formula with parameters in J_κ such that

$$J_\kappa \models \forall\xi < \alpha\ \exists\eta\ \varphi(\xi,\eta).$$

We want to show that the formula $\forall\xi < \alpha\ \exists\eta\varphi$ can be reflected down to a J_γ such that $\gamma \in X$. Indeed assume otherwise. That means that

$$\forall\gamma < \kappa\ \exists\xi < \alpha(\gamma \in X \to J_\gamma \models \forall\eta \neg\varphi(\xi,\eta))$$

For each $\xi < \alpha$ consider the set

$$X_\xi = \{\gamma < \kappa : J_\gamma \models \forall\eta \neg\varphi(\xi,\eta)\} .$$

It is clear that the family $<X_\xi : \xi < \alpha>$ is a $\Delta_1(J_\kappa)$ cover of $X \setminus (\alpha+1)$. It follows that there exists $\xi_0 < \alpha$ such that the set $X_{\xi_0} \cap S^n_\kappa$ is cofinal in κ. Let $A = X_{\xi_0} \cap S^n_\kappa$. If $\gamma \in A$, then $\forall\eta < \gamma\ J_\gamma \models \neg\varphi(\xi_0,\eta)$. But $A \subseteq S^n_\kappa$ and hence for all $\gamma \in A$, $\forall\eta < \gamma\ J_\kappa \models \neg\varphi(\xi_0,\eta)$. Since A is cofinal in κ it follow that $J_\kappa \models \forall\eta \neg\varphi(\xi_0,\eta)$, which is a contradiction.

Theorem 4.5. If κ has the Δ_1 c.p. on S^n_κ, then κ is $(\forall^b\Sigma_{n+2} \wedge \Pi_{n+2})$-reflecting on S^n_κ.

Proof. For any Π_{n+2} formula φ with parameters from J_κ but no free variables consider the set $X_\varphi = \{\gamma \in S^n : J_\gamma \models \varphi\}$. The ordinal κ is Σ_{n+1} admissible and therefore if $J_\kappa \models \varphi$ then the set X_φ is closed and unbounded in κ. The proof is based on the following.

Claim. If C is a $\Pi_n(J_\kappa)$ club subset of S^n_κ and $\langle C_\xi : \xi < \alpha \rangle$ $(\alpha < \kappa)$ is any $\Delta_1(J_\kappa)$ cover of C, then there exists $\xi_0 < \alpha$ such that the set $C_{\xi_0} \cap S^n_\kappa$ is cf in κ.

Proof of claim. Suppose that C, $\langle C_\xi : \xi < \alpha \rangle$ are as in the claim and let θ be a $\Pi_n(J_\kappa)$ formula which defines C in J_κ. For $\xi < \alpha$ define the set X_ξ as follows:

$$\beta \in X_\xi \longleftrightarrow \beta \in C_\xi \vee J_\beta \models \exists\rho \ (\theta(\rho) \wedge \forall\delta > \rho \ \theta(\delta) \wedge \psi(\xi,\rho)),$$

where ψ is the Σ_1 formula which defines $\rho \in C_\xi$ in J_κ. It is clear that $\langle X_\xi : \xi < \alpha \rangle$ is $\Delta_1(J_\kappa)$ uniformly in $\xi < \alpha$. We next show that $\langle X_\xi : \xi < \alpha \rangle$ covers S^n_κ. Indeed take any $\beta \in S^n_\kappa$. If $\beta \in C$, then $\beta \in C_{\xi_0}$, for some $\xi_0 < \alpha$ and so $\beta \in X_{\xi_0}$. If on the other hand $\beta \notin C$ then since C is closed, $\beta \cap C$ has a greatest member ρ. Since $\rho \in C$ there exists $\xi_0 < \alpha$ such that $\rho \in C_{\xi_0}$. It follows from the definition of X_{ξ_0} that $\beta \in X_{\xi_0}$. This shows that $\langle X_\xi : \xi < \alpha \rangle$ covers S^n_κ. But κ has the Δ_1 c.p. on S^n_κ and hence there exists $\xi_0 < \alpha$ such that $X_{\xi_0} \cap S^n_\kappa$ is cf in κ. Now it can be shown that $C_{\xi_0} \cap S^n_\kappa$ is cf in κ. Indeed take any $\beta_0 < \kappa$. By the cf of C, there exists $\beta_1 > \beta_0$ such that $\beta_1 \in C$. Since $X_{\xi_0} \cap S^n_\kappa$ is cf in κ one can find $\beta_2 \in X_{\xi_0} \cap S^n_\kappa$ such that $\beta_2 > \beta_1$. But by definition of X_{ξ_0}, either $\beta_2 \in C_{\xi_0}$, in which case the result follows, or else there exists $\beta_1 \leq \beta_3 < \beta_2$ such that $\beta_3 \in C_{\xi_0} \cap C$. It follows that $\beta_3 \in C_{\xi_0} \cap S^n_\kappa$, and hence the proof of the claim is complete.

Now using the claim we can complete the proof of the theorem. It is enough to show that for any Π_{n+2} formula φ such that $J_\kappa \models \varphi$, the ordinal κ is $\forall^b\Sigma_{n+2}$-reflecting on X_φ. But for any such φ, the set X_φ is closed and unbounded. So the result follows immediately from the claim and Lemma 4.4.

In the presence of the power set axiom one can obtain the following.

Corollary 4.6. If $J_\kappa \models$ PSA and κ satisfies the Δ_1 c.p. on S^n_κ, then is Π_{n+3}-reflecting.

Proof. By theorem 4.5. such a κ must be $(\forall^b\Sigma_{n+2} \wedge \Pi_{n+2})$-reflecting. It follows from Theorem 2.4. and the Remark 2.5. following it, that κ is Π_{n+3}-reflecting.

The above corollary can also be restated in terms of the partition theoretic notation of Chapter 3. Namely,

Corollary 4.7. If $J_\kappa \models PSA$ and $\kappa \to {}^{\Delta_1}(\kappa\text{-}S^n_\kappa)^1_{<\kappa}$, then κ is Π_{n+3}-reflecting.

Remark 4.8. It should be noted that "κ is Σ_{n+1} admissible" does not imply "κ is Π_{n+3}-reflecting". The former notion is equivalent to κ" is Π_{n+2}-reflecting on S^n_κ" by Theorem 4.3 of [K1]. Moreover the partition relation $\kappa \to {}^{\Delta_1}(\kappa\text{-}S^n)^1_{<\kappa}$, in contrast to the partition relation $\subseteq \kappa \to {}^{\Delta_1}(\kappa\text{-}S^n)^1_{<\kappa}$, is weaker than Σ_{n+2} admissibility (See Theorme 3.11). This is not hard to see; for example if there were an $\alpha > \kappa$ such that $J_\kappa \prec_{n+1} J_\alpha$ then $\kappa \to {}^{\Delta_1}(\kappa\text{-}S^n)^1_{<\kappa}$, but such a κ does not need to be Σ_{n+2} admissible.

(5) DEFINABLE HOMOGENEOUS SETS FOR PARTITIONS OF FINITE EXPONENT GREATER THAN ONE

Given any class Φ of $\in$-formulas and any partition $f:\kappa \to \alpha$ of κ into α many pieces $(\alpha < \kappa)$ such that f is $\Phi(J_\kappa)$ definable, if there is a homogeneous set for f, then there is one of the form $f^{-1}[\{\xi\}]$, where $\xi < \alpha$. Clearly the definability of a homogeneous set of this form is simply related to the definability of the given partition of κ. However when one considers partitions of higher exponents (i.e., partitions of $[\kappa]^2$, $[\kappa]^3, \ldots$ etc.) the study fo the definability of the homogeneous sets of a given partition is much more complicated. Work on this question was originally done by Specker (see [Sp]) and later by Jockusch (see [J]) and Yates (see [Y]) for the case $\kappa = \omega$. Some results related to this question had also been obtained in [K2]. So in the present section we will concentrate on the definability of the homogeneous set in definable partitions of $[\kappa]^m$ (where $m \geq 2$). It should be pointed out that our work in the present section has been greatly influenced by the corresponding work of Jockusch (see ([J]) for $\kappa = \omega$.

The first result will be of much use later and is known in the literature as the limit lemma (see for example Lemma 1.7 in [S2]).

Lemma 5.1. (Limit Lemma). If κ is Σ_n admissible anf $f:\kappa^m \to 2$ is $\Sigma_{n+1}(J_\kappa)$, then there exists a $\Sigma_n(J_\kappa)$ function $g:k^{m+1} \to 2$ such that $f(\bar{\alpha}) = \lim_{\alpha \to \kappa} g(\bar{\alpha},\sigma)$, for all $\bar{\alpha} < \kappa$, i.e., $\forall\bar{\alpha} < \kappa\ \exists\beta < \kappa\ \forall\sigma \geq \beta\ g(\bar{\alpha},\sigma) = f(\bar{\alpha})$.

Proof: Since f is $\Sigma_{n+1}(J_\kappa)$ there exists a $\Sigma_{n-1}(J_\kappa)$ relation R such that

$$f(\bar{\alpha}) = \delta \leftrightarrow \exists\xi\forall y \quad R(\xi,y,\bar{\alpha},\delta)$$

Define $g:\kappa^{m+1} \to 2$ as follows

$$g(\bar{a},\sigma) = \begin{cases} 0 & \text{if } \exists\xi < \sigma\ [\forall y \in J_\sigma\ R(\xi,y,\bar{a},0)\ \wedge \\ & \qquad \wedge\ \forall\rho \le \xi\ \exists y \in J_\sigma \neg R(\rho,y,\bar{a},1)] \\ 1 & \qquad \text{o.w.} \end{cases}$$

It is clear that g is $\Delta_n(J_\kappa)$ definable, because κ is Σ_n admissible. One has to use the Σ_n admissibility of κ once more to show that for all $\bar{a} < \kappa$

$$f(\bar{a}) = \lim_{\sigma\to\kappa} g(\bar{a},\sigma).$$

Indeed let $\bar{a} < \kappa$. Suppose $f(\bar{a}) = 0$. Let $\xi_0 < \kappa$ such that $\forall y\ R(\xi_0,y,\bar{a},0)$. Since $f(\bar{a}) \neq 1$, $\forall\xi\ \exists y \neg R(\xi,y,\bar{a},1)$. It follows that $\forall\rho \le \xi_0\ \exists y \neg R(\rho,y,a,1)$. Using the Σ_n admissibility of κ we can find $\beta < \kappa$ such that

$\forall\rho \le \xi_0\ \exists y \in J_\beta \neg R(\rho,y,\bar{a},1)$. It follows that for all $\sigma > \max\{\xi_0,\beta\}$, $g(\bar{a},\sigma) = 0$. If on the other hand $f(\bar{a}) = 1$, then there exists $\xi_0 < \kappa$ such that $\forall y\ R(\xi_0,y,\bar{a},1)$. As before it is true that $\forall\rho \le \xi_0\ \exists y \in J_\beta \neg R(\rho,y,a,0)$ for some $\beta < \kappa$. It follows that for all $\sigma > \max\{\xi_0,\beta\}$, $g(\bar{a},\sigma) = 1$. This completes the proof.

Remark 5.2. It should be pointed out that a modulus function $h:\kappa^m \to \kappa$ (namely a function which satisfies

$$\forall\bar{a} < \kappa\ \forall\sigma \ge h(\bar{a})\quad g(\bar{a},\sigma) = f(\bar{a}))$$

can be defined by: $h(\bar{a})$ = least β such that

$$(\exists\xi < \beta\ \ \forall y\ R(\xi,y,\bar{a},0)\ \wedge\ \forall\rho \le \xi\ \exists y \in J_\beta \neg R(\rho,y,\bar{a},1))\ \vee$$

$$\vee\ (\exists\xi < \beta\ \forall y\ \ R(\xi,y,\bar{a},1)\ \wedge\ \forall\rho \le \xi\ \exists y \in J_\beta \neg R(\rho,y,\bar{a},0)).$$

Using Σ_n-collection one can see that h is $C_n(J_\kappa)$ definable.

Remark 5.3. One can use the idea in the proof of the limit lemma to prove the following more general

Lemma. If κ is Σ_n admissible and $f:\kappa^m \to \gamma$ $(\gamma < \kappa)$ is a $\Sigma_{n+1}(J_\kappa)$ definable function, then there is a $\Sigma_n(J_\kappa)$ function $g:\kappa^{m+1} \to \gamma$ such that

$$f(\bar{a}) = \lim_{\sigma\to\kappa} g(\bar{a},\sigma)\ .$$

Proof (outline). If f is defined by: $f(\bar{a}) = \delta$ if and only if $\exists\xi\forall y\ R(\xi,y,\bar{a},\delta)$, define $R'(\bar{a},\sigma,\delta)$ if and only if

$$\exists\xi < \sigma\ [\forall y \in J_\sigma\ R(\xi,y,\bar{a},\delta)\ \wedge$$
$$\forall\delta' < \gamma\ \forall\rho \le \xi\ \exists y \in J_\sigma(\delta' \neq \delta \to \neg R(\rho,y,\bar{a},\delta'))].$$

It can be checked that for any $\bar{a},\sigma$ there is at most one δ such that $R'(\bar{a},\sigma,\delta)$. Now define $g:\kappa^{m+1} \to \gamma$ by

$$g(a,\sigma) = \begin{cases} \delta & \text{if } R'(\bar{a},\sigma,\delta) \\ 0 & \text{if } \forall\delta < \gamma \neg R'(\bar{a},\sigma,\delta). \end{cases}$$

The proof can be completed as before.

The limit lemma can be used to prove the following stepping up lemma as a corolla

Corollary 5.4. (Stepping up lemma). For all $\alpha < \kappa$, $m \geq 1$

$$\kappa \xrightarrow{\Sigma_n} (\kappa)^{m+1}_{\alpha} \Longrightarrow \kappa \xrightarrow{\Sigma_{n+1}} (\kappa)^{m}_{\alpha}.$$

Proof. Given a $\Sigma_{n+1}(J_\kappa)$ definable function $f:[\kappa]^m \to \alpha$, apply the limit lemma to find a $\Sigma_n(J_\kappa)$ definable function $g:[\kappa]^{m+1} \to \alpha$ such that for all $\bar{\beta} < \kappa$ $f(\bar{\beta}) = \lim_{\sigma\to\kappa} g(\bar{\beta},\sigma)$. But using the hypothesis, g has a homogeneous set I of order type κ; i.e., there exists $\xi < \alpha$ such that $g[[I]^{m+1}] = \{\xi\}$. Passing to the limit it is easy to see that $f[[I]^m] = \{\xi\}$.

The limit lemma can also be used to give an alternative proof of a result of [K3] concerning definable Ramsey ordinals.

Corollary 5.5.

$$\kappa \xrightarrow{\Sigma_1} (\kappa)^{<\omega}_2 \Longrightarrow \kappa \xrightarrow{\Sigma_\omega} (\kappa)^{<\omega}_{<\kappa}$$

Proof. Assume $\kappa \xrightarrow{\Sigma_1} (\kappa)^{<\omega}_2$.

Claim. $\kappa \xrightarrow{\Sigma_\omega} (\kappa)^{<\omega}_2$.

To prove the claim it is enough to show that for all $n \geq 1$, if $\kappa \xrightarrow{\Sigma_n} (\kappa)^{<\omega}_2$ then $\kappa \xrightarrow{\Sigma_{n+1}} (\kappa)^{<\omega}_2$. Indeed given $f:[\kappa]^{<\omega} \to 2$ which is $\Sigma_{n+1}(J_\kappa)$ definable, use the proof of the limit lemma to find a function $g:[\kappa]^{<\omega} \times \kappa \to 2$ which is $\Sigma_n(J_\kappa)$ definable and satisfies $f(\cdot) = \lim_{\sigma\to\kappa} g(\cdot,\sigma)$.

Define a $\Sigma_n(J_\kappa)$ function $h:[\kappa]^{<\omega} \to 2$ such that

$$h(\emptyset) = 0 \quad \text{and}$$

$$h(\xi_1,\ldots,\xi_{m+1}) = g((\xi_1,\ldots,\xi_m),\xi_{m+1}) \quad \text{if } m \geq 0.$$

It is clear that any homogeneous set for h which is cofinal in κ is also homogeneous for f, which establishes the claim.

It remains to show $\kappa \xrightarrow{\Sigma_\omega} (\kappa)^{<\omega}_{<\kappa}$. So suppose $\alpha < \kappa$ and $f:[\kappa]^{<\omega} \to \alpha$ is $\Sigma_\omega(J_\kappa)$. Define $g:[\kappa]^{<\omega} \to 2$ as follows:

For any $\xi_1,\ldots,\xi_{2m+1} < \kappa$

$$g(\emptyset) = 0$$

$$g(\xi_1) = 0$$

and for $m \geq 1$

$$g(\xi_1,\dots,\xi_{2m}) = \begin{cases} 0 & \text{if } f(\xi_1,\dots,\xi_m) \ge f(\xi_{m+1},\dots,\xi_{2m}) \\ 1 & \text{if } f(\xi_1,\dots,\xi_m) < f(\xi_{m+1},\dots,\xi_{2m}) \end{cases}$$

$$g(\xi_1,\dots,\xi_{2m+1}) = \begin{cases} 0 & \text{if } f(\xi_1,\dots,\xi_m) \le f(\xi_{m+1},\dots,\xi_{2m}) \\ 1 & \text{if } f(\xi_1,\dots,\xi_m) > f(\xi_{m+1},\dots,\xi_{2m}) \end{cases}$$

Then g is $\Sigma_\omega(J_\kappa)$ definable, and by the claim, g has a homogeneous set I of order-type κ. It can be seen that for all $r \ge 0$, $g[[I]^r] = 0$ (to show that $g[[I]^{2m}] = g[[I]^{2m+1}] = 0$ for $m \ge 1$, enumerate I in increasing order as $\{\eta_\gamma : \gamma < \kappa\}$ and then split I into $\bigcup\{A_\beta : \beta < \kappa\}$, where $A_\beta = \{\eta_{m\beta},\dots,\eta_{m\beta+m-1}\}$). It is then straightforward to see that I is homogeneous for f, which completes the proof.

Our method of proof of $\kappa \to^{\Sigma_\omega} (\kappa)^{<\omega}_2 \Longrightarrow \kappa \to^{\Sigma_\omega} (\kappa)^{<\omega}_{<\kappa}$ would plainly also yield $\kappa \to (\kappa)^{<\omega}_2 \Longrightarrow \kappa \to (\kappa)^{<\omega}_{<\kappa}$. This is shown by Jech (see [Je, Lemma 29.9]); his proof would not work for $\Sigma_\omega(J_\kappa)$ partitions.

Corollary 5.5 shows that the existence of "recursively Ramsey" ordinals, i.e., ordinals κ satisfying $\kappa \to^{\Sigma_1} (\kappa)^{<\omega}_2$, is not provable in ZFC, since if $\kappa \to^{\Sigma_\omega} (\kappa)^{<\omega}_{<\kappa}$ then by Theorem 3.2. κ is Σ_ω admissible, and by Theorem 3.4. of [K1], $J_\kappa \models$ PSA.

Theorem 4.6 of [K3], the proof of which relied on an analysis of indiscernibles, actually gives a stronger version of Corollary 5.5, namely

$$\kappa \to^{\Sigma_1^-} (\kappa)^{<\omega}_2 \Longrightarrow \kappa \to^{\Sigma_\omega} (\kappa)^{<\omega}_{<\kappa} ,$$

where for any class Φ of $\in$-formulas $\kappa \to^{\Phi^-} (\kappa)^{<\omega}_2$ means the same as $\kappa \to^{\Phi} (\kappa)^{<\omega}_2$, except that the partitions are $\Phi(J_\kappa)$ with no parameters, instead of $\Phi(J_\kappa)$. Corollary 5.6 remedies this deficiency in Corollary 5.5. Its proof again makes use of the limit lemma.

<u>Corollary 5.6</u>.

$$\kappa \to^{\Sigma_1^-} (\kappa)^{<\omega}_2 \Longrightarrow \kappa \to^{\Sigma_1} (\kappa)^{<\omega}_2 .$$

<u>Proof</u>. Assume $\kappa \to^{\Sigma_1^-} (\kappa)^{<\omega}_2$. Then $\kappa \to^{\Sigma_\omega^-} (\kappa)^{<\omega}_2$; the proof is the same as that of the claim in the proof of Corollary 5.5, except that we use a parameter-free version of the limit lemma.

For $z \in [\kappa]^{<\omega}$, say $z = (z_1,\dots,z_m)$, let

$$F_1(z) = (m)_2$$

$$F_2(z) = \begin{cases} \langle z_1,\dots,z_{(m)_1}\rangle & \text{if } (m)_1 \ge 1 \\ \emptyset & \text{if } (m)_1 = 0 \end{cases}$$

$$F_3(z) = \begin{cases} \langle z_{(m)_1},\dots,z_{(m)_1+(m)_2}\rangle & \text{if } (m)_2 \geq 1 \\ \emptyset & \text{if } (m)_2 = 0 \end{cases}$$

Here $(\ )_1$ and $(\ )_2$ are the inverses of the ordinal pairing function of the Preliminaries. Note that for all $s,t < \omega$, $s+t \leq \langle s,t\rangle$. Let h be the canonical Σ_1 Skolem function for J_κ; h is $\Sigma_1(J_\kappa)$ with no parameters (see Lemma 32 of [D2]).

Let $f:[\kappa]^{<\omega} \to 2$ be defined by

$$f(x) = i \Longleftrightarrow J_\kappa \models \varphi(a,x,i)$$

where $\varphi(y,x,i)$ is Σ_1 with no parameters, and $a \in J_\kappa$ is a fixed parameter. Now let ψ be a Σ_1 formula with no parameters such that for all $z,y \in [\kappa]^{<\omega}$,

$$J_\kappa \models \psi(y,x,i) \longleftrightarrow \varphi(h(F_1(y),F_2(y)),x,i).$$

Define $g:[\kappa]^{<\omega} \to 2$ by

$$g(z) = \begin{cases} 0 & \text{if } J_\kappa \models \psi(F_2(z),F_3(z),0) \\ 1 & \text{otherwise} \end{cases}$$

Then g is $\Sigma_\omega(J_\kappa)$ with no parameters, and so it has a homogeneous set I of order-type κ. Let $M = h[\omega \times [I]^{<\omega}]$. Then since I is (essentially) closed under ordered pairs, $[I]^{<\omega} \subseteq M \prec_1 J_\kappa$ (see Lemma 35 of [D2]). As I has order-type κ, by the condensation lemma there is a collapsing isomorphism $\pi:M \cong J_\kappa$. If we let $I' = \pi[I]$, then

$$h[\omega \times [I']^{<\omega}] = \pi(h[\omega \times [I]^{<\omega}]) = J_\kappa,$$

by a standard fine-structural argument (see for instance the proof of Claim 2, Theorem 39 in [D2]). I plainly has order-type κ, and is moreover homogeneous for g; to see this take $z' \in [I']^{<\omega}$. Then $z' = \pi(z)$ for some $z \in [I]^{<\omega}$ and

$$\begin{aligned} g(z') = 0 &\Longleftrightarrow J_\kappa \models \psi(F_2(z'),F_3(z'),0) \\ &\Longleftrightarrow J_\kappa \models \psi(F_2(\pi(z)),F_3(\pi(z)),0) \\ &\Longleftrightarrow M \models \psi(F_2(z),F_3(z),0) \\ &\Longleftrightarrow J_\kappa \models \psi(F_2(z),F_3(z),0) \\ &\qquad g(z) = 0. \end{aligned}$$

Since $J_\kappa = h[\omega \times [I']^{<\omega}]$, there are $r < \omega$ and $y' \in [I']^{<\omega}$ such that $a = h(r,y')$. Moreover there is $y = \langle y_1,\dots,y_s\rangle \in [I']^{<\omega}$ such that $r = F_1(y)$ and $y' = F_2(y)$. So for $x \in [\kappa]^{<\omega}$, $i < 2$,

$$f(x) = i \Longleftrightarrow J_\kappa \models \psi(y,x,i).$$

Let $I'' = I' \setminus (y_s+1)$.

Claim. I'' is homogeneous for f.

Suppose $x = \langle x_1,\ldots,x_t\rangle$, $x' = \langle x_1',\ldots,x_t'\rangle \in [I'']^{<\omega}$. Let $m = \langle s,t\rangle$ $(m < \omega)$. Choose $z_{s+t+1},\ldots,z_m \in I''$ satisfying $\max(x_t,x_t') < z_{s+t+1} < \ldots < z_m$. Then

$$\begin{aligned} f(x) = 0 &\iff J_\kappa \models \psi(y,x,0) \\ &\iff g(y_1,\ldots,y_s,x_1,\ldots,x_t,z_{s+t+1},\ldots,z_m) = 0 \\ &\iff g(y_1,\ldots,y_s,x_1',\ldots,x_t',z_{s+t+1},\ldots,z_m) = 0 \\ &\iff J_\kappa = \psi(y,x',0) \\ &\iff f(x') = 0. \end{aligned}$$

This establishes the claim, and the proof is complete.

Another application of the proof of the limit lemma is the following.

Corollary 5.7. For all $\alpha < \kappa$ and $m \geq 1$,

$$\kappa \to^{\Sigma_1} (\kappa - S_\kappa^{n-1})_\alpha^{m+1} \Longrightarrow \kappa \to^{\Sigma_{n+1}} (\kappa)_\alpha^m$$

Proof. Given a function $f:[\kappa]^m \to \alpha$ which is $\Sigma_{n+1}(J_\kappa)$ definable, consider the relation R which was defined in Theorem 5.1. Alter the definition of g by relativizing R to J_σ. The resulting function, say $h:[\kappa]^m \times \kappa \to \alpha$, will be $\Sigma_1(J_\kappa)$ definable. Moreover we have

$$\forall\beta \in [\kappa]^m \;\; \exists\sigma < \kappa \; \forall\rho \geq \sigma(\rho \in S_\kappa^{n-1} \to g(\bar\beta,\rho) = f(\bar\beta))$$

The rest of the proof is easy.

The next corollary gives some information about the definability of the homogeneous set. This result, which is essentially due to Jockusch, will be extended in Theorem 5.9.

Corollary 5.8. There is no ordinal κ such that $\kappa \to^{\Sigma_n} (\kappa - \Sigma_n)_2^2$.

Proof. Assume on the contrary that there is an ordinal κ which satisfies the above partition relation. Clearly such a κ must be Σ_{n+1} admissible. Using the Σ_{n+1} admissibility of κ it is easy to see that there exists a $\Delta_{n+1}(J_\kappa)$ set $A \subseteq \kappa$ such that neither A nor $\kappa\setminus A$ can include a $\Sigma_n(J_\kappa)$ definable set which is cofinal in κ. (One uses the classical construction of a bi-immune set, e.g. see Theorem III; 8.2 in [R]). Let $f:\kappa \to 2$ be the characteristic function of A which is $\Sigma_{n+1}(J_\kappa)$. By the limit lemma there exists a $\Sigma_n(J_\kappa)$ definable function $g:\kappa \times \kappa \to 2$ such that $f(\alpha) = \lim_{\sigma\to\kappa} g(\alpha,\sigma)$. Let h be the restriction of g to $[\kappa]^2$. But using the fact that A is Σ_n-bi-immune, it is easy to see that h cannot have a homogeneous set which is $\Sigma_n(J_\kappa)$ definable and cofinal in κ.

So in view of Corollary 5.8. for any Σ_{n+1} admissible ordinal κ there exists a $\Sigma_n(J_\kappa)$ function $h:[\kappa]^2 \to 2$ which has no $\Sigma_n(J_\kappa)$ definable, homogeneous set cofinal in κ. So the natural question is whether the homogeneous set in the above result can be assumed to be $\Sigma_{n+1}(J_\kappa)$ definable. This is answered in the theorem below which is based on Theorem 3.1. of [J].

Theorem 5.9. If κ is Σ_{n+1} admissible and $J_\kappa \models$ PSA, then there is a $\Sigma_n(J_\kappa)$ function $h:[\kappa]^2 \to 2$ which has no $\Sigma_{n+1}(J_\kappa)$ definable, homogeneous set cofinal in κ.

Proof. Let $<\theta_\alpha : \alpha < \kappa>$ be a $\Delta_1(J_\kappa)$ enumeration of the Σ_{n-1} $\in$-formulas with four free variables and parameters from J_κ. Define a $\Sigma_n(J_\kappa)$ function $g:\kappa \times \kappa \times \kappa \to 2$ as follows

$$g(\alpha,\gamma,\sigma) = \begin{cases} 0 & \text{if} \quad \exists\xi < \sigma[\forall y \in J_\sigma\ \theta_\alpha(\xi,y,\gamma,0) \wedge \\ & \qquad \wedge \forall\rho \le \xi\ \exists y \in J_\sigma \neg\theta_\alpha(\rho,y,\gamma,1)] \\ 1 & \text{otherwise} \end{cases}$$

(Note the above g is a uniform version of the function g defined in the limit lemma). Next we use Σ_n recursion to define a $\Sigma_n(J_\kappa)$ function $f:[\kappa]^2 \to \kappa$. We define $f(\alpha,\beta)$ by recursion on α for each fixed β; so assume $<f(\rho,\beta):\rho < \alpha>$ have been defined. Let $Z_{\alpha,\beta} = \cup\{(f(\rho,\beta))_1, (f(\rho,\beta))_2 : \rho < \alpha\}$. Let $R(\alpha,\beta,\gamma)$ abbreviate $\gamma \notin Z_{\alpha,\beta} \wedge g(\alpha,\gamma,\beta) = 0$. If there exists $\gamma < \beta$ such that $R(\alpha,\beta,\gamma)$, then let γ_1 be the least such γ; otherwise $\gamma_1 = \beta$. Similarly if there exists $\gamma < \beta$ such that $\gamma \ne \gamma_1 \wedge R(\alpha,\beta,\gamma)$, then let γ_2 be the least such γ; otherwise $\gamma_2 = \beta$. Finally we define $f(\alpha,\beta) = <\gamma_1,\gamma_2>$. The required partition $h:[\kappa]^2 \to 2$ is now given by

$$h(\alpha,\beta) = \begin{cases} 0 & \text{if} \quad \exists\gamma < \beta((f(\gamma,\beta))_1 = \alpha) \\ 1 & \text{otherwise} \end{cases}$$

Next it will be shown that h has no unbounded $\Delta_{n+1}(J_\kappa)$ definable homogeneous set (clearly this implies easily that h cannot have an unbounded $\Sigma_{n+1}(J_\kappa)$ definable homogeneous set). Assume on the contrary that there exists an unbounded $\Delta_{n+1}(J_\kappa)$ definable homogeneous set I for the partition h. There exists a Σ_{n-1} formula θ with four free variables and parameters in J_κ such that

$$\gamma \in I \longleftrightarrow J_\kappa \models \exists\xi\ \ \forall y\theta(\xi,y,\gamma,0)$$

$$\gamma \notin I \longleftrightarrow J_\kappa \models \exists\xi\ \ \forall y\theta(\xi,y,\gamma,1)$$

(one takes the formula which defines the characteristic function of the set I). Clearly $\theta \equiv \theta_\alpha$, for some $\alpha < \kappa$. Take any $\beta > \alpha$ and define $\bar{f}_\beta:\alpha \times 2 \to Z_{\alpha,\beta}$ by $\bar{f}_\beta(\gamma,i) = (f(\gamma,\beta))_{i+1}$; it is clear that $\bar{f}_\beta$ is onto and $\Sigma_n(J_\kappa)$ definable. Using Σ_n admissibility and the fact that J_κ satisfies the schema of Δ_n-separation it follows that $\bar{f}_\beta \in J_\kappa$. Since J_κ is Σ_{n+1} admissible and satisfies the power set axiom, there exists an initial segment H of I such that $H \in J_\kappa$ and $J_\kappa \models |H| > |\alpha \times 2|$. Hence there exists $\beta_0 < \kappa$ such that $H = I \cap \beta_0$. Let $\ell:\kappa \to 2$ be the characteristic function of I which is $\Sigma_{n+1}(J_\kappa)$ definable. It follows from the proof of the limit lemma, Remark 5.2., and the Σ_{n+1} admissibility of κ that there exists $\beta_1 < \kappa$ such that

$$\forall\gamma < \beta_0 \ \forall\sigma > \beta_1 (g(\alpha,\gamma,\sigma) = \ell(\gamma)).$$

Now let $\beta \geq \max\{\beta_0,\beta_1,\alpha\}$. We know that $J_\kappa \models |Z_{\alpha,\beta}| < |H|$. Hence there exist γ_1,γ_2 distinct ordinals in $H \setminus Z_{\alpha,\beta}$. It follows that $g(\alpha,\gamma_1,\beta) = g(\alpha,\gamma_2,\beta) = 0$. Hence both $(f(\alpha,\beta))_i$, $i = 1,2$, are less than β_0 and must belong to H. However by definition of h, $h((f(\alpha,\beta))_{i+1},\beta) = i, i = 0,1$. But if $\beta \in I$ and $\beta \geq \beta_1$, then I cannot be homogeneous for h, which is a contradiction. This completes the proof of the theorem.

The above result can easily be extended to partitions with exponent greater than 2. In fact it is known that if $\kappa \xrightarrow{\Sigma_n} (\kappa)^3_2$, then $J_\kappa \models$ PSA (see theorem 3.4. in [Kl]); hence we can drop the hypothesis $J_\kappa \models$ PSA.

Corollary 5.10. If $m \geq 1$, then there is no ordinal κ such that $\kappa \xrightarrow{\Sigma_n} (\kappa - \Sigma_{n+m+1})^{m+2}_2$.

Proof. This is immediate using Theorem 5.9., the stepping up lemma (Corollary 5.4.) and the fact that $\kappa \xrightarrow{\Sigma_n} (\kappa)^3_2 \Longrightarrow J_\kappa \models$ PSA.

Next we want to show that Theorem 5.9. is the best possible, i.e., it will be shown that under certain conditions on κ the homogeneous set can be $\Sigma_{n+2}(J_\kappa)$ definable. This is based on Lemma 5.11., which is nothing else but a definable analogue of König's infinity lemma. Recall from [Kl] that a Σ^κ_n tree is a tree $T = \langle\kappa,<_T\rangle$ of height κ such that if $T_\alpha = \alpha^{th}$ level of T, then the predicates $<_T, \{(\xi,\alpha):\xi \in T_\alpha\}$ are $\Sigma_n(J_\kappa)$ and $\forall\alpha < \kappa\ (T_\alpha \in J_\kappa)$. The height of ξ, denoted by $ht(\xi)$, is the unique ordinal $\alpha < \kappa$ such that $\xi \in T_\alpha$.

Lemma 5.11. If κ is Σ_{n+1} admissible and $cf^{n+j}\kappa = \omega$, then every Σ^κ_n tree has a $\Delta_{n+j}(J_\kappa)$ definable branch of height κ (note $j \geq 2$).

Proof. Let $T = \langle\kappa, <_T\rangle$ be a Σ_n tree. Since $cf^{n+j}\kappa = \omega$, there exists an increasing $\Sigma_{n+j}(J_\kappa)$ sequence $\langle\kappa_r : r < \omega\rangle$ of ordinals $< \kappa$ such that $\kappa_0 = 0$ and $\bigcup\{\kappa_r : r < \omega\} = \kappa$. Define a $\Sigma_{n+j}(J_\kappa)$ relation $R(m,x,y)$ by

$$y \in T_{\kappa_{m+1}} \wedge x <_T y \wedge \{z \in \kappa : y <_T z\} \text{ is cf in } \kappa.$$

(Notice that $j \geq 2$, because κ is Σ_{n+1} admissible and $cf^{n+j}\kappa = \omega$). As in König's lemma one can prove, using the Σ_{n+1} admissibility of κ, the following.

Claim. There is $x_0 \in T_0$ such that $\{z \in \kappa : x_0 <_T z\}$ is cf in κ; moreover for all $m < \omega$,

$$(\forall x \in T_{\kappa_m})[\{z \in \kappa : x <_T z\} \text{ cf}\kappa \rightarrow \exists y\ R(m,x,y)].$$

Using the Σ_{n+j}-selection theorem, one can find a $\Sigma_{n+j}(J_\kappa)$ function G such that for all $m < \omega$ and all x

$$\text{(i)} \quad \mathrm{dom}\, G = \{(m,x) : \exists y\, R(m,x,y)\}$$

$$\text{(ii)} \quad \exists y\, R(m,x,y) \longleftrightarrow R(m,x,G(m,x)).$$

Now define $f:\omega \to \kappa$ by recursion:

$$f(0) = x_0$$

$$f(m+1) = G(m,f(m)).$$

It is easy to see that f is $\Sigma_{n+j}(J_\kappa)$ definable. Note that $\forall m < \omega\; f(m) \in T_{\kappa_{m+1}}$. Finally let $B = \{x \in \kappa : \exists m < \omega\; x <_T f(m)\}$. Then B is a branch of height κ. B is plainly $\Sigma_{n+j}(J_\kappa)$ definable; it is in fact $\Delta_{n+j}(J_\kappa)$, since

$$x \notin B \longleftrightarrow \exists m < \omega \quad (\mathrm{ht}(x) < \kappa_m \wedge x \not<_T f(m)),$$

so that B is $\Pi_{n+j}(J_\kappa)$ definable.

Remarks 5.12. (i) It is interesting to note that Lemma 5.11 is valid only for κ-finite trees (i.e., trees such that $\forall\alpha < \kappa\; T_\alpha \in J_\kappa$); otherwise one can easily construct $\Delta_1(J_\kappa)$ trees which have branches of height κ, but which have no branches of height κ which are members of $J_{\kappa+}$ (κ^+ = next admissible ordinal $>\kappa$). One can do this using the Barwise compactness theorem.

(ii) The hypothesis $\mathrm{cf}^{n+j}\kappa = \omega (j \geq 2)$ in Lemma 5.11 is also essential. Indeed if $V = L$ and $\kappa = \omega_1$ = the first uncountable ordinal, then one can prove that there is a Σ_1^κ tree which has no branch of height κ (one uses Jensen's diamond principle- see Theorem 3.8 in [D1]).

(iii) If $V = L$ and $J \prec J_{\omega_1}$ $(\kappa < \omega_1)$, let $T = \langle\omega_1, <_T\rangle$ be the $\Delta_1(J_\kappa)$ tree which has no infinite branch of height ω_1 (see (ii) above). If $T' = \langle\kappa, <_T\rangle$ is the restriction of the tree T to κ, then one can easily verify that T' is a $\Delta_1(J_\kappa)$ tree on the countable Σ_ω admissible ordinal κ which has no $\Sigma_\omega(J_\kappa)$ branch of height κ.

Theorem 5.13. If κ is Σ_{n+1} admissible, $J_\kappa \models$ PSA and $\mathrm{cf}^{n+j}\kappa = \omega$, then $\kappa \to^{\Sigma_n} (\kappa\text{-}\Delta_{n+j})^2_2$ (note $j \geq 2$).

Proof. Given any partition $h:[\kappa]^2 \to 2$ which is $\Sigma_n(J_\kappa)$ one defines as usual the Σ_n^κ tree $T = \langle\kappa, <_T\rangle$ which is associated with h. (See Theorem 7.2. in [K1]). By Lemma 5.11 T has a branch B which is of height κ and $\Delta_{n+j}(J_\kappa)$ definable. The function $g:B \to 2$ defined by $g(\alpha) = h(\alpha,\beta)$, some $\beta \in B$ such that $\beta\; {}_T\!> \alpha$, is well defined and $\Delta_{n+j}(J_\kappa)$ definable. Clearly there exists an $i < 2$ such that $I = g^{-1}[\{i\}]$ is cf in κ. It follows that I is homogeneous for h and $\Delta_{n+j}(J_\kappa)$ definable.

Finally we would like to show that the hypothesis $cf^{n+j}\kappa = \omega$ of Theorem 5.13 is essential. This will follow easily from Theorem 5.14, which is based on a theorem of Jensen about the characterization of weak compactness in L (see Theorem 11.1 in [D1]). At this point we should like to thank P. Welch for a very fruitful discussion which enabled us to observe Theorem 5.14.

Theorem 5.14. If $V = L$ and κ is a regular uncountable cardinal, then

$$\kappa \to^{\Sigma_2} (\kappa)^2_2 \Longrightarrow \kappa \text{ is } \Pi^1_1 \text{ reflecting.}$$

Proof. This is based on Theorem 11.1 in [D1]. Assume that κ is not Π^1_1 reflecting. There exists a first order formula $\varphi(R)$ with parameters in J_κ, where R is a new unary predicate, such that $\forall R \subseteq \kappa\ J_\kappa \models \varphi(R)$ and $\forall \alpha < \kappa\ \exists R \subseteq \alpha\ J_\alpha \models \neg\varphi(R)$. As in Theorem 11.1 in [D1] one defines a $C_1(J_\kappa)$ stationary set E such that for all $\alpha < \kappa$, $\alpha \cap E$ is not stationary in α. Using this set E one can define a $\diamondsuit_\kappa(E)$-sequence $\langle S_\alpha : \alpha \in E\rangle$ which is $\Delta_2(J_\kappa)$ definable (see Theorem 9.1 in [D1]). This is next used to define a Souslin tree of height κ which is $\Delta_2(J_\kappa)$ definable (see Theorem 11.2 in [D1]). However since κ is a regular uncountable cardinal, the last statement implies that $\kappa \not\to^{\Sigma_2} (\kappa)^2_2$. This completes the proof of the theorem.

Now assuming the existence of an inaccessible cardinal we can get the desired counterexamples.

Corollary 5.15. If there exists an inaccessible cardinal then there are ordinals such that the following hold:

(i) α is countable (so that $cf\alpha = \omega$)

(ii) $\alpha \to^{\Sigma_\omega} (\alpha)^2_2$

(iii) $\alpha \not\to^{\Sigma_2} (\alpha\text{-}\Sigma_\omega)^2_2$.

Proof. Let κ be the last inaccessible cardinal in L. Then κ is not Π^1_1-reflecting in L. By Theorem 5.14 inside L,

$$L \models \kappa \not\to^{\Sigma_2} (\kappa)^2_2.$$

In particular, $\kappa \not\to^{\Sigma_2} (\kappa\text{-}\Sigma_\omega)^2_2$. Now let α be a countable ordinal such that $J_\alpha \prec J_\kappa$. J_α satisfies Σ_α-collection and the power set axiom. It follows from Theorem 7.3 in [Ki] that the partition relation $\alpha \to^{\Sigma_\omega} (\alpha)^2_2$ holds. On the other hand, since $J_\alpha \prec J_\kappa$ we must have $\alpha \not\to^{\Sigma_2} (\alpha\text{-}\Sigma_\omega)^2_2$. This completes the proof.

REFERENCES

[B] J. Barwise, Admissible Sets and Structures, Springer Verlag, Heidelberg, 1975.

[D1] K. Devlin, Aspects of Constructibility, Springer Verlag Lecture Notes in Mathematics 354, 1973.

[D2] K. Devlin, An Introduction to the Fine Structure of the Constructible Hierarchy (Results of Ronald Jensen), in J. E. Fenstad and P. G. Himnam, eds., Generalize Recursion Theory, North-Holland 1974, pp. 123-163.

[Je] T. Jech, Set Theory, Academic Press, 1978.

[J] C. Jockusch, Ramsey's Theorem and Recursion Theory, Journal of Symbolic Logic, 1972, pp. 268-280.

[K-K] M. Kaufmann and E. Kranakis, Definable Ultrapowers and Ultrafilters over Admissible Ordinals, to appear in Zeitschrift für Mathematische Logik and Grundlagen der Mathematik.

[K1] E. Kranakis, Reflection and Partition Properties of Admissible Ordinals, Annals of Mathematical Logic, 1982, pp. 213-242.

[K2] E. Kranakis, Stepping Up Lemmas in Definable Partitions, JSL, 1984, pp. 22-31.

[K3] E. Kranakis, On Definable Ramsey and Definable Erdös Ordinals, to appear in Archiv für Mathematische Logik und Grundlagenforschung, 23/3-4,1983, pp. 115-12

[P] I.C.C. Phillips, Definability Theory for Σ_n Admissible Ordinals with Particular References to Partition Relations and End Extensions, Oxford D. Phil Thesis, 1983.

[R-A] W. Richter and P. Aczel, Inductive Definitions and Reflecting Properties of Admissible Ordinals, in J. E. Fenstad and P. G. Hinman, eds., Generalized Recursion Theory, North-Holland, 1974, pp. 301-381.

[R] H. Rogers Jr., Recursive Functions and Effective Computability, McGraw-Hill, New York, 1967.

[S1] S. Simpson, Short Course on Admissible Set Theory, in J. E. Fenstad, R. O. Gand and G. E. Sacks, eds., Generalized Recursion Theory II, North Holland, 1978, pp. 355-390.

[S2] S. Simpson, Degree Theory on Admissible Ordinals, in J. E. Fenstad and P. G. Hinman, eds., Generalized Recursion Theory, North-Holland, 1974, pp. 165-
19 195.

[S3] S. Simpson, Σ_2 Admissibility, Handwritten Notes.

[Sp] E. Specker, Ramsey's Theorem does not hold in Recursive Set Theory, in R. O. Gandy and C.M.E. Yates, eds., Logic Colloquium '69, North-Holland, 1971, pp. 439-442.

[Y] C.E.M. Yates, A Note on Arithmetical Sets of Indiscernibles, in R. O. Gandy and C.E.M. Yates, Logic Colloquium '69, North-Holland, 1971, pp. 443-451.

ELIMINATION OF QUANTIFIERS FOR THE THEORY OF ARCHIMEDEAN ORDERED DIVISIBLE GROUPS IN A LOGIC WITH RAMSEY QUANTIFIERS

Wolfgang Lenski
Mathematisches Institut
Universität Heidelberg
Im Neuenheimer Feld 294
D-6900 Heidelberg 1

1. Introduction

Throughout this paper n and m are used as natural numbers. For $n > 1$ let Q^n_α be the quantifier (often called Ramsey quantifier) investigated by Magidor and Malitz (MM), here taken in the $\aleph_0$-interpretation Q^n_0: $L(Q^n_0)$ is the logic obtained from first-order logic by adding the following formation rule: If φ is a formula and $x_1,\ldots,x_n$ are distinct variables then $Q^n_0 x_1,\ldots,x_n \varphi$ is a formula. The quantifier Q^n_0 is given the following meaning in a structure $\mathfrak{A}$: $\mathfrak{A} \models Q^n_0 x_1,\ldots,x_n \varphi[b_1,\ldots,b_m]$ iff there is a subset M of A (the universe of $\mathfrak{A}$) with power $\aleph_0$ such that for all distinct $a_1,\ldots,a_n \in M$ $\mathfrak{A} \models \varphi[a_1,\ldots,a_n,b_1,\ldots,b_m]$. Q^1_α can be identified with the quantifier "There exist $\aleph_\alpha$-many ..." (which is mostly written Q_α), and $L^{<\omega}_\alpha$ is the logic with all the quantifiers Q^n_α , $n<\omega$.

Szmielew stated in (Sz) a set of basic sentences for the elementary theory of abelian groups in the language L' with the non-logical symbols '+','-','o' and '$p^n|$' for every prime p such that elimination of quantifiers is possible for each complete extension and showed the decidability of the elementary theory of abelian groups. In (Ba1) Baudisch proved an analoguous result for the theory T' of all abelian groups in $L'(Q^1_\alpha)$ extending the set of Szmielew basic sentences. In (Ba2) he showed that in the theory of all abelian groups in $L^{<\omega}_\alpha$ all quantifiers Q^n_α are eliminable by Q^1_α. This result yields the decidability of the theory of all abelian groups with Ramsey quantifiers since the elimination procedure is effective.

Let now DOG be the first-order theory of divisible, <u>ordered</u> abelian groups in the language $L=\{+,-,o,<\}$. Recall that an ordered abelian group is archimedean ordered iff for all $o<a<b$ there is a m such that $b<ma$. A theorem of Hölder states that every archimedean ordered group is isomorphic to an ordered subgroup of the reals. It follows from the theorem of Löwenheim-Skolem-Tarski that there is no first-order theory T in L such that Mod(T) is exactly the class of all archimedean ordered groups. This situation changes in a logic L* obtained from first-order logic by adding at least one of the quantifiers Q_o^n, $n>1$. Within L* we <u>can</u> distinguish an archimedean ordered group from a non-archimedean ordered group.

In this paper it will be shown that in the theory of archimedean ordered divisible groups in $L_o^{<\omega}$ all Ramsey quantifiers Q_o^n are eliminable. Using the same arguments it can be shown that the theory of archimedean ordered divisible groups admits elimination of quantifiers in $L(Q_o^n)$ for an arbitrary, fixed n, $n>1$. The logic $L_o^{<\omega}$ is considered instead of $L(Q_o^n)$ only for technical reasons.

2. Preliminaries

'⊣' marks the end of a proof.

Let $\mathfrak{A}$ be a L-structure, let L(A) be the language L expanded by symbols $\dot{c}$ for every element $c\in A$. With $(\mathfrak{A},A)$ we denote the L(A)-structure which arises from $\mathfrak{A}$ by interpreting the symbols $\dot{c}$ by c.

With t,t_i we denote terms of L. If there is no danger of confusion, we write t instead of $t^{\mathfrak{A}}$, too. Writing $k_1x_1+k_2x_2+\ldots+k_nx_n<t$ with k_i integers, we do allow some, but not all k_i to be o where $ox_i:=o$, $i=1,\ldots,n$.

<u>Definition 2.1</u>: Let t be a term of L, $n>o$.

t is called a x_n-term if t is a term of the form $k_1x_1+k_2x_2+\ldots+k_nx_n$.

Every atomic formula of $L_o^{<\omega}$ is in Mod(DOG) equivalent to a formula of the form $t_1=t_2$ resp. $t_1<t_2$ where t_1 is a x_n-term for a suitable $n<\omega$ and the variables x_i, $i<\omega$, do not occur in t_2. Note that t may well contain other variables e.g. y.

Every negated atomic formula is in Mod(DOG) equivalent to a disjunction of atomic formulas: $\neg t_1=t_2$ is in Mod(DOG) equivalent to $t_1<t_2 \vee t_2<t_1$, and $\neg t_1<t_2$ is in Mod(DOG) equivalent to $t_1=t_2 \vee t_2<t_1$ which formula we denote by $t_2\leq t_1$. Instead of $t_1<t_2$ we sometimes write $t_2>t_1$. Every quantifier free formula of $L_0^{<\omega}$ is therefore in Mod(DOG) equivalent to a disjunction of conjunctions of atomic formulas.

We use $\exists z<y\varphi$ as an abbreviation for $\exists z(\ z<y\wedge\varphi\ )$, and correspondingly we use $\exists z>y\varphi$, $\forall z<y\varphi$, and $\forall z>y\varphi$. $\exists! x\varphi(x)$ is given the meaning $\exists x\varphi(x)\wedge\forall y\forall z(\ \varphi(y)\wedge\varphi(z)\rightarrow y=z\ )$. As a simplification we write $Q_0^n x_1<\ldots<x_n\varphi$ instead of $Q_0^n x_1,\ldots,x_n(\ x_1<\ldots<x_n\rightarrow\varphi\ )$.

Given a set I, we let $[I]^n$ denote the set of all subsets of I with exactly n elements

Finally we write $\forall$x resp. $\exists$x instead of "For all x ..." resp. "There exists a x.."

We now turn to the characterization of the class of all archimedean ordered divisible groups in $L_0^{<\omega}$.

<u>Lemma 2.1</u>: Let $\mathfrak{A}$ be an ordered abelian group, let φ be the sentence

$\forall y\forall z(\ o<y\rightarrow\neg Q_0^2 x_1<x_2(\ o<x_1\wedge x_2<z\wedge y<x_2-x_1\ )\ )$.

Then $\mathfrak{A}$ is archimedean ordered iff $\mathfrak{A}\models\varphi$.

Proof: Let φ be as in the lemma.

"$\Rightarrow$": Let $\mathfrak{A}$ be an archimedean ordered group and suppose that $\mathfrak{A}\models\neg\varphi$. Then there exist $a,b\in A$, $a<b$, and a set $M\subseteq A$ with $|M|>\aleph_0$ such that for all $c_1,c_2\in M$ with $c_1<c_2$ $a<c_2-c_1$, $o<c_1$, and $c_2<b$. Choose $m<\omega$. Then there exist $c_0,\ldots,c_m\in M$, $o<c_0<\ldots<c_m$, such that $a<c_j-c_i$, $i<j\leq m$. Consequently $ma<(c_1-c_0)+(c_2-c_1)+\ldots+(c_m-c_{m-1})=c_m-c_0<c_m<b$. ↯

"$\Leftarrow$": Suppose $\mathfrak{A}$ is not archimedean ordered, $\mathfrak{A}\models\varphi$. Then there exist $a,b\in A$, $o<a<b$, such that for all n $na<b$. $M=\{2na\mid o<n<\omega\}$ is an infinite subset of A. For $2ka,2la\in M$, $k>l$, is $2la>o$ and $2ka-2la=2(k-l)a>a$. Since $na<b$ for all n, we get $c<b$ for all $c\in M$ and $\neg\varphi$ is valid in $\mathfrak{A}$. ↯ ⊣

Let φ be as in lemma 2.1. Define ADOG:=DOG $\cup\{\varphi\}$ in $L_0^{<\omega}$, that is ADOG is the theory of archimedean ordered divisible abelian groups in $L_0^{<\omega}$. It will be shown that ADOG admits elimination of quantifiers in $L_0^{<\omega}$.

Now let φ be a quantifier free formula of L. We have to show that in Mod(ADOG) $\exists x\varphi$ and $Q_0^n x_1,\ldots,x_n\varphi$ are respectively equivalent to a quantifier free formula of L.

Since DOG admits elimination of quantifiers in L (see e.g. (W)), it suffices to show that in Mod(ADOG) every formula of the form $Q_0^n x_1,\dots,x_n\varphi$ is equivalent to a formula of L, i.e. a formula not containing any Ramsey quantifier Q_0^n, $n<\omega$. In general the formula φ is a disjunction of conjunctions of atomic formulas, and the quantifier Q_0^n does not distribute over disjunctions in general. However, a slight generalization of a lemma of Cowles ((Co), p.68) allows us to concentrate on formulas φ of the form $x_1<\dots<x_n\to\psi$, ψ being a conjunction of atomic formulas.

<u>Lemma 2.2</u> (Cowles): Let L be a language with a 2-place relation symbol $<$, T a theory in $L_0^{<\omega}$ such that the axioms of a linear ordering for $<$ are derivable from T, and let φ_i be quantifier free formulas, $i=1,\dots,k$.
Then each formula of the form $Q_0^n x_1,\dots,x_n(\ \varphi_1\vee\dots\vee\varphi_k\ )$ is in Mod(T) equivalent to $\bigvee_{f\in F(S(n),k)} Q_0^n x_1<\dots<x_n(\ \bigwedge_{s\in S(n)} \varphi^s_{f(s)}\ )$,
where $S(n)$ is the symmetric group on the set $\{1,\dots,n\}$

$F(S(n),k)$ is the set of functions from $S(n)$ into $\{1,\dots,k\}$

and φ^s is the formula obtained from φ by replacing each of the variables x_i by $x_{s(i)}$, $i=1,\dots,n$.

Proof: Construct a function f as required in the lemma by several applications of the theorem of Ramsey (see Chang-Keisler (CK), p.145). For a detailed proof see Cowles (Co), p.68.⊣

As a further step we can assume that the conjunction ψ of atomic formulas does not contain an equality:

<u>Lemma 2.3</u>: Let t_1 be a term of L in which the variables $x_1,\dots,x_n$ do not occur, let t be a x_n-term and let φ be a formula of the form $Q_0^n x_1<\dots<x_n(\ t=t_1\wedge\psi\ )$, ψ a formula of $L_0^{<\omega}$.
Then φ is in Mod(ADOG) equivalent to $\neg o=o$ if at least one of the variables $x_1,\dots,x_n$ does occur in t, and equivalent to $t=t_1\wedge Q_0^n x_1<\dots<x_n(\psi)$ if not.

Proof: Let t,t_1,φ be as in lemma 2.3.
Clearly φ is in Mod(ADOG) equivalent to $t=t_1\wedge Q_0^n x_1<\dots<x_n(\ \psi\ )$ if none of the variables $x_1,\dots,x_n$ actually occurs in t resp. t_1.

Else observe that for suitable, distinct $a_1,\dots,a_n,b_1,\dots,b_n$ $t(a_1,\dots,a_n)\neq t(b_1,\dots,b_n)$ and therefore $t(a_1,\dots,a_n)\neq t_1$ or $t(b_1,\dots,b_n)\neq t_1$.⊣

Now we turn to the structure of the archimedean ordered groups.

Theorem 2.4 (Hölder): Every archimedean ordered group is isomorphic to an ordered subgroup of the reals $\mathbb{R}$.

For a proof see Fuchs (F), p.45. Hence all archimedean ordered groups are abelian.

Definition 2.2: Let $\varphi(x_1,\dots,x_n)$ be a formula of $L(A)$, $\mathfrak{A}\models$ ADOG.
A subset $H\subseteq A$ is called a homogeneous set of solutions for φ in $\mathfrak{A}$, if whenever $a_1,\dots,a_n\in H$, $a_i<a_j$ for $o<i<j\leq n$, then $(\mathfrak{A},A)\models\varphi(a_1,\dots,a_n)$.

From now on we suppose that the variables x_k, $k<\omega$, do not occur in the terms t,t_i.

Lemma 2.5: ADOG admits elimination of quantifiers in $L_0^{<\omega}$ iff each formula of the form $Q_0^n x_1<\dots<x_n(\bigwedge_{i=1}^{m} k_{1,i}x_1+\dots+k_{n,i}x_n<t_i)$ is in $\mathbb{R}$ equivalent to a formula of L.

Proof: Let ψ be $\bigwedge_{i=1}^{m} k_{1,i}x_1+\dots+k_{n,i}x_n<t_i$, let φ be $Q_0^n x_1<\dots<x_n(\psi)$.
According to lemma 2.2, lemma 2.3 and theorem 2.4, we have to show that for every (not necessarily quantifier free) formula φ^* of L the following equivalence holds: $\mathfrak{A}\models\varphi\leftrightarrow\varphi^*$ for all $\mathfrak{A}\models$ ADOG, $\mathfrak{A}\subseteq\mathbb{R}$, iff $\mathbb{R}\models\varphi\leftrightarrow\varphi^*$. The part "$\Rightarrow$" is obvious.
"$\Leftarrow$": Let $\mathfrak{A}$ be a model of ADOG, $\mathfrak{A}\subseteq\mathbb{R}$. Since DOG admits elimination of quantifiers, each formula of $L(A)$ is satisfied in $(\mathbb{R},A)$ iff it is satisfied in $(\mathfrak{A},A)$. If H is an infinite homogeneous set of solutions for ψ in $\mathfrak{A}$ then H is an infinite homogeneous set of solutions for ψ in $\mathbb{R}$, too. Let H be a homogeneous set of solutions for ψ in $\mathbb{R}$. If H contains an interval $I\subseteq R$, then $H\cap A$ is an infinite homogeneous set of solutions for ψ in $\mathfrak{A}$. Otherwise H is a discrete set and contains either an infinite descending sequence or an infinite ascending sequence. Consider the case when H contains an infinite ascending sequence $S=(a_j)_{j<\omega}$ (the case when H contains an infinite descending sequence is similar). Since $\mathfrak{A}\models$ ADOG, A is dense in R and thus there exist $b_j\in A$ such that $a_j<b_j<a_{j+1}$, $j<\omega$. We prove that $B:=\{b_{2j}\mid j<\omega\}$ is an infinite homogeneous set of solutions for ψ in $\mathfrak{A}$. Observe that there are always two elements of S between every two elements of B:

$a_{2j} < b_{2j} < a_{2j+1} < a_{2(j+1)} < b_{2(j+1)} < a_{2j+3}$, $j<\omega$. Pick $i \leq m$, $o<i$, and $j(1),\ldots,j(n)<\omega$, $j(1)<\ldots<j(n)$. Now we compare $k_{1,i}b_{2j(1)}+\ldots k_{n,i}b_{2j(n)}$ with $k_{1,i}a_1^*+\ldots+k_{n,i}a_n^*$ where $a_l^*=a_{2j(l)}$ or $a_l^*=a_{2j(l)+1}$ depending on the sign of $k_{l,i}$, $l=1,\ldots,n$: Let for $l=1,\ldots,n$ $c_{2j(l)}=\max(k_{l,i}a_{2j(l)}, k_{l,i}a_{2j(l)+1})$. Hence $k_{1,i}b_{2j(1)}+\ldots+k_{n,i}b_{2j(n)} \leq c_{2j(1)}+\ldots c_{2j(n)} < t_i$, and this completes the proof.⊣

Thus to show that ADOG admits elimination of quantifiers it will be sufficient to focus on the problem in the reals.

In some cases however it will be much more suitable to consider formulas φ that are conjunctions of formulas of the form $k_1x_1+\ldots+k_nx_n \leq t$ instead of conjunctions of formulas of the form $k_1x_1+\ldots+k_nx_n < t$. This is made possible by

Lemma 2.6: Let φ be a formula of L.
Then each formula of the form $Q_o^n x_1<\ldots<x_n(\ k_1x_1+\ldots+k_nx_n<t \wedge \varphi)$ is in $\mathbb{R}$ equivalent to $Q_o^n x_1<\ldots<x_n(\ k_1x_1+\ldots+k_nx_n \leq t \wedge \varphi)$.

Proof: The part "$\rightarrow$" is trivial.
"$\leftarrow$": Let ψ be $k_1x_1+\ldots+k_nx_n \leq t$ and suppose $(\mathbb{R},R) \models Q_o^n x_1<\ldots<x_n(\ \psi \wedge \varphi\ )$. Then there exists an infinite homogeneous set H of solutions for $\psi \wedge \varphi$ in $\mathbb{R}$. If H contains an interval $I \subseteq R$ then select a discrete subset $J \subset I \subseteq H$. To complete the proof we take the same argument as in lemma 2.5.⊣

An analogous argument yields a proof of the following corollary, too.

Corollary 2.7: Let ψ be a formula, let $k_j=o$, $j \leq i \leq n$.
Then $Q_o^n x_1<\ldots<x_n(\ k_1x_1+\ldots+k_ix_i<t \wedge \psi\ )$ is in $\mathbb{R}$ equivalent to $Q_o^n x_1<\ldots<x_n(\ k_1x_1+\ldots+k_{j-1}x_{j-1}+k_{j+1}x_j+\ldots+k_ix_{i-1}<t \wedge \psi)$.

3. Elimination of the quantifier Q_o^2

In this section it will be proved that in Mod(ADOG) the quantifier Q_o^2 is eliminable or equivalently that ADOG admits elimination of quantifiers in $L(Q_o^2)$. By lemma 2.5 and lemma 2.6 this will be done by showing that each formula of the form $Q_o^2x_1<x_2(\varphi)$, φ a conjunction of formulas of the form $k_1x_1+k_2x_2 \leq t$, is in $\mathbb{R}$ equivalent to a formula of L.

To show that, we transfer the problem to subsets of $R^{2+} := R^2 \setminus \{(x_1,x_2) \mid x_1 > x_2\}$. Each conjunction φ of formulas of the form $k_1x_1+k_2x_2 \leq t$ determines a subset K of R^{2+}. The existence of an infinite homogeneous set of solutions for φ in $\mathbb{R}$ becomes equivalent to the existence of a suitable subset of K.

Definition 3.1: i) A subset T of R^{2+} is called convex, if whenever $(a_1,a_2),(b_1,b_2) \in K$ then for all $r \in [0,1]$ $(r(a_1-b_1)+b_1, r(a_2-b_2)+b_2) \in K$.
ii) A subset T of R^{2+} is called closed, if it contains all its accumulation points.

Lemma 3.1: Let φ be $\bigwedge_{i=1}^{m} k_{1,i}x_1+k_{2,i}x_2 \leq t_i$, let K_φ be the set of all pairs (a_1,a_2), $a_1,a_2 \in R$, $a_1 < a_2$, that satisfy φ.
Then K_φ is closed, convex, and enclosed by the straight lines $g_i = \{(x_1,x_2) \in R^2 \mid k_{1,i}x_1+k_{2,i}x_2 = t_i\}$ and $\{(x_1,x_2) \in R^2 \mid x_1 = x_2\}$, $i=1,\ldots,m$.

The proof is clear.

Definition 3.2: A subset $K \subseteq R^{2+}$ is called triangular, if for all a_1,a_2,a_3,a_4:
i) whenever $(a_1,a_2) \in K$ and $(a_2,a_3) \in K$ then $(a_1,a_3) \in K$
ii) whenever $(a_1,a_2) \in K$ and $(a_1,a_3) \in K$ and $a_2 < a_3$ then $(a_2,a_3) \in K$
iii) whenever $(a_1,a_2) \in K$ and $(a_3,a_4) \in K$ and $a_1 < a_3$ then $(a_1,a_3) \in K$

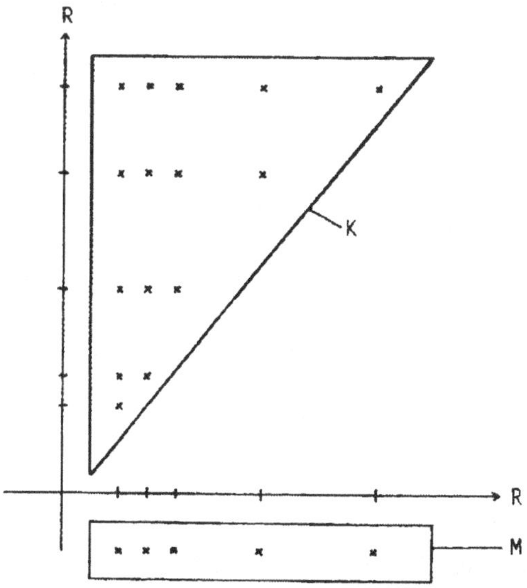

Diagram 3.1

The triangular subsets of K_φ, φ a conjunction of formulas of the form $k_1x_1+k_2x_2\leq t$, are exactly those which can be transferred to homogeneous sets of solutions for φ in $\mathbb{R}$.

Lemma 3.2: Let φ and K_φ be as in lemma 3.1.

i) Let M be a homogeneous set of solutions for φ in $\mathbb{R}$.
   Then $H:=\{(x_1,x_2) \mid x_1,x_2 \in M,\ x_1<x_2\}$ is triangular and $H\subseteq K_\varphi$.

ii) Let H be a triangular subset of K_φ.
   Then $M:=\{x \mid$ there exists y such that $(x,y)\in H$ or $(y,x)\in H\}$ is a homogeneous set of solutions for φ in $\mathbb{R}$.
   If H is infinite then M is infinite and $H=\{(x,y) \mid x,y\in M,\ x<y\}$.

iii) Let H be a triangular subset of K_φ, $(a,b)\in H$,
   $H_1=\{(x,y)\in H \mid y\leq a\}$, $H_2=\{(x,y)\in H \mid a\leq x\}$.
   Then H_1 and H_2 are triangular, and H is infinite iff H_1 or H_2 is infinite.

Diagram 3.1 represents a triangular subset K of R^{2+} together with the corresponding homogeneous set of solutions M for an appropriate formula φ of L.

Proof of lemma 3.2: i) Conditions i)-iii) of definition 3.2 are easily verified.
ii) Pick $a_1,a_2\in M$. We have to show: $(a_1,a_2)\in H$. The proof is straightforward.
iii) Clearly H_1 and H_2 are triangular and $H_1,H_2\subseteq H$.
Let H be infinite and assume H_1 and H_2 are finite. According to definition 3.2,iii) there are only finitely many $(x,y)\in H$ with $x<a$ and only finitely many $(x,y)\in H$ with $a<x$. Hence there are distinct a_i, $(a,a_i)\in H$ for $i<\omega$. From definition 3.2,ii) we get $(a_i,a_j)\in H_2$ or $(a_j,a_i)\in H_2$ for $i<j<\omega$. ↯ ⊣

Corollary 3.3: Let $a,b\in R$, $a<b$, $M=\{(x,y) \mid a<x<y<b\}$.
Then M is triangular.

Proof: Conditions i)-iii) of definition 3.2 are easily checked.

Definition 3.3: Let M be a subset of R^{2+}.

i) M is called triangularly bounded iff
   $\exists x\forall y>x\exists z\forall u>z(\ (y,u)\notin M\ )$ and $\exists x\forall u<x\exists y\forall z<y(\ (z,u)\notin M\ )$.

ii) M is called triangularly unbounded iff M is not triangularly bounded.

Let M be a subset of R^{2+}. M is triangularly bounded iff there are vertical lines arbitrarily far at right for which no upper end-segment is contained in M, and there are horizontal lines arbitrarily far below for which no left-side end-segment is contained in M.

Lemma 3.4: Let M be a subset of R^{2+}, $M \neq \emptyset$, $a \in R$.

i) If M is triangularly bounded and $K \subseteq M$ then K is triangularly bounded.

ii) If M is triangularly unbounded then $M \cap \{(x_1,x_2) \mid x_1 < x_2 < a\}$ is triangularly unbounded or $M \cap \{(x_1,x_2) \mid a < x_1 < x_2\}$ is triangularly unbounded.

Proof: i) is obvious.

ii) Let M be a subset of R^{2+}, $a \in R$, M triangularly unbounded.

Suppose $\forall x \exists y > x \forall z \exists u > z(\ (y,u) \in M\ )$. Especially $\forall x > a \exists y > x \forall z \exists u > z(\ (y,u) \in M\ )$.

Since $M \subseteq R^{2+}$, $M \cap \{(x_1,x_2) \mid a < x_1 < x_2\}$ is triangularly unbounded.

Similarly we get that $M \cap \{(x_1,x_2) \mid x_1 < x_2 < a\}$ is triangularly unbounded, supposing $\forall x \exists u < x \forall y \exists z < y(\ (z,u) \in M\ )$.$\dashv$

Lemma 3.5: Let M be a subset of R^{2+}, $M \neq \emptyset$, let for all $x \in R$ $(x,x) \notin M$, let M be convex and closed.

Then M is triangularly unbounded iff M contains an infinite triangular subset.

Proof:"$\Rightarrow$": Let M be as in the lemma, M triangularly unbounded.

Suppose $\forall x \exists y > x \forall z \exists u > z(\ (y,u) \in M\ )$. Consequently, there exists $a > o$ such that $\forall z \exists u > z(\ (a,u) \in M\ )$.

Construct an increasing sequence $(a_i)_{i<\omega}$ of elements of R such that for all $i < j < \omega$ $(a_i,a_j) \in M$ and for all a_i, $i < \omega$, $\forall z \exists u > z(\ (a_i,u) \in M\ )$ as follows: Define $a_o := a$.

Now suppose a_i are defined for all $i \leq j < \omega$. According to the assumption, there exists $b > a_j$ such that $\forall z \exists u > z(\ (b,u) \in M\ )$. Define $a_{j+1} := b$.

Let $i < j+1$. Since $\forall z \exists u > z(\ (a_i,u) \in M\ )$, there exists $c > a_{j+1}$ such that $(a_i,c) \in M$.

From $(a_i,a_j) \in M$, $(a_i,c) \in M$, $c > a_{j+1} > a_j$, we get by convexity of M $(a_i,a_{j+1}) \in M$.

The conditions i)-iii) of definition 3.2 are easily verified for the set $A = \{(a_i,a_j) \mid i < j < \omega\}$, and therefore A is triangular.

Next suppose $\forall x \exists u < x \forall y \exists z < y(\ (z,u) \in M\ )$. Similarly, we construct a descending sequence $(a_i)_{i<\omega}$ such that $A=\{(a_j,a_i) \mid i<j<\omega\}$ is triangular. This completes "$\Rightarrow$".

"$\Leftarrow$": Let $H\subseteq M$, H infinite and triangular. Pick $(a,b)\in H$ and define $H_1=\{(x,y)\in H \mid a\leq x\}$, $H_2=\{(x,y)\in H \mid y\leq a\}$. According to lemma 3.2,iii), H_1 and H_2 are triangular and H_1 is infinite or H_2 is infinite.

First assume H_1 is infinite.

Define $P_1:=\{x \mid$ there is a y such that $(x,y)\in H_1$ or $(y,x)\in H_1\}$. If follows from lemma 3.2,ii) that P_1 is infinite.

Claim: P_1 is unbounded from above.

Proof of the claim: Assume there exists a $x\in R$ such that $y<x$ for all $y\in P_1$. Since $a\leq y$ for all $y\in P_1$ and P_1 is infinite, there is a point of accumulation c of P_1, $a\leq c\leq x$. Without loss of generality we can assume that c is a point of accumulation from below (the case when c is a point of accumulation from above is similar). Then there exists $(c_i)_{i<\omega}$ such that $c_i<c_j<c$ for $i<j<\omega$ and c is a point of accumulation for $\{c_i \mid i<\omega\}$. Consequently, (c,c) is a point of accumulation for $\{(c_i,c_{i+1}) \mid i<\omega\}$. Since $(c_i,c_{i+1})\in H_1\subseteq M$ for $i<\omega$ and M is closed, we get $(c,c)\in M$ which is a contradiction. ↯

Hence P_1 is unbounded from above. By definition, H_1 is unbounded, that is for all x, $x>a_1$, there exists $y>x$, $y\in P_1$, and for all z there exists $u>z$, $u\in P_1$, such that $(y,u)\in H_1$. Hence $\forall x\exists y>x\forall z\exists u>z(\ (y,u)\in H_1\ )$. Since $H_1\subseteq M$, M is triangularly unbounded.

Now assuming H_2 is infinite we prove by an analogous argument that M is triangularly unbounded in this case, too.⊣

Remark: The proof of "$\Rightarrow$" does not use the assumption $\forall x\in R(\ (x,x)\notin M\ )$.

It remains to investigate whether a set $M\subseteq R^{2+}$ which meets $\{(x,x) \mid x\in R\}$ in exactly one point contains an (infinite) triangular set.

Lemma 3.6: Let $M\subseteq R^{2+}$, M convex and closed, let $\exists! x(\ (x,x)\in M\ )$, $(a,a)\in M$, for all $z<a$ let $M\cap\{(x_1,x_2) \mid x_1<x_2\leq z\}$ be triangularly bounded, for all $z>a$ let $M\cap\{(x_1,x_2) \mid z\leq x_1<x_2\}$ be triangularly bounded.

Then M contains an infinite triangular subset H iff one of the following conditions holds:

i) $\exists z<a\forall u>z(\ u<a\rightarrow(u,a)\in M\wedge\exists y<a(\ (u,y)\in M\ )\ )$

ii) $\exists z>a\forall u<z(\ a<u\rightarrow(a,u)\in M\wedge\exists y(\ (u,y)\in M\ )$.

Conditions i),ii) state that at least in a neighborhood of (a,a) M contains a (convex) triangular (c,b),(c,a),(a,a) for $c<b<a$ resp. (a,a),(a,b),(c,b) for $a<c<b$.

Proof of lemma 3.6: Let M be as in the lemma.

"$\Rightarrow$": Let M contain an infinite triangular subset H.

Define $H_1:=\{(x_1,x_2)\in H \mid x_2<a\}$, $H_2:=\{(x_1,x_2)\in H \mid a<x_1\}$. From lemma 3.2,iii), H_1 is infinite or H_2 is infinite. For example consider the case when H_1 is infinite; the case when H_2 is infinite is similar.

Because for all $z<a$ $M_z:=M\cap\{(x_1,x_2) \mid x_1<x_2\leq z\}$ is triangularly bounded and for all $x\in R$ $(x,x)\notin M_z$, we get from lemma 3.5 that M_z does not contain an infinite triangular subset. Hence for all $z<a$ $H_1\cap\{(x_1,x_2) \mid x_1<x_2\leq z\}$ is finite. Since H_1 is infinite and $H_1=H_1\cap\{(x_1,x_2) \mid x_1<x_2\leq a\}$, (a,a) is an accumulation point of H_1 and it is the only one. Pick $(a_1,a_2)\in H_1$.

Assume $(a_1,a)\notin M$.

Let b be maximal such that $(a_1,b)\in M$ (b exists since M is convex and contains all its accumulation points). Again by convexity, we get $b<a$. Because (a,a) is a point of accumulation of H_1, there are $(b_1,b_2)\in H_1$ with $b<b_1$. Since H_1 is triangular, $(a_1,b_1)\in H_1\subseteq M$ and $b_1>b$. ↯

Consequently $(a_1,a)\in M$.

Now assume conditions i) and ii) of lemma 3.6 do not hold.

Then there exists u, $a_1<u<a$, such that $(u,a)\notin M$ or for all $y<a$ $(u,y)\notin M$. Since $(a_1,a)\in M$, $(a,a)\in M$, $a_1<u<a$ and M convex, $(u,a)\in M$. Therefore $(u,y)\notin M$ for all $y<a$. But because (a,a) is an accumulation point, there is a $(c_1,c_2)\in H_1$ with $a_2,u<c_1$. The straight line from (a_1,a_2) to (c_1,c_2) meets the straight line $\{(x,y) \mid x=u\}$ in v. We have $a_1<u<c_1$, $a_2<c_1<c_2<a$, and from this $v<a$. From the convexity of M we get $(u,v)\in M$. ↯ This completes "$\Rightarrow$".

"$\Leftarrow$": For example consider the case when condition ii) is satisfied; the case when condition i) is satisfied is similar.

Condition ii) states: $\exists z>a\forall u<z(\ a<u\rightarrow(a,u)\in M\wedge\exists y(\ (u,y)\in M\ ))$.

Let $b>a$ such that for all u, $a<u<b$, $(a,u)\in M$ and $\exists y(\ (u,y)\in M\ )$. Pick c,c', $a<c'<b$, such that $(a,c')\in M$ and $(c',c)\in M$, and let g be the straight line from (a,a) to (c',c). (Cf. diagram 3.2). Since M is convex, $g\subseteq M$. Hence for all d,

$a < d < c'$, there exists d', $a \leq d' < d$, such that $(d',d) \in g$.

Define a descending sequence $(c_i)_{i<\omega}$ such that $a < c_i < b$ as follows: $c_0 := c'$.

Now suppose that c_i, $a < c_i \leq c_0$, are defined for all $i \leq j$, $j < \omega$. Then there exists c_j' such that $(c_j', c_j) \in g$. Define $c_{j+1} := c_j'$. Since $(a, c_0) \in M$, $(c_1, c_0) \in M$, and M is convex, we get $(u, c_0) \in M$ for all $a \leq u < c_1$; especially there is $(c_{j+1}, c_0) \in M$ and $(c_{j+1}, c_j) \in M$. Again, from the convexity of M $(c_{j+1}, c_i) \in M$ for all $i \leq j$. Therefore the set $H := \{(c_j, c_i) \mid i < j < \omega\}$ is triangular and infinite, and $H \subseteq M$. ⊣

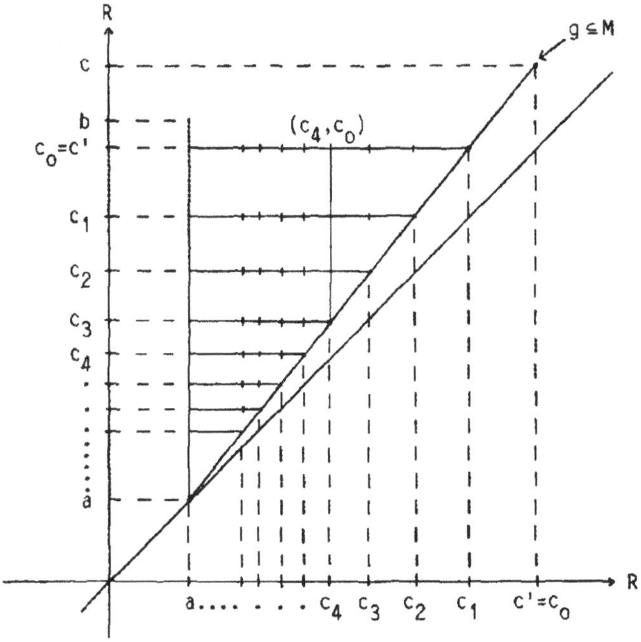

Diagram 3.2

Now we are able to prove that the quantifier Q_0^2 is eliminable in Mod(ADOG).

Theorem 3.7: Let $\varphi(x_1,x_2) \equiv \bigwedge_{i=1}^{m} k_{1,i}x_1 + k_{2,i}x_2 \leq t_i$, let

$\psi_1 \equiv \exists x_1 \exists x_2 (x_1 < x_2 \wedge \varphi(x_1,x_1) \wedge \varphi(x_2,x_2) \wedge \varphi(x_1,x_2))$,

$\psi_2 \equiv \forall x \exists y > x \forall z \exists u > z \varphi(y,u)$,

$\psi_3 \equiv \forall x \exists u < x \forall y \exists z < y \varphi(z,u)$,

$\psi_4 \equiv \exists! x \varphi(x,x) \wedge \forall x (\varphi(x,x) \rightarrow \exists z < x \forall u > z (u < x \rightarrow \varphi(u,x) \wedge \exists y < x \varphi(u,y)))$,

$\psi_5 \equiv \exists! x \varphi(x,x) \wedge \forall x (\varphi(x,x) \rightarrow \exists z > x \forall u < z (x < u \rightarrow \varphi(x,u) \wedge \exists y \varphi(u,y)))$.

Then $Q_0^2 x_1 < x_2 \varphi$ is in $\mathbb{R}$ equivalent to $\psi_1 \vee \psi_2 \vee \psi_3 \vee \psi_4 \vee \psi_5$.

Proof: Let φ be as in the theorem. In conformity with lemma 3.1, φ determines uniquely a convex subset $K=K_\varphi$ of R^{2+} such that K contains all its accumulation points and $(a,b)\in K$ iff $(\mathbb{R},R)\models\varphi(a,b)$.

"$\Rightarrow$": Suppose $(\mathbb{R},R)\models Q_0^n x_1<\ldots<x_n\varphi$.

According to lemma 3.2,i) , K contains an infinite triangular subset H.

Now suppose $(\mathbb{R},R)\models\neg\exists x_1\exists x_2(\ x_1<x_2\wedge\varphi(x_1,x_1)\wedge\varphi(x_2,x_2)\wedge\varphi(x_1,x_2)\ )$ and $(\mathbb{R},R)\models\neg\forall x\exists y>x\forall z\exists u>z\varphi(y,u)\ \wedge\ \neg\forall x\exists u<x\forall y\exists z<y\varphi(z,u)$.

Especially K is triangularly bounded. We show

i) $(\mathbb{R},R)\models\exists!x\varphi(x,x)$

ii) Let $a\in R$, $(\mathbb{R},R)\models\varphi(a,a)$. Then $(\mathbb{R},R)\models\exists z<a\forall u>z(\ u<a\rightarrow\varphi(u,a)\wedge\exists y<a\varphi(u,y)\ )$
or $(\mathbb{R},R)\models\exists z>a\forall u<z(\ a<u\rightarrow\varphi(a,u)\wedge\exists y\varphi(u,y)\ )$.

$(\mathbb{R},R)\models\exists x\varphi(x,x)$. Otherwise we get from lemma 3.5 that K is triangularly unbounded. ↯

Now assume $(\mathbb{R},R)\models\exists x_1\exists x_2(\ \varphi(x_1,x_1)\wedge\varphi(x_2,x_2)\wedge x_1<x_2\ )$.

Let $a,b\in R$, $a<b$, $(a,a)\in K$ and $(b,b)\in K$. Because K is convex and contains an infinite triangular subset, there exist $c_1,c_2\in R$, $c_1<c_2$, such that $(c_1,c_1),(c_2,c_2),(c_1,c_2)\in K$. ↯ Therefore $(\mathbb{R},R)\models\exists!x\varphi(x,x)$.

Let $a\in R$, $(a,a)\in K$. According to lemma 3.4,i), every subset of K is triangularly bounded, and "$\Rightarrow$" is completed by lemma 3.6.

"$\Leftarrow$": We distinguish three cases.

<u>Case 1</u>: $(\mathbb{R},R)\models\psi_1$.

Since K is convex, we get for suitable $a,b\in R$, $a<b$, $\{(x_1,x_2)\mid a<x_1<x_2<b\}\subseteq K$ that is (see corollary 3.3) K contains an infinite triangular subset and the proof is completed (cf. lemma 3.2,ii)).

<u>Case 2</u>: $(\mathbb{R},R)\models\psi_2\vee\psi_3$ or $(\mathbb{R},R)\models\psi_4\vee\psi_5$, and K is triangularly unbounded.

Then K is triangularly unbounded. Because the proof of "$\Rightarrow$" of lemma 3.5 does not use the assumption $\forall x(\ (x,x)\notin K\ )$, K contains an infinite triangular subset, and this completes the proof.

<u>Case 3</u>: $(\mathbb{R},R)\models\psi_4\vee\psi_5$, and K is triangularly bounded.

Then we get a proof from lemma 3.6. ⊣

Hence we have the result that Mod(ADOG) admits elimination of quantifiers in $L(Q_0^2)$.

4. Elimination of quantifiers

In this section we will prove that ADOG admits elimination of quantifiers in $L_0^{<\omega}$. By lemma 2.5, we have to show that each formula φ of $L_0^{<\omega}$ is in $\mathbb{R}$ equivalent to a quantifier free formula of $L_0^{<\omega}$. This will be proved by induction on the "Q_0^n-complexity" $n(\varphi)$ of φ which is the maximal m such that Q_0^m occurs in φ. In section 3 we proved the assumption in the case $n(\varphi)=2$. Now assuming $n(\varphi)>2$, we will show that there is a formula φ^* of $L_0^{<\omega}$ with $n(\varphi^*)<n(\varphi)$, such that φ^* is in $\mathbb{R}$ equivalent to φ. To this end we consider formulas of the form $Q_0^n x_1<\ldots<x_n\psi$, ψ a conjunction of formulas of the form $k_1x_1+\ldots+k_nx_n<t$. Then we reduce step by step the number of such formulas $k_1x_1+\ldots+k_nx_n<t$ within the scope of Q_0^n for which all "coefficients" k_i do not vanish, $i=1,\ldots,n$ (the absolute number of atomic formulas in the scope of Q_0^n may well increase!). This procedure leads to the distinction of several cases depending on the sign of k_1,k_n, and $k_{n-1}+k_n$. Note finally that the reduction of the Q_0^n-complexity does <u>not</u> work for the case n=2 as it stands, because additional formulas of the form $k_1x_1+k_2x_2<c$ are essentially involved in the reduction procedure. For the following lemmata we suppose that k_i are integers, $k_1\neq o\neq k_n$ and $k_{n-1}\neq o$.

<u>Lemma 4.1</u>: Let ψ be a formula of L, t a term of L, in both of which the variable y does not occur, let $k_n>o$.
Then $Q_0^n x_1<\ldots<x_n(\ k_1x_1+\ldots+k_nx_n<t\wedge\psi\ )$ is in $\mathbb{R}$ equivalent to
$\exists y Q_0^n x_1<\ldots<x_n(\ x_n<y\wedge k_1x_1+\ldots+k_{n-1}x_{n-1}<t-k_ny\wedge\psi\ )$.

Proof: Let ψ, k_n be as in the lemma.
"$\Rightarrow$": Suppose $(\mathbb{R},R)\models Q_0^n x_1<\ldots<x_n(\ k_1x_1+\ldots+k_nx_n<t\wedge\psi\ )$.
Then there exists an infinite homogeneous set H of solutions for $k_1x_1+\ldots+k_nx_n<t\wedge\psi$ in R. Pick $a_1,\ldots,a_n\in H$, $a_1<\ldots<a_n$. Since $k_1a_1+\ldots+k_na_n<t$, $k_na_n<t-(\ k_1a_1+\ldots+k_{n-1}a_{n-1}\ )$. Hence given $a_1,\ldots,a_n\in H$, $a_1<\ldots<a_n$, $t-(\ k_1a_1+\ldots+k_{n-1}a_{n-1}\ )$ is an upper bound for all k_na, $a\geq a_n$, $a\in H$. Because $k_n>o$, H is bounded, and so there exists in R a least upper bound c. It follows $a_n\leq c\wedge k_1a_1+\ldots+k_{n-1}a_{n-1}+k_nc\leq t\wedge\psi$ for all $a_1,\ldots,a_n\in H$, and part "$\Rightarrow$" is accomplished by lemma 2.6.

"$\Leftarrow$": Pick $c \in R$ such that there exists an infinite homogeneous set H of solutions for $x_n < y \wedge k_1x_1+\ldots+k_{n-1}x_{n-1} < t-k_ny \wedge \psi$. Since $k_n > o$, $k_1a_1+\ldots+k_na_n < k_1a_1+\ldots+k_{n-1}a_{n-1}+k_nc < t$ for all $a_1,\ldots,a_n \in H$, $a_1 < \ldots < a_n$.⊣

Lemma 4.2: Let ψ be a formula of L, t a term of L, in both of which the variable y does not occur, let $k_1 < o$.
Then $Q_o^nx_1 < \ldots < x_n(\ k_1x_1+\ldots+k_nx_n < t \wedge \psi)$ is in $\mathbb{R}$ equivalent to $\exists y Q_o^n x_1 < \ldots < x_n(\ y < x_1 \wedge k_2x_2+\ldots+k_nx_n < t-k_1y \wedge \psi\ )$.

The proof of lemma 4.2 is essentially the same as for lemma 4.1.⊣

Lemma 4.3: Let $c \in R$, $H_c^+ = \{x \in R \mid c < x\}$, $H_c^- = \{x \in R \mid x < c\}$.

i)(a) If H is unbounded from above, then there exists an infinite homogeneous subset $H_o \subseteq H \cap H_c^+$ of solutions for $2x_1-x_2 < c$ such that H_o is unbounded from above, too.

(b) If H, $H \subseteq H_c^+$, has c as an accumulation point, then there exists an infinite homogeneous subset $H_o \subseteq H$ of solutions for $2x_1-x_2 < c$ with c as an accumulation point.

(c) Let H, $H \subseteq H_c^+$, be a homogeneous set of solutions for $2x_1-x_2 < c$. If b is an accumulation point for H, then b=c.

ii)(a) If H is unbounded from below, then there exists an infinite homogeneous subset $H_o \subseteq H \cap H_c^-$ of solutions for $x_1-2x_2 < -c$ such that H_o is unbounded from below,too.

(b) If H, $H \subseteq H_c^-$, has c as an accumulation point, then there exists an infinite homogeneous subset $H_o \subseteq H$ of solutions for $x_1-2x_2 < -c$ with c as an accumulation point.

(c) Let H, $H \subseteq H_c^-$, be a homogeneous set of solutions for $x_1-2x_2 < -c$. If b is an accumulation point for H, then b=c.

Proof: Let c, H_c^+, H_c^- be as in the lemma.

i)(a): Let H be unbounded from above. Construct an infinite increasing sequence $(a_i)_{i<\omega}$ as follows: Since H is unbounded from above, $H \cap H_c^+ \neq \emptyset$. Pick $a_o \in H \cap H_c^+$ arbitrarily. Suppose a_i are defined for all $i \leq j$, $j < \omega$. Again since H is unbounded from above, there exists $a^* \in H \cap H_c^+$, $a^* > 2a_j - c$. Define $a_{j+1} := a^*$.
Now $H_o = \{a_j \mid j < \omega\}$ is an infinite homogeneous set of solutions for $2x_1-x_2 < c$, $H_o \subseteq H \cap H_c^+$.

i)(b): Let $H \subseteq H_c^+$, c an accumulation point for H, let $(c_i)_{i<\omega}$ be an infinite decreasing sequence with limit o. Construct an infinite decreasing sequence $(b_i)_{i<\omega}$ such that for all $i<j<\omega$ $2b_i-b_j<c$ and $c<b_j<c+c_j$ as follows: Pick $b_0 \in H$, $b_0>c$, $b_0<c+c_0$. Suppose b_i are defined for all $i \leq j$, $j<\omega$. Since $c<b_j$ and c is an accumulation point for H, $2c<c+b_j$ and there exists $b \in H$, $c<b<b_j$, $b<c+c_{j+1}$, with $2b<c+b_j$. Define $b_{j+1}:=b$. Hence $2b_{j+1}-b_i \leq 2b_{j+1}-b_j<c$, $i \leq j$.

Therefore $H_0=\{b_j \mid j<\omega\}$ is an infinite homogeneous set of solutions for $2x_1-x_2<c$ with c as an accumulation point, $H_0 \subseteq H$.

i)(c): Let H, $H \subseteq H_c^+$, be a homogeneous set of solutions for $2x_1-x_2<c$, let b be an accumulation point for H.

Assume $b-c=e>o$.

For any elements $a_1,a_2 \in H$ with $b<a_1<b+e$ and $a_1<a_2$ we get $a_2>2a_1-c>2b-c=b+e$. Thus there can be at most one element a of H satisfying $b<a<b+e$. Consequently $H_1=H \cap \{x \in R \mid x<b\}$ is infinite and b is an accumulation point for H_1. Since $b>c$, there exist $a_1,a_2 \in H_1$ with $2a_1>b+c$ and $a_1<a_2$. Therefrom $a_2>2a_1-c>b+c-c=b$ ↯.

Hence b=c and "⇒" is proved.

ii) The proof of ii) is similar.⊣

The inequalities $2x_1-x_2<c$ resp. $x_1-2x_2<-c$ in their special properties as states in lemma 4.3 will serve as codes for those properties of homogeneous sets of solutions for $k_1x_1+\ldots+k_nx_n<t \wedge \psi$ that are essential in this context, namely whether they contain accumulation points or not resp. whether they are bounded from above/from below or not.

<u>Lemma 4.4</u>: Let ψ be a formula of L, t a term of L in both of which the variables y,z do not occur, let $k_1>o$, $k_n<o$, let

$\varphi_1 \equiv z<x_1 \wedge 2x_1-x_2<y$,

$\varphi_2 \equiv x_n<y \wedge x_1-2x_2<-z$

$\varphi_3 \equiv y<x_1 \wedge x_n<z \wedge 2x_1-x_2<y \wedge k_nx_n<t-(k_1+\ldots+k_{n-1})y$,

$\varphi_4 \equiv y<x_1 \wedge x_n<z \wedge x_1-2x_2<-z \wedge k_1x_1<t-(k_2+\ldots+k_n)z$,

$\varphi_i^* \equiv Q_0^n x_1<\ldots<x_n(\varphi_i \wedge \psi)$ for $1 \leq i \leq 4$,

$\varphi^* \equiv \varphi_1^* \vee \varphi_2^* \vee \varphi_3^* \vee \varphi_4^*$.

Then $Q_0^n x_1<\ldots<x_n(k_1x_1+\ldots+k_nx_n<t \wedge \psi)$ is in $\mathbb{R}$ equivalent to $\exists y \exists z(y<z \wedge \varphi^*)$.

Proof: "$\Rightarrow$": Suppose $(\mathbb{R},R)\models Q_0^n x_1<\ldots<x_n(\ k_1x_1+\ldots+k_nx_n<t\wedge\Psi\ )$.

Let H be an infinite homogeneous set of solutions for $k_1x_1+\ldots+k_nx_n<t\wedge\Psi$.

Case 1: H is unbounded from above.

Pick $c,d\in R$, $c<d$. According to lemma 4.3,i)(a), there exists an infinite homogeneous subset $H_0\subseteq H\cap H_d^+$ of solutions for $2x_1-x_2<c$. Since $H_0\subseteq H$, H_0 is an infinite homogeneous set of solutions for Ψ, too. Hence H_0 is an infinite homogeneous set of solutions for $d<x_1\wedge 2x_1-x_2<c\wedge\Psi$. Consequently $(\mathbb{R},R)\models\exists y\exists z(\ y<z\wedge\varphi_1^*\ )$.

Case 2: H is unbounded from below.

Using lemma 4.3,ii)(a), we get $(\mathbb{R},R)\models\exists y\exists z(\ y<z\wedge\varphi_2^*\ )$ by an argument analogous to the one in the first case.

Case 3: H has an accumulation point c, and c is a point of accumulation from above.

Assume there is $a\in H$, $c<a$, with $k_na>t-(k_1+\ldots+k_{n-1})c$. Define $k:=|k_1|+\ldots+|k_{n-1}|$. Then there is an $e>0$ such that $k_na-ke>t-(k_1+\ldots+k_{n-1})c$. Since c is an accumulation point, there exist $b_1,\ldots,b_{n-1}\in H$, $b_1<\ldots<b_{n-1}<c+e$, $b_{n-1}<a$. From that

$k_1b_1+\ldots+k_{n-1}b_{n-1}+k_na=(k_1+\ldots+k_{n-1})c+k_1(b_1-c)+\ldots+k_{n-1}(b_{n-1}-c)+k_na$

$\geq(k_1+\ldots+k_{n-1})c-|k_1|(b_1-c)-\ldots-|k_{n-1}|(b_{n-1}-c)+k_na>(k_1+\ldots+k_{n-1})c-ke+k_na>t.$ ↯

Thus $k_na\leq t-(k_1+\ldots+k_{n-1})c$ for all $a\in H$, $c<a$. According to lemma 2.6 and lemma 4.3,i)(b), there exists $d\in R$, $c<d$, and an infinite homogeneous set H_0 of solutions for $c<x_1\wedge x_n<d\wedge 2x_1-x_2<c\wedge k_nx_n<t-(k_1+\ldots+k_{n-1})c\wedge\Psi$.

Hence $(\mathbb{R},R)\models\exists y\exists z(\ y<z\wedge\varphi_3^*\ )$.

Case 4: H has an accumulation point c, and c is a point of accumulation from below.

We get $(\mathbb{R},R)\models\exists y\exists z(\ y<z\wedge\varphi_4^*\ )$ by arguments similar to those used in case 3.

Since $\exists y\exists z(\ y<z\wedge\varphi^*\ )$ is equivalent to

$\exists y\exists z(\ y<z\wedge\varphi_1^*\ )\vee\exists y\exists z(\ y<z\wedge\varphi_2^*\ )\vee\exists y\exists z(\ y<z\wedge\varphi_3^*\ )\vee\exists y\exists z(\ y<z\wedge\varphi_4^*\ )$, "$\Rightarrow$" is completed.

"$\Leftarrow$": We have to show $(\mathbb{R},R)\models Q_0^n x_1<\ldots<x_n(\ k_1x_1+\ldots+k_nx_n<t\wedge\Psi\ )$, distinguishing four cases.

Case 1: $(\mathbb{R},R)\models\exists y\exists z(\ y<z\wedge Q_0^n x_1<\ldots<x_n(\ z<x_1\wedge 2x_1-x_2<y\wedge\Psi\ )\ )$.

There exist $c,d\in R$, $c<d$, and an infinite homogeneous set H of solutions for $d<x_1\wedge 2x_1-x_2<c\wedge\Psi$. According to lemma 4.3,i)(c), H does not have a point of accumulation. So H is unbounded from above. Construct an increasing sequence $(H_i)_{i<\omega}$ of finite subsets of H as follows: Pick $a_1,\dots,a_{n-1}\in H$, $a_1<\dots<a_{n-1}$, and define $H_0:=\emptyset$, $H_1:=\{a_1\}$, $H_2:=\{a_1,a_2\}$,..., $H_{n-1}:=\{a_1,\dots,a_{n-1}\}$.

Supposing H_j has been defined, $n-1\leq j<\omega$, let

$S=\{(b_1,\dots,b_{n-1}) \mid b_1,\dots,b_{n-1}\in H_j,\ b_1<\dots<b_{n-1}\}$. For each $s\in S$ define $b_s:=k_1b_1+\dots+k_{n-1}b_{n-1}$ and $b:=\max(\ b_s \mid s\in S\ )$. Since $k_n<o$ and H is unbounded from above, there exists $a\in H$, $a>x$ for all $x\in H_j$, with $k_na<t-b$. Define $H_{j+1}:=H_j\cup\{a\}$.

By the construction, $H^*:=\bigcup_{i<\omega}H_i$ is an infinite homogeneous set of solutions for $k_1x_1+\dots+k_nx_n<t\wedge\Psi$.

Case 2: $(\mathbb{R},R)\models\exists y\exists z(\ y<z\wedge Q_0^n x_1<\dots<x_n(\ x_n<y\wedge x_1-2x_2<-z\wedge\Psi)\ )$.

A construction analogous to case 1 yields a proof for case 2, too.

Case 3: $(\mathbb{R},R)\models\exists y\exists z(y<z\wedge Q_0^n x_1<\dots<x_n(y<x_1\wedge x_n<z\wedge 2x_1-x_2<y\wedge k_nx_n<t-(k_1+\dots+k_{n-1})y\wedge\Psi))$.

There exist $c,d\in R$, $c<d$, and an infinite homogeneous set H of solutions for $c<x_1\wedge x_n<d\wedge 2x_1-x_2<c\wedge k_nx_n<t-(k_1+\dots+k_{n-1})c\wedge\Psi$.

Since H is infinite and $c<x<d$ for all $x\in H$, H has an accumulation point b. By lemma 4.3,i)(c), b=c. Construct a decreasing sequence $(a_i)_{i<\omega}$ of elements of H as follows: Pick $a_0\in H$ arbitrarily.

Suppose a_i are defined for all $i\leq j$, $j<\omega$. From the assumption we get $k_na_j<t-(k_1+\dots+k_{n-1})c$. Since c is a point of accumulation for H, there is an $e>o$ such that for all $b_1,\dots,b_{n-1}\in H$, $b_1<\dots<b_{n-1}<c+e$, $k_na_j<t-(k_1b_1+\dots+k_{n-1}b_{n-1})$. From this $k_1b_1+\dots+k_{n-1}b_{n-1}+k_na_i\leq k_1b_1+\dots+k_{n-1}b_{n-1}+k_na_j<t$ for all $i\leq j$ and $b_1,\dots,b_{n-1}\in H$, $b_1<\dots<b_{n-1}<c+e$. Pick $a\in H$, $c<a<c+e$, $a<a_j$, and define $a_{j+1}:=a$.

Hence $H_0:=\{a_i \mid i<\omega\}$ is an infinite homogeneous set of solutions for $k_1x_1+\dots+k_nx_n<t\wedge\Psi$, $H_0\subseteq H$.

Case 4: $(\mathbb{R},R)\models\exists y\exists z(y<z\wedge Q_0^n x_1<\dots<x_n(y<x_1\wedge x_n<z\wedge x_1-2x_2<-z\wedge k_1x_1<t-(k_2+\dots+k_n)z\wedge\Psi))$

The construction in case 4 is essentially the same as the one in case 3.⊣

Theorem 4.5: ADOG admits elimination of quantifiers in $L_0^{<\omega}$.

Proof: Since DOG admits elimination of quantifiers in L, it suffices to show that each formula of $L_0^{<\omega}$ is in Mod(ADOG) equivalent to a formula of L (which need not be quantifier free). Moreover, by lemma 2.5 it suffices to show that each formula of $L_0^{<\omega}$ is in $\mathbb{R}$ equivalent to a formula of L.

Now we define for each formula φ of $L_0^{<\omega}$ the natural number $n(\varphi)$ as the maximal m such that Q_0^m occurs in φ. $n(\varphi)$ is called the "Q_0^n-complexity" of φ. We will prove by induction on the Q_0^n-complexity of formulas that each formula φ of $L_0^{<\omega}$ is in $\mathbb{R}$ equivalent to a formula of L. By induction on formulas and according to lemma 2.2 and lemma 2.3, we can assume without loss of generality $\varphi \equiv Q_0^n x_1 < \ldots < x_n \psi$, ψ a conjunction of formulas of the form $k_1x_1+\ldots+k_nx_n<t$, t a term of L in which the variables $x_1,\ldots,x_n$ do not occur. Using lemma 2.6, theorem 3.10 yields a proof for $n(\varphi)=2$. Next suppose that we have proved the assumption for all formulas ψ with $n(\psi)<m$, $m>2$. Let $\varphi \equiv Q_0^m x_1 < \ldots < x_m (\bigwedge_{i=1}^{n} k_{1,i}x_1+\ldots+k_{n,i}x_n<t_i)$. We prove the assumption by induction on the number $q(\varphi)$ of inequalities of the form $k_1x_1+\ldots+k_mx_m<t$ in the scope of Q_0^m for which all $k_i \neq o$, $i=1,\ldots,m$.

Let $q(\varphi)=o$. Then by corollary 2.7, φ is in $\mathbb{R}$ equivalent to a formula of the form $Q_0^m x_1 < \ldots < x_m \psi$, ψ a conjunction of formulas of the form $k_1x_1+\ldots k_{m-1}x_{m-1}<t$. Hence φ is in $\mathbb{R}$ equivalent to $\varphi^* \equiv Q_0^{m-1} x_1 < \ldots < x_{m-1}\psi$. Since $n(\varphi^*)<n(\varphi)$, φ^* is in $\mathbb{R}$ equivalent to a formula of L. Hence φ is in $\mathbb{R}$ equivalent to a formula of L.

Now suppose that each formula ψ with $n(\psi)=m$ and $q(\psi)<h$ is in $\mathbb{R}$ equivalent to a formula of L. Let φ be a formula of $L_0^{<\omega}$ of the form given above, $n(\varphi)=m$, $q(\varphi)=h$. Then by lemma 4.1 resp. lemma 4.2 resp. lemma 4.4, φ is in $\mathbb{R}$ equivalent to a formula φ^* all subformulas ψ of which have $q(\psi)<q(\varphi)$ and $n(\psi)\leq n(\varphi)$. Thus by induction on formulas, φ^* is in $\mathbb{R}$ equivalent to a formula of L. This completes the proof.⊣

Since DOG is decidable in L, it follows from theorem 4.5 together with the constructive nature of the proof

Corollary 4.6: ADOG is decidable in $L_0^{<\omega}$.

References

(B) Barwise,K.J. (ed.): Handbook of mathematical logic
North-Holland Publ. Co., Amsterdam-London, 1977

(Ba1) Baudisch,A.: Elimination of the quantifier Q in the theory of abelian groups
Bull. de l'Acad. Polon. Sci.,
Ser. Sci. Math. Astronom. Phys. 24 (1976), 543-549

(Ba2) Baudisch,A.: Decidability of the theory of abelian groups with Ramsey quantifiers
Bull. de l'Acad. Polon. Sci.,
Ser. Sci. Math. Astronom. Phys. 25 (1977), 733-739

(CK) Chang,C.C. & Keisler,H.J.: Model theory
North-Holland Publ. Co., Amsterdam-London, 1973

(Co) Cowles,J.: The theory of archimedean real closed fields in logics with Ramsey quantifiers
Fund. Math. 103 (1979), 65-76

(Cu) Courant,R.: Vorlesungen über Differential- und Integralrechnung (2 Bände)
Springer Verlag, Heidelberg-Berlin-New York, 1971/72

(F) Fuchs,L.: Partially ordered algebraic systems
Pergamon Press, Oxford, 1963

(MM) Magidor,M. & Malitz,J.: Compact extensions of L(Q) (part 1a)
Ann. Math, Logic 11 (1977), 217-261

(Mo) Mostowski,A.: On a generalization of quantifiers
Fund. Math. 44 (1957), 12-36

(Sc) Schmitt,P.H.: Model theory of ordered abelian groups
Habilitationsschrift Univ. Heidelberg, 1982

(Sz) Szmielew,W.: Elementary properties of abelian groups
Fund. Math. 41 (1955), 203-271

(W) Weispfenning,V.: Elimination of quantifiers for certain ordered and lattice-ordered abelian groups
Bull. Soc. Math. Belg. 33 (1981), 131-155

A PROOF-THEORETIC APPROACH TO NON STANDARD ANALYSIS (CONTINUED)

By

Shih-Chao Liu

This talk is based upon my previous paper [1], namely "A proof-theoretic approach to non-standard analysis with emphasis on distinguishing between constructive and non-constructive results" in The Kleene Symposium, 1980. I first give a sketch of the main points of that paper but in a newly reorganized form with some new materials which were not contained in [1], and then add more remarks and reflections on this subject. I think that such a proof-theoretic approach to non-standard analysis is worth to be further developed either theoretically or in its applications.

We start with the assumption that the axiomatic set theory ZF is consistent. For each intuitive natural number i, let $\ulcorner i \urcorner$ to denote the set definable in ZF according to the following recursion: $\ulcorner 0 \urcorner = 0$, $\ulcorner i+1 \urcorner = \ulcorner i \urcorner \cup \{\ulcorner i \urcorner\}$. By Coedel's technique and the fact that the law of excluded middle holds in ZF we can show that there exists a set definable in ZF, which we denote by ∞, such that (i) ZF $\vdash \infty \in \omega$, and (ii) $\infty \leqslant \ulcorner i \urcorner$ is unprovable in ZF for $i = 0, 1, \ldots$. Hence ZF can be extended to a consistent theory ZF* by adding as new axioms the sentences $\ulcorner 0 \urcorner < \infty$, $\ulcorner 1 \urcorner < \infty$, $\ulcorner 2 \urcorner < \infty$, ... Since the real field R, the rational field Q, the set of integers C, the set of natural numbers ω and the following relation and operations $x < y$, $x+y$, $x \cdot y$, $|x|$, $-x$, $1/x$ are all definable in ZF, they are also definable in the extension ZF* by the same formulas as in ZF. We further introduce the notation δ by $\delta = 1/\infty$. By non-standard anslysis I mean the discipline of metamathematics about ZF* especially when the notion of convergence defined below in terms of infinitesimal such as δ for real functions (namely functions from R to R) is involved.

1 Metadefinition. Let η be a definable set in ZF* such that ZF* $\vdash \eta \in R$. We say that η is an infinity if ZF* $\vdash \ulcorner i \urcorner < |\eta|$ for $i = 0, 1, \ldots$; an infinitesimal if ZF* $\vdash |\eta| < 1/\ulcorner i \urcorner$ for $i = 1, 2, \ldots$. Accordingly, ∞ is an infinity and δ is an infinitesimal.

2 Metadefinition. Let $f(x_1, \ldots, x_m)$ be a formula and $F(x_1, \ldots, x_m)$ be a function definable in ZF*. We say that $F(x_1, \ldots, x_m)$ is infinitesimal for $x_1, \ldots, x_m$ satisfying $f(x_1, \ldots, x_m)$ if ZF* $\vdash$ $\forall x_1 \ldots x_m(f(x_1, \ldots, x_m) \rightarrow |F(x_1, \ldots, x_m)| < 1/\ulcorner i \urcorner)$ for $i = 1, 2, \ldots$.

We say that the function $F(x_1, \ldots, x_m, t)$ converges uniformly as $t \rightarrow 0$ with respect to $x_1, \ldots, x_m$ satisfying $f(x_1, \ldots, x_m)$ if for any two infinitesimals η , $\sigma \neq 0$, the difference $F(x_1, \ldots, x_m, \eta) - F(x_1, \ldots, x_m, \sigma)$ is always infinitesimal for $x_1, \ldots, x_m$ satisfying $f(x_1, \ldots, x_m)$.

3 Metadefinition. Suppose $G(x)$ is a function definable in ZF* such that F* $\vdash \forall x(x \in R \rightarrow G(x) \in R)$ and that (a,b) is an interval where a, b are two efinable sets in ZF* such that ZF* $\vdash a \in R \ \& \ b \in R \ \& \ a < b$. ow let $F(x, t)$ denote the function $(G(x+t) - G(x))/t$. We say that (x) is differentiable in (a, b) if $F(x, t)$ converges uniformly as $\rightarrow 0$ with respect to x satisfying $x \in (a, b)$. In case that $G(x)$ s differentiable in (a, b) we use $dG(x)/dx$ to denote $F(x, \delta)$ as function of the variable x ranging over (a, b).

4 Metadefinition. Suppose $G(x)$ is a real function definable in F* on the real line, and $[a, b]$ is a closed interval. We let $x'y$ enote the value of x at y in case x is a function and $y \in dom(x)$. et $P(x, y, t)$ be the set function definable in ZF* such that $(x, y, t) = \{\xi\colon \xi$ is an increasing funciton & $range(\xi) \subset R$ & $om(\xi) \in \omega$ & $\forall s(s \in dom(\xi) - 1 \rightarrow \xi'(s+1) - \xi's \leqslant t)$ & $\xi'0 = x$ & $'(dom(\xi) - 1) = y\}$. We may call $P(x,y,t)$ the set of all partitions over the nterval $[x,y]$ with mesh $\leq t$. We say that $G(x)$ is integrable over $[a,b]$ if or any positive infinitesimal η we have that

$$\sum_{s<dom(\mu)-1} (\mu'(s+1)-\mu's) \cdot G(\mu'(s+1)) - \sum_{s<dom(\nu)-1} (\nu'(s+1)-\nu's) \cdot G(\nu'(s+1))$$

s infinitesimal for all x,y,μ,ν satisfying $[x, y] \subset [a, b]$ & $\in P(x, y, \eta)$ & $\nu \in P(x, y, \eta)$. In case that $G(x)$ is integrable ver $[a, b]$ we use $M(x, y, t)$ to denote the function

$$\sum_{s<N(y-x,|t|)} |t| \cdot G(x+|t| \cdot (s+1)) + (y-x-|t| \cdot N(y-x, |t|)) \cdot G(y)$$

here $N(x,t)$ is a function such that ZF* $\vdash 0 < x \ \& \ 0 < t \rightarrow N(x,t) \in \omega$ $t.N(x,t) \leq x < t.(N(x,t) + 1)$ and immediately see that $M(x,y,t)$ converges uniformly as $t \to 0$ with respect to x,y satisfying $x<y$ & $[x,y] \subset [a,b]$. Now we define $S_x^y G(t)dt$ as a function of x, y such that ZF* $\vdash \forall x,y(x,y \in [a, b] \to ((x < y \to S_x^y G(t)dt = M(x, y, \delta))$ & $(y < x \to S_x^y G(t)dt = -M(y, x, \delta))$ & $(x = y \to S_x^y G(t)dt = 0)))$. Thus when G(x) is integrable over [a, b] and [c, d] is any interval such that ZF* $\vdash [c, d] \subset [a, b]$ we see that G(x) is also integrable over [c, d] and $S_c^d G(t)dt = M(c, d, \delta)$.

After these definitions are given we can derive many theorems including the usual theorems concerning differentiation and integration. For illustration, we cite the following 4 metatheorems.

5 Metatheorem. Suppose G(x) is a real function definable in ZF* on the real line and suppose G(x) is differentiable over the interval (a, b) where b-a may be an infinity. If G(x) satisfies the following three conditions: (i) G(x) has a finite bound in (a, b) in the sense that for some i, ZF* $\vdash x \in (a, b) \to |G(x)| < \ulcorner i \urcorner$, (ii) dG(x)/dx has a finite bound in (a, b), and (iii) 1/G(x) has a finite bound in (a, b), then 1/G(x) is also differentiable over (a, b) and $d(1/G(x))/dx - (-dG(x)/dx)/(G(x))^2$ is infinitesimal for $x \in (a, b)$. (Note. The conditions (i), (ii), (iii) are sufficient but not necessary. Sometimes (i) can be dropped as is illustrated by our example below in 10 when we take $G(x) = x^2$ in $(\infty - \ulcorner 5 \urcorner, \infty$

6 Metatheorem. Suppose G(x) is integrable over [a, b]. Then $S_x^y G(t)dt - [S_z^y G(t)dt + S_x^z G(t)dt]$ is infinitesimal for all $x,y,z \in [a, b]$ satisfying $x < z < y$. In particular, when c is a definable set such that ZF* $\vdash a < c$ & $c < b$ we have that $S_a^b G(t)dt - [S_c^b G(t)dt + S_a^c G(t)dt]$ is infinitesimal.

7 Metatheorem. Suppose that $G(x)$ is a real function which is ontinuous uniformly in $[a, b]$ in the sense that for any infinitesimal , the function $G(x) - G(y)$ is infinitesimal for all $x,y \in [a, b]$ atisfying $|x - y| < \eta$. If $b-a$ is a finite number, i.e. for some i, F* $\vdash |b-a| < \ulcorner i \urcorner$, then $G(x)$ is integrable over $[a, b]$.

8 Metatheorem. Suppose $G(x)$ is differentiable in (a, b) and $G(x)/dx$ has a finite bound in (a, b). We further suppose that $G(x)$ s integrable over $[a, b]$ (here $b-a$ may not be finite). Then $S_a^x G(t)dt$ s differentiable and $d(S_a^x G(t)dt)/dx - G(x)$ is infinitesimal for $\in (a, b-\delta)$. (Our method of proof of this theorem will be illustrated elow in an example in 10.)

9 Metatheorem. Suppose $G(x)$ is differentiable in (a, b) and $G(x)/dx$ is infinitesimal for $x \in (a, b)$. If $b-a$ is finite, then $(a) - G(x)$ is infinitesimal for $x \in (a, b)$.

Suppose $I(x)$ is differentiable of degree δ^{n+1} in (a, b) in the ense that for any two infinitesimals $\eta, \sigma \neq 0$ we always have F* $\vdash \forall x(x \in (a, b) \rightarrow |(I(x+\eta) - I(x))/\eta - (I(x+\sigma) - I(x))/\sigma| \leqslant \ulcorner i \urcorner . \delta^{n+1})$ or some fixed i. Further we suppose that $dI(x)/dx$, namely $(I(x+\delta) - (x))/\delta$, is infinitesimal of degree δ^{n+1} in the sense that ZF* $\vdash \forall x(x \in$ $a, b) \rightarrow dI(x)/dx \leqslant \ulcorner i \urcorner \cdot \delta^{n+1})$ for some fixed i. If $b-a \leqslant \infty^n$, then $(a) - I(x)$ is infinitesimal for $x \in (a, b-\delta)$. (An example will be iven below in 10.)

10 Examples. Let $F(x) \overset{df.}{==} \infty + x$. Then by definition $F(x)$ is

differentiable in the interval $(-\infty, \infty)$. Since $F(x)$ is continuous uniformly in $[0, 1]$, by Metatheorem 7 we see that $F(x)$ is integrable over $[0, 1]$. Then by Matatheorem 8, $S_0^x F(t)dt$ is differentiable and $d(S_0^x F(t)dt)/dx - F(x)$ is infinitesimal for x in the interval $(1/\ulcorner i \urcorner, \ulcorner 1 \urcorner - (1/\ulcorner i \urcorner))$. Now let $G(x) \overset{\text{df.}}{==} 1/x^2$. When we use $L(x, t)$ to denote $(G(x+t) - G(x))/t$ we see that $ZF^* \vdash L(x, t) = (-(2.x + t))/((x+t)^2.x^2)$. Hence for any two infinitesimals $\eta, \sigma \neq 0$, we see that $L(x, \eta) - L(x, \sigma)$ is infinitesimal for all x in the interval $(\infty - \ulcorner 5 \urcorner, \infty)$. Then by definition, $G(x)$ is differentiable and $dG(x)/dx - 0$ is infinitesimal for x in the interval $(\infty - \ulcorner 5 \urcorner, \infty)$. It can be verified that $F(x) + G(x)$, namely the function $\infty + x + 1/x^2$, is differentiable and $d(F(x) + G(x))/dx - (dF(x)/dx + dG(x)/dx)$ is infinitesimal for x in $(\infty - \ulcorner 5 \urcorner, \infty)$.

Now we are to give an informal proof for that $G(x)$ is integrable over $[\ulcorner 1 \urcorner, \infty]$. For any positive infinitesimal η we need to show in ZF^* that $(x < y \ \& \ [x, y] \subset [\ulcorner 1 \urcorner, \infty] \ \& \ \mu \in P(x, y, \eta) \ \& \ \nu \in P(x, y, \eta)) \rightarrow |\sum_{s<\mathrm{dom}(\mu)-1} (\mu'(s+1) - \mu's).\ G(\mu'(s+1)) -$

$$\sum_{s<\mathrm{dom}(\nu)-1} (\nu'(s+1) - \nu's) \cdot G(\nu'(s+1))| < 1/\ulcorner i \urcorner \quad \text{for all } i = 1, 2, \ldots$$

Let i be fixed but arbitrary. We can find j such that $1/(2^{j-1}) < 1/(9.i)$. Then we have three cases. (i) $x < y \leqslant \ulcorner 2^j \urcorner$, (ii) $x < \ulcorner 2^j \urcorner < y$, (iii) $\ulcorner 2^j \urcorner \leqslant x < y$. For illustration we only consider the case (ii). Since $\mu'0 = \nu'0 = x$, and $\mu'(\mathrm{dom}(\mu)-1) = \nu'(\mathrm{dom}(\nu)-1) = y$, we can find p and q such that $\mu'p \leqslant \ulcorner 2^j \urcorner < \mu'(p+1)$, $\nu'q \leqslant \ulcorner 2^j \urcorner < \nu'(q+1)$. Since $G(x) = 1/(x^2)$ is continuous uniformly in $[\ulcorner 1 \urcorner, \infty]$, and since $\mu'(p+1) - \ulcorner 2^j \urcorner < \mu'(p+1) - \mu'p \leq \eta$, hence $G(\ulcorner 2^j \urcorner) - G(\mu'(p+1))$ is infinitesimal. Thus we have

$$(1)\quad \Big|\sum_{s<\mathrm{dom}(\mu)-1}(\mu'(s+1)-\mu's)\cdot G(\mu'(s+1))-\Big(\sum_{s<p}(\mu'(s+1)-\mu's)\cdot G(\mu'(s+1))+$$

$$(\ulcorner 2^j\urcorner-\mu'p)\cdot G(\ulcorner 2^j\urcorner)+(\mu'(p+1)-\ulcorner 2^j\urcorner)\cdot G(\mu'(p+1))+\sum_{p<s<\mathrm{dom}(\mu)-1}(\mu'(s+1)-$$

$$\mu's)\cdot G(\mu'(s+1))\Big)\Big| = (\ulcorner 2^j\urcorner-\mu'p).|G(\ulcorner 2^j\urcorner)-G(\mu'(p+1))| < 1/\ulcorner(3.i)\urcorner .$$

Similarly we have

$$(2)\quad \Big|\sum_{s<\mathrm{dom}(\nu)-1}(\nu'(s+1)-\nu's)\cdot G(\nu'(s+1))-\Big(\sum_{s<q}(\nu'(s+1)-\nu's)\cdot G(\nu'(s+1))+$$

$$(\ulcorner 2^j\urcorner-\nu'q)\cdot G(\ulcorner 2^j\urcorner)+(\nu'(q+1)-\ulcorner 2^j\urcorner)\cdot G(\nu'(q+1))+\sum_{q<s<\mathrm{dom}(\nu)-1}(\nu'(s+1)-$$

$$\nu's)\cdot G(\nu'(s+1))\Big)\Big| < 1/\ulcorner(3.i)\urcorner.$$

Since $\ulcorner 2^j\urcorner - \ulcorner 1\urcorner$ is finite, by Metatheorem 7 we have that $G(x)$ is integrable over $[\ulcorner 1\urcorner, \ulcorner 2^j\urcorner]$. Hence

$$(3)\quad \Big|\Big(\sum_{s<p}(\mu'(s+1)-\mu's)\cdot G(\mu'(s+1)) + (\ulcorner 2^j\urcorner-\mu'p)\cdot G(\ulcorner 2^j\urcorner)\Big) -$$

$$\Big(\sum_{s<q}(\nu'(s+1)-\nu's)\cdot G(\nu'(s+1)) + (\ulcorner 2^j\urcorner-\nu'q)\cdot G(\ulcorner 2^j\urcorner)\Big)\Big| < 1/\ulcorner(9.i)\urcorner.$$

Since $y \leqslant \infty$, there is $z \in \omega$ such that $2^z \leqslant y < 2^{z+1} - 1$. Since $G(x)$ is decreasing we can see that

$$(4)\quad \Big|\Big((\mu'(p+1)-\ulcorner 2^j\urcorner)\cdot G(\mu'(p+1))+\sum_{p<s<\mathrm{dom}(\mu)-1}(\mu'(s+1)-\mu's)\cdot G(\mu'(s+1))\Big)\Big|$$

$\leq \sum_{c\leq s<d} G(s)$ where c stands for $\ulcorner 2^j\urcorner$ and d for 2^{z+1}

$$\leq \sum_{j\leq s<z+1} G(2^s)\cdot 2^s$$

$$= \sum_{j\leq s<z+1} 1/(2^s) \qquad (G(x) = 1/(x^2))$$

$$\leq 1/\ulcorner(2^{j-1})\urcorner$$

$$\leq 1/\ulcorner(9.i)\urcorner .$$

Similarly we have

$$(5)\quad |((\nu'(q+1)-\ulcorner 2^j \urcorner)\cdot G(\nu'(q+1)) + \sum_{q<s<dom(\nu)-1} (\nu'(s+1) - \nu's)\cdot G(\nu'(s+1)))| < 1/\ulcorner(9.i)\urcorner .$$

By the above arguments (1) - (5) it easily follows from $(x < y$ & $[x,y] \subset [1, \infty]$ & $\mu \in P(x,y,\eta)$ & $\nu \in P(x,y,\eta))$ that

$$\left| \sum_{s<dom(\mu)-1} (\mu'(s+1)-\mu's)\cdot G(\mu'(s+1)) - \sum_{s<dom(\nu)-1} (\nu'(s+1)-\nu's)\cdot G(\nu'(s+1))\right| < 1/\ulcorner i \urcorner .$$

Hence $G(x)$ is integrable over $[\ulcorner 1 \urcorner, \infty]$. Since $G(x)$ is differentiable and $dG(x)/dx = ((1/(x+\delta)^2 - 1/(x^2))/\delta$ has a finite bound in $(\ulcorner 1 \urcorner, \infty)$, hence by Metatheorem 8, $S_1^x G(t)dt$ is differentiable and $d(S_1^x G(x)dt)/dx - G(x)$ is infinitesimal in $(\ulcorner 1 \urcorner, \infty-\delta)$.

Let $G(x)$ still denote $1/(x^2)$. Let σ denote δ^{n+1} and let $S_1^x G(t)\sigma t$ denote $\sum_{s<N(x-1,\sigma)} \sigma.G(1 + \sigma.(s+1)) + (x - 1 - \sigma.N(x-1, \sigma))\cdot G(x)$. And let $I(x)$ to denote $1/x + S_1^x G(t)\sigma t$. We want to show that $I(x)$ is differentiable of degree δ^{n+1} in $(1, \infty^n)$, and $dI(x)/dx$ is infinitesimal of degree δ^{n+1} in $(1, \infty^n)$. Some of our arguments used here can also be used to prove the above metatheorem 8.

Let $\eta \neq 0$ be any infinitesimal. We assume $0 < \eta$ (the case that $\eta < 0$ can be considered similarly). We have two cases: (i) $1 + \sigma.N(x-1, \sigma) \le x < x+\eta < 1 + \sigma.(N(x-1,\sigma) + 1)$. (ii) $x < 1 + \sigma.(N(x-1, \sigma) + 1) \le x+\eta$. In case (i) we have $N(x+\eta-1, \sigma) = N(x-1, \sigma)$ and also the following:

$$(1)\quad S_1^{x+\eta} G(t)\sigma t - S_1^x G(t)\sigma t = (x+\eta - 1 - \sigma.N(x+\eta-1, \sigma))\cdot G(x+\eta) - (x - 1 - \sigma.N(x-1, \sigma))\cdot G(x)$$

$$= (x - 1 - \sigma.N(x-1, \sigma))\cdot (G(x+\eta) - G(x)) + \eta.G(x+\eta).$$

(2) $|(x - 1 - \sigma.N(x-1, \sigma)) \cdot (G(x+\eta) - G(x))|$

$< \sigma.|(G(x+\eta) - G(x))|$ (since $(x - 1 - \sigma.N(x-1, \sigma)) < \sigma$)

$= \sigma.|((G(x+\eta) - G(x))/\eta)|.\eta$ where $(G(x+\eta) - G(x))/\eta$ has finite bound for $x > 1$ because $G(x) = 1/(x^2)$ is differentiable for $x > 1$ and $dG(x)/dx = (-2x + \delta)/((x+\delta)^2 \cdot x^2)$ has a finite bound for $x > 1$.

(3) $\eta G(x+\eta) = \eta/(x+\eta)^2$

$\leqslant \eta\cdot(1/(x+\eta) \cdot (1/x)$

$= 1/x - 1/(x+\eta)$.

From (1), (2) and (3) it follows that (since in case (i) $\eta < \sigma$)

(4) $(S_1^{x+\eta}G(t)\sigma t - S_1^{x}G(t)\sigma t)/\eta + (1/(x+\eta)-1/x)/\eta$ is infinitesimal of degree σ, namel δ^{n+1} for $x > 1$.

Now consider the case (ii). In this case we have $N(x-1, \sigma) + 1 \leq N(x+\eta-1, \sigma)$ and we assume that $N(x-1, \sigma)+1 \leq N(x-1, \sigma)+z = N(x+\eta-1, \sigma)$. where $z \in \omega$. In this case we also have the following:

(5) $S_1^{x+\eta}G(t)\sigma t - S_1^{x}G(t)\sigma t$

$$= \sigma\cdot G(1+\sigma(N(x-1,\sigma)+1)) + \sum_{N(x-1,\sigma)<s<N(x-1,\sigma)+Z} \sigma\cdot G(1+\sigma(s+1)) +$$

$$(x+\eta-1 - \sigma N(x+\eta-1,\sigma)) \cdot G(x+\eta) - (x-1- \sigma N(x-1,\sigma)) \cdot G(x)$$

$$= (x-1- \sigma N(x-1,\sigma))(G(1+ \sigma(N(x-1,\sigma)+1))-G(x)) +$$

$$(\sigma-(x-1- \sigma N(x-1,\sigma))) \cdot G(1+\sigma(N(x-1,\sigma)+1))+$$

$$\sum_{N(x-1,\sigma)<s<N(x-1,\sigma)+Z} \sigma G(1+\sigma(s+1)) + (x+\eta-1 - \sigma N(x+\eta-1,\sigma)) \cdot G(x+\eta).$$

(6) $|(x-1-\sigma.N(x-1,\sigma))\cdot(G(1+\sigma.(N(x-1,\sigma)+1)) - G(x))|$

$\leqslant \sigma.|((G(1+\sigma.(N(x-1,\sigma)+1)) - G(x))/(1+\sigma.(N(x-1,\sigma)+1) - x))|.$

$.|(1+\sigma.(N(x-1,\sigma)+1) - x)/\eta|.\eta$

$\leqslant \sigma.\ulcorner i \urcorner.\eta$ for some fixed i (since $G(x)$ evidently is differentiabl for $x > 1$ and $dG(x)/dx$ has a finite bound $\ulcorner 3 \urcorner$ for $x < 1$. We have $1+\sigma(N(x-1,\sigma)+1) - x \leqslant \eta$ because in case (ii) $1+\sigma(N(x-1,\sigma)+1) \leqslant 1 + \sigma(N(x-1,\sigma)+z) \leqslant x+\eta$.).

(7) $|(\sigma-(x-1-\sigma.N(x-1,\sigma)))\cdot G(1+\sigma.(N(x-1,\sigma)+1))|$

$\leqslant (1+\sigma.(N(x-1,\sigma)+1)-x)\cdot(1/(1+\sigma.(N(x-1,\sigma)+1)))\cdot(1/x)$

$= 1/x - 1/(1+\sigma\cdot(N(x-1,\sigma)+1))$ (Since $G(x)$ is $1/x^2$)

Similarly we have

(8) $$\sum_{N(x-1,\sigma)<s<N(x-1,\sigma)+Z} \sigma G(1+\sigma(s+1))$$

$$\leqslant \sum_{N(x-1,\sigma)<s<N(x-1,\sigma)+Z} (1/(1+\sigma s) - 1/(1+\sigma(s+1))).$$

(9) $(x+\eta-1 - \sigma N(x+\eta-1,\sigma))\cdot G(x+\eta)$

$\leqslant 1/(1 + \sigma N(x+\eta-1,\sigma)) - 1/(x+\eta)$

$= 1/(1+\sigma(N(x-1,\sigma)+Z)) - 1/(x+\eta).$

From (5), (6), (7), (8) and (9) it follows that for $1 < x$,

(10) $S_1^{x+\eta} G(t)\delta t - S_1^x G(t)\delta t \leqslant \eta(\sigma\cdot\ulcorner i \urcorner + (1/x - 1/(x+\eta))/\eta)$ or

$(S_1^{x+\eta} G(t)\sigma t - S_1^x G(t)\sigma t)/\eta + (1/(x+\eta) - 1/x)/\eta \leqslant \sigma\cdot\ulcorner i \urcorner$

By (10) and (4) we concluded that for any positive infinitesimal η we have

(11) $(S_1^{x+\eta}G(t)\sigma t - S_1^{x}G(t)\sigma t)/\eta + (1/(x+\eta) - 1/x)/\eta$ is infinitesimal of degree σ, namely $\delta^{\eta+1}$ for $1 < x$. (11) also holds for any negative infinitesimal η. Hence $1/x + S_1^{x}G(t)\sigma t$, namely $I(x)$, is differentiable of degree δ^{n+1} for $x > 1$ and $I(x)/dx$ is infinitesimal of degree δ^{n+1} for $x > 1$. By metatheorem 9, $I(\ulcorner 1 \urcorner) - I(x)$ is infinitesimal for $x \in (\ulcorner 1 \urcorner, \infty^{n})$, i.e. $S_1^{x}G(t)\sigma t - (\ulcorner 1 \urcorner - 1/x)$ is infinitesimal for $x \in (\ulcorner 1 \urcorner, \infty^{n})$.

11 Remark. The above examples show some theorems concerning differentiation and integration which definitely will not appear in standard analysis; they belong to our non-standard analysis proper. For one more example we mention Dirac Delta function which can be treated adequately in non-standard analysis. In [6] we can find many references concerning Dirac Delta functions. We may expect that there will appear a lot of functions and theorems which are in the non-standard analysis proper and which are interesting.

12 Remark. Though ∞ is a real number in R and in particular, a natural number in ω, it can be used neither for counting nor for measurement. ∞ is, indeed, only a number by name. We may say that ∞ is a non-constructive number or that ∞ is an ideal object. In my previous paper mentioned above I have defined that a constructive real number is a set b definable in ZF* such that ZF* $\vdash b \in R$ and for each intuitive natural number i one can effectively find j, k such that ZF* $\vdash |b - \ulcorner j \urcorner / \ulcorner k \urcorner| < 1/\ulcorner i \urcorner$ or ZF* $\vdash |b - -\ulcorner j \urcorner / \ulcorner k \urcorner| < 1/\ulcorner i \urcorner$. An interesting non-constructive real number can be defined as follows. By some technique in proof theory we can find a formula $A(x)$ such that for any consistent extension ZF** whose axioms are recursively

enumerable there is an intuitive natural number f for which $A(\ulcorner f \urcorner)$ is undecidable in ZF**. Then by the law of excluded middle which holds in ZF* we have ZF* $\vdash \forall x(x \in \omega \rightarrow E!y((y = 0 \ \& \ A(x)) \vee (y = 1 \ \& \ \neg A(x))$. This theorem defines a function $G(x)$ in ZF* such that ZF* $\vdash \forall x(x \in \omega \rightarrow (G(x) = 0 \leftrightarrow A(x)) \ \& \ (G(x) = \ulcorner 1 \urcorner \leftrightarrow \neg A(x)))$. Then $\sum_{s < \infty} G(s) \cdot (1/3^s)$ is a finite real number but it is not constructive. For, to assume that it is a constructive real number will lead to the conclusion that $A(\ulcorner i \urcorner)$ is decidable in ZF* for $i = 0, 1, \ldots,$. This contradicts the fact that $A(\ulcorner f \urcorner)$ is undecidable in ZF* for some intuitive natural number f. Let us call a function definable in ZF* constructive if for any argument which is constructive real number the function always take a constructive real number as its value. Since there is no general method for determining whether any function definable in ZF* is a constructive real function, an object of our proof-theoretic theory of non-standard analysis is to find some sufficient conditions for a function definable in ZF* to be a constructive real function.

13 Metadefinition. Let $F(x, t)$ be a real function definable in ZF*, and (a, b) be an interval. We say that $F(x, t)$ converges constructively uniformly as $t \rightarrow 0$ with respect to x in (a, b) if for any intuitive j we can effectively find an intuitive i such that ZF* $\vdash \forall x(x \in (a, b) \rightarrow \forall y,z(0 < |y| < 1/\ulcorner i \urcorner \ \& \ 0 < |z| < 1/\ulcorner i \urcorner \rightarrow |F(x,y) - F(x, z)| < 1/\ulcorner j \urcorner))$.

14 Metatheorem. Suppose $F(x, t)$ converges constructively uniformly as $t \rightarrow 0$ with respect to x in (a, b). Then $F(x, t)$ also converges uniformly as $t \rightarrow 0$ with respect to x in (a, b) in the sense of Definition 2. Suppose further that $F(x, t)$ is a constructive real function. Then $F(x, \delta)$ is also a real constructive function.

We have shown that constructively uniformly continuous functions are useful for defining constructive real functions by means of integration. We have also given examples of function which is uniformly continuous but is not constructively uniformly continuous (in my previous paper [1]).

15 Remark. A merit of our theory of non-standard analysis is that it takes as its universe of discourse the class of all sets rather than some small parts of this class, say, the superstructures V(R) over R and V(R*) over R* which have been used by the modelists in developing their theory of non-standard analysis. For example we may take Keisler's book [2]. Recently we find three different theories which develop non-standard analysis also in the whole universe of sets. These works are by Nelson in [3], Hrbacek in [4] and Vopenka in [5]. However, these approaches are all different from our's; they are all model-theoretically motivated.

Besides, our theory can develop straightforwardly and speedily. We may expect that the proof-theoretic approach to non-standard analysis would become more popular sooner or later. On the other hand, our theory do not have a carefully formulated 'Leibniz Principle' as appears in the literature of 'conventional' non-standard analysis. However, in taking infinities and infinitesimals as fixed numbers which obey the general laws just as the ordinary finite numbers do (indeed, the infinities and infinitesimals obey all the general laws that can be proved in ZF because any general law, for example, $x+y = y+x$ which is provable in ZF must also be provable in ZF*) our theory of analysis conforms to the idea which must have been in Leibniz' mind when he suggested that the real numbers should be extended to a larger system which has the same elementary properties but contains infinitesimals. If this idea of Leibniz is what is referred to when 'Leibniz Principle' is talked about, then our theory of non-standard analysis already complies with Leibniz Principle merely without explicitly stating it.

16 Remark. When Goedel first discovered his example of undecidable sentence it seemed that the mathematical circle considered the discovery a crisis. However, after the elapse of 50 years we may view the matter from a different angle. In a sense we can take undecidable sentences as our asset rather than liability. For, when a sentence which is undecidable in a certain consistent theory appears we may choose to obtain a new consistent theory by adding the undecidable sentence or alternatively its negation to the original theory. In general, the new theory so obtained contains a lot of new theorems. In case the undecidable sentence added is of the form $E!xf(x)$ the resulting new theory must contain a new object defined by $f(x)$ which may be an 'ideal' object but which may also be useful. Some times, in this way we may be able to obtain a new theory which proves very interesting and useful as is illustrated by our example ZF* which provides us with a whole theory of non-standard analysis. With a view to getting new interesting consistent theories by adding undecidable sentences to old theories we may expect to have still more work to be done in proof theory.

Postscript. Now I add this paragraph to discuss briefly how the non-standard definitions of differentiability and integrability in this paper are related to the usual standard ones. We first note that the notion of improper integral in standard analysis is already subsumed, as a special case, under our non-standard notion of integrability (Example 10). On the other hand, let us suppose that for any standard real function (namely, function definable in ZF) we define differentiability and integrability in exactly the same manner as in a standard calculus text book except that the notion of limit adopted is in the following sense: ℓ is the limit of $F(x)$ as $x \to 0$ if for each i there is a j such that $ZF \vdash 0 < x < \ulcorner j \urcorner \to |F(x) - \ell| < \ulcorner i \urcorner$. It can be verified that if a function is differentiable (or integrable) in an interval in such a standard sense, then so it is also in the non-standard sense as defined above in 3 or 4. (compare this with Metatheorem 14 above) Thus in some sense we can say that the non-standard analysis of this paper is a generalization of the conventional standard analysis.

Acknowledgement. The author wishes to express his thanks to Professor Michael M. Richter for informing of the workg done by Nelson, Hrbacek and Vopenka and for other helps; also to a referee of this paper for kindly making many corrections and for suggesting a discussion about how the non-standard definitions given in this paper are related to the corresponding usual ones.

References

1. Shih-Chao Liu, A proof-theoretic approach to non-standard analysis with emphasis on distinguishing between constructive and non-constructive results, The Kleene Symposium (1980), pp.391-414, North-Holland Publishing Company.

2. H. J. Keisler, Foundations of infinitesimal calculus, (1976), Prindle, Weber & Schmidt, Boston, MA.

3. E. Nelson, Internal set theory, Bull. AMS. Vol. 83(1977), pp.1165-1198.

4. K. Hrbacek, Axiomatic foundations for non-standard analysis, Fund. Math. vol. XCVIII (1978), p.1-19.

5. P. Vokenka, Mathematics in the alternative set theory, Leipzig 1979.

6. M. Richter, Ideale Punkte, Monaden und Nichtstandard-Methoden, VIEWEG 1982.

Institute of Mathematics
Academia Sinica
Nankang, Taipei, Taiwan 115
Republic of China

INTERPRETATIONS AND THE MODEL THEORY OF THE CLASSICAL GEOMETRIES*

Kenneth L. Manders
Philosophy, University of Pittsburgh
Pittsburgh, PA 15260 USA

The main goal of this article is to show that, for suitable choices of primitives, projective geometry is the model completion of affine geometry. Thus the geometers' predilection for a projective setting is analogous to the preference for working over the algebraic or real closure of an initially given field. Of course, we have to specify and qualify to make our thesis precise, but the choices which work are natural ones in a model theoretic investigation of projective closure.

We start from sufficient conditions for an inverse pair of interpretations to transfer model completeness from one theory to another, (1.2-3), and check whether traditional constructions among geometries and fields satisfy these (2-3.1). For example, a projective plane with collinearity relation is model complete iff the underlying (skew) field is model complete. Then we study the relations of projective closure and hyperplane removal to affine embeddings, with conclusions about existential undefinability of affine parallelism (3.2) and model companions (3.4-9); using a finer, algebraic, characterization of affine embeddings (3.10), we treat amalgamation and model completions (3.9-12). Model companions of affine spaces over ordered fields would have to be projective betweenness spaces, which we study in §4, reducing to questions about ordered fields with additional primitives. A discussion of our choice of primitives in §5 leads into rather puzzling philosophical questions about the semantics of algebraic geometry.

For basic definitions and properties of model completeness, model companions and completions, we refer to Macintyre's Handbook article [M]. We develop the theory for spaces of arbitrary fixed finite dimension $n \geqslant 2$ over infinite skew (i.e., not necessarily commutative) fields, giving a parallel treatment of the ordered and non-ordered cases. Commutativity isn't missed, but there are some differences between dimension 2 and higher dimensions. For example, collinearity embeddings between n-dimensional spaces need not have n-dimensional image for $n>2$. To recover this desirable feature one must instead take co-hyperplanarity as a primitive (§3.13). This is at variance with usage in the literature on logical foundations of geometry ([T], [Sch]), which has been primarily concerned with questions of axiomatizability and decidability; problems concerning morphisms are more sensitive to choices of primitive notions.

---

*Research supported by a NATO Postdoctoral Fellowship in Science.

1. MUTUAL INTERPRETABILITY AND TRANSFER OF MODEL COMPLETENESS.

The mutual interpretabilities between the classical geometries and field theories have not been used to transfer model completeness; indeed, Diller [D] pointed out the failure of model completeness for affine betweenness planes over real-closed fields, and Szczerba [Sz3] denies that transfer of model completeness arises from mutual interpretability, giving a counterexample constructed by P. Tuschik. Nonetheless, most classical geometries are model complete over model complete field theories; and for suitable choices of primitive notions. We give sufficient conditions on a mutual interpretability to guarantee transfer of model completeness; these may be seen to account for model completeness when it occurs among classical geometries.

1.1. To establish notation and a semantic point of view, we recall the notion of interpretation, following e.g. [Sz3], [Sz4]. Given a σ_1-structure $\underline{M}$, satisfying a theory T_{12} to be described, we construct a σ_2-structure $I_{12}(\underline{M})$: (i) The domain is given, for some fixed n, by σ_1-formula $\psi_d(x_1 \ldots x_n)$, and $\exists\bar{x}\psi_d(x)$ is in T_{12}; or rather, by the quotient of the preceding by σ_1-formula $\psi_=(\bar{x},\bar{y})$, with "$\psi_=$ defines an equivalence relation" in T_{12}; or there could be more than one domain of this kind (with $\exists\bar{x}\psi_d(\bar{x})$ in T_{12} for at least one); (ii) Each k-ary symbol R in σ_2 is given an interpretation on the domain(s) by σ_1-formula $\psi_R(\bar{x}_1,\ldots,\bar{x}_k)$; and "$\psi_=$ defines a congruence relation w.r.t. ψ_R" is in T_{12}; (iii) all σ_1-formulas in the above may have extra free variables $\bar{y}$; then the construction, indicated as $I_{12}(\underline{M},\bar{p})$, is determined by $\underline{M}$ together with a choice of parameters $\bar{p}$ from $\underline{M}$ for $\bar{y}$; the admissible parameter choices are given by σ_1-formula $\psi_p(\bar{y})$, with $\exists\bar{y}\psi_p(\bar{y})$ in T_{12}; the preceding requirements on $\underline{M}$ must be relativised to ψ_p in this case, e.g. we want $\forall\bar{y}[\psi_p(\bar{y}) \to \exists\bar{x}\psi_d(\bar{x},\bar{y})]$ in T_{12}.

Thus the various formulas ψ together determine a model theoretic operation

$$I_{12}: \mathrm{Mod}_{\sigma_1}(T_{12}) \to \mathrm{Str}(\sigma_2),$$

as well as the familiar syntactic translation operation

$$I^*_{12}: \mathrm{Form}(\sigma_2) \to \mathrm{Form}(\sigma_1);$$

for any $\underline{M} \in \mathrm{Mod}_{\sigma_1}(T_{12})$, $\varphi \in \mathrm{Form}(\sigma_2)$ and parameters $\bar{p}$ from $\underline{M}$, these satisfy

$$I_{12}(\underline{M},\bar{p}) \models \varphi \iff (\underline{M},\bar{p}) \models I^*_{12}(\varphi).$$

We need a few elementary observations. If $I_{12}: \mathrm{Mod}_{\sigma_1}(T_{12}) \to \mathrm{Str}(\sigma_2)$ and $I_{23}: \mathrm{Mod}_{\sigma_2}(T_{23}) \to \mathrm{Str}(\sigma_3)$ are interpretations, so is the composite interpretation

$I_{12} \circ I_{23}$: $\mathrm{Mod}_{\sigma_1}(T_{13}) \to \mathrm{Str}(\sigma_3)$ given by translating I_{23}, e.g.

$$T_{13} = T_{12} \cup \{\forall \bar{y}[\psi_p^{12}(\bar{y}) \to I_{12}^*(T_{23})]\},$$

$$\psi_p^{13}(\bar{y},\bar{z}_1,\ldots,\bar{z}_t) =_{def} \psi_p^{12}(\bar{y}) \wedge I_{12}^*(\psi_p^{23}(z_1,\ldots,z_t)), \text{ etc.}$$

If j: $\underline{M} \to \underline{M}'$ is an elementary embedding and I_{12} an interpretation, as above, then for any fixed $\bar{p}$ from $\underline{M}$ we have an induced elementary embedding

$$I_{12}(j,\bar{p}): I_{12}(\underline{M},p) \to I_{12}(\underline{M}',j(\bar{p}))$$

and this action respects composition $j_1 \circ j_2$ and identity map. For $I_{12}(\underline{M},\bar{p})$ is obtained from $(\underline{M},\bar{p})$ by σ_1-definable constructions, so given that j is σ_1-elementary, these constructions act in the same way on (sequences of) elements of $(\underline{M},\bar{p})$ as on their j-images, (sequences of) elements of $(j(\underline{M}),j(\bar{p}))$ within $(\underline{M}',j(\bar{p}))$. If $T_1 \supseteq T_{12}$ is a σ_1-theory, I_{12} is a <u>Δ-interpretation in</u> T_1 iff the domain and parameter formulas of I_{12} are equivalent to existential formulas in T_1, and all other formulas of I_{12} are equivalent in T_1 both to an existential and to a universal formula. By the same reasoning as just above, we have that if I_{12} is a Δ interpretation in T_1, then any σ_1-embedding j: $\underline{M} \to \underline{M}'$ of models of T_1 induces a σ_2-embedding $I_{12}(j,\bar{p})$: $I_{12}(\underline{M},\bar{p}) \to I_{12}(\underline{M}',j(\bar{p}))$. The composition of Δ interpretations is a Δ interpretation. As a general notion, if i: $\underline{M} \to \underline{M}'$ and j: $\underline{N} \to \underline{N}'$ are σ-embeddings, i and j are <u>elementarily equivalent embeddings</u> iff $\underline{M}'$ and $\underline{N}'$ are elementarily equivalent in the language σ augmented by a unary predicate designating the image of i resp. j. Now we have

1.2. <u>Theorem</u>. Let I_{12}: $\mathrm{Mod}_{\sigma_1}(T_{12}) \to \mathrm{Str}(\sigma_2)$, I_{21}: $\mathrm{Mod}_{\sigma_2}(T_{21}) \to \mathrm{Str}(\sigma_1)$ be interpretations, $T_1 \supseteq T_{12}$, $T_2 \supseteq T_{21}$ a σ_1 resp. σ_2-theory, such that $I_{12}(\mathrm{Mod}(T_1)) \subseteq \mathrm{Mod}(T_2)$ and T_2 is model complete. Then T_1 is model complete if

(i) I_{12} is a Δ interpretation in T_1, and

(ii) for any embedding j: $\underline{M} \to \underline{M}'$ of models of T_1 which under the composite interpretation $I_{12} \circ I_{21}$: $\mathrm{Str}(\sigma_1) \to \mathrm{Str}(\sigma_1)$ induces an embedding of the images for some choice of parameters from $\underline{M}$, some such embedding is elementarily equivalent to j.

Proof: We must show that any embedding j: $\underline{M} \to \underline{M}'$ of models of T_1 is elementary. As I_{12} is Δ in T_1, it induces a σ_2-embedding j': $I_{12}(\underline{M},\bar{p}) \to I_{12}(\underline{M}',j(\bar{p}))$. As these structures satisfy the model complete T_2, j' is in fact elementary. But then j' induces an elementary embedding under I_{21} (with possibly further parameters from $I_{12}(\underline{M},\bar{p})$, which may be traced back to $\underline{M}$). By (ii), this elementary embedding is elementarily equivalent to j, which therefore must also be elementarily. □

1.3. Note that condition (i) is necessary, (ii) possibly not. In our applications,

(ii) will always arise from a much stronger condition, if it holds at all. Consider

$$(*)\quad\begin{cases}\text{For any } \underline{M} \models T_1\text{, any parameters } \bar{p} \text{ for } I_{12} \text{ from } \underline{M}\text{, there exist parameters } \bar{q}\\ \text{for } I_{21} \text{ from } I_{12}(\underline{M},\bar{p})\text{, tracing back to } \bar{p}' \text{ on } \underline{M} \text{ via } I_{12}\text{, such that}\\ \qquad I_{21}(I_{12}(\underline{M},\bar{p}),\bar{q}) \simeq \underline{M}\\ \text{and such an isomorphism is explicitly definable on } \underline{M} \text{ with parameters}\\ \bar{p}'' \supseteq \bar{p},\bar{p} \text{ by a formula } \psi_{\simeq}(x,\bar{x},\bar{y}'') \text{ independent of } \underline{M},\bar{p}''.\end{cases}$$

This makes sense, as by the definition of interpretation, any isomorphism as in (*) is indeed a relation on $\underline{M}$; by Beth's theorem (*) therefore follows from the existence on $\mathrm{Mod}(T_1)$ of at most one such isomorphism of $\underline{M}$ with prescribed behavior on some sufficiently long finite sequence of parameters $\bar{p}''$. But in applications, we can easily write down $\psi_{\simeq}$ directly. When (*) holds, we say that (I_{12},I_{21}) is a <u>definably inverse pair of interpretations</u> (<u>on</u> T_1).

However, (*) does not entail condition (ii), even given the other hypotheses of theorem 1.2.(2.6.2.). Rather, we must require that $\psi_{\simeq}$ be existential, and give an existential σ_1-formula $\psi_{p''}(\bar{y}'')$ such that

$$(**)\qquad T_1 \models \exists\bar{y}''\ \psi_{p''}(\bar{y}'') \wedge \forall\bar{y}''[\psi_{p''}(\bar{y}'') \rightarrow {}''\psi_{\simeq}\colon \underline{M} \simeq I_{21}(I_{12}(\underline{M},\bar{p}),\bar{q})''].$$

Then the isomorphism defined by $\psi_{\simeq}$ commutes with the embedding $j\colon \underline{M} \to \underline{M}'$ of models of T_1, which entails (ii); we say that (I_{12},I_{21}) is an <u>$\exists$-definably inverse pair of interpretations on</u> T_1.

These strong definability conditions give a syntactically explicit transfer of model completeness: Assuming the conditions of theorem 1.2, we have, for any σ_2-formula φ

$$T_2 \models I^*_{21}(\varphi) \leftrightarrow \chi_{\forall}, \chi_{\exists}$$

by model completeness; as I_{12} maps $\mathrm{Mod}(T_1)$ into $\mathrm{Mod}(T_2)$, this gives

$$T_1 \models I^*_{12}(I^*_{21}(\varphi)) \leftrightarrow I^*_{12}(\chi_{\forall}),\ I^*_{12}(\chi_{\exists}).$$

Because (I_{12},I_{21}) are definably inverse (with parameter formula satisfying **), $\varphi(x_1\ldots x_t)$ is equivalent in T_1 to both of

$$\forall\bar{y}''\ \forall\bar{x}_1\ldots\bar{x}_t[(\psi_{p''}(\bar{y}'') \wedge \bigwedge_i \psi_{\simeq}(x_i,\bar{x}_i,\bar{y}'')) \rightarrow I^*_{12}(I^*_{21}(\varphi))],$$

$$\exists\bar{y}''\ \exists\bar{x}_1\ldots\bar{x}_t[\psi_{p''}(\bar{y}'') \wedge \bigwedge_i \psi_{\simeq}(x_i,\bar{x}_i,\bar{y}'') \wedge I^*_{12}(I^*_{21}(\varphi))].$$

As I_{12} is Δ in T_1 and $\psi_{p''},\psi_{\simeq}$ are existential, substitution of $I^*_{12}(\chi_{\forall})$ resp. $I^*_{12}(\chi_{\exists})$ in these expressions give a universal resp. existential equivalent of φ in T_1.

1.4. We indicate a generalisation. Let a <u>$\forall_n$-embedding</u> be an embedding preserving

$\forall_n$-formulas, a Δ_n-interpretation in T be one in which domain and parameter formulas are $\exists_n$ and all others have both $\exists_n$ and $\forall_n$ equivalents in T; and let T have prefix dimension $\leqslant n$ iff every $\forall_n$-formula is equivalent to an $\exists_n$-formula in T iff every $\forall_n$-embedding of models of T is elementary. The proof of theorem 2.1 shows: If T_2 has prefix dimension $\leqslant m$, (ii) holds at least for $\forall_m$-embeddings j under $I_{12} \circ I_{21}$, and I_{12} is a Δ_n-interpretation, then T_1 has prefix dimension $\leqslant m+n-1$. Again, the transfer of prefix bound is explicit if this version of (ii) is replaced by (*,**) for $\exists_{n+m-1}$-formulae $\psi_{\simeq}, \psi_{p''}$.

2. SOME CLASSICAL GEOMETRICAL CONSTRUCTIONS.

We review a number of constructions relating rings and classical geometries as indicated in the diagram, to settle terminology and notation, but mainly to obtain the information about syntactic form of the constructions needed for later applications. For a parallel discussion (of Euclidean geometry), giving full detail, see [Sch]§3.25-67. (The reader might prefer to go directly to §3, referring back to the items here as needed.)

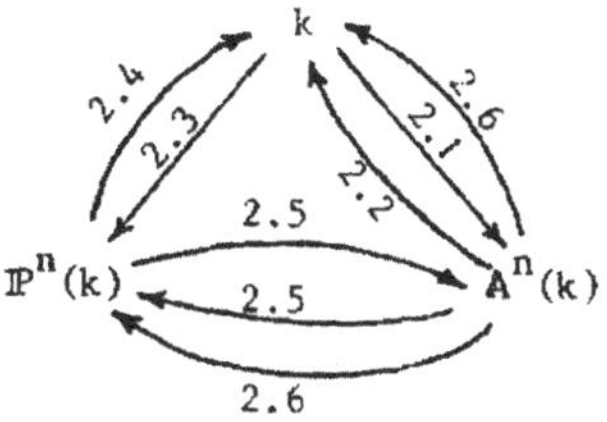

Here k is a (not necessarily commutative) field, $\mathbb{A}^n(k)$ affine space, $\mathbb{P}^n(k)$ projective space over k. We give parallel treatments for (a) k in language of rings with 1, called linear case; (b) k in language of ordered rings with 1, called ordered case, i.e. here we deal with additional geometrical primitives to correspond to the order on the field.

2.1. Affine n-space $\mathbb{A}^n(k)$, k (skew) field, in the language $\{0,1,+,\cdot\}$ of rings, $n \geqslant 2$.

2.1.1. Linear case: domain k^n. By quantifier free solvability criteria (Bourbaki, [Bo], §6.10), we have the relations, for $\bar{x}_0,\ldots,\bar{x}_{n+1} \in k^n$, and any $m \leqslant n$: $C_m\bar{x}_0,\ldots\bar{x}_{m+1}$ iff $\bar{x}_0,\ldots\bar{x}_{m+1}$ lie in an m-dimensional linear left subspace.
Of course, $C_0\bar{x}_0\bar{x}_1$ iff $\bar{x}_0 = \bar{x}_1$; C_n holds universally, and C_1 is the familiar relation of collinearity. $C_1,\ldots,C_{n-1}$ are interdefinable; in particular it will suf-

fice to take C_{n-1} as primitive, for we have, for $m \leq n-2$ and distinct $x_0,\ldots x_{m+1}$

$$C_m x_0 \ldots x_{m+1} \text{ iff } \forall x_{m+2}\, C_{m+1} x_0 \ldots x_{m+1} x_{m+2}$$
$$\text{iff } \exists yz[\sim C_{m+1} x_0 \ldots x_m yz \wedge C_{m+1} x_0 \ldots x_{m+1} y \wedge C_{m+1} x_0 \ldots x_{m+1} z] .$$

Conversely, there is a positive existential definition of C_{i+1} from C_i for $0<i<n-1$: let π range over permutations of $\{0,\ldots,i+2\}$;

$$C_{i+1} x_0 \ldots x_{i+2} \text{ iff } \exists y \exists \pi\, [C_i\, x_{\pi(0)} \ldots x_{\pi(i)} y \wedge C_1 y x_{\pi(i+1)} x_{\pi(i+2)}];$$

but for $m>1$, C_m is not definable by a universal formula in $C_1,\ldots,C_{m-1}$ (3.13). Because of this collinearity would not be suited to our purposes as a unique primitive for $n>2$; unless explicitly noted otherwise, the primitive of $\mathbb{A}^n(k)$ is C_{n-1}. This gives a Δ interpretation in any case.

2.1.2. Occasionally, parallelism is used as a primitive in $\mathbb{A}^n(k)$, cf. [Sz4] for $n=2$:

$$xy \parallel \bar{u}\bar{v} \text{ iff } (x_i-y_i)(u_j-v_j) = (x_j-y_j)(u_i-v_i) \text{ for each } 1 \leq i<j \leq n.$$

From this, C_1 is definable

$$C_1 xyz \text{ iff } xy \parallel xz,$$

but for $n \geq m \geq 2$, C_m is not definable from $\parallel$ by a universal formula.
We will often use $\parallel$ as a defined expression, for $n=2$ deleting C_2,

$$xy \parallel uv \text{ iff } (C_1 xyu \wedge C_1 xyv) \vee \{C_2 xyuv \wedge \forall z[C_2 xyuz \rightarrow \sim(C_1 xyz \wedge C_1 uvz)]\}$$

(coplanar non-intersection); but $\parallel$ is not definable from C_{n-1} by an existential formula (3.2).

2.1.3. Ordered case: k in language of ordered rings $\{0,1,+,\cdot,\leq\}$. As primitive, add the ternary betweenness relation,

$$B\bar{x}\bar{y}\bar{z} \text{ iff } C_1 \bar{x}\bar{y}\bar{z} \wedge \forall j \leq n\ 0 \leq x_j-y_j \leq x_j-z_j \text{ or } 0 \geq x_j-y_j \geq x_j-z_j .$$

So this remains a Δ interpretation. For $n=2$, B may replace C_1 as a primitive. (In the Polish tradition, B is taken as the unique primitive for all n. Cf. [ST], [K]. For us, this is as inadequate as using only C_1 in 2.1.1.)

2.1.4. The theory $Th(\{\mathbb{A}^n(k) : k \text{ skew field}\})$ turns out to be the theory of Desarguean affine n-spaces; axioms could be worked out from the axioms of Blumenthal [BlM], Chs. IV, V. (Given for $n=2$, these are: there is a unique line through any two given points, a unique parallel to a given line through a given point, and a four-point: a quadruple of points no 3 of which are collinear; and two universal axioms which are affine versions of the projective Desargues, cf. [G]§3.1. Adequacy of the axioms is shown by carrying out coordinate ring construction (2.2.1), co-ordinatisation (2.2.4) and verifying that a skew field and an isomorphism arise.) Additional axioms on B in the ordered case: see [HD], pp. 20, 40, 149-150.

2.2. Affine Coordinate ring.

2.2.1. linear case. parameters $0, e_1, e_2$.

$\psi_p(x,y,z)$ iff $\sim C_1xyz$, $\psi_d(x)$ iff C_1x0e_1, $\psi_=(x,y)$ iff $x=y$;

$\psi_0(x)$ iff $x=0$, $\psi_1(x)$ iff $x=e_1$;

$\psi_+(x,y,z)$ iff $\exists u\ xu \parallel 0e_2 \wedge e_2u \parallel 0e_1 \wedge pe_2 \parallel uz$;

$\psi_\cdot(x,y,z)$ iff $\exists u\ xu \parallel e_1e_2 \wedge C_10e_2u \wedge e_2y \parallel uz$.

As the point u in the last two formulas is uniquely determined in Desarguean affine n-spaces by the conditions given, these formulas have universal equivalents. Thus we have a Δ interpretation (in Desarguean affine n-spaces), assuming $\parallel$ as a primitive. Assuming C_{n-1} as primitive, this gives a Δ_2 interpretation. For these spaces, with $n \geqslant 2$ throughout, the construction gives a skew field, see [Bl]. (It is shown there that a minor generalisation of the construction already gives a $\{+,\cdot\}$-algebra without the Desargues axioms; but then most skew field identities may fail.)

2.2.2. ordered case. As above, but we also recover the field order from B, by defining the positive elements

$\psi_P(x)$ iff $B0xe_1 \vee B0e_1x$.

So still a Δ resp. Δ_2 interpretation; applied to an ordered Desarguean affine geometry it gives an ordered skew field.

2.2.3. coordinate isomorphism.
Starting from an ordered or unordered skew field, form affine n-space, and then construct the affine coordinate ring, with parameters $0=\bar{0}$, $e_1=(1,0\ldots0)$, $e_2=(0,1,0\ldots0)$. The resulting ring isomorphism, $x \mapsto (x,0\ldots0)$ is evidently uniformly $\exists$-definable on the original skew field, i.e. we have a $\exists$-definably inverse pair of interpretations. Note: the interpretations themselves are not both Δ!

2.2.4. coordinatisation isomorphism.
Starting from an ordered or unordered Desarguean affine n-space $\underline{A}$, execute the coordinate ring construction and then the affine n-space construction. We uniformly define an isomorphism between $\underline{A}$ and the result, with parameters $0,e_1,\ldots,e_n$ ($0,e_1,e_2$ coinciding with those of the coordinate ring construction) such that $\sim C_{n-1}0e_1\ldots e_n$; $\psi_\simeq(x,x_1\ldots x_n,y_0,\ldots,y_n)$ iff $\forall i$ "x_i is obtained from x by success-

ive projections parallel to $y_0y_1,\dots,y_0y_n$, but not y_0y_i -ultimately giving a point on y_0y_i- followed by a projection to y_0y_1 parallel to y_iy_1."
As the intermediate points in the construction are uniquely determined by quantifier free conditions (in $\|, C_{n-1}, C_1$), we have an $\exists$-definition in terms of $\{C_{n-1},\|\}$; in terms of C_{n-1} alone, this becomes $\exists_2$. So this is a $\exists$ resp. $\exists_2$-definably inverse pair of interpretations.

2.3. Projective n-space $\mathbb{P}^n(k)$ over (ordered; skew) field k, $n\geq 1$.

2.3.1. linear case. Compose the interpretation $k \to \mathbb{A}^{n+1}(k)$ with the following construction $\mathbb{A}^{n+1}(k) \to \mathbb{P}^n(k)$: Domain $\mathbb{A}^{n+1}$, subject to restriction $\psi_d(\bar{x})$ iff $\bar{x}\neq\bar{0}$ and reduction modulo (left) linear equivalence.

$$\psi_{=}(x,y) \quad \text{iff} \quad \bar{x}\neq 0 \wedge \bar{y}\neq\bar{0} \wedge C_1\bar{0}\bar{x}\bar{y}.$$

As relational primitive: C_{n-1}

$$\psi_{C_{n-1}}(\bar{x}_0,\dots,\bar{x}_n) \quad \text{iff} \quad C_n\ \bar{x}_0,\dots,\bar{x}_n,\bar{0}$$

for which $\psi_{=}$ defines a congruence. This gives a Δ interpretation. The comments under 2.1.1. concerning the relations $C_0,\dots,C_n$ continue to hold in $\mathbb{P}^n(k)$.

2.3.2. ordered case. We add the quaternary relation of collinear separation S: Sxyuv iff x,y,u,v are collinear, and the pairs (x,y),(u,v) of points separate each other on the line. A line of $\mathbb{P}^n$ is a plane through $\bar{0}$ in $\mathbb{A}^{n+1}$, its points are lines through zero in that plane, for the intersection points of these lines with any line in the plane not through zero or parallel to the four given ones, "Bxuy $\wedge$ Buyx or any cyclic permutation" gives S.
This is a Δ definition, as there are always such intersection points, any such set will work, and all give the same result.

2.3.3. The theory Th($\{\mathbb{P}^n(k)$: k skewfield$\}$) turns out to be the theory of <u>Desarguean projective n-spaces</u> cf. [G]§6.1-2. for axioms in the linear case, and [HD], p. 150 for the order axioms.

2.3.4. These constructions work for any Desarguean affine (n+1)-space $\underline{A}^{n+1}$: Replace $\bar{0}$ by a parameter $p\in\underline{A}^{n+1}$ (as we didn't otherwise refer to the coordinates).

2.4. <u>Projective coordinate ring</u>

2.4.1. linear case.

Parameters $0,e_1,e_2, u_1,u_2 : \psi_p(\sim)$ iff "$C_1 0e_1e_2 \wedge \forall i[C_1 0e_iu_i \wedge 0 \neq u_i \neq e_i]$; $\psi_d(x)$ iff $C_1 0e_1 x \wedge x \neq u_1$. From here on, the construction coincides with the affine coordinate ring construction, replacing everywhere

'ab ∥ cd' by "ab and cd have a common intersection with u_1u_2".

Thus the interpretation becomes Δ in C_{n-1}: any such intersection is uniquely determined by a,b,c,d,u_1,u_2, and may hence be introduced interchangeably by universal or existential quantification. For very clear analysis, see Garner [G], §3.3-3.4. The construction gives a skew field iff applied to a Desarguean projective space; the the skew field is unique up to isomorphism; indeed, for fixed $0,e_1,u_1$ the skew field is uniquely determined regardless of the choice of e_2,u_2 (satisfying ψ_p).

2.4.2. ordered case. As above, but recover the skew field order by defining the positive elements by

$$\psi_{Pos}(x) \quad \text{iff} \quad S0xe_1u_1 \vee S0e_1xu_1,$$

so again a Δ interpretation in Desarguean projective space.

2.4.3. coordinate ring isomorphism.
Entirely as in the affine case (2.2.3) we obtain an $\exists$-definable ring isomorphism by a suitable choice of parameters in composing projective n-space construction with projective coordinate ring construction, so this pair is $\exists$-definably inverse.

2.4.4. coordinatisation isomorphism.
The construction is described in [G], p. 93 (for $n=2$). The isomorphism itself is a Δ-interpretation; it makes (proj. coord. ring, $\mathbb{P}^n$) into an $\exists$-definably inverse pair.

2.5. Projective closure and hyperplane removal.

2.5.1. projective closure of affine n-space $\mathbb{A}^n = \mathbb{A}^n(k)$.
Adjoining the hyperplane $\mathbb{P}^{n-1}(k)$ of directions in $\mathbb{A}^n(k)$ (2.3.1) to $\mathbb{A}^n$, with as new hyperplanes (i) the $\mathbb{P}^{n-1}$ adjoined and (ii) the hyperplanes of $\mathbb{A}^n$ with the directions of all lines there in adjoined, we obtain a structure which is canonically isomorphic to $\mathbb{P}^n(k)$ as constructed in (2.3.1) from $\mathbb{A}^{n+1}(k)$.

2.5.2. Given $\mathbb{P}^n = \mathbb{P}^n(k)$, we may take the C_{n-1}-substructure obtained by removing any hyperplane from $\mathbb{P}^n$. The result is isomorphic to $\mathbb{A}^n(k)$.

2.5.3. These operations are definably inverse interpretations; the definitions

are easily worked out from those of the constructions. If we adjoin a hyperplane
to A^n and then remove a different one, $\psi_{\cong}$ must describe on A^n an automorphism
of P^n taking one hyperplane to the other.

2.5.4. The ordered case gives no further difficulties, see [HD], p. 150.

2.6. General affine coordinate ring and coordinatisation.

2.6.1. linear case.
Given a Desarguean affine n-space $A^n = A^n(k)$, we could execute the projective
coordinate ring construction in the projective closure $P^n(k)$. If none of the
parameters $0, e_1 e_2 u_1 u_2$ lie in the infinite hyperplane, the construction restricts
to a Δ interpretation on A^n. This was developed (for the ordered case) by
Szczerba [Sz1], [ST]. We do not recover k, for one point is missing in $A^n(k)$
(the infinite point on $0e_1$). But k can be recovered by then applying the additive
analog of the quotient ring construction ("algebra of differences"), a Δ interpretation without parameters.

    Szczerba gives a Δ construction ψ of coordinates in A^n, again the restriction
of projective coordinatisation, which is an isomorphic embedding of A^n in $P^n(k)$,
that is, with properties analogous to $\psi_{\cong}$ in (1.3. *,**), except that the embedding
is onto the complement of a hyperplane in $P^n(k)$; the construction uses a system
of parameters $0, e_1, \ldots, e_n, u_1, \ldots, u_n$ as for projective co-ordinatisation. See [ST],
p. 166-167 (only $n=2$ is explicitly given). The arguments require that lines contain sufficiently many points; perhaps $\geqslant 5$ would do, we will simply require lines
to be infinite.

2.6.2. Using additional parameters which characterise the missing hyperplane, and
applying the appropriate projective transformation η (definable over k from these
parameters) to the coordinates produced by ψ, we obtain the commuting diagram

$$A^n \xrightarrow{\;2.6.1.\;} k \xrightarrow{\;2.1.1.\;} A^n(k)$$
$$\eta \circ \psi$$

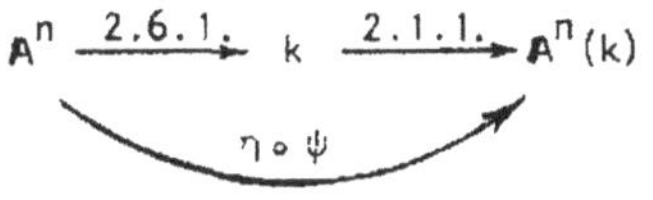

which satisfies (1.3. *), and $\eta \cdot \psi$ is an isomorphism defined by a Δ condition.
Nonetheless, (1.3. **) may not be satisfied in this way: Given an affine embedding
$A^n(k) \rightarrow A^n(K)$, the parameters needed for η in $\eta \circ \psi(A^n(K))$ need not belong to
$A^n(k)$ at all; this precludes the existence of an $\exists$-formula $\psi_{\bar{p}}$, with the properties in (1.3. **). This occurs, for example, for the embedding of (3.3), regard-
less of whether k,K are model complete.

2.6.3. ordered case.

Using B as in (2.2.2), the general affine coordinate ring construction, gives the underlying ordered (skew) field k. But now Szczerba and Tarski realise a great gain: The field is already obtained from its positive elements, which are the points between 0 and u_1; and all arithmetic structure on these is determined by line constructions (the ordinary projective ones w.r.t. $0, e_1 e_2, u_1 u_2$) which remain entirely within the triangle $0u_1u_2$. So this Δ coordinate ring construction reconstructs k from any convex subset of $\mathbb{A}^n$. (That is, containing any points of $\mathbb{A}^n$ between contained points.) See [ST], §2, or [Sz1]; and also [Sp].

Szczerba defines a theory WGA_n, weak general (n-dimensional desarguean) ordered affine geometry, in which the coordinate ring construction is defined, and in which (the restriction of) projective coordinatisation determines an embedding of any model of WGA_n with coordinate ring k onto a nonempty convex open n-dimensional subset of $\mathbb{P}^n(k)$, preserving C_{n-1} and pairwise collinear separation S (which may be defined from B). We define the <u>general projective closure</u> of a model of WGA_n to be this embedding; it agrees with the ordinary projective closure on $\mathbb{A}^n(k)$. It may be observed that the general projective closure is a Δ interpretation.

Both [ST] and [Sz1] include in WGA_n the extension axiom

$$\forall xy\, \exists z\, [z \neq y \wedge Bxyz].$$

But all steps with existential import in general affine coordinate ring and coordinatisation constructions require the determination of intersections (C_1, C_{n-1}) within the closed convex hull of the coordinatisation parameters and point being co-ordinatised; these constructions remain Δ in WGA_n without the extension axiom. What fails, of course, are the openness statements in the representation theorems, but we will have no real use for these. For the reader's convenience, we stick to WGA_n; but it will be understood how the results are to be modified to accommodate the weaker theory.

3. MODEL THEORETIC CONSEQUENCES.

3.1. From this syntactic information about interpretations and inverse pairs of interpretations we have some immediate consequences. Let us set, for any theory T of skew fields, $\mathbb{P}^n(T) = Th(\{\mathbb{P}^n(k) : k \models T\})$, $\mathbb{A}^n(T)$ similarly, where the primitives are C_{n-1}, together with the relevant order primitive in the ordered case. Let $n \geq 2$.

<u>Proposition</u> (i) $\mathbb{P}^n(T)$ is model complete iff T is; (2.4.3)

(ii) $\mathbb{A}^n(T)$ is model complete iff T is, <u>if the quaternary parallelism</u>

relation is added to the language;

(iii) If T is model complete, $\mathbb{A}^n(T)$ has prefix dimension 2. (2.2.4)

Here (iii) follows directly from (ii), as we must simply substitute a universal formula for each occurrence of '$\|$' in the equivalent universal and existential formulas of (ii).

3.2. We show that in any infinite affine n-space, the relation $\|$ has no nonempty existentially definable subset; so (2.2) and 3.1 (ii) - (iii) are best possible. Existential undefinability of $\|$ was already shown by Diller [D] in the ordered commutative case, cf. (3.3).

Suppose in $\underline{A}^n$, ab $\|$ cd and these satisfy $\exists\bar{x}\phi(\bar{x},\bar{y})$, ϕ quantifierfree and $\underline{A}^n \models \forall\bar{y}[\exists\bar{x}\phi(\bar{x},\bar{y}) \rightarrow y_1y_2 \| y_3y_4]$. Then in the projective closure $\underline{P}^n$ of $\underline{A}^n$, we have an intersection e of ab with cd, and $\bar{x}$ such that $\phi(\bar{x}$,abcd). By infiniteness of $\underline{A}^n$, we may choose a hyperplane H in $\underline{P}^n$ not containing any of a,b,c,d,e,$\bar{x}$; in the ordered case we also require H to lie outside the convex hull in $\underline{A}^n$ of a,b,c,d,$\bar{x}$. Removing H from $\underline{P}^n$, we obtain $\underline{B}^n \cong \underline{A}^n$ in which ab intersects cd in e, but still $\phi(\bar{x}$,abcd), because a,b,c,d,$\bar{x}$ satisfy the same C_{n-1} and B relations as in $\underline{A}^n$.

□

3.3. Any $\{C_{n-1},B\}$-structure is also a $\{C_{n-1},S\}$-structure, by the definition: Sxyuv iff x,y,u,v are pairwise distinct, collinear, and

$$Bxuy \leftrightarrow (Bxyz \vee Bzxy).$$

An <u>embedding</u> between an ordered affine and projective space (of the same fixed dimension n) will be understood to be a $\{C_{n-1},S\}$ embedding between the affine $\{C_{n-1},S\}$-structure determined in this way and the projective structure. With these understandings in the ordered case, we have

<u>Corollary</u>. Let $n \geq 2$ and k an infinite (possibly ordered or skew) field. For any embedding $\mathbb{A}^n(k) \rightarrow \mathbb{P}^n(k)$ there is an embedding $\mathbb{P}^n(k) \rightarrow \mathbb{A}^n(K)$, for some K elementarily extending k, such that the composition map j: $\mathbb{A}^n(k) \rightarrow \mathbb{A}^n(K)$ is a $\{C_{n-1},B\}$-embedding (in the ordered case); no two parallels have parallel images under j.

Proof: The hyperplane removal argument of (3.2) shows that the diagram of $\mathbb{P}^n(k)$, together with the B-diagram on the image of $\mathbb{A}^n(k)$ is consistent with Th$(\mathbb{A}^n(k))$, which by (2.4.4) is $\mathbb{A}^n$(Th(k)). □

The argument in fact justifies taking K an ultrapower of k.
But no ring embedding $k \rightarrow K$ induces j via the Δ interpretation $\mathbb{A}^n$(2.1.1-3); for then parallelism would be preserved. Further examination of the argument of (3.2) would show how to preserve parallelism in n-i directions and destroy it in the i

remaining (independent) ones. But a logic-free algebraic approach (3.10) gives this, together with tighter control of $k \to K$. Direct algebraic constructions have been given by Szczerba (cf. [Sch] §6, 59-61) and (for the commutative, ordered, plane case) by Diller [D].

3.4. model companions, linear case.

A theory T is model consistent with a theory T^* if every model of T may be embedded in a model of T^*. T^* is the model companion of T if T^* is model complete and mutually model consistent with T.

Theorem. Let $n \geqslant 2$, and T, T^* theories of infinite skew fields. Then

(a) $\mathbb{P}^n(T)$ is the model companion of $\mathbb{A}^n(T)$ iff T is model complete.

(b) $\mathbb{P}^n(T^*)$ is the model companion of $\mathbb{A}^n(T)$ (and $\mathbb{P}^n(T)$) iff T^* is the model companion of T.

Proof: equivalence of model completeness is (3.1.i), so we need only the equivalences between model consistency statements. These are easily worked out using the constructions of §2 ; that an embedding $\mathbb{A}^n(k) \to \mathbb{P}^n(K)$ induces an embedding $k \to K$ uses general affine coordinatisation (2.6), or (3.8).

□

3.5. In the ordered case, the claims of theorem (3.4) may only be made by stretching the accepted definition of model consistency: the projective and affine order primitives are different, so we do not really have embeddings. Nor can we force the issue by extending the definition of betweenness to projective closures of affine spaces (putting the point at infinity at an extreme point of each line); for while this can be done consistently with the universal theory of $\{C_{n-1}, B\}$ affine n-space -this is what implicitly happens in the embedding of (3.3)- the resulting structures simply are not model complete.

Still, (3.3) gives more than just mutual model consistency of $\mathbb{P}^n$ and $\mathbb{A}^n$ as $\{C_{n-1}, S\}$-structures: it allows the basic alternating chain arguments derived from mutual model consistency (see [M], §3.3) to be made between projective and affine structures, with the extension chain on the affine side consisting of $\{C_{n-1}, B\}$-embeddings, and B,S consistently related as in (3.3). One alternative in this situation is to name this type of mutual model consistency, intermediate between mutual model consistency in $\{C_{n-1}, S\}$ and in $\{C_{n-1}, B\}$, weak mutual model consistency; and correspondingly to speak of weak model companions and completions. By the familiar alternating chain argument, weak model companions and completions (in language $\{C_{n-1}, S\}$, w.r.t. the definition of S in (3.3)) are unique. We do this for the rest of this section. In particular, the argument of (3.4) gives

Theorem. Let $n \geqslant 2$, and T, T^* theories of ordered (skew) fields. Then

(a) $\mathbb{P}^n(T)$ is the weak model companion of $\mathbb{A}^n(T)$ iff T is model complete.

(b) $\mathbb{P}^n(T^*)$ is the weak model companion of $\mathbb{A}^n(T)$ iff T^* is the model companion of T.

On the other hand, we see that if there is a model companion of $\mathbb{A}^n$ in the $\{C_{n-1},B\}$ language, it too is weakly mutually model consistent with the model complete $\mathbb{P}^n$'s, and by the alternating chain companion structures would have to be model complete <u>projective</u> spaces. As model companion of $\mathbb{A}^n$'s, they would have to satisfy the extension axiom $\forall xy\, \exists z \neq y\; Bxyz$. Such <u>projective betweenness spaces</u> satisfying the extension axiom were discovered by Szczerba [Sz2] in another context. We study them further in §4.

3.6. Weak general affine geometry. [Sz1]

For any theory T of <u>ordered</u> (skew) fields, let

$$WA^n(T) = Th(\{\underline{A} \models WGA_n;\ \underline{A} \text{ has coordinate ring } k \models T\})$$
$$= WGA_n \cup I^*(T)$$

where I is the weak general affine coordinate ring construction (2.6); taking parameters into account, $I^*(T)$ abbreviates

$$\{\forall \bar{y}[\psi_p(\bar{y}) \rightarrow I^*(\phi)]: \phi \in T\}.$$

The arguments of (3.2-5) may be extended to this more general class of theories. First, projective closure may be replaced by general projective closure (2.6): any model $\underline{A} \models WA^n(T)$, say with coordinate ring k, is embedded as a $\{C_{n-1},S\}$-structure in $\mathbb{P}^n(k) \models \mathbb{P}^n(T)$. Second, the supplement for the ordered case in (3.3) may be recovered by a consideration about finitely generated convex subsets of $\mathbb{P}^n$ (We avoid defining these, as we use only very simple properties, and anyhow only need to consider images of finitely generated convex subsets of $\underline{A}$: smallest sets containing the generators and containing any point between two contained points.)

<u>Lemma</u>. Any finitely generated convex subset of $\mathbb{P}^n$ misses a hyperplane of $\mathbb{P}^n$.

Proof: induction on n, trivial for $n=1$. For $n>1$, take any point x not in the subset; (a hyperplane through one of the generating points must contain such x, for if no hyperplane through generator g contains a point not in the set, the generator may be deleted); projecting to $\mathbb{P}^{n-1}$ through x we obtain a convex subset of $\mathbb{P}^{n-1}$ with the same generators. This misses a hyperplane of $\mathbb{P}^{n-1}$, with inverse image a hyperplane of $\mathbb{P}^n$ outside the set. □

<u>Theorem</u>. Let $n \geq 2$, and T,T^* be theories of ordered (skew) fields.

(a) $\mathbb{P}^n(T^*)$ is the weak model companion of $WA^n(T^*)$ iff T^* is model complete.

(b) $\mathbb{P}^n(T^*)$ is the weak model companion of $WA^n(T)$ iff T^* is the model companion of T.

3.7. Hyperbolic geometry ([ST], Example 6.3).

For ordered (skew) fields k, $n\geqslant 2$, let $H^n(k)$ be the (result of) the Δ interpretation which gives the restriction of $\mathbb{A}^n(k)$ to the interior of the unit hypersphere; and for theories T of ordered (skew) fields, let $H^n(T)$ be the associated theory of restricted affine planes, defined as in (3.1). These are just the affine reducts of Klein models of hyperbolic geometry over ordered (skew) fields k, and their theories. They satisfy the axioms of WGA_n. Thus Theorem (3.6) extends to $H^n(T), H^n(T^*)$ once we show that $\mathbb{P}^n(T)$ is model consistent with $H^n(T)$ for any theory T of ordered skew fields. But this is a trivial refinement of the ordered case of (3.3): once we have removed the hyperplane from $\mathbb{P}^n(k)$ avoiding the convex hull of the finitely many given points, we may find a hypersphere in $\mathbb{A}^n(k)$ including the given points; so the image of $\mathbb{P}^n(k)$ in $\mathbb{A}^n(K)$ is within a hypersphere of $\mathbb{A}^n(K)$.

In particular, real projective space is the weak model companion of real affine hyperbolic space. But this is a red herring, as the linear (C_{n-1}) structure of $\mathbb{P}^n$ is Euclidean. The proper conclusion is that while the affine structure fully determines the structure of hyperbolic space in terms of first-order definability or invariance under automorphisms, the congruence hyperbolic structure is not robust under affine morphisms, and should be studied with an additional primitive.

3.8. Let an affine space (for implicitly fixed dimension n) be (a) in the linear case, $\mathbb{A}^n(k)$ for some (skew) field k, or (b) in the ordered case, a model of WGA_n, hence necessarily with coordinate ring an ordered (skew) field; in language C_{n-1} resp. $\{C_{n-1}, B\}$.

Theorem. Let $j: \underline{A} \to \underline{A}'$ be an embedding of affine spaces with all lines infinite. Then j extends uniquely to an embedding $\hat{j}: \hat{A} \to \hat{A}'$ of the (weak) projective closures (up to an automorphism of $\underline{\hat{A}}'$ over $\underline{\hat{A}}$).

We originally obtained a self-contained synthetic proof of this result- even for $j: \mathbb{A}^n(k) \to \mathbb{A}^n(k')$ it is not trivial, for if j does not preserve parallelism, affine coordinatisation of $\mathbb{A}^n$ does not commute with j. The theorem has also been shown by Carter and Vogt [CV]. for the linear plane case.

The arguments for $n=2$ and $n>2$ are genuinely different, at least in our proof, as $n=2$ uses converse Desargues and $n>2$ a projection technique, both of which fail in the other case.

For brevity, we use another approach here, unfortunately not self-contained: The general projective closure (2.6) is a Δ interpretation. Therefore, j simply induces the embedding of general projective closures! (1.1) (Ordinary projective closure is not Δ; it implicitly determines the location of the hyperplane at infinity.)

3.9. A prime model of a theory is one which is embeddable in all other models of the theory; a prime model of T over a structure M is a prime model of T ∪ Diagram(M) Here all primitives on M should be among those of T; if M is ordered affine and T is ordered projective, we make the convention that Diagram(M) be taken in the $\{C_{n-1}, S\}$-language following (3.3).

Corollary. Let $n \geq 2$; T, T^* theories of infinite (possibly ordered or skew) fields.

(a) $\mathbb{P}^n(k)$ is the prime model of the theory of Desarguean projective spaces over any affine space with (general affine) coordinate ring k.

(b) If there is a prime model of T^* over any model k of T, then (and only then) is there a prime model of $\mathbb{P}^n(T^*)$ over any affine space with coordinate ring a model of T.

Proof: general affine coordinatisation on the affine space A agrees with projective coordinatisation on the overlying $\mathbb{P}^n(K)$ and induces the factorisation of any given embedding into $A \to \mathbb{P}^n(k) \to \mathbb{P}^n(K)$, where in (b), k is the prime model of T over the coordinate ring of A. Conversely in (b), $k \to K$ induces $\mathbb{A}^n(k) \to \mathbb{P}^n(K)$, and using the prime projective space $\mathbb{P}^n(K')$ over $\mathbb{A}^n(k)$ in $\mathbb{P}^n(T^*)$ this factors and induces $K' \to K$, so K' is prime over k.

□

The model completion T^* of T is a model companion such that $T^* \cup$ Diagram(M) is complete for any $M \models T$. In our ordered case, we speak of weak model completion, continuing to take Diagram(M) in language $\{C_{n-1}, S\}$. Thus a (weak) model companion of T with prime models over models of T is the (weak) model completion.

Corollary. n, T, T^* as above.

(a) $\mathbb{P}^n(T)$ is the model completion of $\mathbb{A}^n(T)$ (the weak model completion of $\mathbb{A}^n(T)$ and $W\mathbb{A}^n(T)$) iff T is model complete.

(b) $\mathbb{P}^n(T^*)$ is the model completion of $\mathbb{A}^n(T)$ (the weak completion of $\mathbb{A}^n(T)$ and $W\mathbb{A}^n(T)$) if T^* is the model completion of T and T^* has a prime model over any model of T.

With more detailed analysis (partially avoidable in the linear case) we will eliminate the bothersome restriction in (b). But all examples of model completions of commutative field theories appear to be covered by the present statement. In particular, the theory of $\mathbb{P}^n(\mathbb{C})$ is the model completion of the theory of Pappian affine n-spaces (= $\mathbb{A}^n$ (commutative fields)); in the ordered case, $\mathbb{P}^n(\mathbb{R})$ is the weak model completion of Pappian ordered affine n-spaces, and of Pappian weak general affine n-spaces, also without extension axiom. This last result is quite strong; the topological primitive B allows a local result -the models of $WA^n(\mathbb{R})$ are convex open n-dimensional neighborhoods of $\mathbb{A}^n_n(\mathbb{R})$;- this suggests looking for conditions under which projective closure commutes with glueing in real manifolds, so as to consider model completion phenomena among such topological structures.

3.10. We now study embeddings $\mathbb{A}^n(k) \to \mathbb{A}^n(K)$, $\mathbb{P}^n(k) \to \mathbb{A}^n(K)$ algebraically. Assume first we have $\mathbb{P}^n(K)$ coordinatised, with parameters $0 = (1,\bar{0})$, $e_i = (1:\bar{0}:1_i:\bar{0})$, $u_i = (0:\bar{0}:1_i:\bar{0})$, $i = 1\ldots n$. Suppose we want to fix $(0,\bar{e})$, but move u_i to u_i^*, where $u_i^* = u_i$ or $u_i^* = (1:\bar{0}:\alpha_i:\bar{0})$, $\alpha_i \neq 0,1$. This is accomplished by the map $\eta(\theta_1,\ldots,\theta_n)$:

$$\begin{cases} (\pi_0:\ldots\pi_n) \mapsto (\pi_0 + \Sigma\pi_i.(\Theta_i - 1): \ \ldots : \pi_i\Theta_i : \ldots), \text{ all } \pi_i \in k, \\ \Theta_i = 1, \text{ if } u_i^* = u_i, \\ \Theta_i = \alpha_i(\alpha_i - 1)^{-1} = 1 + (\alpha_i - 1)^{-1}, \text{ if } u_i^* = (1:\bar{0}:\alpha_i:\bar{0}). \end{cases}$$

Here $1,\Theta_1,\ldots,\Theta_n$ are the eigenvalues associated with eigenvectors $0,e_1\ldots e_n$ of a K-linear transformation of $\mathbb{A}^{n+1}(K)$ which induces η, which is therefore evidently an automorphism of $\mathbb{P}^n(K)$. Conversely, for any $\Theta_1\ldots\Theta_n \neq 0$ in K one solves for x_i's such that $\eta(\Theta_1\ldots\Theta_n)$ produces a transformation of the type described initially.

Also, $\eta(\theta_1,\ldots,\theta_n)^{-1} = \eta(\theta_1^{-1},\ldots,\theta_n^{-1})$, and $\theta_i^{-1} - 1 = -\alpha_i^{-1}$.

Now let j: $\underline{A} \to \underline{A}^*$ be an arbitrary embedding of affine n-spaces. Choosing parameters 0, $\bar{e}$ in $\underline{A}$ and forming projective closures $\underline{P}$, $\underline{P}^*$, we obtain projective parameters $0,\bar{e},\bar{u}$, u_i the ideal point on $0e_i$ in $\underline{P}$ (resp. u_i^* in $\underline{P}^*$). We have a diagram

$$\begin{array}{ccccccc} (\underline{A},0,\bar{e}) & \xrightarrow[(2.5)]{} & (\underline{P},0,\bar{e},\bar{u}) & \xrightarrow[(2.4)]{\quad} & (k,0,e_1,+,\cdot) & \longrightarrow & (\mathbb{P}^n(k),0,\bar{e},\bar{u}) \\ \Big\downarrow j & & \Big\downarrow \hat{j}\ (3.8) & & \Big\downarrow \hat{\hat{j}}\ (2.4,1.1) & & \Big\downarrow \hat{\hat{j}}\ (2.4.3,1.1) \\ (\underline{A}^*,0,\bar{e}) & \xrightarrow[(2.5)]{} & (\underline{P}^*,0,\bar{e},\hat{j}(\bar{u})) & \xrightarrow[(2.4.)]{\quad} & (k,0,e_1,+,\cdot) & \longrightarrow & (\mathbb{P}^n(k),0,\bar{e},\hat{j}(\bar{u})) \end{array}$$

In general, $\hat{j}(\bar{u}) \neq \bar{u}^*$(3.3). We may describe $\hat{j}$ as obtained from $(\underline{A},0,\bar{e})$ and an embedding $\hat{\hat{j}}$: $k \to K$, k its (projective) coordinate ring, by removal of a suitable hyperplane, spanned by points $\bar{u}^*$, say with coordinates $u_i^* = (1:\bar{0}:\alpha_i:\bar{0})$ w.r.t. $0,\bar{e},\hat{j}(\bar{u})$ if $u_i^* \neq \hat{j}(u_i)$. If we had an embedding $\underline{P} \to \underline{A}^*$ of a projective n-space $\underline{P}$, the same analysis would hold after picking a full parameter set $0,\bar{e},\bar{u}$ in $\underline{P}$.

Theorem. Let $n \geq 2$, $\underline{A} = \mathbb{A}^n(k)$, $\underline{A}^* = \mathbb{A}^n(K)$; k,K infinite (possibly ordered or skew) fields; notation further as above.

1. Relative to a choice $(0,\bar{e})$ of affine parameters, the embeddings $\underline{A} \to \underline{A}^*$ are exactly the maps obtained from embeddings $k \to K$ by removal of the hyperplane $\eta(\Theta_1 \ldots \Theta_n)[\bar{u}] = \bar{u}^*$ from $\mathbb{P}^n(K)$, for $\alpha_1 \ldots \alpha_n \in K$ such that (writing $k^* = k \setminus 0$)
   (i) linear case: $\sum_{i=1}^{n} p_i(\alpha_i^{-1}) \notin k^*$, all $p_1,\ldots,p_n \in k$;
   (ii) ordered case: the Θ_i all belong to the k-monad of 1 in K (i.e., no Θ_i is bounded away from 1 in K by an element of k),

   up to automorphisms induced by those of K over the given embedding.
2. An embedding $\underline{A} \to \underline{A}^*$ extends to $\mathbb{P}^n(k) \to \underline{A}^*$ iff also
   (iii) $1, \alpha_1^{-1}, \ldots \alpha_n^{-1}$ are left k-linearly independent.

Proof: (i) given $\underline{A} \to \mathbb{P}^n(k) \to \mathbb{P}^n(K)$, the hyperplane $H(\bar{u}^*)$ spanned by $\bar{u}^* = \eta[\bar{u}]$ misses $\underline{A}$ iff $\eta^{-1}[\underline{A}]$ misses $H(\bar{u})$ iff for all $p_1 \ldots p_n \in k, p_0 \in k^*$, $p_0 - \sum_{i=1}^{n} p_i \alpha_i^{-1} \neq o$ (because $\Theta_i^{-1} - 1 = \alpha_i^{-1}$). (ii) In the ordered case, we obtain a B-embedding iff no points of $\underline{A}$ are separated by $H(\bar{u})$ and $H(\bar{u}^*)$ iff each line of $\underline{A}$ intersects $H(\bar{u}^*)$ at a k-infinite parameter value $\lambda \in K$ (in the formula for $\lambda - 1$ in 4.2, with $p_c \neq o$) iff all α_i are k-infinite iff all θ_i are in the monad of 1 in K. (iii) For $(\underline{P}, 0, \bar{e}, \bar{u}) \to \underline{A}^*$, $H(\bar{u}) \cap \underline{P}$ should also have η^{-1}-image missing $H(\bar{u})$. η^{-1} maps $(0: \ldots p_i \ldots)$ to $H(\bar{u})$ iff $\sum_{i=1}^{n} p_i \alpha_i^{-1} = o$; given (i), this fails iff $\Sigma p_i \alpha_i^{-1} \notin k$ iff (iii) holds. □

For example, we have $\mathbb{P}^i(\mathbb{R}) \to \mathbb{A}^i(\mathbb{C})$, $i = 1,2$, but $\mathbb{P}^3(k) \to \mathbb{A}^3(K)$ requires an extension of degree at least 3. For the unordered plane case, a complete analysis of these embeddings including the exceptional cases for small finite planes is given by Carter and Vogt [CV].

3.11. A theory T has <u>amalgamation</u> if any two extensions of a model of T to models of T have a common extension to a model of T such that the embeddings commute. Suppose we are given an ∃-definably inverse pair of interpretations (1.3) on T_1, <u>both</u> of which are Δ interpretations, in T_1 resp. $T_2 = Th(I_{12}(Mod(T_1)))$. Then T_1 has amalgamation iff T_2 has amalgamation; for we may convert the given extensions in (say) T_1 to extensions in T_2; amalgamate there, and the amalgamating extensions induce T_1 extensions, which by ∃-definably inverseness serve to solve the original amalgamation problem. This gives all but (1,2 ⇒ 3) of

<u>Theorem</u>. For a theory T of (ordered; skew) fields, $n \geq 2$, the following are equivalent:

1. T has amalgamation

2. $\mathbb{P}^n(T)$ has amalgamation
3. $\mathbb{A}^n(T)$ has amalgamation.

Proof: (1) $\Rightarrow$ (3). Given embeddings of affine spaces, say

$$\mathbb{A}^n(k) \to \mathbb{A}^n(k_1),\ \mathbb{A}^n(k) \to \mathbb{A}^n(k_2),$$

we may amalgamate the field extensions induced by taking the projective closures (3.8) of the given affine embeddings, and a system of coordinatisation parameters $0, e_1, \dots e_n$ from $\mathbb{A}^n(k)$, $u_1, \dots, u_n$ from $\mathbb{P}^n(k)$, the points at infinity on $0e_1, \dots, 0e_n$. By (3.10), the given affine embeddings arise from the canonical ones over the field extensions by†automorphisms determined by $\Theta^1 = (\Theta_1^1, \dots \Theta_n^1)$ resp. Θ^2, satisfying (3.10.i) or (3.10.ii), w.r.t. $k \to k_1$ resp. $k \to k_2$. Making sure that the amalgamating field K contains an element t

(i) in neither k_1 nor k_2, and

(ii) in the ordered case, in the k_1 and the k_2-monad of 1 in K,

both of which may certainly be accomplished by moving to an elementary extension of the amalgamating field if necessary, and embedding $\mathbb{A}^n(k_1) \to \mathbb{A}^n(K)$ resp. $\mathbb{A}^n(k_2) \to \mathbb{A}^n(K)$ by the canonical embedding followed by†the automorphism determined by $((\Theta_1^1)^{-1}t, \dots, (\Theta_n^1)^{-1}t)$, resp. $((\Theta_1^2)^{-1}t, \dots, (\Theta_n^2)^{-1}t)$, the amalgamation is accomplished by (3.10. i,ii). □

†suppression of the hyperplane determined by

3.12. If T* is the model companion of T, then T* is the model completion of T iff T has amalgamation, as one verifies easily from mutual model consistency; and <u>weak</u> mutual model consistency (3.5) in our ordered case gives exactly what is needed for the "weak" analog of this argument (with T in language $\{C_{n-1}, B\}$ for amalgamation). Thus (3.9) may be completed:

<u>Corollary</u>. n,T,T* as in (3.9).

<u>$\mathbb{P}^n(T^*)$ is the (weak) model completion of $\mathbb{A}^n(T)$ iff T* is the model completion of T.</u>

Wheeler [W] has shown that for a <u>universal</u> theory of (ordered or unordered) commutative fields with amalgamation, existential closure coincides with real resp. algebraic closure; therefore these cases are already covered by (3.9). It would be interesting to have an example of a model completion of some theory of non commutative skew fields; especially one where the model completion does not have prime models over the models of the original theory. (The first type probably arises by the Δ interpretations associated with finite dimensional algebras over commutative fields; the second type could require a genuinely new example.) While these results on model completions may apply in relatively few cases, certainly the discussion of model companions (3.4) is widely applicable; Macintyre, McKenna and van den Dries [MMD] quote a result from McKenna's thesis, that there are very many model complete theories of fields.

Van den Dries has asked whether, say in the result that projective planes over algebraically closed commutative fields are the model completion of affine planes over commutative fields, the latter theory could be weakened. The most attractive candidate would of course be the universal part of that theory. But a model completion of a universal theory has elimination of quantifiers ([M], p. 155), and $\mathbb{P}^n$ (alg. closed) does not, in this language. In the ordered case, one does a bit better, getting most results for weak general affine geometry.

3.13. We show that in any infinite Desarguean affine or projective n-space, the relation $C_{i+1}, i+1<n$, has no nonempty subset universally definable in $C_i,..C_1$, so that we must indeed take C_{n-1} as a primitive if model completeness is to be preserved. The argument, for $i=1$, was used in the affine case by Kordos [K] to show non $\forall\exists$-axiomatisability of the lower dimension axiom for $n>2$.

First consider $\mathbb{P}^n(k)$, and suppose $\sim C_{i+1}\, a_0 \ldots a_{i+2}$, and these satisfy $\exists\bar{x}\,\phi(\bar{x}\bar{y})$, ϕ quantifierfree, and $\mathbb{P}^n \models \forall\bar{y}[\exists\bar{x}\,\phi(\bar{x}\bar{y}) \rightarrow \sim C_{i+1}\,\bar{y}\,]$, ϕ in $C_i,..C_1$, or in the ordered case, $C_i,..C_1$ and S. By infiniteness, we may choose a point $p\in\mathbb{P}^n$ which lies neither (i) on a line through any two of $a_0,\ldots a_{i+2}, x_1\ldots,x_t$; nor (ii) on an (n-1) hyperplane spanned by n of these points. Then projection through p gives a $\{C_{n-2},S\}$-homomorphism $\mathbb{P}^n(k) \rightarrow \mathbb{P}^{n-1}(k)$ which is injective on the given points $a_0,\ldots,x_t$. Proceeding inductively, we repeat this until we obtain a $\{C_i,S\}$-homomorphism on $\mathbb{P}^{i+1}(k)$, then embed as a subspace in $\mathbb{P}^n(k)$. For the images $\hat{\bar{x}},\hat{\bar{a}}$, we have

$$\phi(\hat{\bar{x}},\hat{\bar{a}}) \wedge C_{i+1}\,\hat{\bar{a}}.$$

a contradiction.

Next, for $\mathbb{A}^n(k)$, we embed in $\mathbb{P}^n(k)$ and proceed as above, removing a hyperplane avoiding the given points in $\mathbb{P}^{i+1}(k)$ to get the image in $\mathbb{A}^{i+1}(k)$. In the ordered case, we must make all our choices (p, and the hyperplane to be removed) so as to avoid the image of the convex hull in $\mathbb{A}^n(k)$ of the given points; this is conveniently done using the fact that some hyperplane in projective space always avoids a finitely generated convex set (3.6); p can be chosen in such a hyperplane, and in the end it may be removed.

3.14. Note that (3.11) applies to skew fields, for they amalgamate. [C]
The alternating chain intended in (3.5), referred to in §4: all $k_i \models T^*$.

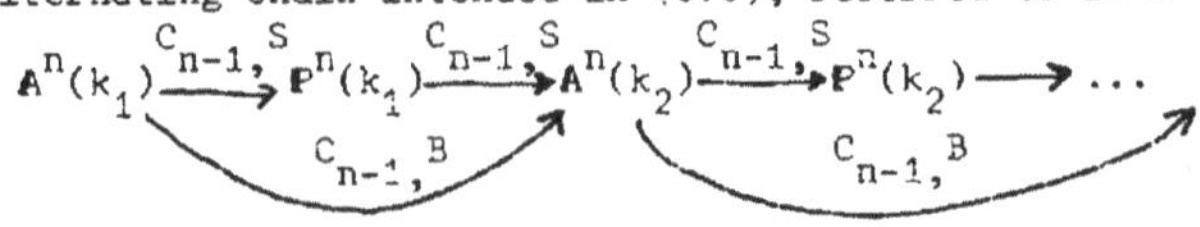

$\mathbb{A}^n(k_1)$
C_{n-1}, S
$\mathbb{P}^n(k_1)$
C_{n-1}, S
$\mathbb{A}^n(k_2)$
C_{n-1}, S
$\mathbb{P}^n(k_2)$
...
C_{n-1}, B
C_{n-1}, B

4. PROJECTIVE BETWEENNESS SPACES.

The alternating chains argument suggested in (3.5) would show that the model companions of affine ordered spaces $\mathbb{A}^n(T)$ in the language $\{C_{n-1},B\}$ would have to contain at least, for some model complete theory of ordered skew fields T^*,

(i) $\mathbb{P}^n(T^*)$, in language $\{C_{n-1},S\}$,

(ii) The $\forall\exists$-part of $\mathbb{A}^n(T^*)$, in language $\{C_{n-1},B\}$,

and the definition of S in terms of B; by that argument, all this is mutually model consistent with $\mathbb{A}^n(T)$.

4.1. In order to investigate model completeness of this theory, we develop the representation theory for <u>projective betweenness n-spaces</u>, defined by

4.1.1. Desarguean projective n-space axioms (C_{n-1} only),

4.1.2. $\forall xyz\ C_1xyz \leftrightarrow Bxyz \vee Byzx \vee Bzxy$,

4.1.3. order axioms for B:

$\forall xy\ Bxyx \rightarrow x=y$, $\forall xyzu\ Bxyz \wedge Byzu \wedge y\neq z \rightarrow Bxyu$,

$\forall xyzu\ Bxyz \wedge Bxyu \wedge x\neq y \rightarrow Byzu \vee Byuz$,

4.1.4. Pasch, which by 4.1.1. may be written

$\forall xyzuvw\ Bxyz \wedge Buvy \wedge C_1wvz \wedge C_1xwu \rightarrow Bwvz \wedge Bxwu$,

4.1.5. $\forall xy\exists z\ Bxyz \wedge z\neq y$ (extension axiom).

C_1 is an abbreviation if $n>2$. 4.1.1-4 together with the definition of S from B (3.3) entail the theory of Desarguean projective n-space in $\{C_{n-1},S\}$. Verifying the ordinary $\forall\exists$-form of Pasch ([Sz1]A5), one sees that 4.1.1-5 entail the axioms of WGA_n and its Euclidean strengthening WEA_n (op. cit.).

4.2. If $\underline{A}$ is an affine space (3.8), $\underline{A} \models WGA_n$ (even without extension axiom), with coordinate ring k, then there is a $\{C_{n-1},B\}$-embedding $\underline{A} \rightarrow \mathbb{A}^n(K)$ for some K elementarily extending k by (3.6). Thus if $\underline{A} = \underline{P}^n_B$ is a projective betweenness space, we may regard $\underline{A}$ as $\mathbb{P}^n(k)$ endowed with a B-relation by the $\{C_{n-1},\text{-}S\}$-embedding $\mathbb{P}^n(k) \rightarrow \mathbb{P}^n(K)$ followed by removal of a hyperplane H, as in (3.10).

Choosing a set of parameters $(0,\bar{e},\bar{u})$ in $\underline{A}$, the lines $0e_iu_i$ intersect H in points $(1:\bar{0}:\alpha_i:\bar{0})$, $\alpha_i \in K\setminus k$. Defining Θ_i from α_i as in (3.10), we have (3.10.iii), that $1,\alpha_1^{-1},\ldots\alpha_n^{-1}$ must be left k-linearly independent.

However, the extension axiom for $\underline{P}^n_B$ requires now that no point in $\underline{P}^n_B$ be the last point in $\underline{P}^n_B$ on any line in $\underline{P}^n_B$. Choose a pair of points $p,r\in\mathbb{P}^n(k)$, $p\notin k.r$ i.e., distinct, say

$p = (p_0:\ldots p_i\ldots)$, $p_0 = \sum_{i>0} p_i$ (so $p\in H(\bar{e})$, hyperplane through $\bar{e}$),

$$r = (0 : \ldots r_i \ldots) \qquad \text{so } r \in H(\bar{\alpha}) ,$$

and some canonical normalisation fixing the ratio $p_i:r_i$. Then we parametrize the line pr of $\underline{P}^n_B$ in $\mathbb{P}^n(K)$ as

$$x = p + (\lambda-1)r \quad , \lambda-1 \in \mathbb{P}^1(K)$$

with H in particular given by

$$x_i = e_i + (\alpha_i - 1)u_i \qquad i = 1\ldots n,$$

and the line pr intersects H for

$$\lambda-1 = (p_0 - \Sigma p_i \alpha_i^{-1}) \cdot (\Sigma r_i \alpha_i^{-1})^{-1}.$$

which is defined and not in k by k-linear independence. The extension axiom is satisfied on pr iff $\mathbb{P}'(k)$ contains no point closest to $\lambda-1$.

This holds for all pr iff

(i) α_i is <u>archimedean</u>, $i = 1\ldots n$ $(\exists \gamma_1 \gamma_2 \in k: 0 < \gamma_1 < |\alpha_i| < \gamma_2)$, and

(ii) for any $q = (q_0 : \ldots : q_n) \in \mathbb{P}^n(k)$,

$q_0 + \Sigma q_i \alpha_i^{-1}$ is bounded away from 0 by an element of k.

We abbreviate the latter condition as: $1, \alpha_1 \ldots \alpha_n$ are <u>strongly left k-linearly</u> independent. Violation of (i) violates the extension axiom on $e_i u_i$; given (i), violation of (ii) for $q = p$ or $q = r$ violates the axiom on pr; conversely, if (i,ii) hold, and $\ell \in \mathbb{P}'(k)$ is closest to $\lambda-1$, then $\ell \neq 0, \infty$, and $\Sigma r_i \alpha_i^{-1}$ is archimedean, $\ell \sim \lambda-1$ $(\forall \varepsilon > 0 \text{ in } k\, |\ell - (\lambda-1)| < \varepsilon)$, so

$$\ell \cdot \Sigma r_i \alpha_i^{-1} \sim p_0 - \Sigma p_i \alpha_i^{-1}$$

$$0 \sim p_0 - \Sigma(\ell r_i + p_i)\alpha_i^{-1}, \text{ contradicting (ii).}$$

4.3. Conversely, let an extension $k \to K$ be given, with $\alpha_1 \ldots \alpha_n \in K$ satisfying (4.2. i,ii); define $\mathbb{P}^n_B(k, \alpha_1 \ldots \alpha_n)$ as $\mathbb{P}^n(k)$ endowed with the B-relation induced by the maps $\mathbb{P}^n(k) \to \mathbb{A}^n(K) \to \mathbb{P}^n(K)$ obtained from the canonical embedding $\mathbb{P}^n(k) \to \mathbb{P}^n(K)$ by removal of the hyperplane H determined by $\alpha_1 \ldots \alpha_n$ as in (4.2). First (4.2. ii) guarantees that H misses $\mathbb{P}^n(k)$; then this does make $\mathbb{P}^n(k)$ into a $\{C_{n-1}, B\}$-structure. As a $\{C_{n-1}, B\}$-substructure of $\mathbb{A}^n(K)$, it satisfies the universal axioms (4.1.2-4). As argued in (4.2), conditions (i,ii) now guarantee the extension axiom (4.1.5).

The models of WGA_n (resp. WGA_n without extension axiom) are the n-dimensional open B-convex (resp. n-dimensional B-convex) $\{C_{n-1}, B\}$-substructures of suitable $\underline{P}^n_B$, by the initial argument of 4.2, valid in this generality, together with the corresponding result of [Sz1] for $\{C_{n-1}, S\}$-embeddings in $\mathbb{P}^n(k)$. So the same analysis applies to these theories, replacing (4.2i,ii) in the case without extension axiom by the weaker (3.10. iii)

(iii) $1, \alpha_1^{-1} \ldots \alpha_n^{-1}$ are left-k-linearly independent.

<u>Theorem</u>. Let $n \geq 2$; k,K ordered skew fields.

(1) The projective betweenness n-spaces are exactly the structures $\mathbb{P}^n_B(k,\alpha_1 \ldots \alpha_n)$ obtained as above from $k \to K, \alpha_1 \ldots \alpha_n$ satisfying (4.2. i,ii).

(2) The models of WGA_n are exactly the n-dimensional B-convex open substructures of these.

(3) The models of WGA_n without extension axiom are exactly the n-dimensional B-convex substructures of $\mathbb{P}^n_B(k,\alpha_1 \ldots \alpha_n)$ obtained as above from $k \to K, \alpha_1 \ldots \alpha_n$ satisfying (iii).

Parts (1),(2) follow closely the representation theorems [Sz2, Thm. 3.1], [Sz1, Thm. 8]; however, the statements and proofs of these results are incorrect in that the author substitutes the weaker (iii) for (4.2. ii), inferring the extension axiom from (iii) by the lemma [Sz2. 1.1] that k-rational combinations of archimedean elements are archimedean. (This fails for differences.)

4.4. The representation theorem does not give the type of connection between skew fields and spaces which allows transfer of model completeness following §1. Nor could there be such a connection, for the coordinate rings of projective betweenness spaces do not form a first-order (EC_Δ) class: $\mathbb{R}$ cannot be the coordinate ring of a projective betweenness space (4.5), but by an alternating chain

$$\mathbb{A}^n(\mathbb{R}) \to \mathbb{P}^n(\mathbb{R}) \to \mathbb{A}^n(K_1) \to \mathbb{P}^n(K_1) \to \mathbb{A}^n(K_2) \ldots$$

we obtain a projective betweenness space with coordinate ring an elementary extension of $\mathbb{R}$.

The point is that projective betweenness spaces are correlated with ordered skew fields <u>with extra structure</u>; for example, after coordinatisation, the betweenness relation itself corresponds to some 3n-ary relation $\underline{R}_B$ on the coordinate ring. This gives a Δ construction of ordered skew fields-with-R_B from projective betweenness spaces such that the $\psi_{\simeq}$ of projective coordinatisation makes an $\exists$-definably inverse pair. (The construction of projective betweenness space from ordered skew field with R_B is evident.) Hence the model complete projective betweenness spaces correspond to the model complete skew fields-with-R_B satisfying the relevant conditions: the R_B-translations of (4.1.2-4.1.5).

4.5. This seems too complicated for useful analysis, and so we extract an additional axiom from our original goal of characterising model companions of theories $\mathbb{A}^n(T)$. In this case, we may replace R_B by something simpler, the cuts $\kappa_1 \ldots \kappa_n$ in k of $\alpha_1 \ldots \alpha_n$ from (4.2).

Given $\underline{P}^n_B$, we may choose the parameters $0,\bar{e},\bar{n}$ such that $B0e_iu_i$. Then the Δ coordinate ring construction may be extended to give $(k,\kappa_1 \ldots \kappa_n)$, by the Δ

definitions

$$a < \kappa_i \quad \text{iff} \quad B0u_i(1:\bar{0}:a:\bar{0}), \qquad (*)$$

where the coordinate term is eliminable in favor of a Δ description in terms of the Δ formula ψ_α. However, in general the structure $(K,\kappa_1\ldots\kappa_n)$ does <u>not</u> determine the B relation of $\underline{P}^n_B$.

From this definition of $<_i$ we see that by the extension axiom (4.1.5) the κ_i are <u>k-finite</u>, <u>open</u> cuts: They partition k into two nonempty intervals, both of which are open $(-\infty,\kappa_i),(<_i,\infty)$. Szczerba points out by this argument that no such structure has coordinate ring $k = \mathbb{R}$. If $\underline{P}^n_B$ satisfies the model companion of an $\mathbb{A}^n(T)$, the cuts κ_i are also <u>Dedekind</u>: their breadth is no greater than (hence, equals) that of the cut at 0 in k, i.e. $\{y-x:\ x<\kappa_i<y\}$ is downward cofinal in $(0,1]$ of k. (Perhaps any model complete projective betweenness space has all such B-definable cuts Dedekind ?) To see this, consider the alternating chain from mutual model consistency with $\mathbb{A}^n(T)$:

$$\ldots \to (\underline{P}^n_B(k_i),0,\bar{e},\bar{u}) \to \mathbb{A}^n(k_i') \to \underline{P}^n_B(k_{i+1}) \to \ldots$$

The images of $(0,\bar{e},\bar{u})$ in $\mathbb{A}^n(k_i')$ give rise to cuts $\kappa_1\ldots\kappa_n$ in k_i' by $(*)$; but these cuts simply indicate the position w.r.t. $0,\bar{e},\bar{u}$ of the infinite hyperplane in $\mathbb{P}^n(k_i')$, and hence are not open in k_i'. But then they must be Dedekind. Taking unions of chains,

$$\cup\,\underline{P}^n_B(k_i) \simeq \cup\,\mathbb{A}^n(k_i') \models \text{all } \kappa_i \text{ are Dedekind w.r.t. } 0,\bar{e},\bar{u},$$

as this statement is $\forall\exists$ and true at each i. By model completeness, the statement holds of $\underline{P}^n_B(k_0)$, which could have been chosen arbitrarily.

But if the cuts $\bar{\kappa} = \kappa_1\ldots\kappa_n$ in k are Dedekind, there is at most one way to make $\mathbb{P}^n(k)$ into a projective betweenness space with the κ_i given by $(*)$. For if there were two distinct such B relations on $\mathbb{P}^n(k)$, say

$$B \text{ given by } \alpha_1'\ldots\alpha_n' \text{ in } K',\ k\to K',\ \alpha_i \text{ in } \kappa_i\ \forall i$$

$$B' \text{ given by } \alpha_1''\ldots\alpha_n'' \text{ in } K'',\ k\to K'',\ \alpha_i'' \text{ in } \kappa_i\ \forall i$$

then without loss of generality we may assume that $k\to K'$ is an elementary extension, and therefore we may obtain a common extension $k\to K$ containing the α_i' and the α_i''. But now each $\alpha_i'-\alpha_i''\in K$ is k-infinitesimal because κ_i is Dedekind. Consider any line pr of $\mathbb{P}^n(k)$ as in (4.2). As both $\Sigma r_i(\alpha_i')^{-1}$ and $\Sigma r_i(\alpha_i'')^{-1}$ are k-finite nonzero, the parameter difference of the intersection points of H' resp. H'' with pr

$$(\lambda'-1)-(\lambda''-1) = (p_0-\Sigma p_i(\alpha_i')^{-1})(\Sigma r_i(\alpha_i')^{-1})^{-1} - (p_0-\Sigma p_i(\alpha_i'')^{-1})\cdot(\Sigma r_i(\alpha_i'')^{-1})^{-1}$$

is k-infinitesimal; but both $\lambda'-1$ and $\lambda''-1$ differ from any value in k by a k-finite amount by the extension axiom, and so give the same B. Note that this last argument also shows that if the cuts $\kappa_1 \ldots \kappa_n$ determined in a projective betweenness space w.r.t. one set $0, \bar{e}, \bar{u}$ of parameters are Dedekind, this will hold w.r.t. any other set $0, \bar{e}', \bar{u}'$ as well; thus we can speak of Dedekind projective betweenness spaces. This condition is expressable as an $\forall\exists$ statement in $\{C_{n-1}, B\}$.

This argument for uniqueness of projective betweenness spaces over k with given Dedekind cuts $\bar{\kappa}$ also shows that for Dedekind projective betweenness spaces with or without the extension axiom, strong left k-linear independence of $1, \alpha_1 \ldots \alpha_n$ in $\kappa_1 \ldots \kappa_n$ in K extending k is a property of $\kappa_1 \ldots \kappa_n$ in k; all extensions $k \to K$ and $\alpha_i \in K$ equally give dependence or independence in this sense. We say that $1, \kappa_1 \ldots \kappa_n$ are k-linearly independent cuts iff for any $(q_0 : \ldots : q_n) \in \mathbb{P}^n(k)$ the linear form $q_0 + \Sigma q_i x_i^{-1}$ is uniformly bounded away from 0 in some k-finite neighborhood of $(\kappa_1, \ldots, \kappa_n)$ in k^n; more formally iff k satisfies

$$\forall q_0 \ldots q_n \neq (0 \ldots 0)\ \exists\delta\ \exists x_i^- < \kappa_i < x_i^+ < 0,\ i = 1 \ldots n,$$

$$\forall i \forall y_i \in \{x_i^-, x_i^+\}\ q_0 + \Sigma q_i y_i^{-1} \text{ is bounded away from zero by } \delta.$$

This condition clearly entails the extension axiom in any structure $\underline{P}^n_B$ giving $(k, \bar{\kappa})$ which satisfies (4.1.1-4.1.4); if the κ_i are all Dedekind it is also necessary for the extension axiom, for if it were to fail for $\bar{\kappa}, q_0 \ldots q_n$, then some extension $k \to K$ has α_i in κ_i, $i = 1 \ldots n$ such that $q_0 + \Sigma q_i \alpha_i^{-1} = 0$, and then this linear form must be k-infinitesimal or zero for any $\alpha_1 \ldots \alpha_n$ in $\bar{\kappa}$ in any K extending k.

Augment the language of ordered rings by unary predicates for the cuts $\kappa_1 \ldots \kappa_n$; we continue to write these as '$x < \kappa_i$'. The theory of Dedekind $\bar{\kappa}$-independently ordered (skew) fields is the theory of ordered (skew) fields together with the statements that the κ_i are k-finite open k-linearly independent Dedekind cuts. By (4.3), projective betweenness spaces are obtained from such $(K, \bar{\kappa})$ by expansion by a suitable relation R_B. We now see that there is exactly one such expansion of a Dedekind $\bar{\kappa}$-independently ordered skew field, and so by Beth's theorem R_B is explicitly first-order definable in this theory. The diagram

$$(k, \bar{\kappa}) \to (k', \bar{\kappa}') \to (K, \alpha_1 \ldots \alpha_n);\ \alpha_i \text{ in } \kappa_i' \text{ so in } \kappa_i,$$

where $(K, \bar{\kappa}) \to (K', \bar{\kappa}')$ is an extension of Dedekind $\bar{\kappa}$-independently ordered skew fields, then shows that this definition has both an $\exists$ and a $\forall$ equivalent in this theory. (The definition could be worked out explicitly from the formula for the parameter value λ of $pr \cap H$ in (4.2), but we won't need it.) Augmenting the Δ interpretation $\mathbb{P}^n(k)$ by this definition, we have a Δ interpretation $\mathbb{P}^n_B(k, \bar{\kappa})$ con-

<u>structing Dedekind projective betweenness spaces from Dedekind $\bar{\kappa}$-independently ordered skew fields. This and the Δ interpretation $(k,\bar{\kappa})$ of projective coordinate ring construction augmented by (*) form an $\exists$-definably inverse pair in either order w.r.t. the appropriate projective ψ_{α}</u> (2.4.2-3).

For any theory T of ordered (skew) fields, let T_D = T + "Dedekind $\bar{\kappa}$-independently ordered". For any theory T' of Dedekind $\bar{\kappa}$-independently ordered skew fields, let $\mathbb{P}^n_B(T') = Th(\{\mathbb{P}^n_B(k,\bar{\kappa}) : (k,\bar{\kappa}) \models T'\})$. By alternating chains between $\mathbb{P}^n(T)$ and $\mathbb{A}^n(T)$ with limits giving projective betweenness spaces over the limit fields, $\mathbb{A}^n(T)$ is mutually model consistent with $\mathbb{P}^n_B(T_D)$, and for T_1, T_2 skew field theories, $(T_1)_D$ is model consistent with $(T_2)_D$ iff T_1 is model consistent with T_2. Together with the conclusions underlined above, this gives

<u>Theorem</u>.

1. The Dedekind projective betweenness spaces are exactly those obtained as $\mathbb{P}^n_B(k,\bar{\kappa})$ from Dedekind $\bar{\kappa}$-independently ordered skew fields $(k,\bar{\kappa})$.
2. $\mathbb{P}^n_B(k,\bar{\kappa})$ is model complete iff $(k,\bar{\kappa})$ is, for any Dedekind $\bar{\kappa}$-independently ordered skew fields $(k,\bar{\kappa})$.
3. $\mathbb{P}^n_B(T')$ is the model companion of $\mathbb{A}^n(T)$ iff T' is the model companion of T_D.
4. If $T \mapsto T^*$ is the Kaiser Hull operator, and T a theory of ordered skew fields, $(T^*)_D \subseteq (T_D)^*$.

Perhaps equality holds in (4.). In particular, one would conjecture that if T is the theory of real-closed fields, T_D is model complete. One should also be able to tell which real-closed fields can be expanded to models of T_D. By alternating chains between $\mathbb{P}^n(T)$ and $\mathbb{A}^n(T)$, this holds at least for any real-closed field $\underline{M} = \bigcup \underline{M}_i$ which can be obtained as a union of an increasing chain of extensions such that $\underline{M}_{i+1}$ contains $\underline{M}_i$-infinitesimals, i.e. $\underline{M}$ has an infinite decreasing chain of Archimedean classes. Finally, does one obtain prime models-over given affine spaces $\mathbb{A}^n(k)$-of the model companion of $\mathbb{A}^n(T)$ (assuming that it exists)?

5. DISCUSSION: THE ROLE OF PROJECTIVE CLOSURE.

Justifying our geometrical primitives leads to philosophical questions. Our remarks below, exceedingly exploratory in character, are intended to set the stage not for mathematical advance, but for philosophical catching up: The traditional semantic theory implicit in classical languages and Tarski semantics, which fit so closely with pure algebra (as in Robinson's work) appears unsuited for an "intrinsic" presentation of algebraic geometry (except via the artifice of first-order set theory).

An investigation of classical geometries which is so very sensitive to choices of primitive notions in a formal description as the preceding may seem surprising, or even somewhat suspect. I surmise that such intuitions rest on the notion, which goes back to the influence of Klein's "Erlanger Programm", that geometrical properties are intrinsic and in no way depend on choices of primitives or description.

Of course such views, while illuminating, are not unassailable; limitations of Klein's automorphism group based classification program have long been understood; foundations of geometry include a host of different conceptions, each illuminating a distinctive realm of geometrical investigation. From the logician's point of view, one would want to add: (1) definability classification gives positive information, e.g. on constructibility from given primitives, which it is in the very nature of Klein's method to ignore; (2) experience in model theory, as in algebra, shows the importance of morphisms between structures. And morphisms bring out definability distinctions, as was recognized in model theory from the very beginning, and comes out above most starkly in (3.7): automorphisms of hyperbolic planes may be characterisable as collinearity automorphisms, but collinearity embeddings lose the metric structure. Just so for affine parallelism structure.

Our particular choice of morphisms -hence of primitive notions- was made in order to obtain a formal counterpart to the commonplace intuition of the naturalness of projective geometry and projective closure - the desirability of considering projective versions of geometrical objects obtained in other contexts as a first step in classification and understanding of the original objects. To give just two random examples: (i) The headings of the motivational sections of [CK]: "§2. Real Projective Space - the Unifier. §3. Complex Projective Space - the Great Unifier." (ii) From the introduction to [Seg]: "I shall show how these results can be completed and given a simple form, when the Galois spaces are considered from the projective point of view instead of the affine one." (p. 129). A similar phenomenon has long been recognised in the movement to the algebraic closure (real closure) of fields in (real) algebraic geometry; Robinson developed the notion of model completion as a formal model for these relationships. In the body of the paper, we have seen how to formally assimilate projective closure to these other cases. Now we ask, in a more philosophical and speculative mood, to what extent these formal features of the situation may be taken to account for the mathematical usefulness of moving to a projective setting.

5.1. The homogenisation of equations arising in projective closure -both $X^2+Y^2 = 1$ and $X^2-Y^2 = 1$ become $U^2+V^2 = W^2$- is a process of existential closure (adjoining solutions) which leads to unification - hyperbolas are circles intersecting the line at infinity in two points. The further unification in the complex case alluded to in the headings of [Cℓ] is similarly due to adjoining solutions in the algebraic closure. So this fits rather nicely, but in fact points out an inadequacy in the concept of model completion: Homogenisation gives unification regardless of whether the ground field is model complete. So we would want a concept of relative model completion: a closure operation which eliminates additional quantifier complexity of definable sets contributed by a construction [say $\mathbb{A}^n(\cdot)$] regardless of the base field. This is what our arguments show of projective closure.

5.2. What appears to go far deeper is that projective closure, at least to $\mathbb{P}^n(\mathbb{C})$ or $\mathbb{P}^n(\mathbb{R})$, is a compactification. Compactness is a crucial property in the study of geometric structures by analytic methods. Consideration of examples quickly shows that the power of the method of compactification in geometric problems does not lie in compactification per se —e.g., in the point-set topological sense— but rather in the choice of a compactification intimately related to the geometric structures at hand. Indeed, one would expect something like this from a naive point of view, as study of a compactification (that is, a new structure) appears unlikely to give information about the original structure unless some mechanism for transfer of information back to that structure is available. It is therefore an interesting logical problem to give a formal description of such compactifications which would illuminate how they function in geometrical reasoning. Here we consider the correlation between such compactifications and existential closure. Such connections are especially intriguing because existential closure is a first-order phenomenon whereas compactness is restricted to topological fields such as $\mathbb{R}$ and $\mathbb{C}$.

In classical geometry, as considered in the body of the paper, such compactifications typically have to do with extending the action of a group which becomes the automorphism group of the resulting structure. This is brought out very clearly in [Bℓa] §1-3, first for projective transformations by extension of continuous bijective collinearity-preserving maps between affine open regions, and then for circle geometries of Möbius, Laguerre and Lie. (This last, largest group requires a two-sheeted cover of the euclidean plane circles, ramified in the circles of radius 0 and one ideal point.) Another such "classical" example is the conformal compactification of Minkowski space-time (e.g., [We], p. 38). Immensely more subtle are the compactification problems in modern algebraic geometry, such as those for moduli spaces. The question of what objects (singular curves) to add in order to obtain a parametric definition of all nonsingular curves of a given type as a projective variety now becomes very intricate, see [Mu 1]; [Mu 2], p. 182. Again a group action plays a crucial, though somewhat different role in the construction, in which compactification is eventually achieved as closure to a projective variety.

The "logical" point is, that in each case, one is extending an algebraic action of an algebraic group, which from our point of view might be read as: a group of maps whose graphs are explicitly definable on the geometrical space by a formula with parameters. This is at least clear in the case of $\mathbb{A}^n \to \mathbb{P}^n$, where we can coordinatise explicitly by $\psi_{\simeq}$, and obtain a Δ definition with parameters giving the graph of all projective transformations. It follows that points wit-out image or preimage satisfy a universal formula, and adding ideal points to function as such images or preimages is an obvious case of existential closure. So far, it remains mysterious, (a) how to deal with multiply-sheeted coverings instead of simple adjunctions of points, (b) what to say about definability on nontrivial algebraic varieties (e.g. moduli spaces) where no intrinsic primitives are in evidence. We return to the last point in (6.4); but one is at least tempted to speculate at this point that something akin to existential closure is involved in these compactifications, and that some kind of definability control from below, as in construction of a prime model over a given model, plays a role in the transferability of information about the compactification to the original structure.

5.3. The algebraic geometers in fact have an algebraic (as opposed to topological) analysis of the effect of "projective closure" of varieties. Just as the Hausdorff property of varieties over $\mathbb{C}$ is analogous to the condition that the diagonal map

$$(id_V, id_V) : V \to V \times V$$

be closed in the Zariski topology (where closed sets are solution sets of polynomial systems, at least locally), compactness has an algebraic analogue: V is complete iff any projection map

$$V \times W \to W$$

is closed (w.r.t. the Zariski topology). Now compactification may be compared with suitable conservative embedding in a complete variety; as all projective varieties are complete, projective closure will constitute such an embedding. Catch: once we have an embedding as a suitably definable (quasi projective) subset of some $\mathbb{P}^n$.

From a logical point of view, it is not clear how to evaluate this. The algebraic character of the notion of completeness leads one to suspect it is close to first-order. Indeed, the proof of completeness of projective varieties recently given by Van den Dries [VdD] makes explicit that a logical property is involved, namely a positive quantifier elimination - but in a language based of field primitives rather than geometric primitives. Also, the Lie geometry example (or Mumford's observation, that well-behaved moduli spaces can only be obtained for "polarised" abelian varieties, [Mu 2] p. 97) suggests that a general model theoretic understanding of projective closure requires a more structured relationship than simple embedding, e.g., capturing a notion of many-sheeted coverings extending a given definable group action.

.4. This brings us back to the original conceptual problems of this discussion -the choice of geometrical primitive notions. In the context of algebraic geometry, which is certainly an interesting and most important one, this problem appears quite intractable in traditional model theoretic terms, if one is serious about geometrical notions. In part, the difficulty seems parallel to one encountered in philosophy of physics. For on the one hand, the geometers appear convinced that they are dealing and must deal with intrinsically and inherently geometrical notions- just as the physicist must be taken to be studying an intrinsically physical world ultimately independent of mathematical objects, mathematical quantities and functions. On the other hand, just as the physicist has no other formulation of his theory except in terms of such mathematical objects, the algebraic geometer actually studies objects defined in terms of polynomial rings over fields and entities derived from these. To avoid making assertions which depend on these geometrically non-intrinsic objects, or specific embeddings of varieties in an ambient space such as $\mathbb{A}^n(k)$, $\mathbb{P}^n(k)$ one studies equivalence classes of varieties under some notion of isomorphism (birational equivalence, proper birational equivalence) and attempts to discover structural invariants w.r.t. the equivalence relation. This contrasts sharply with the simple cases of classical geometries dealt with in earlier sections, where the intrinsic geometrical structure is spelled out in advance by the explicit choice of primitives.

This raises the critical question -just as for physics- whether the intuition that one is studying intrinsic geometric objects which are independent of representation used to study them can be fully justified. A simple-minded (but ambitious) way to tackle this is to try to spell out intrinsic geometrical primitives on specific varieties just as for $\mathbb{A}^n$ and $\mathbb{P}^n$ (which after all are just the most trivial algebraic varieties.) This is not to be confused with defining a given variety V in $\mathbb{P}^n$ in the language of $\mathbb{P}^n$ -this is simple, we just translate the algebraic definition of V back via the coordinate ring interpretation; but gives us V as embedded in $\mathbb{P}^n$, which fixes quantities such as the degree of homogeneous defining polynomials, which are not intrinsic geometric structure (birationally invariant) of V.

Given a variety V of dimension $n \geqslant 2$, let us say a surface, taken as a point set in some $\mathbb{P}^n(k)$, we now look for some primitive geometrical relations on V. Experience in algebraic geometry seems to suggest that we should look at curves on V; perhaps select one or more algebraic families and look at their intersection behavior. Thus one would start with primitives such as

$R^i x_0 \ldots x_{n_i}$: x_0 lies on the curve determined by its points $x_1 \ldots x_{n_i}$ (among the i-th family of curves)

This gives us one or more incidence relations. We would like to recover the base field k from this (provided that V contains enough points over k), and some sort of coordinatisation isomorphism. It is not clear how one should accomplish these things, as the intersection behavior of our families of curves could be quite complicated. Presumably, one might need additional primitives ? Nor is it clear how the families of curves should be selected in the first place, and to what extent this could be done uniformly for different V. In higher dimensions, one would expect to look at families of subvarieties of codimension 1; rather more complicated behavior is to be anticipated.

Regardless of the outcome of such attempts -they do not seem altogether without hope- one may have one's doubts about the semantic analysis of the objects of algebraic geometry provided thereby. The modern language of schemes reflects, via the mechanism of base change, a view of geometrical objects as functors rather than solution sets of equations over fields, much as one might view projective geometry as the construction $\mathbb{P}^n$ of (2.3) rather than $\mathbb{P}^n(\mathbb{C})$ or even all $\mathbb{P}^n(k)$ together with their embeddings. The difference hardly comes out as long as one is interested in a single fixed rich base field such as $\mathbb{R}$ or $\mathbb{C}$, as in classical algebraic geometry.

One difference between the conceptions becomes visible when one considers our embeddings $\mathbb{P}^n(k) \to \mathbb{A}^n(K)$: this makes perfect sense to a classical model theorist, but looks very strange to an algebraic geometer. If varieties such as $\mathbb{P}^n$ and $\mathbb{A}^n$ are functors, and embeddings relations between functors, then embeddings will mean: of varieties over the same base field, uniformly depending on that base field. As Jan Denef pointed out, there are "intrinsic geometric reasons" why one could not have an embedding $\mathbb{P}^n \to \mathbb{A}^n$: Over $\mathbb{A}^n$, there is a large ring of regular functions, whereas over $\mathbb{P}^n$ there are very few. If we had an embedding -functorial and always over the same base field; but this is automatically understood- the regular functions would restrict. Contradiction.

5.5. In summary, we seem to have at least the following difficulties. (i) While the formalism of modern algebraic geometry is perfectly clear mathematically, we do not see how to design a well-adapted "logic" (language, with accompanying semantic theory) which would give a descriptively close fit with the mathematical concepts in question--unless, again, indirectly via formalisation in some general framework such as set theory. And if we are really forced to modify our semantic conceptions, the philosopher (!) wants to understand the nature of the change, and what has forced it. Why must even innovative attempts to use Tarski semantics, say with unobvious but geometrically intrinsic primitives, break down in describing modern algebraic geometry? Here the philosopher seeks his own, critical, understanding of what the geometer has grasped. (ii) To what extent does the formal notion of model completion capture what is important in general about moving to projective settings?

As I hope the above will have brought out, these questions have significant mathematical components (though the necessary definitions are unavailable). Their motivation is largely non-mathematical.

ACKNOWLEDGEMENT

I very much thank *Jan Denef*, *Justus Diller*, *Lou van den Dries*, and *Frans Oort* for stimulating and helpful conversations. The research and writing was done while I enjoyed the hospitality of the Mathematical Institute at Utrecht; I especially thank *Sophie van Sterkenburg* there for the typing.

REFERENCES

[Bla] W. Blaschke, Differential-Geometrie III. Springer, 1929.

[Blu] L.M. Blumenthal, A Modern View of Geometry. W.H. Freeman, 1961. (Dover reprint, 1980).

[Bo] N. Bourbaki, Algèbre, Ch. II.

[C] P.M. Cohn, The embedding of firs in skew fields. Proc. Lond. Math. Soc. 23 (1971), 193-213.

[Cl] C.H. Clemens, A Scrapbook of Complex Curve theory. Plenum, 1980.

[CV] D. Carter & A. Vogt, Collinearity-preserving functions between Desarguean-planes. Memoirs AMS 235 (1980).

[D] J. Diller, Nicht-persistenz der Parallelität in affinen Ebenen. Zeitschr.f. Math. Logik und Grundlagen d. Math. 15 (1969), 431-33.

[G] L.E. Garner, An Outline of Projective Geometry. North-Holland, 1981.

[HD] G. Hessenberg & J. Diller, Grundlagen der Geometrie. W. de Gruyter, 1967.

[K] M. Kordos, On the syntactic form of dimension axiom for affine geometry. Bull. Acad. Pol. Sci. 17 (1969), 833-37.

[M] A. Macintyre, Model Completeness. In: Handbook of Mathematical Logic, J. Barwise ed. North-Holland, 1977.

[MMD] A. Macintyre, K. McKenna, L. van den Dries, Elimination of Quantifiers in Algebraic Structures. Advances in Mathematics 47 (1983), 74-87.

[Mu1] D. Mumford, Stability of projective varieties. L'Enseignement Math. 23 (1977), 39-110.

[Mu2] D. Mumford, Geometric Invariant Theory, 2nd ed. Springer, 1982.

[Sch] W. Schwabhäuser, II. Metamathematische Betrachtungen. In [SST].

[Seg] B. Segre, Geometry and algebra in Galois spaces. Abh. Math. Sem. Univ. Hamburg 25 (1962), 129-139.

[Sp] E. Sperner, Zur Begründung der Geometrie im begrenzten Ebenenstück. Halle a.d.S.: Niemeyer, 1938. (cf. Zentralblatt 19 (1938), p. 179)

[ST] L. Szczerba, A. Tarski, Metamathematical discussion of some affine geometries. Fund. Math. 104 (1974), 155-192.

[Sz1] L. Szczerba, Weak general affine geometry. Bull. Acad. Pol. Sci. 20 (1972), 752-61.

[Sz2] ——— , A paradoxical model of euclidean affine geometry. Bull. Acad. Pol. Sci. 20 (1972), 845-51.

[Sz3] ——— , Interpretability of Elementary Theories. In: Butts & Hintikka (eds.), Logic, Foundations of Mathematics and Computability Theory, 129-45.

[Sz4] ——— , Interpretations with parameters. Zeitschr.f. math. Logik und Grundlagen d. Math. 26 (1980), 35-39.

[T] A. Tarski, What is elementary geometry? In: Henkin, Suppes, Tarski (eds.), The Axiomatic Method. North-Holland, 1959, 16-29.

[VdD] L. van den Dries, Some applications of a model theoretic fact to (semi-) algebraic geometry. Indag. Math. 44 (1982), 397-401.

[W] W. Wheeler, Amalgamation and elimination of quantifiers for theories of fields. Proc. AMS 77 (1979), 243-50.

[We] R. O. Wells, Complex geometry in mathematical physics. Presses Univ. Montreal, 1982.

ADDENDA

[SST] W. Schwabhauser, W. Szmielew, A. Tarski, Metamathematische Methoden in der Geometrie. Springer, 1983.

It is consistent with the description in Zentralblatt that Sperner's early work [SP] already contains results on general affine coordinatisation, for which we have referred to [ST]. We have not been able to consult [SP].

ON CANTOR-BENDIXSON SPECTRA CONTAINING (1,1) - I (°)

Annalisa Marcja
Università degli Studi di Trento
Dipartimento di Matematica

Carlo Toffalori
Università degli Studi
di Firenze
Istituto Matematico "U.Dini"

§ 0. Introduction

The general problem we are interested in is the classification of the countable complete, quantifier eliminable theories by the Boolean algebras of parametrically definable subsets of their models (cf. [MM1]). Owing to the obvious difficulty presented by the problem, a step by step approach is needed. A first result in this direction was the introduction and the study of the concept of pseudo $\aleph_0$-categorical theory [MM1], [MT]; but, for example, theorem 2.1 in [MT] proves that it is impossible to distinguish stable, superstable or unstable theories by simply considering the Boolean algebras of definable subsets of their countable models. For this reason $\aleph_1$-Boolean spectra were introduced in [MM2] .

(°) Work performed under the auspices of Italian C.N.R. (G.N.S.A.G.A.)

On the other hand the situation is quite different for ω-stable theories, for which the analysis of the Boolean algebras of definable subsets of countable models seems to be very natural.

In fact we have that T is ω-stable if and only if for every $M \models T$, $|M| = \aleph_0$ the Boolean algebra of (parametrically) definable subsets of M (B(M)) is superatomic, and every countable superatomic Boolean algebra is characterized, up to isomorphisms, by its Cantor-Bendixson type. More precisely:

For every Boolean algebra B, a sequence of ideals $I_\lambda(B)$ of B can be defined for every ordinal λ in this way:

1. $I_0(B) = \{0\}$;
2. $I_1(B)$ = ideal of finite elements of B;
3. $I_{\nu+1}(B)$ = preimage in B (in the natural epimorphism) of $I_1(B/I_\nu(B))$;
4. $I_\lambda(B) = \bigcup_{\nu<\lambda} I_\nu(B)$ if λ is a limit ordinal.

If B is superatomic then there exists an ordinal μ such that $I_\mu(B)=B$ and the least such μ is a successor ordinal. Then we define:

α_B = the greatest ordinal ν such that $I_\nu(B) \neq B$.

d_B = number of atoms of $B/I_{\alpha_B}(B)$.

In a similar way we can define α_b, d_b for every $b \in B$. We have the following.

i. If B is countable then $\alpha_B < \omega_1$;

ii. $d_B < \omega$;

iii. for every ordered pair (α,d) with $1 \leq \alpha < \omega_1$, $1 \leq d < \omega$ there is a countable superatomic Boolean algebra B such that $(\alpha,d) = (\alpha_B,d_B)$;

iv. for every countable superatomic Boolean algebras B_1, B_2, B_1 is isomorphic to B_2 if and only if $(\alpha_{B_1},d_{B_1}) = (\alpha_{B_2},d_{B_2})$ (see [D]).

Let T now be an ω-stable theory (complete, Q.E.), M a countable model of T; we set $(\alpha_M,d_M) = (\alpha_{B(M)},d_{B(M)})$ and we say that:

* α_M is the Cantor-Bendixson rank (CB-rank) of M.
* d_M is the Cantor-Bendixson degree (CB-degree) of M.
* (α_M,d_M) is the Cantor-Bendixson type (CB-type) of M.

We define <u>Cantor-Bendixson spectrum</u> of T (CB-Spec T) the set

$$\{(\alpha_M,d_M) \; : \; M \models T, \; |M| = \aleph_0\}$$

ordered lexicographically.

CB-Spec T has a minimal pair, corresponding to the prime model of T and a maximal pair corresponding to the countably saturated model $\bar{M}$ of T. Moreover $\alpha_{\bar{M}}$ coincides with the Morley rank α_T of T and $d_{\bar{M}}$ coincides with the Morley degree d_T of T (see [B1]). We will say that (α_T,d_T) is the Morley type of T.

Then, as a first approach to our general problem, it turns out to be natural to study all possible CB-spectra of ω-stable theories.

Of course it makes sense only observing that not every subset of $(\omega_1 - \{0\}) \times (\omega-\{0\})$ admitting a maximal element, can be a CB-spectrum (see [T]). The most restrictive condition seems to be $(1,1) \in$ Spec T; for this reason in the next step our analysis will be confined to

CB-spectra containing (1,1). In fact, either (1,1) is the only pair in CB-Spec T (and hence T is $\aleph_1$-categorical), or there is a second pair (α,d) in it. In this case, what are the other possible pairs in the spectrum? The presence of the pair (1,1) seems to introduce strong limitations to the entire spectrum.

In this paper we consider the case in which $(\alpha,d) = (1,n+1)$ in order to illustrate the necessary techniques, useful also in the general case, involving normalization in Lachlan's sense (cf. [L_1], [L_2]) and forking (see [S], [LP]). We will show that $\{(1,k\,n+1) : k \geq 0\} \cup \{(2,1)\} \subseteq$ CB Spec T and that the prime model is the unique model of type (1,1). We will deal with the general case ($\alpha \geq 2$) in a forthcoming paper.

The paper is organized as follows.

- In section 1 we give some algebraic examples of ω-stable theories whose CB-spectra contain (1,1).
- In section 2 we prove a "weak" version of the finite equivalence relation theorem for CB-rank.
- Finally, in section 3, we investigate ω-stable theories T whose CB-spectra contain (1,1), (1,n+1) as minimal pairs.

We assume that the reader is familiar with the concepts of theory of models. The notation will be the usual one.

We wish to thank Gregory Cherlin for his valuable contributions and Leo Harrington for having suggested Lemma 3.2.

§ 1. Some examples

1.1 Groups

Let G be an infinite ω-stable group. Then:

- (1,1) $\in$ CB-Spec Th(G) if and only if G is abelian and either $G = Q^{(\alpha)} \oplus (\bigoplus_{p \text{ prime}} Z(p^{\infty})^{(\alpha p)})$ where α_p is finite for every prime p or $G = Z(p)^{(\alpha)}$ where p is prime and $\alpha \geq \aleph_0$.

Observe that if (1,1) $\in$ CB-Spec Th(G) then CB-Spec Th(G) = {(1,1)} so Th(G) is pseudo $\aleph_0$-categorical and hence $\aleph_1$-categorical (cf. [MT] : a theory T is said to be pseudo $\aleph_0$-categorical if, for every countable models M, N of T, B(M) $\cong$ B(N); hence, if T is ω-stable, then T is pseudo $\aleph_0$-categorical if and only if |CB-Spec T| = 1).

The above classification is implicit in [R]; we give here a sketch of the proof.

It is obvious that, for every countable abelian group G corresponding to the previous classification, G has CB-type (1,1). Conversely, let G be an ω-stable infinite group such that (1,1) $\in$ CB-Spec Th(G); we can assume without loss of generality that $|G| = \aleph_0$ and G has CB-type (1,1).

We have:

(1.1,a) G has no proper definable infinite subgroup.

(1.1,b) G is abelian, because every ω-stable infinite group admits an infinite abelian definable subgroup ([C], p. 67).

(1.1,c) By the Macintyre classification of ω-stable abelian groups ([M]) it is $G = D \oplus H$, where D is divisible and H has finite exponent m. It follows that D = mG, then by (1.1,a) we have two cases:

Case 1 : G = D, H = (0), so G is divisible. In this case, we obtain readily

$$G = Q^{(\alpha)} \oplus (\bigoplus_{p\ \text{prime}} Z(p^{\infty})^{(\alpha_p)})$$

where α_p is finite for every prime p.

Case 2 : D = (0), G = H has finite exponent; by (1.1,a) and the Prüfer theorem we obtain $G = Z(p)^{(\alpha)}$ where p is prime, $\alpha = \aleph_0$.

1.2 Rings

Let R be an infinite ω-stable ring. Then:

- (1,1) ε CB-Spec Th(R) if and only if either R is an algebraically closed field, or R has trivial multiplication, and additive structure corresponding to 1.1.

In particular, if (1,1) ε CB-Spec Th(R), then CB-Spec Th(R) = {(1,1)} so Th(R) is pseudo $\aleph_0$-categorical and hence $\aleph_1$-categorical again. The above result is implicit in [P]. Here is a sketch of the proof. It is obvious that, whenever a ring R corresponds to the previous classification, R has CB type (1,1). Conversely, let R be an ω-stable (infinite) ring such that (1,1) ε CB-Spec Th(R), we can suppose without loss of generality that $|R| = \aleph_0$ and R has CB-type (1,1).

(1.2,a) R has no proper definable infinite ideal.

(1.2,b) The Jacobson radical J(R) of R is 0-definable, then

either $J(R)$ is finite or $R = J(R)$.

(1.2,c) If $J(R)$ is finite, it is easy to see that $J(R) = (0)$, so R is semisimple. Recalling the classification of ω-stable semisimple rings [BR] we have

$$R = F_1^{(n_1)} \times \dots \times F_r^{(n_r)}$$

where, for every $j = 1,\dots,r$, F_j is a finite field or an algebraically closed field, n_j is a positive integer and $F_j^{(n_j)}$ is the complete $n_j \times n_j$ matrix ring over F_j; furthermore F_j is definable in R, so there exists $j=1,\dots,r$ such that F_j is an algebraically closed field and $R = F_j$.

(1.2,d) If $R = J(R)$, then R is nilpotent: it is easy to see that R has nil-exponent 2, so R has a trivial multiplication; moreover the additive group of R corresponds to the classification given in 1.1.

1.3 Differential rings

Let R be a commutative differential ω-stable ring (with identity). Then:

- $(1,1) \in$ CB-Spec Th(R) <u>if and only if</u> R <u>is an algebraically closed field and it has a trivial derivation</u> D (i.e. $D(r) = 0$ <u>for every</u> $r \in R$).

<u>Proof</u>: suppose $|R| = \aleph_0$, R of CB-type $(1,1)$. It follows by 1.2 that R is an algebraically closed field. Let us denote by $C(R)$ the constant subfield of R. If R has characteristic p (p prime), $R=R^p=$

= C(R) so $D \equiv 0$ in R. If R has characteristic 0, C(R) contains the prime subfield of R, so C(R) is an infinite differential subfield of R; it follows that R = C(R) so $D \equiv 0$ in R again.

Looking at the previous examples we notice that we always obtained CB-Spec T = $\{(1,1)\}$. Obviously, there exist ω-stable theories T such that CB-Spec T $\supsetneq \{(1,1)\}$: for example for every ordinal $\alpha (2 \leq \alpha < \omega_1)$ there is an ω-stable theory such that CB-Spec T = $\{(1,1), (\alpha,1)\}$ (cf. [T]). Here we are interested in an ω-stable theory T such that CB-Spec T = $\{(1,kn+1) : k \in N\} \cup \{(2,1)\}$. For example consider the theory T of an equivalence relation E and a bijection f such that:

* E admits exactly n classes with m elements for every positive integer m;
* f induces a cycle of length n on the E-classes with m elements, for every m.

We will prove in section 3 that this example is "typical" in the sense that if (1,1), (1,n+1) $\in$ CB-Spec T and (1,d) $\notin$ CB-Spec T for every d, 1 < d < n+1, then CB-Spec T $\supseteq \{(1,kn+1) : k > 1\} \cup \{(2,1)\}$.

Notice that, when n = 1, the generic ω-stable theory whose CB-spectrum contains (1,1), (1,2) strongly resembles the following well known example [B1] :

the theory T of an equivalence relation admitting exactly one class of n elements for every n > 0.

§ 2. The equivalence relation

We want to prove here a "finite equivalence relation" like theorem for ω-stable theories, relatively to CB-rank.

Theorem 2.1

If T is ω-stable, $M \models T$, M prime over X, CB-type of M = (α, d), then there exists an X-definable finite equivalence relation with exactly d classes of CB-rank α on M.

Proof:

Let $\phi = \phi(v,\bar{c})$ be a formula, $\bar{c}$ in M, such that:

1. CB-type of $\phi = (\alpha,1)$
2. Morley type of ϕ(M-type ϕ) minimal subject to 1.

Therefore,

3. for every formula ψ(of L_M) CB-rank $(\phi\wedge\psi) = \alpha$ if and only if the Morley rank (M-rank) of $\phi \wedge \psi$ is equal to M-rank ϕ.

Let $\phi^*(v,\bar{w})$ a normalization of $\phi(v,\bar{w})$ (in Lachlan's sense) at $p_\circ = tp(\bar{c}/\emptyset)$; it means that if $\bar{c}_1, \bar{c}_2 \models p_\circ$ then

* - M-type $(\phi^*(v,\bar{c}_1) \wedge \phi^*(v,\bar{c}_2))$ = M-type $(\phi^*(v,\bar{c}_1))$ implies $\models \forall v(\phi^*(v,\bar{c}_1) \leftrightarrow \phi^*(v,\bar{c}_2))$, and

* - M-type $(\phi^*(v,\bar{c}_1) \wedge \phi(v,\bar{c}_1))$ = M-type $\phi^*(v,\bar{c}_1)$ = M-type $(\phi(v,\bar{c}_1))$.

We have:

4. if $\bar{c}_1, \bar{c}_2 \models p_\circ$ then CB-rank $(\phi^*(v,\bar{c}_1) \wedge \phi^*(v,\bar{c}_2)) = \alpha$ implies $\models \forall v(\phi^*(v,\bar{c}_1) \leftrightarrow \phi^*(v,\bar{c}_2))$.

(Then we say that ϕ^* is also a CB-normalization). Notice that CB-type $(\phi^*(v,c)) = (\alpha,1)$.

5. Let $p = tp(\bar{c}/X)$. Notice that, if $\bar{c}_1 \models p, \bar{c}_1$ in M, then CB-type $(\phi^*(v,\bar{c}_1)) = (\alpha,1)$, since M is prime over X.

6. Define $E_o(v,w)$: $(\forall \bar{t} \models p)(\phi^*(v,\bar{t}) \leftrightarrow \phi^*(w,\bar{t}))$.

E_o is X-definable as p is isolated. Consider the possible E_o-classes.

If $\bar{c}_1$, $\bar{c}_2 \models p$ we have either CB-rank $(\phi^*(v,\bar{c}_1) \wedge \phi^*(v,\bar{c}_2)) = \alpha$ and hence $\phi^*(M,\bar{c}_1) = \phi^*(M,\bar{c}_2)$, or CB-rank $(\phi^*(v,\bar{c}_1) \wedge \phi^*(v,\bar{c}_2)) < \alpha$

There exist $d(\phi^*) \leq d$ and $\bar{c}_i \models p$, $1 \leq i \leq d(\phi^*)$ such that

CB-rank $(\phi^*(v,\bar{c}_i)) = \alpha$ for every i ,

CB-rank $(\phi^*(v,\bar{c}_i) \wedge \phi^*(v,\bar{c}_j)) < \alpha$ for $i \neq j$.

It follows that E_o has

* $d(\phi^*)$ classes having CB-rank α (and CB-degree 1) corresponding to

$$\phi^*(M,\bar{c}_i) - \bigcup_{j \neq i} \phi^*(M,\bar{c}_j)$$

* a finite number of classes having CB-rank $< \alpha$, corresponding to the meets;

* possibly another class having CB-rank α, corresponding to M-S(M), where

$$S(v) : (\exists \bar{t} \models p)\ \phi^*(v,\bar{t}) .$$

If CB-rank $(M-S(M)) < \alpha$, then $d = d(\phi^*)$: if CB-rank $(M-S(M)) = \alpha$ we continue our analysis, by considering a formula $\phi_1 = \phi_1(v,\bar{b})$ such that

0' - $\phi_1(M,\bar{b}) \cap S(M) = \emptyset$

1' - CB-type $\phi_1 = (\alpha,1)$

2' - Morley type of ϕ_1 minimal subject to 0' and 1'. ∎

§ 3. The (1,1), (1,n+1) case

We are dealing here with an ω-stable theory T whose CB-spectrum contains (1,1), (1,n+1) as minimal pairs.

As T is not $\aleph_0$-categorical we may suppose $S_1(\emptyset)$ infinite (we fix $\bar{c}$ with $tp(\bar{c}/\emptyset)$ isolated and $S_1(\bar{c})$ infinite and we adjoin $\bar{c}$ to the language). Let M_0 now be the prime model of T; M_0 is of CB-type (1,1); for every $x \in M_0$, $tp(x/\emptyset)$ is algebraic and we have that M_0 is the algebraic closure of the empty set and w.l.o.g. the elements of M_0 are constants.

We may suppose there is a finite sequence $\bar{m}$,a such that $M = M_0[\bar{m}]$ has CB-type (1,1) and $M_1 = M[a]$ has CB-type (1,n+1). (By $N[x]$ we denote the prime model over $N \cup \{x\}$).

Furthermore notice that $(1,1) \in$ CB-Spec T implies that the Morley degree d_T of T is 1 (cf. [T] Lemma 1.2). So, for every model M of T, there is exactly one type over M, whose Morley rank is equal to α_T. We call such type the <u>generic type</u> over M, and we denote it by p_M.

When M has CB-type (1,1), p_M is the only non trivial type over M, hence p_M is strongly regular and regular [S]. It follows that for every $M \models T$, p_M is strongly regular and regular.

Theorem 3.1

<u>Let</u> T <u>as above, then</u>:

(3.1.1) M_0 <u>is the only countable model of</u> T <u>whose CB-type is</u> (1,1).

(3.1.2) <u>There exists a</u> 0-<u>definable equivalence relation</u> E <u>on the</u>

<u>models of</u> T <u>such that</u> $M_1 = M_o \cup a/E$.

<u>Proof</u>.

(3.1.2)' Firstly we prove that there exists an M-definable equivalence relation E' such that $M_1 = M \cup a/E'$. In view of (3.1.1) in the following we consider $Th(M_M)$ instead of T, for the sake of simplicity.

Let $\phi = \phi(v,\bar{c})$ be a formula, $\bar{c}$ in M, such that:

1. CB-type $\phi = (1,1)$;
2. $\phi(M[a],\bar{c}) \cap M = \emptyset$;
3. M-type of ϕ minimal subject to 1,2.

Notice that, by Baldwin's characterization of stability [B2], for every L_M-formula $\psi(v,\bar{d})$, $\psi(M[a],\bar{d}) \cap M$ is M-definable, and hence finite or cofinite; then a formula $\phi(v,\bar{c})$ satisfying 1-3 exists.

As in theorem 2.1 we can find a formula $\phi^*(v,\bar{c})$ such that:

4. $\phi^*(v,\bar{w})$ is CB-normal at $p_o = tp(\bar{c}/\emptyset)$ i.e. if $\bar{c}_1, \bar{c}_2 \models p_o$ and CB-rank $(\phi^*(v,\bar{c}_1) \wedge \phi^*(v,\bar{c}_2)) = 1$ then $\models \forall v(\phi^*(v,\bar{c}_1) \leftrightarrow \phi^*(v,\bar{c}_2))$.

5. $\phi^*(v,\bar{c})$ has CB-type (1,1) and hence for every $\bar{c}_1 \models p_o$, CB-rank $(\phi^*(v,\bar{c}_1)) = 1$.

Furthermore, $\phi^*(v,\bar{c})$ is equivalent to a positive Boolean combination of formulas of the form $\phi(v,\bar{d})$ where $tp(\bar{d}/\emptyset) = p_o$, hence $\phi^*(M[a], \bar{c}) \cap M = \emptyset$ so that $\phi^*(M[a],\bar{b}) \cap M = \emptyset$, for every $\bar{b} \models p_o$.

a) Observe that there exists a finite number $\bar{k} \leq n+1$ of sequen ces $\bar{c}_1,\dots,\bar{c}_{\bar{k}} \models p_o$ such that $\phi^*(M[a],\bar{c}_i) \cap \phi^*(M[a],\bar{c}_j)$ is finite for $1 \le i < j \le \bar{k}$; let h be the greatest power of these meets.

b) Let $\psi(\bar{w},a)$ be the formula isolating $tp(\bar{c}/a)$.

We have:

(*) $\models \exists\, \bar{w}\; \psi(\bar{w},a)$

(**) $\bar{b} \models p_o$ for every $\bar{b}$ such that $\models \psi(\bar{b},a)$

Let $\bar{\psi}(\bar{w},a)$ a formula satisfying (*), (**) and such that

$$\phi_o(v,a) : \exists\, \bar{w}(\bar{\psi}(\bar{w},a) \wedge \phi^*(v,\bar{w}))$$

has maximal CB-type (observe that $\phi_o(M[a],a)$ is infinite coinfinite).

Then we prove that for every $a' \in M[a]$, $a' \in M[a]-M$ if and only if $\models \forall\; v(\phi_o(v,a) \leftrightarrow \phi_o(v,a'))$.

In fact if $a' \in M$, then $\phi_o(M[a],a')$ is finite or cofinite, in any case it is different from $\phi_o(M[a],a)$.

Let now $a' \in M[a]-M$, then $tp(a/\emptyset) = tp(a'/\emptyset)$ and the following formulas belong to $tp(a/\emptyset)$:

- $\exists^{!k}\, \bar{w}_i \;(\bigwedge_{1\le i\le k} \bar{\psi}(\bar{w}_i,v) \wedge \bigwedge_{i\ne j} \exists^{\le h} t(\phi^*(t,\bar{w}_i) \wedge \phi^*(t,\bar{w}_j)))$

  for suitable $k \le \bar{k}$.

- $\forall\; \bar{w}_i\; \forall\; \bar{w}_j\; ((\bar{\psi}(\bar{w}_i,v) \wedge \bar{\psi}(\bar{w}_j,v) \wedge \exists^{>h} t(\phi^*(t,\bar{w}_i) \wedge$

  $\wedge\; \phi^*(t,\bar{w}_j))) \;\rightarrow\; \forall\; t(\phi^*(t,\bar{w}_i) \leftrightarrow \phi^*(t,\bar{w}_j)))$.

These formulas belong also to $tp(a'/\emptyset)$, hence there exist $\bar{b}_1,\dots,\bar{b}_k$ such that

* for $i \ne j$ $|\phi^*(M[a],\bar{b}_i) \cap \phi^*(M[a],\bar{b}_j)| \le h$;

* $\models \bar{\psi}(\bar{b}_i,a')$ for every i: as above $\bar{b}_i \models p_o$ by (**) ;

* if $\models \bar{\psi}(\bar{b},a')$ then there exists i such that $|\phi^*(M[a],\bar{b}) \cap \phi^*(M[a],\bar{b}_i)| > h$ and hence $\phi^*(M[a],\bar{b}) = \phi^*(M[a],\bar{b}_i)$.

Let us suppose towards a contradiction that there exists $i = 1,\dots,k$ such that, for every $\bar{c}$, $\models \bar{\psi}(\bar{c},a)$, it is $\phi^*(M[a],\bar{b}_i) \neq \phi^*(M[a],\bar{c})$ (i.e. the intersection is finite); consider the formulas

$$\chi(v,a) \quad \text{isolating} \quad tp(a'/a)$$

$$\psi'(\bar{w},a) \; : \; \bar{\psi}(\bar{w},a) \vee \exists s(\bar{\psi}(\bar{w},s) \wedge \chi(s,a))$$

We have : $\models \exists \bar{w}\, \psi'(\bar{w},a)$ and if $\models \psi'(\bar{b},a)$ then $\models \bar{\psi}(\bar{b},a)$ and $\bar{b} \models p_0$; otherwise there exists $a'' \models tp(a'/a)$ such that $\models \bar{\psi}(\bar{b},a'')$: but then $a'' \notin M$, $tp(a/\emptyset) = tp(a''/\emptyset)$ hence $\bar{b} \models p_0$ again.

Furthermore CB-type of

$$\phi_0'(v,a) \; : \; \exists \bar{w}(\psi'(\bar{w},a) \wedge \phi^*(v,\bar{w}))$$

is greater than CB-type $(\phi_0(v,a))$, as $\phi^*(M[a],\bar{b}_i) \subseteq \phi_0'(M[a],a)$ and $\phi^*(M[a],\bar{b}_i) \not\subseteq \phi_0(M[a],a)$. A contradiction.

It follows that $\phi_0(M[a],a) = \phi_0(M[a],a')$.

In $M[a]$ we define the following equivalence relation E'

$$E'(v,w) \; : \; \forall z(\phi_0(z,v) \leftrightarrow \phi_0(z,w)).$$

E' is M-definable $a/E' = M[a]-M$ while in M E' has infinitely many classes of finite unbounded size and hence, for every n, there is at most a finite number of classes with n elements.

(3.1.1) Suppose, towards a contradiction, that $M_0 \neq M$; we may set

$M = M_0[m]$. Notice that $a \models p_M$; it follows that a and m are M_0-independent and hence $M_0[a]$ and $M_0[m]$ are M_0-independent furthermore $M_0[a] \cong_{M_0} M_0[m]$.

$M_0[a] - M_0 = a/E' \cap M_0[a]$ is $M_0[a]$-definable by Baldwin's theorem [B2]; it follows that $M - M_0$ is M-definable and hence $M - M_0$ is finite. Then $E' \cap M_0^2$ is an equivalence relation which is M_0-definable [B2] and has in M_0 infinitely many classes of finite unbounded size. Consequently $M = M_0$ (otherwise $M - M_0$ is infinite). A contradiction.

(3.1.2) Obvious by (3.1.2)' and (3.1.1). ■

For simplicity we summarize the situation.

a - M_0, the prime model of T, has CB-type (1,1),

b - $M_1 = M_0[a]$ is a prime model extension generated by a new element a,

c - M_1 has CB-type (1,n+1),

d - there is a 0-definable equivalence relation E on the models of T such that $M_1 = M_0 \cup a/E$.

Lemma 3.2

<u>Let</u> T <u>be an</u> ω-<u>stable theory satisfying</u> (a), (b), (d). <u>If</u> M <u>is a model of</u> T, q <u>is the generic type over</u> M, $M' = M[c]$ <u>is the prime model over</u> $M \cup \{c\}$, <u>where</u> c <u>is a realization of</u> q, <u>then</u>:

(3.2.1) <u>for every</u> $b \in M - M_0$, <u>the equivalence class</u> b/E <u>in</u> M <u>is equal to the class</u> b/E <u>in</u> M'.

(3.2.2) $M' - M = c/E$.

Proof. (3.2.1) Recall that p_{M_0} is a regular type, p_M is the non-forking extension of p_{M_0} to M.

Fix $b \in M - M_0$, $b' \in M'$ with $E(b,b')$, and set $q' = tp(b'/M)$; q' is a forking extension of p_{M_0} to M, as the formula $E(v,w)$ is represented in q' and not in p_{M_0}. It follows that q is orthogonal to q', hence q' is not realized in $M'-M$ (see [La], 4.3), so $b' \in M$ as claimed.

(3.2.2) Let (M^{**}, M^*) be an elementary extension of (M_1, M_0) such that $M < M^*$. Let p^* be the generic type over M^*.

We have:

* - $M^{**} = M^* \cup a/E$: in fact the language for pairs of models of T is $L(T) \cup \{P\}$, where P is 1-place relation symbol. Then:

$$(M_1, M_0) \models \forall v \; \forall w(\neg P(v) \rightarrow (\neg P(w) \leftrightarrow E(v,w)))$$

and hence the same sentence holds in (M^{**}, M^*).

* - a realizes p^* in M^*: it is sufficient to prove that $tp(a/M^*)$ does not fork over M_0. Let $\phi(u,\bar{w})$ be a formula of L(T), $\bar{m}$ in M^* such that $\phi(v,\bar{m}) \in tp(a/M^*)$. Therefore $(M^{**},M^*) \models \exists \bar{w}(P(\bar{w}) \wedge \wedge \phi(a,\bar{w}))$ and the same is true in (M_1, M_0). So there exists $\bar{m}_0$ in M_0 such that $\phi(v,\bar{m}_0) \in p_{M_0}$.

Consequently $tp(a/M) = tp(c/M) = q$. Then we can embed M' into M^{**} fixing M and sending c to a.

The image of this map is prime over $M \cup \{a\}$ so that it will be denoted from now on by $M[a]$. We have $M[a] \cap M^* = M$ because a is independent from $M^* - M$ over M.

Concluding, we have an embedding of (M', M) into (M^{**}, M^*).

Hence $(M',M) \models \forall v \forall w (\neg P(v) \to (\neg P(w) \leftrightarrow E(u,w)))$ ■

Looking at the (1,1), (1,n+1) case we can finally prove the following:

Theorem 3.3

Let T be an ω-stable theory satisfying (a)-(d). Then $\{(1,kn+1): k \in N\} \cup \{(2,1)\} \subseteq$ CB-Spec T.

Proof: First of all we show that $(1,2n+1) \in$ CB-Spec T.

Let $M_2 = M_1[b]$, where b is a realization of the generic type p_1 over M_1. It follows that M-rank $(tp(b/M_o))$ = M-rank $(tp(b/M_o \cup \{a\})) = \alpha_T$. So a,b are M_o-independent. By lemma 3.2 we deduce that $M_2 = M_o \cup a/E \cup b/E = M_o[a] \cup M_o[b]$. Hence M_2 has CB-type (1,2n+1).

More generally, for every $k \geq 1$, $0 \leq i \leq k$ we set:

* p_i the generic type over M_i;

* a_i a realization of p_i ;

* $M_{i+1} = M_i[a_i]$.

Proceeding as above, we obtain that M_k has CB-type (1,kn+1).

Finally $(2,1) \in$ CB-Spec T:

Looking at the previous construction we set

* $M_1 = M_o[a_1] = M_o \cup a_1/E$

* $M_{n+1} = M_n[a_{n+1}] = M_o \cup \bigcup_{i=1}^{n+1} a_i/E$ for $n \geq 1$

* $M^* = \bigcup_{n \in N} M_n = M_o \cup \bigcup_{i=1}^{\infty} a_i/E$.

Then:

(1) M^* is a model of T, $M^* > M_n$ for every $n \in N$;

(2) Let $X = \phi(M^*;\bar{m}_r)$ be a definable infinite subset of M^*: there

exists $r \in N$ such that all parameters $\bar{m}_r$ belong to M_r.

(3) Consider $X \cap M_o = \{m \in M_r : M_r \models \phi(m,\bar{m}_r) \wedge \bigwedge_{i=1}^{r} \neg E(m,a_i)\}$.

Case 1 $X \cap M_o$ is finite. Therefore

$$M_r \models \exists^{!q} v(\phi(v,\bar{m}_r) \wedge \bigwedge_{i=1}^{r} \neg E(v,a_i))$$

for a suitable non negative integer q: the same holds in M^*, so $X \cap a_n/E = \emptyset$ for all $n > r$.

Case 2 $X \cap M_o$ is cofinite; proceeding as above we deduce $X \cap a_n/E = a_n/E$ for all $n > r$.

(4) In the case 1, X has CB-rank 1, in the case 2, X' has CB-rank 1. Consequently, M^* has CB-type (2,1). ∎

REFERENCES

[B1] - J.Baldwin, α_T is finite for $\aleph_1$-categorical T. Trans.Amer. Math. Soc. 181 (1973) 37-51.

[B2] - J.Baldwin, Conservative extensions and the two cardinal theorem for stable theories, Fund.Math. 88 (1975), 7-9.

[BR] - J.Baldwin - B.Rose, $\aleph_0$-categoricity and stability of rings. J.Algebra 45 (1977) 1-16.

[C] - G.Cherlin, Stable algebraic theories, Logic Colloquium '78, North Holland, Amsterdam (1979), 53-75.

[D] - G.W.Day, Superatomic Boolean Algebras. Pacific J.Math. 23 (1967) 479-489.

[L1] - A.H.Lachlan, Two conjectures regarding the stability of ω-categorical theories. Fund.Math. 81 (1974) 133-145.

[L2] - A.H.Lachlan, Dimension and totally transcendental theories of rank 2. In "Set theory and hierarchy theory - A memorial tribute to A.Mostowski". Springer Lectur Notes n° 537, 153-183.

[La] - D.Lascar, Ordre de Rudin-Keisler and poids dans les theories stable, Zeitchr.f.Math. Logik und Grundlagen d.Math. 28 (1982) 413-430.

[LP] - D.Lascar - B.Poizat, An introduction to forking. J.Symb. Logic 44 (1979), 330-350.

[M] - A.Macintyre, On ω_1-categorical theories of abelian groups. Fund.Math. 70 (1970) 253-270.

[MM1] - P.Mangani - A.Marcja, Shelah rank for Boolean algebras and some application to elementary theories I. Alg.Universalis 10 (1980) 247-257.

[MM2] - P.Mangani, A.Marcja, $\aleph_1$-Boolean spectrum and stability. Atti Accad.Naz.Lincei 72 (1982) 269-272.

[MT] - A.Marcja - C.Toffalori, On pseudo $\aleph_0$-categorical theories.

Zeitschr.f.Math.Logik und Grundlagen d.Math. (to appear).

[P] - K.P.Podewski, Minimale Ringe. Math.Phys. Semesterbeit 22 (1975) 193-197.

[R] - J.Reineke, Minimale Gruppen. Zeitschr. für Math. Logik 21 (1975) 357-359.

[S] - S.Shelah, Classification theory and the number of non isomorphic models. North Holland, Amsterdam (1978).

[T] - C.Toffalori, Cantor-Bendixson spectra of ω-stable theories (to appear).

ABSTRACT MODEL-THEORY AND NETS OF C*-ALGEBRAS: NONCOMMUTATIVE INTERPOLATION AND PRESERVATION PROPERTIES

Daniele Mundici (National Research Council, GNSAGA Group, and Mathematical Institute "U.Dini" of the University of Florence), Loc.Romola n.76, 50060 Donnini (Florence) Italy

With every compact logic L and set T of types directed by inclusion we associate a net of abelian C*-algebras (1.3-1.5). Nets of C*-algebras are widely used in mathematical physics (2.2). All nets considered in relativistic quantum field theory satisfy a property of statistical independence, known as the Schlieder property, with respect to spacelike separation: it turns out that all nets associated with L satisfy the Schlieder property with respect to disjointness of types (2.3). In quantum statistical mechanics nets of C*-algebras often satisfy a strong form of statistical independence (3.1) and/or a weak form of additivity property (4.1): for nets arising from L the first property is equivalent to Craig interpolation (3.2), and the second property is equivalent to the Feferman Vaught ($\equiv_L$ preservation) property for pairs of structures (4.4). We show that any arbitrary net of abelian C*-algebras with totally disconnected compact Hausdorff spectrum can be obtained in our model-theoretical framework (5.2). We discuss the role of AF C*-algebras in the separable case (5.3-5.5). This paper is a first presentation of a construction which is proposed for further study.

1. Framework

Let $L = L(Q^j)_{j\in J}$ be a logic generated by a set of quantifiers [5]. As usual, ϱ, σ, τ denote (many-sorted, similarity) types, also called vocabularies. We let $Str(\tau)$ denote the class of all structures of type τ. For any structure $\mathfrak{M} \in Str(\tau)$ we let $[\mathfrak{M}] = \{\mathfrak{N} \in Str(\tau) \mid \mathfrak{M} \equiv_L \mathfrak{N}\}$, where as usual $\equiv_L$ is L-elementary equivalence [5]. For each type τ we consider the topological space $S_\tau = \{[\mathfrak{M}] \mid \mathfrak{M} \in Str(\tau)\}$ equipped with the topology generated by the family of all sets of the form $X_\varphi = \{[\mathfrak{M}] \in S_\tau \mid \mathfrak{M} \models_L \varphi\}$, for φ ranging over the set $L(\tau)$ of all sentences in L of type τ. S_τ is a totally disconnected Hausdorff space: the family $\{X_\varphi \mid \varphi \in L(\tau)\}$ is a clopen basis for S_τ, since L has negation and conjunction. If $\sigma \subseteq \tau$ are types then the map $\pi^\tau_\sigma : S_\tau \longrightarrow S_\sigma$ given by $\pi^\tau_\sigma([\mathfrak{M}]) = [\mathfrak{M}\upharpoonright^\tau_\sigma]$ for any $\mathfrak{M} \in Str(\tau)$, is well-defined and continuous, since $\varphi \in L(\sigma)$ implies $\varphi \in L(\tau)$; here $\upharpoonright^\tau_\sigma$ denotes reduct to σ of a structure of type τ, [5]. See [6] for all the topological notions used here.

Let T be a nonempty set of types <u>directed</u> by inclusion, in the sense that for any $\varrho, \sigma \in T$ there is $\zeta \in T$ with $\varrho, \sigma \subseteq \zeta$. Consider the inverse topological system $S_T = \{(S_\tau, \pi^\tau_\sigma) \mid \sigma, \tau \in T, \sigma \subseteq \tau\}$,

where the dependence on L is understood. In our paper "Inverse topological systems and compactness in abstract model-theory" (to appear) we proved the following result, which however, shall never be used in this paper:

<u>Theorem</u>. For any $L = L(Q^j)_{j\in J}$ the following are equivalent:

(i) L is compact;

(ii) each mapping π^τ_σ is closed;

(iii) each space S_τ is locally compact and paracompact, and each clopen $X \subseteq S_\tau$ is of finite type, i.e. there is ϱ finite, $\varrho \subseteq \tau$, with $X = (\pi^\tau_\varrho)^{-1} B$ for some clopen $B \subseteq S_\varrho$.

If, in addition, $|J| \leq \omega$, then the compactness of L is also equivalent to the following:

(iv) for each τ, each clopen subset of S_τ is of finite type, and S_τ is homeomorphic to $\varprojlim S_T$, where $T = \{\sigma \subseteq \tau \mid \sigma \text{ finite}\}$. $\square$

Although this paper is independent of the above theorem, still the latter motivates our concentration on <u>compact</u> logics: indeed, the theorem states that in every noncompact logic several (usually, much weaker) corollaries of compactness are missing; thus, function spaces over the S_τ's are more interesting if L is compact. There are many examples of compact logics in (higher) model-theory, the most notable perhaps being first-order logic, $L_{\omega\omega}$, by Gödel's theorem [2]. Other examples are given by the logic with the cofinality-ω quantifier, and related logics [16]. More examples are found in the realm of enriched, e.g. topological, structures [8]. We shall study below the relationship between compact logics and nets of C*-algebras: see [14] and [17] for background on C*-algebras.

<u>1.0 Definition</u>. [10], [1, p.121]. For I a nonempty set, directed by the partial order relation $\leqslant$, a <u>net</u> $\{A, A_\alpha\}_{\alpha \in I}$ <u>of C*-algebras with the same unit</u> 1, is a family of C*-algebras such that $1 \in A_\alpha \subseteq A_\beta \subseteq A$ for every $\alpha, \beta \in I$ with $\alpha \leqslant \beta$, and A is the norm-closure of $\bigcup_{\alpha \in I} A_\alpha$. □

As usual, I being <u>directed</u> by $\leqslant$ means that for any $\alpha, \beta \in I$ there is $\gamma \in I$ with $\alpha \leqslant \gamma$ and $\beta \leqslant \gamma$. Throughout this paper $\{A, A_\alpha\}_{\alpha \in I}$ will always denote a net of C*-algebras with the same unit; unless otherwise specified, I is a nonempty set directed by the partial order relation $\leqslant$. We shall say for short, <u>the net</u> $\{A, A_\alpha\}_{\alpha \in I}$ without any further specification. There is an extensive literature in mathematical physics on nets of C*-algebras, mainly in connection with quantum field theory and statistical mechanics: see, e.g., the references quoted in [1] and [10]. In this paper it will be shown that nets of C*-algebras also describe logics, and provide a possible framework for noncommutative generalizations of basic concepts from higher-model theory.

<u>1.1 Definition</u>. Given a compact logic L and a type τ, consider the C*-algebra $C(S_\tau)$ of all complex-valued continuous functions $f: S_\tau \longrightarrow \mathbb{C}$. Since L is compact then S_τ is compact, and $C(S_\tau)$ has an identity. Given types $\sigma \subseteq \tau$, define the map $\Phi_\sigma^\tau : C(S_\sigma) \longrightarrow C(S_\tau)$ by

$$(*) \qquad \Phi_\sigma^\tau(g) = g \circ \pi_\sigma^\tau \quad , \text{ for any } g \in C(S_\sigma). \qquad \square$$

<u>1.2 Proposition</u>. Each map Φ_σ^τ is a $*$-isomorphism of $C(S_\sigma)$ into $C(S_\tau)$; the range of Φ_σ^τ is the set of those $f \in C(S_\tau)$ which <u>do not separate π_σ^τ-equivalent points</u> (i.e. those $f \in C(S_\tau)$ such that for any $p,q \in S_\tau$, if $\pi_\sigma^\tau(p) = \pi_\sigma^\tau(q)$ then $f(p)=f(q)$). Furthermore, $\Phi_\sigma^\tau(1_\sigma) = 1_\tau$, where 1_σ and 1_τ are the identities of $C(S_\sigma)$ and $C(S_\tau)$ respectively. Also, for any $\rho \subseteq \sigma \subseteq \tau$ we have $\Phi_\rho^\tau = \Phi_\sigma^\tau \circ \Phi_\rho^\sigma$.

<u>Proof</u>. If $f \in C(S_\tau)$ and f does not separate π_σ^τ-equivalent points, then let $c: S_\sigma \to S_\tau$ be an arbitrary but fixed function such that

$$(1)_c \qquad \pi_\sigma^\tau(c(p)) = p, \quad \forall p \in S_\sigma .$$

Thus, c expands $p = [\mathfrak{M}]$ into some $q = [\mathfrak{N}] \in S_\tau$, $\mathfrak{M} = \mathfrak{N} \restriction_\sigma^\tau$.
Let $g: S_\sigma \to \mathbb{C}$ be defined by

$$(2) \qquad g(p) = f(c(p)), \quad \forall p \in S_\sigma .$$

By our assumptions about f, if c' is another function obeying $(1)_{c'}$, and $g' = f \circ c'$, then $g'=g$; thus g does not depend on c. We now note that the following holds:

$$(3) \qquad f = g \circ \pi_\sigma^\tau .$$

Indeed, for any $q \in S_\tau$ we have $(g \circ \pi_\sigma^\tau)(q) = (f \circ c \circ \pi_\sigma^\tau)(q)$; let $q' = (c \circ \pi_\sigma^\tau)(q)$; then $\pi_\sigma^\tau(q') = (\pi_\sigma^\tau \circ c \circ \pi_\sigma^\tau)(q) = \pi_\sigma^\tau(q)$ by (1), so that $f(q') = f(q)$ and $g \circ \pi_\sigma^\tau = f$, which proves (3).
<u>Claim</u>. The function g is continuous.
We must show that for any closed $F \subseteq \mathbb{C}$, $g^{-1}(F)$, i.e., by (2), $c^{-1}(f^{-1}(F))$ is closed. To this purpose, let $G = f^{-1}(F)$; then $G \subseteq S_\tau$ is closed by the assumed continuity of f; also, whenever $q', q'' \in S_\tau$ and $\pi_\sigma^\tau(q') = \pi_\sigma^\tau(q'')$ we have that $q' \in G$ iff $q'' \in G$, since f does not separate π_σ^τ-equivalent points; we now verify that

$$(4) \qquad c^{-1}(G) = \pi_\sigma^\tau(G).$$

Indeed, $p \in c^{-1}(G) \Longrightarrow q = c(p) \in G \Longrightarrow \pi_\sigma^\tau(q) = \pi_\sigma^\tau(c(p)) = p$, by (1), so that $p \in \pi_\sigma^\tau(G)$. Conversely, $p \in \pi_\sigma^\tau(G) \Longrightarrow$ there is $q \in G$ with $p = \pi_\sigma^\tau(q)$; now, $\pi_\sigma^\tau(q) = p = \pi_\sigma^\tau(c(p))$ by (1), hence $f(c(p)) = f(q)$, whence $c(p) \in G$ by our assumption about f. In definitive, $p \in c^{-1}(G)$; this settles (4). We now see that $c^{-1}(G)$ is closed, by (4), because π_σ^τ is a (continuous and) closed mapping, by the assumed compactness of L, and because G is closed. This proves the claim.

Up to this point we have proved that range(Φ_σ^τ) contains the set $\{f \in C(S_\tau) \mid \pi_\sigma^\tau(q') = \pi_\sigma^\tau(q'') \Longrightarrow f(q')=f(q'')\}$. The reverse inclusion is immediate, noting that for every $g \in C(S_\sigma)$, $g \circ \pi_\sigma^\tau$ is a composition of continuous functions, and does not separate π_σ^τ-equivalent points. The identities $\Phi_\sigma^\tau(1_\sigma) = 1_\tau$ and $\Phi_\varrho^\tau = \Phi_\sigma^\tau \circ \Phi_\varrho^\sigma$ are easily proved; for the latter identity one may recall that $\pi_\varrho^\tau = \pi_\varrho^\sigma \circ \pi_\sigma^\tau$, and then use clause (*) in 1.1 . Finally, to prove that Φ_σ^τ is a *-isomorphism, one immediately verifies using (*) that Φ_σ^τ is one-one (recalling that π_σ^τ is onto), and preserves addition, multiplication, involution and scalar multiplication. □

Let T be a nonempty set of types directed by inclusion. The family $\{C(S_\tau)\}_{\tau \in T}$ of C*-algebras with identity, together with the family $\{\Phi_\sigma^\tau \mid \sigma,\tau \in T,\ \sigma \subseteq \tau\}$, where Φ_σ^τ is the *-isomorphism of $C(S_\sigma)$ into $C(S_\tau)$ of Proposition 1.2, satisfies the conditions of [14, 1.23, p.70-71] . Then we have:

1.3 Proposition. For any compact logic L, and any directed set of types $T \neq \emptyset$, there is a net of C*-algebras $\{A, A_\tau\}_{\tau \in T}$ with the same unit 1, together with a family $\{\Phi_\tau\}_{\tau \in T}$ where each Φ_τ is a *-isomorphism of $C(S_\tau)$ onto A_τ , such that $\Phi_\tau \circ \Phi_\sigma^\tau = \Phi_\sigma$ for all $\sigma, \tau \in T$ with $\sigma \subseteq \tau$. In other words, we have the following commutative diagram:

$$\begin{array}{ccc} C(S_\tau) & \xrightarrow{\ \Phi_\tau\ } & A_\tau \\ \uparrow{\scriptstyle \Phi_\sigma^\tau} & & \uparrow{\scriptstyle \text{inclusion}} \\ C(S_\sigma) & \xrightarrow{\ \Phi_\sigma\ } & A_\sigma \end{array} \tag{1}$$

Proof. [14, 1.23.2] □

Given L and T as above, the "uniqueness" of the pair $\{A, A_\tau\}_{\tau \in T}$, $\{\Phi_\tau\}_{\tau \in T}$ is established by the following proposition, whose proof also can be found in [14, 1.23.2] :

<u>1.4 Proposition.</u> Given L and T as in Proposition 1.3, assume $\{B,B_\tau\}_{\tau\in T}$ is a net of C*-algebras with the same unit 1', together with a family $\{\Psi_\tau\}_{\tau\in T}$ where each Ψ_τ is a $*$-isomorphism of $C(S_\tau)$ onto B_τ, such that $\Psi_\tau \circ \Phi_\sigma^\tau = \Psi_\sigma$ for all $\sigma, \tau \in T$ with $\sigma \subseteq \tau$. Then there is a $*$-isomorphism Λ of A onto B such that for all $\tau \in T$, $\Lambda(A_\tau) = B_\tau$ and $\Lambda | A_\tau = \Psi_\tau \circ \Phi_\tau^{-1}$ (where $|$ denotes restriction). □

In the light of Propositions 1.3 and 1.4, we shall freely speak of <u>the net</u> $\{A,A_\tau\}_{\tau\in T}$ <u>associated with</u> L <u>and</u> T. For the sake of notational simplicity we shall not make explicit mention of the fact that the net also carries a family $\{\Phi_\tau\}_{\tau\in T}$ of $*$-isomorphisms obeying (D1); however, the presence of this family shall always be understood.

Given a compact logic L and a directed set of types $T \neq \emptyset$, let $\{A,A_\tau\}_{\tau\in T}$ be the associated net. For each $\tau \in T$ define the map $\theta_\tau : S_\tau \to PA_\tau$ (where PA_τ is the space of <u>pure states</u> of A_τ, a pure state being an extremal positive linear functional of norm one on A_τ, [14], [17]), by the following clause:

(**) $\quad (\theta_\tau(p))(\Phi_\tau(f)) = f(p)$, for any $p \in S_\tau$, and $f \in C(S_\tau)$.

<u>1.5 Proposition.</u> (i) θ_τ is a one-one map of S_τ onto PA_τ ; (ii) for any $p \in S_\tau$ and $\sigma, \tau \in T$ with $\sigma \subseteq \tau$, $\theta_\sigma(\pi_\sigma^\tau(p)) = (\theta_\tau(p))|$ where $|$ denotes restriction; in other words, we have the following commutative diagram:

<u>(D2)</u>

$$\begin{array}{ccc} S_\tau & \xrightarrow{\ \theta_\tau\ } & PA_\tau \\ \downarrow{\scriptstyle \pi_\sigma^\tau} & & \downarrow{\scriptstyle \text{restriction}} \\ S_\sigma & \xrightarrow{\ \theta_\sigma\ } & PA_\sigma \end{array}$$

<u>Proof.</u> (i) This is an immediate application of the Gelfand isomorphi theorem, [14], [17] : one can identify $P(C(S_\sigma))$ and S_σ, writing $f(p) = p(f)$ for any $p \in S_\sigma$ and $f \in C(S_\sigma)$.

(ii) Let p be an arbitrary but fixed element of S_τ. Let $\sigma, \tau \in T$, with $\sigma \subsetneq \tau$. We have to prove the following:

$$\text{for any } x \in A_\sigma \;, \quad (\Theta_\tau(p))(x) = (\Theta_\sigma(\pi^\tau_\sigma(p)))(x).$$

As a matter of fact, for exactly one $f \in C(S_\sigma)$ we can write $x = \Phi_\sigma(f)$, by Proposition 1.3. Then we have :

$$\begin{aligned}(\Theta_\tau(p))(x) = (\Theta_\tau(p))(\Phi_\sigma(f)) &= (\Theta_\tau(p))(\Phi_\tau(\Phi^\tau_\sigma(f))), \quad \text{by (D1)}\\ &= (\Phi^\tau_\sigma(f))(p), \quad \text{by clause (**) above}\\ &= (f \circ \pi^\tau_\sigma)(p), \quad \text{by (*) in 1.1.}\end{aligned}$$

On the other hand, we also have the following:

$$(\Theta_\sigma(\pi^\tau_\sigma(p)))(x) = (\Theta_\sigma(\pi^\tau_\sigma(p)))(\Phi_\sigma(f)) = f(\pi^\tau_\sigma(p)), \text{ by (**)}.$$

This proves (ii), and completes the proof of the proposition. $\square$

2. The Schlieder property.

2.1 Definition. [1, 2.6.3] , [4, 3.3] . Let $\{A, A_\alpha\}_{\alpha \in I}$ be a net; let $\perp$ be a binary symmetric relation on I. Then the net is said to have the commutation property with respect to $\perp$ iff for all $\alpha, \beta \in I$ with $\alpha \perp \beta$ we have $[A_\alpha, A_\beta] = 0$, i.e., $ab = ba$ whenever $a \in A_\alpha$, $b \in A_\beta$. The net $\{A, A_\alpha\}_{\alpha \in I}$ has the Schlieder property with respect to $\perp$ iff for all $\alpha, \beta \in I$ with $\alpha \perp \beta$, for each $a \in A_\alpha$, $b \in A_\beta$ we have that $ab = 0$ implies $a = 0$ or $b = 0$. □

2.2 Examples. (i) In relativistic quantum field theory one considers the net $\{A, A_\alpha\}_{\alpha \in I}$, where A_α denotes the C*-algebra generated by the observables in the bounded open region α of Minkowski space: see [9], [15], [4] and references quoted therein ; I is the totality of such regions, and A = norm-closure of $\bigcup_{\alpha \in I} A_\alpha$. Write $\alpha \times \beta$ iff α and β are spacelike separated; write $\alpha \times\!\!\times \beta$ iff there is an open neighbourhood N of the origin of $\mathbb{R}^4$ such that $(\alpha + z) \times \beta$ for all $z \in N$. Then Schlieder [15] proved that $\{A, A_\alpha\}_{\alpha \in I}$ has the Schlieder property with respect to $\times\!\!\times$. On the other hand, the commutation property of the net with respect to $\times\!\!\times$ (indeed, with respect to $\times$) is a basic postulate in relativistic quantum field theory [9].

(ii) [1, 2.6.12] Let $\mathfrak{J} \neq \emptyset$ be a set, and I_f the set of finite subsets of $\mathfrak{J}$ directed by inclusion. Assume each $i \in \mathfrak{J}$ is associated with a finite-dimensional Hilbert space H_i; then one can associate with each $\lambda \in I_f$ the C*-algebra $A_\lambda = \mathcal{L}(\bigotimes_{i \in \lambda} H_i)$ of linear operators on the tensor product space $\bigotimes_{i \in \lambda} H_i$: see [17] for tensor products. Write $\lambda_1 \perp \lambda_2$ iff λ_1 and λ_2 are disjoint. Note that if this is the case, then A_{λ_1} is isomorphic to the C*-subalgebra $A_{\lambda_1} \otimes 1\!\!1_{\lambda_2}$ of $A_{\lambda_1 \cup \lambda_2}$, where $1\!\!1_{\lambda_2}$ is the identity operator on H_{λ_2}. Identifying A_{λ_1} and $A_{\lambda_1} \otimes 1\!\!1_{\lambda_2}$ (or, using [14, 1.23.2]) one gets an increasing family of matrix algebras, which we also write $\{A_\lambda\}_{\lambda \in I_f}$. Let A be the norm-completion of $\bigcup_{\lambda \in I_f} A_\lambda$. Then $\{A, A_\lambda\}_{\lambda \in I_f}$ is a net of C*-algebras having both the commutation and the Schlieder property with respect to $\perp$. These nets are important in the study of quantum statistical mechanical spin systems [1, 6.2] .

In this paper the binary relation $\perp$ on the index set I is only assumed to be symmetric: compare with [1, 2.6.3] where additional conditions are imposed on $\perp$. We now turn to nets associated with compact logics (see after Proposition 1.4). Evidently, the commutation property with respect to $\perp$ holds, no matter $\perp$. Concerning the Schlieder property, we have the following:

<u>2.3 Theorem</u>. Let L be a compact logic, T a nonempty set of types directed by inclusion, $\{A, A_\tau\}_{\tau \in T}$ the associated net. For any $\varrho, \sigma \in T$ write $\varrho \perp \sigma$ iff $\varrho \cap \sigma = \emptyset$. Then $\{A, A_\tau\}_{\tau \in T}$ has the Schlieder property with respect to $\perp$.

<u>Proof</u>. Assume $\varrho, \sigma \in T$, $\varrho \perp \sigma$, $a \in A_\varrho$, $b \in A_\sigma$, $a, b \neq 0$, $ab = 0$ (absurdum hypothesis). Without loss of generality, $a, b > 0$. Let $\mathbb{R}^+ = \{z \in \mathbb{R} \mid z > 0\}$. Referring to (D1) in Proposition 1.3, let $g = \Phi_\varrho^{-1}(a) \in C(S_\varrho)$. Then there exists $p \in S_\varrho$ with $g(p) > 0$. For suitable $\mathfrak{M} \in Str(\varrho)$ we can write $p = [\mathfrak{M}] = \{\mathfrak{M}' \in Str(\varrho) \mid \mathfrak{M}' \equiv_L \mathfrak{M}\}$. Let similarly $h = \Phi_\sigma^{-1}(b)$; there is $q \in S_\sigma$ such that $h(q) > 0$, with $q = [\mathfrak{N}]$, for suitable $\mathfrak{N} \in Str(\sigma)$. Let $\mathfrak{B} = \langle \mathfrak{M}, \mathfrak{N} \rangle \in Str(\varrho \cup \sigma)$ be the disjoint pair of $\mathfrak{M}$ and $\mathfrak{N}$, which certainly exists because $\varrho \perp \sigma$. Let $\tau \in T$ be such that $\varrho, \sigma \subseteq \tau$; the existence of τ is ensured by T being directed by inclusion. Expand $\mathfrak{B}$ to some $\mathfrak{D} \in Str(\tau)$ and let $r = [\mathfrak{D}] \in S_\tau$. Note that $p = \pi_\varrho^\tau(r)$, and $q = \pi_\sigma^\tau(r)$. Referring to (D1), let $f = \Phi_\tau^{-1}(a) \in C(S_\tau)$: this makes sense, since $a \in A_\varrho \subseteq A_\tau$ and $\tau \in T$. Then we have $f(r) =$

$(\Phi_\tau^{-1}(a))(r) = (\Phi_\varrho^\tau \circ \Phi_\varrho^{-1}(a))(r)$, by (D1)

$= (\Phi_\varrho^\tau(g))(r) = (g \circ \pi_\varrho^\tau)(r)$, by clause (*) in 1.1.

Now observe that $g(\pi_\varrho^\tau(r)) = g(p) > 0$. Similarly, letting $d = \Phi_\tau^{-1}(b)$, we have $d(r) > 0$. Hence, $f(r) \cdot d(r) > 0$, whence $ab \neq 0$, a contradiction. □

One important aspect of the Schlieder property for nets $\{A, A_\alpha\}_{\alpha \in I}$ is in connection with the notion of <u>statistical independence</u> of algebras of regions $\alpha \perp \beta$. This notion may be given the following precise formulation, [18], [9, p.852] : given arbitrary pure states $\omega_1 \in PA_\alpha$ and $\omega_2 \in PA_\beta$, for all $\gamma \geq \alpha, \beta$ there exists $\omega \in PA_\gamma$ which jointly extends ω_1 and ω_2, in the sense that $\omega | A_\alpha = \omega_1$ and $\omega | A_\beta = \omega_2$.

In relativistic quantum field theory, using the Schlieder property (see example 2.2(i)), Roos [13] proved that any two regions $\alpha \times\!\times \beta$ in Minkowski space are statistically independent: this answers a problem of Haag and Kastler [9, p. 852]. The following result shows that the Schlieder property coincides with the above notion of statistical independence. Note that we are dealing with the general (nonabelian) case:

2.4 Theorem. Assume the net $\{A, A_\alpha\}_{\alpha\in I}$ has the commutation property with respect to $\perp$, for $\perp$ a binary symmetric relation on I. Then the following are equivalent:

(i) the net has the Schlieder property with respect to $\perp$;

(ii) for any two pure states $\omega_1 \in PA_\alpha$, $\omega_2 \in PA_\beta$, if $\alpha \perp \beta$ then for all $\gamma \geq \alpha, \beta$ there is $\omega \in PA_\gamma$ which jointly extends ω_1 and ω_2, i.e., $\omega|A_\alpha = \omega_1$ and $\omega|A_\beta = \omega_2$.

Proof. (i)$\Rightarrow$(ii). [13, Theorem 1 and Proposition 1]. (ii) $\Rightarrow$ (i): The proof is an adaptation of the Appendix of [13], as follows: Assume (i) does not hold, and let $a_1 \in A_\alpha$, $a_2 \in A_\beta$, $a_1, a_2 \neq 0$, $a_1 a_2 = 0$ (absurdum hypothesis). Without loss of generality $a_1 > 0$ and 1 belongs to the spectrum [14], [17] of a_1. Let $A_1 = \overline{\{\mathbb{1}, a_1\}} \subseteq A_\alpha$, where $\mathbb{1}$ is the unit of A and $\overline{\{\mathbb{1}, a_1\}}$ is the C*-subalgebra of A_α generated by $\mathbb{1}$ and a_1. Using the Gelfand map [14], [17] one exhibits $\tilde{\omega} \in PA_1$ such that $\tilde{\omega}(\mathbb{1} - a_1) = \tilde{\omega}((\mathbb{1} - a_1)^2) = 0$. Let $\omega_1 \in PA_\alpha$ be an extension of $\tilde{\omega}$ (the existence of such extension is a well-known fact in C*-algebra theory, [14], [17]). Let $\omega_2 \in PA_\beta$ be such that $\omega_2(a_2) \neq 0$. Let $\gamma \geq \alpha, \beta$: γ exists because I is directed by $\geq$: recall our convention after Definition 1.0. Using hypothesis (ii), let $\omega \in PA_\gamma$ be a joint extension of ω_1 and ω_2. Using the Cauchy-Schwartz inequality we get:

$$0 = \omega((\mathbb{1} - a_1)^2) \cdot \omega((\mathbb{1} + a_2)^2) \geq |\omega((\mathbb{1} - a_1)(\mathbb{1} + a_2))|^2$$

$$= |\omega(\mathbb{1} - a_1 + a_2 - a_1 a_2)|^2 = |\tilde{\omega}(\mathbb{1} - a_1) + \omega_2(a_2) - \omega(a_1 a_2)|^2 > 0,$$

a contradiction. □

3. Schlieder, joint extension and Craig interpolation property.

By Theorem 2.4 the Schlieder property with respect to $\perp$ amounts to postulating the existence of joint extensions of pure states of C*-algebras A_α and A_β whenever $\alpha \perp \beta$. One might be interested in situations where joint extensions exist independently of any binary symmetric relation $\perp$. One obvious necessary condition for joint extensibility of functionals is that they agree on their common domain; when this also happens to be a sufficient condition, we have the strongest possible form of joint extension property. The following is a precise definition.

3.1 Definition. A net $\{A, A_\alpha\}_{\alpha \in I}$ has the joint extension property iff for all $\alpha, \beta, \gamma, \delta \in I$ with $\delta = \inf(\alpha, \beta)$ and $\gamma \geq \alpha, \beta$, every two pure states $\omega_1 \in PA_\alpha$ and $\omega_2 \in PA_\beta$ whose restrictions to A_δ are equal to the same pure state $\omega_0 \in PA_\delta$ have a joint pure extension in A_γ, i.e. there is $\omega \in PA_\gamma$ with $\omega | A_\alpha = \omega_1$ and $\omega | A_\beta = \omega_2$. □

Examples and remarks. In a forthcoming paper we shall investigate the role of the joint extension property in noncommutative nets, with particular reference to nets describing quantum statistical systems. We remark here that all nets described in example 2.2(ii) above have the joint extension property. In this paper we shall relate the joint extension property with the Craig interpolation property: the latter is an important notion in (higher) model-theory [2], [8], [5], [11], [12]. Recall that a logic L has the(Craig) interpolation property iff whenever $\psi_1 \in L(\tau_1)$, $\psi_2 \in L(\tau_2)$ are sentences and $\psi_1 \longrightarrow \psi_2$ is valid in L there is a sentence $\chi \in L(\tau_1 \cap \tau_2)$ such that both $\psi_1 \longrightarrow \chi$ and $\chi \longrightarrow \psi_2$ are valid in L. Craig [3] proved the interpolation theorem for first-order logic. Further examples of compact logics with the interpolation property are known in the realm of logics for enriched structures [8]; compact logics with interpolation have a nice theory: see, e.g. [12]; see [11] for further results in the absence of compactness. The following result shows that the joint extension property is a noncommutative generalization of the Craig interpolation property:

3.2 Theorem. For any compact logic L the following are equivalent:
(i) L has the Craig interpolation property;
(ii) for any set of types $T \neq \emptyset$, closed under intersection and directed by inclusion, the associated net $\{A, A_\tau\}_{\tau \in T}$ has the joint extension property;
(iii) same as (ii), but with T an arbitrary set of finite types.

Proof. (i)⟹(ii). In higher model-theory it is well-known [5] that compactness plus interpolation for L implies the Robinson property for L, e.g., in the following sense [11], [12]:

$$\forall \mathfrak{M}_1 \in \mathrm{Str}(\varrho), \mathfrak{M}_2 \in \mathrm{Str}(\sigma), \text{ if } \xi = \varrho \cap \sigma \text{ and } \mathfrak{M}_1 \restriction^{\varrho}_{\xi} \equiv_L \mathfrak{M}_2 \restriction^{\sigma}_{\xi},$$
$$\text{then } \exists \mathfrak{N} \in \mathrm{Str}(\varrho \cup \sigma) \text{ with } \mathfrak{N} \restriction^{\varrho\cup\sigma}_{\varrho} \equiv_L \mathfrak{M}_1 \text{ and } \mathfrak{N} \restriction^{\varrho\cup\sigma}_{\sigma} \equiv_L \mathfrak{M}_2.$$

Stated otherwise, $\forall p_1 \in S_\varrho$, $p_2 \in S_\sigma$, if $\pi^{\varrho}_{\xi}(p_1) = \pi^{\sigma}_{\xi}(p_2)$ and $\xi = \varrho \cap \sigma$, then $\exists\, q \in S_{\varrho\cup\sigma}$ with $\pi^{\varrho\cup\sigma}_{\varrho}(q) = p_1$ and $\pi^{\varrho\cup\sigma}_{\sigma}(q) = p_2$. This in turn is equivalent to saying that $\forall \varrho, \sigma, \xi$ with $\xi = \varrho \cap \sigma$, and $\zeta \supseteq \varrho, \sigma$, if $\pi^{\varrho}_{\xi}(p_1) = \pi^{\sigma}_{\xi}(p_2)$ then $\exists\, q \in S_\zeta$ with $\pi^{\zeta}_{\varrho}(q) = p_1$ and $\pi^{\zeta}_{\sigma}(q) = p_2$. Assume now $\omega_0 \in PA_\xi$, $\omega_1 \in PA_\varrho$, $\omega_2 \in PA_\sigma$ with $\omega_1 | A_\xi = \omega_2 | A_\xi = \omega_0$, and $\xi = \varrho \cap \sigma$, as in the hypothesis of the joint extension property (where $\xi, \varrho, \sigma \in T$). Using (D2) in Proposition 1.5, let $p_1 = \theta^{-1}_{\varrho}(\omega_1)$, $p_2 = \theta^{-1}_{\sigma}(\omega_2)$, and $p_0 = \theta^{-1}_{\xi}(\omega_0)$. By (D2) we have $\pi^{\varrho}_{\xi}(p_1) = \pi^{\varrho}_{\xi}(\theta^{-1}_{\varrho}(\omega_1)) = \theta^{-1}_{\xi}(\omega_1 | A_\xi) = p_0$. Similarly, $\pi^{\sigma}_{\xi}(p_2) = p_0$. Let $\zeta \in T$ obey $\zeta \supseteq \varrho, \sigma$. By the above formulation of hypothesis (i) in terms of the Robinson property of L, there exists $q \in S_\zeta$ such that $\pi^{\zeta}_{\varrho}(q) = p_1$ and $\pi^{\zeta}_{\sigma}(q) = p_2$. Now, for exactly one $\omega \in PA_\zeta$ we can write $q = \theta^{-1}_{\zeta}(\omega)$. By (D2) we have $\theta^{-1}_{\varrho}(\omega | A_\varrho) = \pi^{\zeta}_{\varrho}(\theta^{-1}_{\zeta}(\omega)) = p_1 = \theta^{-1}_{\varrho}(\omega_1)$, and, similarly, $\theta^{-1}_{\sigma}(\omega | A_\sigma) = \pi^{\zeta}_{\sigma}(\theta^{-1}_{\zeta}(\omega)) = p_2 = \theta^{-1}_{\sigma}(\omega_2)$. Therefore $\omega | A_\varrho = \omega_1$ and $\omega | A_\sigma = \omega_2$, as required to prove that (i) implies (ii).

(ii) $\Longrightarrow$ (i). It is well-known in higher model-theory [5] that for any compact logic L the Robinson property implies the Craig interpolation property. Therefore it suffices to prove that property (ii) implies the Robinson property, e.g., in the first formulation given above. To this purpose, assume ξ, ϱ, σ are types with $\xi = \varrho \cap \sigma$. Assume $\mathfrak{M}_1 \in \mathrm{Str}(\varrho)$, $\mathfrak{M}_2 \in \mathrm{Str}(\sigma)$ with $\mathfrak{M}_1 \upharpoonright^{\varrho}_{\xi} \equiv_L \mathfrak{M}_2 \upharpoonright^{\sigma}_{\xi}$. Write $p_1 = [\mathfrak{M}_1] = \{\mathfrak{N} \in \mathrm{Str}(\varrho) \mid \mathfrak{N} \equiv_L \mathfrak{M}_1\}$. Let $p_0 = \pi^{\varrho}_{\xi}(p_1)$. Write $p_2 = [\mathfrak{M}_2] \in S_\sigma$. Note that $\pi^{\varrho}_{\xi}(p_1) = \pi^{\sigma}_{\xi}(p_2) = p_0$. Using (D2) and letting $\omega_1 = \theta_\varrho(p_1)$, $\omega_2 = \theta_\sigma(p_2)$, $\omega_0 = \theta_\xi(p_0)$, we have $\omega_1 | A_\xi = \omega_0 = \omega_2 | A_\xi$. By hypothesis (ii) applied to the net $\{A_{\varrho \cup \sigma}, A_\varrho, A_\sigma, A_\xi\}$, there exists $\omega \in PA_{\varrho\cup\sigma}$ with $\omega | A_\varrho = \omega_1$ and $\omega | A_\sigma = \omega_2$. Again using (D2) and letting $p = \theta^{-1}_{\varrho\cup\sigma}(\omega)$, one has $\pi^{\varrho\cup\sigma}_{\varrho}(p) = p_1$ and $\pi^{\varrho\cup\sigma}_{\sigma}(p) = p_2$. Write $p = [\mathfrak{N}]$ for some $\mathfrak{N} \in \mathrm{Str}(\varrho \cup \sigma)$. Then $\mathfrak{N}\upharpoonright^{\varrho\cup\sigma}_{\varrho} \equiv_L \mathfrak{M}_1$ and $\mathfrak{N}\upharpoonright^{\varrho\cup\sigma}_{\sigma} \equiv_L \mathfrak{M}_2$, as required to prove (i).

(ii) $\Longrightarrow$ (iii). Trivial.

(iii) $\Longrightarrow$ (i). The same argument used to prove (ii) $\Longrightarrow$ (i) yields, using the weaker hypothesis (iii), the so-called Robinson property for finite types, in the sense that $\forall \xi, \varrho, \sigma$ <u>finite</u> types with $\xi = \varrho \cap \sigma$, $\forall \mathfrak{M}_1 \in \mathrm{Str}(\varrho), \mathfrak{M}_2 \in \mathrm{Str}(\sigma)$, if $\mathfrak{M}_1 \upharpoonright^{\varrho}_{\xi} \equiv_L \mathfrak{M}_2 \upharpoonright^{\sigma}_{\xi}$ then $\exists \mathfrak{N} \in \mathrm{Str}(\varrho \cup \sigma)$ with $\mathfrak{N}\upharpoonright^{\varrho\cup\sigma}_{\varrho} \equiv_L \mathfrak{M}_1$ and $\mathfrak{N}\upharpoonright^{\varrho\cup\sigma}_{\sigma} \equiv_L \mathfrak{M}_2$.

Now this weaker form of Robinson property is well-known [5] to be sufficient to give the interpolation property, in the case of a compact logic L. $\square$

4. Schlieder property, factorization and preservation.

As shown in this section, a second important aspect of the Schlieder property for a net $\{A,A_\alpha\}_{\alpha\in I}$ is in connection with the basic inclusion

$$A_\alpha \subseteq \overline{A_\alpha \vee A_\beta} \supseteq A_\beta ,$$

where $A_\alpha \vee A_\beta$ is the normed involutive subalgebra of A generated by A_α and A_β, and $\overline{A_\alpha \vee A_\beta}$ is the C*-subalgebra of A generated by A_α and A_β. In case $\sup(\alpha,\beta)$ exists, one may also consider the basic inclusion

$$\overline{A_\alpha \vee A_\beta} \subseteq A_{\sup(\alpha,\beta)}.$$

Nets which are well-behaved with respect to the above inclusions will be defined in 4.1 below. Recall that $A_\alpha \otimes A_\beta$ is the algebraic tensor product of A_α and A_β, and $A_\alpha \otimes_{min} A_\beta$ is the completion of $A_\alpha \otimes A_\beta$ with respect to minimal norm [17, §4 IV] .

<u>4.1 Definition</u>. Given a net $\{A,A_\alpha\}_{\alpha\in I}$ and a binary symmetric relation $\perp$ on I, we say that the net has the <u>factorization property with respect to $\perp$</u>, iff for all $\alpha,\beta \in I$ with $\alpha\perp\beta$ the homomorphism

$$\sum_{i=1}^{n} x_i \otimes y_i \in A_\alpha \otimes A_\beta \longmapsto \sum_{i=1}^{n} x_i y_i \in A_\alpha \vee A_\beta$$

can be extended to a C*-isomorphism of $A_\alpha \otimes_{min} A_\beta$ onto $\overline{A_\alpha \vee A_\beta}$.

We also say that $\{A,A_\alpha\}_{\alpha\in I}$ has the <u>union property with respect to $\perp$</u> iff $(I,\leqslant)$ is closed under sup and $A_{\sup(\alpha,\beta)} = \overline{A_\alpha \vee A_\beta}$ for all $\alpha,\beta \in I$ with $\alpha\perp\beta$. □

<u>4.2 Proposition</u>. Given a net $\{A,A_\alpha\}_{\alpha\in I}$ and a binary symmetric relation $\perp$ on I, assume the net to have both the commutation and the Schlieder property with respect to $\perp$. It follows that for all $\alpha,\beta\in I$ with $\alpha\perp\beta$, if either A_α or A_β is abelian, then the homomorphism

$$\sum_{i=1}^{n} x_i \otimes y_i \in A_\alpha \otimes A_\beta \longmapsto \sum_{i=1}^{n} x_i y_i \in A_\alpha \vee A_\beta$$

can be extended to a C*-isomorphism of $A_\alpha \otimes_{min} A_\beta$ onto $\overline{A_\alpha \vee A_\beta}$.

<u>Proof</u>. [17, IV §4 ex.2, p.220] , or [13, Theorem 2] . □

Thus in particular the factorization property with respect to $\perp$ holds in all nets of abelian C*-algebras having the Schlieder property with respect to $\perp$. Concerning nets of nonabelian C*-algebras one immediately sees that the nets described in example 2.2(ii) above obey both the factorization and the union property; see [1, 2.6.12 and 6.2] for details. In this paper we shall relate both these properties to the familiar Feferman-Vaught property in higher model-theory.

4.3 Definition. A logic L has the Feferman-Vaught property (for pairs), for short FVP, iff for all types ϱ, σ with $\varrho \cap \sigma = \emptyset$, if $\mathfrak{M}_1, \mathfrak{N}_1 \in \mathrm{Str}(\varrho)$ and $\mathfrak{M}_2, \mathfrak{N}_2 \in \mathrm{Str}(\sigma)$, with $\mathfrak{M}_1 \equiv_L \mathfrak{N}_1$ and $\mathfrak{M}_2 \equiv_L \mathfrak{N}_2$, then $\langle \mathfrak{M}_1, \mathfrak{M}_2 \rangle \equiv_L \langle \mathfrak{N}_1, \mathfrak{N}_2 \rangle$, where $\langle\ ,\ \rangle$ denotes the disjoint pairing of structures. $\square$

Feferman and Vaught [7] proved that first-order logic has FVP. See [8] for more examples of logics with FVP, in the realm of enriched structures. In model-theoretical terminology, the FVP is a typical preservation property, for it amounts to the requirement that $\equiv_L$ is preserved under formation of pairs of structures.

The following theorem shows the role of the factorization and/or the union property as noncommutative generalizations of the Feferman-Vaught property:

4.4 Theorem. For any compact logic L the following are equivalent:

(i) L has FVP;

(ii) for every set of types $T \neq \emptyset$ closed under union, the associated net $\{A, A_\tau\}_{\tau \in T}$ has the factorization and the union properties with respect to the disjointness relation $\perp$;

(iii) for every set of types $T \neq \emptyset$ closed under union, the associated net has the union property with respect to the disjointness relation $\perp$.

For the proof we prepare the following:

4.5 Lemma. If L is compact and has FVP, then for any ϱ, σ with $\varrho \cap \sigma = \emptyset$ the map $h: S_{\varrho\cup\sigma} \longrightarrow S_\varrho \times S_\sigma$ given by $h(p) = (\pi_\varrho^{\varrho\cup\sigma}(p), \pi_\sigma^{\varrho\cup\sigma}(p))$ for any $p \in S_{\varrho\cup\sigma}$, is a one-one map.

<u>Proof of Lemma.</u> Assume $p'\neq p''$, $p',p''\in S_{\varrho\cup\sigma}$ but $h(p')=h(p'')$, (absurdum hypothesis). Write $p' = [\langle\mathfrak{M}',\mathfrak{N}'\rangle]$, $p'' = [\langle\mathfrak{M}'',\mathfrak{N}''\rangle]$ for suitable structures $\mathfrak{M}',\mathfrak{M}''\in \mathrm{Str}(\varrho)$ and $\mathfrak{N}',\mathfrak{N}''\in \mathrm{Str}(\sigma)$. Then by hypothesis we have $[\mathfrak{M}']=[\mathfrak{M}'']$ and $[\mathfrak{N}'] = [\mathfrak{N}'']$. By FVP, $p'=p''$, a contradiction.

<u>Proof of Theorem.</u> (ii)$\Longrightarrow$(iii). Trivial. (iii)$\Longrightarrow$(ii) is a consequence of Theorem 2.3 and Proposition 4.2. We now prove (i)$\Longrightarrow$(iii) i.e. FVP implies $\overline{A_\varrho \vee A_\sigma} = A_{\varrho\cup\sigma}$ whenever $\varrho\perp\sigma$. In the light of the Stone-Weierstrass theorem [6],[14], it suffices to prove that $\overline{A_\varrho \vee A_\sigma}$ separates the pure states of $A_{\varrho\cup\sigma}$. To this purpose, write $A_{\varrho\cup\sigma} = C(S_{\varrho\cup\sigma})$ using the Gelfand map [14],[17]. Assume p', $p''\in S_{\varrho\cup\sigma}$ and $p'\neq p''$. By Lemma 4.5 either $\pi_\varrho^{\varrho\cup\sigma}(p')\neq \pi_\varrho^{\varrho\cup\sigma}(p'')$ or $\pi_\sigma^{\varrho\cup\sigma}(p')\neq \pi_\sigma^{\varrho\cup\sigma}(p'')$, say without loss of generality $\pi_\varrho(p')\neq\pi_\varrho(p'')$ we drop the superscript $\varrho\cup\sigma$ in $\pi_\varrho^{\varrho\cup\sigma}$ for simplicity. For suitable $\mathfrak{M}'$, $\mathfrak{M}''\in \mathrm{Str}(\varrho)$ we have $\pi_\varrho(p')=[\mathfrak{M}']$ and $\pi_\varrho(p'')=[\mathfrak{M}'']$. Since $[\mathfrak{M}']\neq[\mathfrak{M}'']$ then there is a sentence $\chi\in L(\varrho)$ such that $\mathfrak{M}'\models_L \chi$ and $\mathfrak{M}''\models_L \neg\chi$ where, as usual, $\models_L$ is the satisfaction relation in L [5]. In other words, there is a basic clopen $X\subseteq S_\varrho$ such that $\pi_\varrho(p')\in X$ and $\pi_\varrho(p'')\notin X$. Since S_ϱ is a totally disconnected compact Hausdorff space, there is a $\{0,1\}$-valued function $e\in C(S_\varrho)$ such that $e(\pi_\varrho(p'))=1$ and $e(\pi_\varrho(p''))=0$, see [6]. Now, $e\circ\pi_\varrho\in C(S_{\varrho\cup\sigma})$, as a composition of continuous functions and does not separate π_ϱ-equivalent points in $S_{\varrho\cup\sigma}$ (see Proposition 1.2). Hence, by Proposition 1.2 $e\circ\pi_\varrho\in A_\varrho\subseteq A_{\varrho\cup\sigma} = C(S_{\varrho\cup\sigma})$, and $e\circ\pi_\varrho$ separates p' and p''. Now by the Stone-Weierstrass theorem we conclude that $\overline{A_\varrho \vee A_\sigma} = A_{\varrho\cup\sigma}$.

We now prove that (ii)$\Longrightarrow$(i). Assume (ii) holds and FVP does not hold (absurdum hypothesis). There are types $\varrho\perp\sigma$ and structures $\mathfrak{M}_1$, $\mathfrak{N}_1\in \mathrm{Str}(\varrho)$, $\mathfrak{M}_2$, $\mathfrak{N}_2\in\mathrm{Str}(\sigma)$ with $\mathfrak{M}_1\equiv_L\mathfrak{N}_1$ and $\mathfrak{M}_2\equiv_L\mathfrak{N}_2$ and $\langle\mathfrak{M}_1,\mathfrak{M}_2\rangle\not\equiv_L\langle\mathfrak{N}_1,\mathfrak{N}_2\rangle$. Let $p_1=[\mathfrak{M}_1]\in S_\varrho$ $p_2=[\mathfrak{M}_2]$, $q_1=[\mathfrak{N}_1]$, $q_2=[\mathfrak{N}_2]$, $p=[\langle\mathfrak{M}_1,\mathfrak{M}_2\rangle]$, $q=[\langle\mathfrak{N}_1,\mathfrak{N}_2\rangle]$. We have the following

(1) $p\neq q$, $p_1=\pi_\varrho(p)=\pi_\varrho(q)=q_1$; $p_2=\pi_\sigma(p)=\pi_\sigma(q)=q_2$.

Let $\zeta = \varrho \cup \sigma$. Since S_ζ is a totally disconnected compact Hausdorff space, there is a basic clopen $X \subseteq S_\zeta$ with $p \in X$ and $q \notin X$, hence there is a $\{0,1\}$-valued function $e \in C(S_\zeta)$ such that $e(p) = 1$ and $e(q) = 0$. Identify A_ζ with $C(S_\zeta)$ using the Gelfand map. Define μ, ν pure states in PA_ζ by

(2) $\mu(x) = x(p)$, $\nu(x) = x(q)$, for any $x \in A_\zeta$, i.e.,

$$\mu = \theta_\zeta(p), \qquad \nu = \theta_\zeta(q),$$

referring to clause (**) before Proposition 1.5, and recalling that $A_\zeta = C(S_\zeta)$. Then $\mu(e) = 1$ and $\nu(e)=0$. By our hypothesis (ii), $e \in \overline{A_\varrho \vee A_\sigma} = C(S_\zeta)$ and hence e is the norm-limit of finite sums of the form $\Sigma x_i y_i$ with $x_i \in A_\varrho$ and $y_i \in A_\sigma$. By suitably choosing the x_i and y_i we can impose that $\| e - \sum_{i=1}^{n} x_i y_i \| < 1/3$, hence, a-fortiori, (writing Σ for simplicity), $(\Sigma x_i y_i)(p) \neq (\Sigma x_i y_i)(q)$, i.e., by (2), $\mu(\Sigma x_i y_i) \neq \nu(\Sigma x_i y_i)$, hence by linearity, $\Sigma \mu(x_i y_i) \neq \Sigma \nu(x_i y_i)$. Recalling that A_ζ is abelian and hence that restrictions of pure states are pure states [14], [17], we can apply [17, IV 4.11] to the effect that μ and ν factorize, i.e. we can write $\Sigma \mu(x_i) \cdot \mu(y_i) \neq \Sigma \nu(x_i) \cdot \nu(y_i)$. Using restrictions we can write:

(3) $$\Sigma (\mu \mid A_\varrho)(x_i) \cdot (\mu \mid A_\sigma)(y_i) \neq \Sigma (\nu \mid A_\varrho)(x_i) \cdot (\nu \mid A_\sigma)(y_i).$$

In the light of (D1), Proposition 1.3, there exist uniquely determined $f_i \in C(S_\varrho)$, $g_i \in C(S_\sigma)$ such that

(4) $\Phi_\varrho(f_i) = x_i$, and $\Phi_\sigma(g_i) = y_i$.

Using the fact that, e.g., $\mu \mid A_\varrho = \theta_\varrho \theta_\varrho^{-1}(\mu \mid A_\varrho)$, in the light of (4) and of (**) we can write (3) as follows:

$$\Sigma f_i(\theta_\varrho^{-1}(\mu \mid A_\varrho)) g_i(\theta_\sigma^{-1}(\mu \mid A_\sigma)) \neq \Sigma f_i(\theta_\varrho^{-1}(\nu \mid A_\varrho)) g_i(\theta_\sigma^{-1}(\nu \mid A_\sigma)),$$

and by (D2), Proposition 1.5, we have

$$\Sigma f_i(\pi_\varrho^\zeta \circ \theta_\zeta^{-1}(\mu)) g_i(\pi_\sigma^\zeta \circ \theta_\zeta^{-1}(\mu)) \neq \Sigma f_i(\pi_\varrho^\zeta \circ \theta_\zeta^{-1}(\nu)) g_i(\pi_\sigma^\zeta \circ \theta_\zeta^{-1}(\nu)),$$

i.e., recalling (2), $\Sigma (f_i(\pi_\varrho^\zeta(p)))(g_i(\pi_\sigma^\zeta(p))) \neq \Sigma (f_i(\pi_\varrho^\zeta(q)))(g_i(\pi_\sigma^\zeta(q)))$, which contradicts (1). □

5. From C*-nets back to logics. The role of AF C*-algebras

We now study the problem, posed by one of the referees, whether every inductive limit of commutative C*-algebras with totally disconnected compact Hausdorff spectrum can be obtained by our model-theoretical methods. It turns out that for an arbitrary fixed compact logic L, nets of quotients of the $C(S_\tau)$ yield all such inductive limits (Theorem 5.2). For simplicity we shall work with sentential logic $\mathcal{L}$: the set up of section 1 is easily adapted to $\mathcal{L}$: a type τ (denoted by $\mathcal{I}$ in [2]) is now a set of sentence symbols (propositional variables), and the set $\mathcal{L}(\tau)$ of sentences of type τ is defined as in [2, 1.2.2]. Note that for any $\psi \in \mathcal{L}(\tau)$ there is a finite $\sigma \subseteq \tau$ such that $\psi \in \mathcal{L}(\sigma)$. A structure of type τ is now a function $\mathfrak{M}$ which associates with each sentence symbol $S \in \tau$ one of the truth values 1 (true) or 0 (false): we denote by 2^τ the set of all structures of type τ [2, 1.2.1] . The satisfaction relation $\mathfrak{M} \models \psi$ is defined as in [2,1.2.3] . Note that two structures $\mathfrak{N}$ and $\mathfrak{M}$ satisfy the same sentences iff $\mathfrak{N} = \mathfrak{M}$. Thus in sentential logic $\mathcal{L}$, the $\mathcal{L}$-equivalence relation $\equiv_{\mathcal{L}}$ coincides with the identity relation =. One defines the (Stone) space S_τ exactly as we did in section 1 by equipping the set 2^τ of all ($\equiv_{\mathcal{L}}$-equivalence classes of) structures of type τ with the topology generated by the sets $X_\psi = \{\mathfrak{M} \in 2^\tau \mid \mathfrak{M} \models \psi\}$, for all $\psi \in \mathcal{L}(\tau)$. By the Gödel-Maltsev theorem [2,1.2.1] S_τ is a compact (totally disconnected, Hausdorff) space. A <u>theory</u> in $\mathcal{L}(\tau)$ is a set of sentences of $\mathcal{L}(\tau)$, [2,p.12]. Given a theory $\Gamma \subseteq \mathcal{L}(\tau)$ we let $\mathrm{Mod}\,\Gamma = \{\mathfrak{M} \in 2^\tau \mid \mathfrak{M} \models \psi \text{ for all } \psi \in \Gamma\}$ = the set of models of Γ ; the dependence on τ shall always be clear from the context. Note that $\mathrm{Mod}\,\Gamma$ is a closed, hence compact, subspace of S_τ . We denote by $C(\mathrm{Mod}\,\Gamma)$ the C*-algebra of all complex-valued continuous functions over $\mathrm{Mod}\,\Gamma$. $C(\mathrm{Mod}\,\Gamma)$ is isomorphic to the quotient abelian C*-algebra $C(S_\tau)/J_\Gamma$, where J_Γ is the closed ideal of $C(S_\tau)$ canonically determined by the closed subspace $\mathrm{Mod}\,\Gamma \subseteq S_\tau$ in the hull-kernel correspondence [17, I.8.3] . In symbols:

$$(*) \quad C(\mathrm{Mod}\,\Gamma) \cong C(S_\tau)/J_\Gamma \; ; \quad J_\Gamma = \{f \in C(S_\tau) \mid f(\mathfrak{N})=0 \text{ for all } \mathfrak{N} \in \mathrm{Mod}\,\Gamma\}.$$

<u>5.1 Proposition.</u> For any C*-algebra A the following are equivalent:

(i) A is an abelian C*-algebra with totally disconnected compact Hausdorff spectrum;

(ii) For some type τ and theory $\Gamma \subseteq \mathcal{L}(\tau)$ of sentential logic we have $A \cong C(\mathrm{Mod}\,\Gamma)$;

(iii) for some type τ and closed ideal J of $C(S_\tau)$ we have $A \cong C(S_\tau)/J$.

<u>Proof</u>. (ii) $\Longleftrightarrow$ (iii). Use the hull-kernel correspondence [17,I.8.3] and the correspondence between closed subspaces of S_τ and (sets of models of) theories in $\mathcal{L}(\tau)$: recall (*) above.

(ii) $\Longrightarrow$ (i). As remarked above, $\mathrm{Mod}\,\Gamma$ is a compact totally disconnected Hausdorff subspace of S_τ. Now apply the Gelfand theorem to $A \cong C(\mathrm{Mod}\,\Gamma)$.

(i) $\Longrightarrow$ (ii). Let K be the spectrum of A. Using the Gelfand theorem identify A with the C*-algebra $C(K)$ of complex-valued continuous functions over K. Let λ be the weight of K, i.e. [6] the cardinality of a topological base for K of smallest cardinality. To avoid trivialities assume λ infinite. By [6,6.2.10, 6.2.16] we can identify K with a compact subspace of the Cantor cube 2^λ of weight λ [6,p.115]. Let τ be a type of cardinality λ; one equivalent form of the Gödel-Maltsev compactness theorem for sentential logic [2,1.2.12] states that the Stone space S_τ is (homeomorphic to) the Cantor cube 2^λ: therefore K may be identified with a compact subspace of S_τ. Let $\Gamma \subseteq \mathcal{L}(\tau)$ be a theory such that $K = \mathrm{Mod}\,\Gamma$ (such Γ existing by definition of S_τ); then $A \cong C(\mathrm{Mod}\,\Gamma)$ by the Gelfand theorem.

Note that the above argument also yields that for every type σ of cardinality $\geq \lambda$ there is a theory $\Delta \subseteq \mathcal{L}(\sigma)$ such that $A \cong C(\mathrm{Mod}\,\Delta)$. $\square$

We now consider the problem mentioned at the beginning of this section. In the light of [14, 1.23.2, 1.23.3] we may without loss of generality restrict attention to nets of C*-algebras (see Definition 1.0).

5.2 Theorem. For $\langle I,\leq\rangle$ a nonempty directed set, let $\{A,A_\alpha\}_{\alpha\in I}$ be a net of commutative C*-algebras, each A_α having totally disconnected compact Hausdorff spectrum. It follows that:

(1) there is a set T of types directed by inclusion, and an order-preserving one-one map $\alpha\longmapsto\tau_\alpha$ of $\langle I,\leq\rangle$ onto $\langle T,\subseteq\rangle$,

(2) there is a theory $\Gamma\subseteq\mathcal{L}(\tau)$ of sentential logic, where $\tau=\bigcup_{\alpha\in I}\tau_\alpha$, and a *-isomorphism φ of A onto $C(\mathrm{Mod}\,\Gamma)$,

(3) for each $\alpha\in I$ there is a theory $\Gamma_\alpha\subseteq\mathcal{L}(\tau_\alpha)$ and a *-isomorphism φ_α of A_α onto $C(\mathrm{Mod}\,\Gamma_\alpha)$,

(4) for every α, $\beta\in I$ with $\alpha\leq\beta$ there is a *-isomorphism Ξ_β of $C(\mathrm{Mod}\,\Gamma_\beta)$ into $C(\mathrm{Mod}\,\Gamma)$ and a *-isomorphism Ξ_α^β of $C(\mathrm{Mod}\,\Gamma_\alpha)$ into $C(\mathrm{Mod}\,\Gamma_\beta)$,

such that the following diagram commutes:

$$\begin{array}{lcccc}
 & A & \xrightarrow{\ \varphi\ } & C(\mathrm{Mod}\,\Gamma) & \qquad \Gamma\subseteq\mathcal{L}(\tau)\\
\text{inclusion} & \uparrow & & \uparrow\,\Xi_\beta & \\
 & A_\beta & \xrightarrow{\ \varphi_\beta\ } & C(\mathrm{Mod}\,\Gamma_\beta) & \qquad \Gamma_\beta\subseteq\mathcal{L}(\tau_\beta)\\
\text{inclusion} & \uparrow & & \uparrow\,\Xi_\alpha^\beta & \\
 & A_\alpha & \xrightarrow{\ \varphi_\alpha\ } & C(\mathrm{Mod}\,\Gamma_\alpha) & \qquad \Gamma_\alpha\subseteq\mathcal{L}(\tau_\alpha)
\end{array}$$

In particular we have $A\cong\lim_\sigma\{C(S_\sigma)/J_{\Gamma_\sigma},\Xi_\varrho^\sigma\mid\varrho\subseteq\sigma;\ \varrho,\sigma\in T\}$, identifyin[g] $C(\mathrm{Mod}\,\Gamma_\sigma)$ and $C(S_\sigma)/J_{\Gamma_\sigma}$ in the light of (*) above.

Proof. Observe that the spectrum X of A is a totally disconnected compact Hausdorff space (use [6, 6.2.15, 6.2.10] and the Stone-Weierstrass theorem). Let ν be the weight of X. For each $\alpha\in I$ consider the set $P_\alpha\subseteq I$ given by $P_\alpha=\{\gamma\in I\mid\gamma\leq\alpha\}$. The map η sending each $\alpha\in I$ into P_α is a one-one order-preserving function from $\langle I,\leq\rangle$ into $\langle\text{powerset}(I),\subseteq\rangle$. Let Y be a set of cardinality ν such that $Y\cap I=\emptyset$; the map ϑ sending each $\alpha\in I$ into $P_\alpha\cup Y$ is a one-one order-preserving function from $\langle I,\leq\rangle$ into $\langle\text{powerset}(I\cup Y)$, $\subseteq\rangle$, having the additional property that $|\vartheta(\alpha)|\geq\nu$ for each $\alpha\in I$. It is no loss of generality to assume that for each $\alpha\in I$, $\vartheta(\alpha)$ is a type in sentential logic (the set $I\cup Y$ may be thought of as a set of sentence symbols), so let $\tau_\alpha=\vartheta(\alpha)$, $\bigcup_{\alpha\in I}\tau_\alpha=\tau$, $T=\{\tau_\alpha\mid\alpha\in I\}$.

We then see that ϑ is an order-preserving one-one map of $\langle I,\leq\rangle$ onto $\langle T,\subseteq\rangle$ as in (1). Since $|\tau|\geq\nu$ there is a theory $\Gamma\subseteq\mathcal{L}(\tau)$ and a *-isomorphism Φ of A onto $C(\mathrm{Mod}\Gamma)$, by the final remark in the proof of Proposition 5.1. Similarly, noting that $|\tau_\alpha|\geq\nu$, for each $\alpha\in I$ there is $\Gamma_\alpha\subseteq\mathcal{L}(\tau_\alpha)$ and a *-isomorphism Φ_α of A_α onto $C(\mathrm{Mod}\Gamma_\alpha)$. Defining now $\Xi_\beta = \Phi\circ\Phi_\beta^{-1}$, $\Xi_\alpha^\beta = \Phi_\beta\circ\Phi_\alpha^{-1}$ $(\alpha\leq\beta,\ \alpha,\beta\in I)$ we see that the theories Γ, Γ_α, together with *-isomorphisms Φ, Φ_β, Ξ_β, Ξ_α^β defined above obey (2)-(4), and make the above diagram commute. The last statement is immediately proved from (*) and Proposition 5.1. □

In the above theorem no restriction is imposed on the weight of spectra and on the cardinality of types and theories. However in applications one frequently encounters separable C*-algebras and countable types and theories: see Example 2.2(ii), and the characterization (i)⟺(iv) of compact logics generated by countably many quantifiers, in the theorem at the beginning of section 1. In this respect we have:

5.3 Proposition. For any commutative C*-algebra A the following are equivalent:

(i) A has second countable totally disconnected compact Hausdorff spectrum;

(ii) A is an AF C*-algebra, i.e., A is the inductive limit of an increasing sequence of finite dimensional C*-algebras, all with the same unit;

(iii) there is a countable type τ and a theory $\Gamma\subseteq\mathcal{L}(\tau)$ in sentential logic such that $A\cong C(\mathrm{Mod}\,\Gamma)$.

Proof. (i)⟺(ii). By the Gelfand theorem the spectrum X of A is locally compact Hausdorff, and A has a unit iff X is compact. X is homeomorphic to the space prim(A) of primitive ideals in A with the hull-kernel (Jacobson) topology (see, e.g., [31, 4.1.2]). Now apply [21, 3.1] or [32, Lemma 1]. (i)⟺(iii). By inspection of the proof of Proposition 5.1. □

5.4 Examples and Problems. (i) AF (approximately finite dimensional) C*-algebras were introduced by Bratteli in [20] generalizing work done by Glimm and Dixmier. For an account see the monograph [27]. As remarked by Effros [25, p. 31], «If we regard C*-algebras as " noncommutative topological spaces ", the AF C*-algebras would seem to comprise the "zero-dimensional" spaces.» . In particular, properties of the zero-dimensional [6, 6.2.10] second countable Stone spaces of compact logics may be investigated in a noncommutative context using AF C*-algebras. For example, let us agree to say that an AF C*-algebra A has the lattice property (l.p.) iff its family of projections with the natural order is a lattice. Using a result in [29] Lazar proved in [28] that (l.p.) holds in A iff the collection of finite dimensional *-subalgebras of A is directed by inclusion. In the restricted framework of compact logics, when $A = C(S_\tau)$, projections may be identified with classes of models of sentences of type τ, and (l.p.) is then a reformulation of closure under disjunction and conjunction.

(ii) Given a closed two sided ideal J of a C*-algebra A, let $q: A \longrightarrow A/J$ be the quotient map. Consider the so called lifting problem whether each projection $P \in A/J$ has the form $P = q(Q)$ for some projection $Q \in A$. Brown (see [24]) gave a positive solution to this problem, assuming J is the inductive limit of a sequence of finite dimensional C*-algebras (obviously, without a common unit). In the restricted case when A arises from a compact logic L, recalling (*), the lifting problem amounts to settling the following simple corollary of the compactness of L: Let $\Delta_1, \Delta_2 \supseteq \Delta$ be theories of type σ such that for all $\mathfrak{M} \in \mathrm{Mod}\Delta$ we have $\mathfrak{M} \in \mathrm{Mod}\Delta_1$ iff $\mathfrak{M} \notin \mathrm{Mod}\,\Delta_2$. Then there is a sentence χ such that $\mathrm{Mod}\Delta_1 = \mathrm{Mod}\,\Delta \cap \mathrm{Mod}\{\chi\}$. For the case $\mathrm{Mod}\Delta$ = all structures of type σ, this result can be found in [2, 1.2.15] and [2, exercise 2.1.13(ii)], concerning sentential logic and first-order logic respectively. The proof is the same for any compact logic.

(iii) In the above examples results about nonabelian (AF) C*-algebras imply (but had not been conjectured as generalizations of) simple facts in model theory. The following is an example of a problem in C*-algebra theory arising from a sequence of results in (abstract) model theory. Given C*-algebras $A_1, A_2 \subseteq A$ with a common unit and with A generated by $A_1 \cup A_2$, let us agree to say that the triple $\langle A_1, A_2, A\rangle$ has the intermediate projection property (i.p.p.) iff for every two projections $P \in A_1$ and $R \in A_2$ with $P \leq R$ there is a projection $Q \in A_1 \cap A_2$ such that $P \leq Q \leq R$. By an obvious adaptation of the argument yielding the equivalence of Craig interpolation and Robinson consistency for compact logics [5, 7.1.5] one can easily show that in case A is abelian, $\langle A_1, A_2, A\rangle$ has (i.p.p.) iff the net $\{A_1 \cap A_2,\ A_1,\ A_2,\ A\}$ has the joint extension property (3.1). Using (3.2) and [2, exercise 1.2.7, theorem 2.2.20] one can exhibit examples $\langle A_1, A_2, A\rangle$ having (i.p.p.); on the other hand, from any compact logic not satisfying Craig interpolation (see, e.g., [16] and [5, 7.1.3(c)]) one can extract counterexamples to (i.p.p.). One can now pose the <u>problem</u> of generalizing the above characterization of (i.p.p.) to larger classes of C*-algebras.

(iv) To give another example of a problem in C*-algebra theory arising from model-theory, referring to Proposition 5.3, let us agree to say that an abelian AF C*-algebra A has the decidable theory property (d.t.p.) iff $A \cong C(\mathrm{Mod}\,\Delta)$ for some <u>decidable</u> theory Δ in sentential logic, i.e., a theory $\Delta \subseteq \mathcal{L}(\tau)$ such that there is a Turing machine deciding whether or not an arbitrary sentence $\psi \in \mathcal{L}(\tau)$ is a consequence of Δ [2, p.11] . A moment's reflection shows that (d.t.p.) is non-trivial: indeed, the well-known fact that there are uncountably many nonisomorphic countable Boolean algebras implies (by the Stone representation theorem and [6, 6.2.16]) that there are uncountably many non-homeomorphic compact subspaces of the Cantor set, hence (by the Gelfand theorem and Proposition 5.3) uncountably many nonisomorphic abelian AF C*-algebras. On the other hand, there are only countably many decidable theories $\Delta \subseteq \mathcal{L}(\tau)$; thus only countably many isomorphism classes of abelian AF C*-algebras satisfy (d.t.p.). One can pose the problem of giving an internal characterization of (d.t.p.) in terms of the recursion-theoretic properties of the combinatorial counterparts of AF C*-algebras, e.g., their Bratteli diagrams. Note that examples of AF C*-algebras in the literature often have an effective presentation via Bratteli diagrams, [20], [27], [30], [28].

Abelian and nonabelian AF C*-algebras are also useful to give a model-theoretical characterization of a larger class of abelian C*-algebras than in Proposition 5.3:

5.5 Theorem. For any C*-algebra A the following are equivalent:

(i) A is abelian, separable and has a unit;

(ii) there is an AF C*-algebra B such that A = center of B = $= \{b \in B \mid bx = xb \text{ for all } x \in B\}$;

(iii) there are theories $\{\Gamma_i\}_{i\in I}$ and Γ in sentential logic, such th

(1) $A \cong \{f \in C(\mathrm{Mod}\,\Gamma) \mid f(\mathfrak{M}) \neq f(\mathfrak{N}) \Rightarrow \exists\, i \in I$ with $\mathfrak{N} \in \mathrm{Mod}\,\Gamma_i$ and $\mathfrak{M} \notin \mathrm{Mod}\,\Gamma_i$

(2) $\Gamma \subseteq \Gamma_i \subseteq \mathcal{L}(\tau)$ for all $i \in I$; τ is countable;

(3) $\mathrm{Mod}\,\Gamma_i \cap \mathrm{Mod}\,\Gamma_j = \emptyset$ whenever $i \neq j$; $\mathrm{Mod}\,\Gamma_i \neq \emptyset$ for all $i \in I$;

(4) $\bigcup\{\mathrm{Mod}\,\Gamma_i \mid i \in I\} = \mathrm{Mod}\,\Gamma$;

(5) I is either countable, or has the cardinality of the continuum.

Proof. (i) $\Longrightarrow$ (ii) This is the main theorem in [22] ; see also the proof in [27, 5.35] based on the theory of spectral spaces. (ii)$\Longrightarrow$(i) is trivial.

(iii) $\Longrightarrow$ (i) By hypotheses (2)-(4) the family $\{\mathrm{Mod}\,\Gamma_i \mid i \in I\}$ is a partition of $\mathrm{Mod}\,\Gamma$, and each $\mathrm{Mod}\,\Gamma_i$, as well as $\mathrm{Mod}\,\Gamma$, are closed subspaces of the Stone space S_τ, the latter being compact (Hausdorff) by the Gödel-Maltsev theorem. By hypothesis (1), A may be identified with the C*-subalgebra of $\mathrm{Mod}\,\Gamma$ given by those functions which do not separate equivalent points with respect to the equivalence relation given by the above partition. Since $|\tau| \leq \omega$, then $\mathrm{Mod}\,\Gamma$ is second countable, hence both $C(\mathrm{Mod}\Gamma)$ and its subalgebra A are separable (see [17, exercise 5, p.21] and [6, 4.1.16]).

(i)$\Longrightarrow$(iii) By [32, Lemma 2] or [22, Lemma 1] there is an <u>abelian</u> AF C*-algebra $B \supsetneq A$. By Proposition 5.3 we can write $A \subsetneq B = C(\mathrm{Mod}\,\Gamma)$ for some $\Gamma \subseteq \mathcal{L}(\tau)$, τ countable. Define on $\mathrm{Mod}\Gamma$ the equivalence relation $\approx$ given by the following stipulation: $\mathfrak{M} \approx \mathfrak{N}$ iff $f(\mathfrak{N})=f(\mathfrak{M})$ for all $f \in A$. For each $\mathfrak{M} \in \mathrm{Mod}\Gamma$ let $\langle\mathfrak{M}\rangle = \{\mathfrak{N} \in \mathrm{Mod}\Gamma \mid \mathfrak{N} \approx \mathfrak{M}\} = \bigcap_{f\in A}\{\mathfrak{N} \in \mathrm{Mod}\,\Gamma \mid f(\mathfrak{N})=f(\mathfrak{M})\}$: then $\langle\mathfrak{M}\rangle$ is a closed, hence compact, subspace of $\mathrm{Mod}\Gamma$. Let I be the set of $\approx$-equivalence classes over $\mathrm{Mod}\Gamma$; for each $i \in I$ let $\Gamma_i \subseteq \mathcal{L}(\tau)$ be a theory such that $\mathrm{Mod}\,\Gamma_i = i$; Γ_i exists, because i $(=\langle\mathfrak{M}\rangle)$ is a closed subspace of $\mathrm{Mod}\Gamma$. Without loss of generality, since $\mathrm{Mod}\Gamma_i \subseteq \mathrm{Mod}\,\Gamma$, we may further assume $\Gamma_i \supseteq \Gamma$. Now Γ together with the Γ_i , $i \in I$, obey (2)-(4). Define the C*-algebra $W \subseteq C(\mathrm{Mod}\,\Gamma)$ by $W = \{f \in C(\mathrm{Mod}\,\Gamma) \mid f(\mathfrak{M}) \neq f(\mathfrak{N})$ implies $\langle\mathfrak{M}\rangle \neq \langle\mathfrak{N}\rangle\}$. By the Stone Weierstrass theorem, W=A and $\mathrm{Mod}\Gamma/\approx$ = spectrum(A). Thus A satisfies (1). To finally prove (5), note that $|I|$ is the cardinality of the spectrum X of A. Since A is abelian separable and has a unit, then X is second countable compact Hausdorff [17, exercise 5 p.21], [14, 1.2.1] . By standard topological facts [6,1.3.8, 4.2.8, 4.3.28], X can be equipped with a complete metric so as to become a complete metric separable space; a routine application of the Cantor Bendixson theorem [6,4.5.5(b)] now yields the desired conclusion (5). This completes the proof of the theorem. □

See [23, 8.3] for one more characterization of abelian separable C*-algebras with unit, in terms of spectral spaces. There is a well-established theory of closed *-derivations in large classes of abelian C*-algebras, see, e.g. [26] and references therein . Since by a result of Sakai (see [19,p.261 and Theorem 1]) there is no nonzero closed *-derivation in abelian C*-algebras with totally disconnected spectrum, then any useful model-theoretical characterization of abelian separable C*-algebras with nontrivial derivations shall follow the lines of 5.5 .

References.

[1] Bratteli O., Robinson D.W., Operator Algebras and Quantum Statistical Mechanics, I,II, Springer, Berlin, 1979.

[2] Chang C.C., Keisler H.J., Model Theory, North-Holland, Amsterdam, second edition 1977.

[3] Craig W., Three uses of the Herbrand-Gentzen theorem in relating model-theory and proof-theory, J.Symb.Logic,22 (1957) 269-285.

[4] Driessler W., Comments on lightlike translations and applications in relativistic quantum field theory, Commun.Math.Phys.,44 (1975) 133-141.

[5] Ebbinghaus H.D., Chapter II, in: Model-Theoretic Logics, J.Barwise S.Feferman, Editors, Perspectives in Mathematical Logic, Springer, Berlin 1984, to appear.

[6] Engelking R., General Topology, Monografie Matematyczne, Tom 60, PWN-Polish Scientific Publishers, Warszawa 1977.

[7] Feferman S., Vaught R.L., The first-order properties of algebraic systems, Fund.Math., 47 (1959) 57-103.

[8] Flum J., Ziegler M., Topological Model Theory, Lecture Notes in Mathematics, Springer, Berlin 1980.

[9] Haag R., Kastler D., An algebraic approach to quantum field theory J.Math.Phys., 5.7 (1964) 848-861.

[10] Haag R., Kadison R.V., Kastler D., Nets of C*-algebras and classification of states, Commun.Math.Phys.,16 (1970) 81-104.

[11] Mundici D., Compactness, interpolation and Friedman's third problem Annals Math.Logic, 22 (1982) 197-211.

[12] Mundici D., Duality between logics and equivalence relations, Transactions A.M.S., 270 (1982) 111-129.

[13] Roos H., Independence of local algebras in quantum field theory, Commun.Math.Phys., 16(1970) 238-246.

[14] Sakai S., C*-algebras and W*-algebras, Springer, Berlin 1971.

[15] Schlieder S., Einige Bemerkungen über Projektionsoperatoren, Commun Math.Phys., 13 (1969) 216-225.

[16] Shelah S., Generalized quantifiers and compact logics, Transactions A.M.S., 204 (1975) 342-364.

[17] Takesaki M., Theory of Operator Algebras I, Springer, Berlin 1979.

[18] Turumaru T., On the direct product of operator algebras IV, Tôhoku Math.J., 8 (1956) 281-285.

Additional references.

[19] Batty C.J.K., Unbounded derivations of commutative C*-algebras, Commun.Math.Phys., 61 (1978) 261-266.

[20] Bratteli O., Inductive limits of finite dimensional C*-algebras, Transactions A.M.S., 171 (1972) 195-234.

[21] Bratteli O., Structure spaces of approximately finite dimensional C*-algebras, Journal of Funct.Anal., 16 (1974) 192-204.

[22] Bratteli O., The center of approximately finite-dimensional C*-algebras, Journal of Funct.Anal., 21 (1976) 195-202.

[23] Bratteli O., Elliott G.A., Structure spaces of approximately finite dimensional C*-algebras, II, Journal of Funct.Anal., 30 (1978)74-82.

[24] Choi M.D., Lifting projections from quotient C*-algebras, Journal of Operator Theory, 10 (1983) 21-30.

[25] Effros E.G., On the structure theory of C*-algebras: some old and new problems. Proceedings of Symp.in Pure Math., A.M.S., vol.38 (1982) part 1, 19-35.

[26] Goodman F.M., Closed derivations in commutative C*-algebras, Journal of Funct.Anal., 39 (1980) 308-346.

[27] Hofmann K.H., Thayer F.X., Approximately finite-dimensional C*-algebras, Dissertationes Mathematicae (Rozprawy Mat .), 174 (1980) 64 pp.

[28] Lazar A.J., AF algebras with a lattice of projections, Math.Scand., 50 (1982) 135-144.

[29] Lazar A.J., AF algebras with directed sets of finite dimensional *-subalgebras, Transactions A.M.S., 275 (1983) 709-721.

[30] Lazar A.J., Taylor D.C., Approximately finite dimensional C*-algebras and Bratteli diagrams, Transactions A.M.S.259 (1980) 599-619.

[31] Pedersen G.K., C*-algebras and their Automorphism Groups, Academic Press, London (1979).

[32] Thayer F.X., The Weyl-von Neumann theorem for approximately finite C*-algebras, Indiana Math.J., 24(1975) 875-877.

A CONTRIBUTION TO NONSTANDARD TERATOLOGY[1]

Roman Murawski

Institute of Mathematics
A.Mickiewicz University
Poznań, Poland

By PA we denote Peano arithmetic formalized in the first order language L(PA) with constants 0,S,+,· and based on the usual Peano's axioms including the axiom scheme of induction. A_2^- will denote the second order arithmetic (cf. [1]) formalized in a two-sorted language (individual and set variables) and based on Peano's axioms (but with the single induction axiom instead of the axiom scheme of induction), extensionality and comprehension scheme.

In last years there was rather great interest in the problem of connections between PA and A_2^-, in particular in the problem of expandability of nonstandard models of PA to models of A_2^- or its sub- or supertheories (cf. e.g. our survey paper [6]). Recall the appropriate definition.

DEFINITION. A model $M \models PA$ is said to be A_2^--<u>expandable</u> iff there is a family $\mathcal{X}_M \subseteq \mathcal{P}(M)$ such that $(\mathcal{X}_M, M, \in) \models A_2^-$.

Though we know rather a great deal about A_2^--expandability, not too much is known about possible A_2^--expansions of models of PA (the known criterion of expandability due to Schlipf and independently G.Wilmers gives no such information - cf. [6] part III, [2]).

In this paper we shall consider the problem of so called trace expansions. Given an A_2^--expansion $\mathfrak{A} = (\mathcal{X}_M, M, \in)$ of a model $M \models PA$ we can consider submodels $I \subseteq M$ and trace structures generated by I, namely

---

[1] A part of results of this paper was obtained when I was a fellow of the Alexander von Humboldt Foundation and worked at the University of Heidelberg (West Germany) under the direction of Professor Gert H. Müller. I express here my warm appreciation to him.

the structures $(\mathcal{X}_M \cap I, I, \in)$ where $\mathcal{X}_M \cap I = \{Y \subseteq I: (EX \in \mathcal{X}_M)(Y = X \cap I)\}$. We can now look for I's such that the trace structure is a model of A_2^-.

For special $\mathcal{X}_M$, namely for $\mathcal{X}_M = Def(M)$ (= the family of definable - with parameters - subsets of M) this construction was considered by L.Kirby and K.McAloon (cf. [2]). Recall first the following

DEFINITION. Let Q_1 and Q_2 be two families of initial segments of a given model $M \models PA$. We say that Q_1 and Q_2 are <u>symbiotic</u> iff for every $a, b \in M$, $a \leq b$:

$$(EI)_{Q_1}(a \in I < b) \text{ iff } (EJ)_{Q_2}(a \in J < b).$$

L.Kirby and K.McAloon have proved now the following

THEOREM 1 (Kirby, McAloon). Let T be recursively axiomatizable such that $T \vdash A_{pr}$. Then for any countable $M \models PA$ the families

$$\{I \subseteq_e M: I \text{ is T-expandable}\}$$

and

$$\{I \subseteq_e M: (Def(M) \cap I, I, \in) \models T\}$$

are symbiotic.

The construction of trace expansions was considered in a general setting in [9]. We have introduced there a version $\underset{\sim}{A}_2^-$ of A_2^- defined as follows. Its language is $L(PA) \cup \{\tilde{S}\} \cup \{E\}$ where $\tilde{S}$ is a new unary and E a new binary predicate ($\tilde{S}$ = the set of (codes of) sets, E = the membership relation). Axioms of $\underset{\sim}{A}_2^-$ are simply appropriate translations of the axioms of A_2^-. The theories A_2^- and $\underset{\sim}{A}_2^-$ are of course mutually interpretable. Any model of $\underset{\sim}{A}_2^-$ has the form $(M, \tilde{S}, E)$ where $M \models PA$, $\tilde{S} \subseteq M$ is a set of (codes of) sets and $E \subseteq M \times \tilde{S}$ is a membership relation. Every such a model gives a regular model of A_2^-, i.e. a model of the form $(\mathcal{X}_M, M, \in)$, where $M \models PA$, $\mathcal{X}_M \subseteq \mathcal{P}(M)$ and $\in$ is the real membership relation. Indeed we ought to take $\mathcal{X}_M = \{X_a: a \in \tilde{S}\}$, where $X_a = \{x \in M: xEa\}$. On the other hand any model $\mathfrak{A} = (\mathcal{X}_M, M, \in)$ of A_2^- such that $card(\mathcal{X}_M) \leq card(M)$ gives a model of $\underset{\sim}{A}_2^-$. It can be easily seen that for any model $M \models PA$: M is A_2^--expandable iff M is extendable to a model of $\underset{\sim}{A}_2^-$.

Consider now a countable recursively saturated model M of $(PA)^{ZF}$ (in fact it is enough to assume that M is a model of a theory like $(PA)^{A_2^- + Con(A_2^-)}$). Hence M is A_2^--expandable and has an expansion $(\mathcal{X}_M, M, \in)$ with the property:

$$(\mathcal{X}_M, M, \in) \underset{F}{\cong} (M, \tilde{S}, E),$$

where $\tilde{S} = M$ and E is inductive (i.e. substitutable into the axiom scheme of induction). Using now a special indicator (constructed in [9]) we get the following

THEOREM 2. The model M has arbitrarily large and $2^{\aleph_0}$ A_2^- - expandable initial segments $I \leq_e M$ such that

$$(\mathcal{X}_M^I \cap I, I, \epsilon) \models A_2^-,$$

where $\mathcal{X}_M^I$ is the family of all sets $X \in \mathcal{X}_M$ such that $F(X) \in I$.

COROLLARY 3. The family of initial segments of the model M described above whichare A_2^--expandable and the family of initial segments I of M such that $(\mathcal{X}_M^I \cap I, I, \epsilon) \models A_2^-$ are symbiotic.

Using substitutable satisfaction classes (for the definition cf.[4], see also [7] and [2]) we get similar results for elementary initial segments.

THEOREM 4. Let M be a countable recursively saturated model of $(PA)^{ZF}$ and let $(\mathcal{X}_M, M, \epsilon)$ be its A_2^--expansion such that

$$(\mathcal{X}_M, M, \epsilon) \underset{F}{\cong} (M, \tilde{S}, E)$$

where $\tilde{S} = M$ and E is inductive. Then M has arbitrarily large and $2^{\aleph_0}$ A_2^--expandable elementary initial segments I such that

$$(\mathcal{X}_M^I \cap I, I, \epsilon) \prec (\mathcal{X}_M, M, \epsilon).$$

In this paper we consider the "negative" aspect of the trace construction, namely we shall prove, generalizing the result of Mostowski [5], that for some A_2^--expandable nonstandard models $M \models PA$ and for some A_2^--expansions $\mathfrak{A}$ of them there are an elementary extension M_1 of M and an A_2^--expansion $\mathfrak{B} = (\mathcal{X}_{M_1}, M_1, \epsilon)$ of M_1 such that $(\mathcal{X}_{M_1} \cap M, M, \epsilon)$ is not a model of A_2^-.

The contents of the main result justifies the title of the paper. Since we study certain "pathological" features of models of arithmetic, hence the word "teratology" is appropriate. It comes from the Greek: téras - monster and lógos -word, science and means a science studying development defects and monstrosities and abnormalities by people and animals (first of all by the vertebrates). It describes substantial deviations from the average constitution of organisms of a given species, discovers ways of abnormal development of embryos and searches for causes of abnormalities.

Our construction will use substitutable satisfaction classes. We shall consider not expandability to models of A_2^- but to certain extension A_2^* of A_2^- obtained by adding an arithmetical version of the axiom of constructibility CONSTR. This is no loss of generality since for a model $M \models PA$: M is A_2^--expandable iff M is A_2^*-expandable (cf. [8]).

MAIN THEOREM. Let M be a countable nonstandard model of PA which is expandable to a model of $A_2^* = A_2^- + CONSTR$ having a full substitutable satisfaction class. Then there exist an extension M_1 of M and an A_2^--expansion $\mathfrak{F} = (\mathfrak{X}_{M_1}, M_1, \epsilon)$ of M_1 such that

$$M_1 \equiv M$$

and

$$(\mathfrak{X}_{M_1} \cap M, M, \epsilon) \text{ non} \models A_2^-.$$

Proof of Main Theorem

Let $\mathfrak{A}$ be an A_2^*-expansion of M and S a full substitutable satisfaction class over $\mathfrak{A}$. We can assume that the set-universe of $\mathfrak{A}$ is countable. In fact the following general fact holds: any countable model of PA which has an expansion to a model of A_2^* having a satisfaction class with some elementary properties has also a countable A_2^* - expansion having a satisfaction class with these same properties.

The needed structure $\mathfrak{F}$ will be a particular nonstandard Skolem ultrapower of the model $\mathfrak{A}$ (such ultrapowers were considered by Kotlarski in [3]). To simplify our considerations we shall treat the satisfaction class S not as a set of triples (a,b,X), where $\mathfrak{A} \models Fm[a]$ (i.e. a is a formula in the sense of $\mathfrak{A}$, possibly nonstandard), b and X are valuations for number and set variables resp. (using a pairing function we can consider formulas with one number and one set variables only). Using certain coding of sets of $\mathfrak{A}$ by numbers of $\mathfrak{A}$ (for example similar to the one used in [5]) and a pairing function $\langle \cdot,\cdot \rangle$ we can treat S simply as a set of numbers of $\mathfrak{A}$.

Let S-Def($\mathfrak{A}$) be a family of all S-definable subsets of M with parameters, i.e.

$$X \in S\text{-Def}(\mathfrak{A}) \quad \text{iff} \quad (Ea,b)_M[\mathfrak{A} \models \text{"a is a formula in variables } v_0, v_1\text{"} \;\&\; (x)_M(x \in X \equiv (\mathfrak{A},S) \models \underline{S}(\langle a,\langle x,b\rangle\rangle))].$$

We say that the set X is S-definable in $\mathfrak{A}$ by a with a parameter b. Notice that S-Def($\mathfrak{A}$) is a Boolean algebra under the usual operations. Let $\mathcal{F}$ be an ultrafilter in S-Def($\mathfrak{A}$). The nonstandard Skolem ultrapower S-($\mathfrak{A}^M/\mathcal{F}$) is now defined in the following way: Let Tm be the set of all terms (in the sense of $\mathfrak{A}$!) with one number-variable, i.e. all elements of $|\mathfrak{A}|$, which satisfy in $\mathfrak{A}$ the formula representing the set of terms of one free number-variables with parameters. We define an equivalence relation:

$$t_1 \sim t_2 \quad \text{iff} \quad \{x \in M: (\mathfrak{A}, S) \models \underline{S}(\langle \ulcorner t_1(v) = t_2(v)\urcorner, x\rangle)\} \in \mathcal{F}$$

for $t_1, t_2 \in$ Tm. In Tm/$\sim$ we define now for any relation R of the language $L(A_2^-)$:

$$R(\tilde{t}_1, \ldots, \tilde{t}_k) \quad \text{iff} \quad \{x \in M: (\mathfrak{A}, S) \models \underline{S}(\langle \ulcorner R(t_1(v), \ldots, t_k(v)\urcorner, x\rangle)\} \in \mathcal{F}.$$

It can be easily seen that S-($\mathfrak{A}^M/\mathcal{F}$) is a structure for $L(A_2^-)$ and that Łoś's Theorem holds (here we use the fact that $\mathfrak{A} \models A_2^*$, i.e. that we have a definable well-ordering of the universe and hence built - in Skolem functions). Hence S-($\mathfrak{A}^M/\mathcal{F}$) is elementarily equivalent to $\mathfrak{A}$. The structure S-($\mathfrak{A}^M/\mathcal{F}$) is isomorphic to a regular model $\mathfrak{X} = (\mathcal{X}_{M_1}, M_1, \in)$. Hence $\mathfrak{X} \models A_2^*$, $\mathfrak{X} \equiv \mathfrak{A}$ and $M_1 = M$.

Consider now the family $\mathcal{X}_{M_1} \cap M$ of subsets of M.

LEMMA A. The family $\mathcal{X}_{M_1} \cap M$ consists of sets $R_{\mathcal{F}}(X)$ where X ranges over S-Def($\mathfrak{A}$) and

$$R_{\mathcal{F}}(X) = \{x \in M: \{y \in M: \mathfrak{A} \models \text{"}\langle y, x\rangle \in X\text{"}\} \in \mathcal{F}\}.$$

P r o o f. First observe that for an $Y \in \mathcal{X}_{M_1}$, or more exactly for an equivalence class t/$\sim$ where $t \in$ Tm and t/$\sim$ gives the set Y, we have

$$Y \cap M = \{x \in M: \{y \in M: (\mathfrak{A}, S) \models \underline{S}(\langle \ulcorner v_1 \in t(v_2)\urcorner, \langle x, y\rangle\rangle)\} \in \mathcal{F}\}.$$

This follows from the definition of S-($\mathfrak{A}^M/\mathcal{F}$) and the usual properties of ultrapowers. Consider now a nonstandard formula

$$\varphi(v_0) = (Ev_1)[t((v_0)_0) = v_1 \ \& \ (v_0)_1 \in v_1].$$

(Since t may be a nonstandard term we must not write φ as above - i.e. in such a way as if φ were a standard formula but we ought to define its code showing how it is obtained from the code of the term t. Nevertheless the way in which we have written φ, though not completely correct and precise, shows better its structure and meaning.)

We have $(\mathfrak{A},S) \models \underline{S}(\langle \ulcorner\varphi\urcorner, \langle x,y\rangle\rangle)$ iff $(\mathfrak{A},S) \models$ "$y \in t(x)$". Take the set $X = \{z \in M: (\mathfrak{A},S) \models \underline{S}(\langle \ulcorner\varphi\urcorner, z\rangle)\}$. Of course $X \in$ S-Def($\mathfrak{A}$) and for $x \in M$:

$$\begin{aligned} x \in R_{\mathcal{F}}(X) &\equiv \{y \in M: \mathfrak{A} \models \langle y,x\rangle \in X\} \in \mathcal{F} \\ &\equiv \{y \in M: (\mathfrak{A},S) \models \text{"}x \in t(y)\text{"}\} \in \mathcal{F} \\ &\equiv x \in Y \cap M. \end{aligned}$$

Conversely, suppose now that $X \in$ S-Def($\mathfrak{A}$). By the assumption that the satisfaction class S is substitutable and hence in particular the comprehension scheme for atomic formulas containing the predicate $\underline{S}$ holds we have that there is a set Y in $\mathfrak{A}$ such that

$$\langle x,y\rangle \in X \equiv \mathfrak{A} \models \text{"}\langle x,y\rangle \in Y\text{"}.$$

De ine now a sequence $s: M \longrightarrow |\mathfrak{A}|$ with the property:

$$\mathfrak{A} \models \text{"}x \in s(y) \equiv \langle y,x\rangle \in Y\text{"}.$$

We do it using the comprehension axiom. Namely we have

$$\mathfrak{A} \models (y)(Z)(ET)(u)[u \in T \equiv \langle y,u\rangle \in Z].$$

Take now as Z our set Y. So there is a term t such that

$$\begin{aligned} (\mathfrak{A},S) \models \underline{S}(\langle \ulcorner v_0 \in t(v_1)\urcorner, \langle x,y\rangle\rangle) \quad &\text{iff} \quad \mathfrak{A} \models \langle y,x\rangle \in Y \\ &\text{iff} \quad \langle y,x\rangle \in X. \end{aligned}$$

Let Y_1 be an element of $\mathcal{X}_{M_1}$ given by the term t. Then

$$\begin{aligned} x \in Y_1 \cap M &\equiv \{y \in M: (\mathfrak{A},S) \models \underline{S}(\ulcorner x \in t(y)\urcorner)\} \in \mathcal{F} \\ &\equiv \{y \in M: \langle y,x\rangle \in X\} \in \mathcal{F} \\ &\equiv x \in R_{\mathcal{F}}(X). \end{aligned}$$

Q.E.D.

By the definition of S-Def($\mathfrak{A}$) we can associate with every set X belonging to S-Def($\mathfrak{A}$) a number $a_X \in M$ (call it a code of X) - more exactly a_X will be a sequence number consisting of a (nonstandard) formula and a sequence of number- and set-parameters.

LEMMA B. The set C of codes of elements of S-Def($\mathfrak{A}$) is definable in M by a formula with bounded quantifiers.

Using the assumption that S is a substitutable satisfaction class and hence in particular S satisfies the minimum principle we can choose for any $X \in$ S-Def($\mathfrak{A}$) its smallest code. Call such a code the distinguished code.

LEMMA C. The set C^* of distinguished codes of elements of S-Def($\mathfrak{A}$) is definable in (M,S) by a Σ_1^0 formula of the language $L(PA) \cup \{\underline{S}\}$, where $\underline{S}$ is a unary predicate being interpreted as the set S.

P r o o f.

$$a \in C^* \equiv (a \in C \;\&\; (x)_M[x < a \longrightarrow (Ey)_M[S(\langle (a)_0, \langle y, (a)_1, \dots, (a)_{lh(a) \dot- 1}\rangle\rangle) \equiv$$
$$\equiv \neg\, S(\langle (x)_0, \langle y, (x)_1, \dots, (x)_{lh(x) \dot- 1}\rangle\rangle)]]).$$

Q.E.D.

Our next purpose is to construct an ultrafilter $\mathcal{F}$ in S-Def($\mathfrak{A}$) such that $(\mathcal{X}_{M_1} \cap M, M, \in)$ will not be a model of A_2^-, where $(\mathcal{X}_{M_1}, M_1, \in)$ is a regular model of A_2^- isomorphic to S-$(\mathfrak{A}^M/\mathcal{F})$.

For any $a \in M$ denote by $\bar{a}$ and a' the uniquely determined elements of M with the properties: $M \models \bar{a}, a' \geq 0 \;\&\; a' < 2^{\bar{a}} \;\&\; a = 2^{\bar{a}} + a'$. Let

$$D_a = \{x \in M:\ M \models (x \geq 1) \;\&\; (x \equiv a')(\mathrm{mod}\ 2^{\bar{a}})\}.$$

The sets D_a form a full binary tree under inclusion. The maximal element of it is $D_1 = M - \{0\}$, the immediate successors of D_a are D_b and D_c where $M \models [b = 2^{\bar{a}+1} + a' = 2^{\bar{a}} + a]$, $M \models [c = 2^{\bar{a}+1} + 2^{\bar{a}} + a' = 2^{\bar{a}+1} + a]$. Define now in (M,S) a function f:

$$f(0) = 1$$
$$f(x+1) = \begin{cases} 2^x + f(x), & \text{if } x \notin S, \\ 2^{x+1} + f(x), & \text{if } x \in S. \end{cases}$$

Of course $M \models 2^x \leq f(x) < 2^{x+1}$. Let $\mathcal{F}_0 = \{D_{f(x)}:\ x \in M\}$.

LEMMA D. $\mathcal{F}_0$ is a basis of a filter.

LEMMA E. There exists a function definable in (M,S) which enumerates the distinguished codes of members of $\mathcal{F}_0$.

P r o o f. By the definition of the set D_a there exists a formula $\varphi(v_0, v_1)$ of L(PA) such that for $x, y \in M$:

$$x \in D_y \equiv M \models \varphi(v_0, v_1)[x, y].$$

Using coding of finite sequences of M and sequence-constructors we can find an explicite definition of the function f, i.e. a formula $\psi(v_0, v_1)$ in $L(PA) \cup \{\underline{S}\}$ such that for any $a, b \in M$

$$b = f(a) \equiv M \models \psi(v_0, v_1)[a, b].$$

Consider now the formula

$$\chi(v_0,v_1) \equiv: (w)[\psi(v_1,w) \longrightarrow \varphi(v_0,w)].$$

we have for any $x,y \in M$:

$$x \in D_{f(y)} \equiv M \models \chi(v_0,v_1)[x,y].$$

Let $\ulcorner\chi\urcorner$ be a Gödel number of χ. The distinguished code of $D_{f(a)}$ is now

$$\delta(a) = \min y: y \in C^* \& (x)_M[S(\langle(y)_0,\langle x,(y)_1,\ldots,(y)_{lh(y)\dot{-}1}\rangle\rangle) \equiv$$
$$\equiv S(\langle\ulcorner\chi\urcorner,\langle x,a\rangle\rangle)].$$

Q.E.D.

LEMMA F. The set $X_0 = \{\langle x,y\rangle \in M: M \models y>0 \& x \in D_y\}$ is an element of S-Def($\mathfrak{A}$) such that for every ultrfilter $\mathcal{F} \supseteq \mathcal{F}_0$, the set S is definable in $(M,R_{\mathcal{F}}(X_0))$ by a formula of L(PA) with a parameter $R_{\mathcal{F}}(X_0)$.

P r o o f. By the definition of X_0 we have for $x \in M$:

$$x \in R_{\mathcal{F}}(X_0) \equiv \{y \in M: \langle y,x\rangle \in X_0\} \in \mathcal{F} \equiv \{y \in M: y \in D_x\} \in \mathcal{F} \equiv D_x \in \mathcal{F}.$$

We show now that $Im(f) = R_{\mathcal{F}}(X_0)$ where Im(f) is the image of f. This can be proved in a similar way as in [5] but we must work inside M. Having this equality we can define S in (M,Im(f)), namely we have

$$x \in S \equiv (M,Im(f)) \models (Ex_1)(Ex_2)[x_1 \in Im(f) \& x_2 \in Im(f) \&$$
$$\& 2^x \leqslant x_1 < 2^{x+1} \leqslant x_2 < 2^{x+2} \&$$
$$\& x_2 - x_1 = 2^{x+1}].$$

Q.E.D.

LEMMA G. There exists an ultrafilter $\mathcal{F} \supseteq \mathcal{F}_0$ such that each set of the form $R_{\mathcal{F}}(X)$ for $X \in$ S-Def($\mathfrak{A}$) is arithmetically definable in the structure $(M,R_{\mathcal{F}}(X_0))$.

P r o o f. Using Lemma E we show that there is an ultrafilter $\mathcal{F} \supseteq \mathcal{F}_0$ such that the set of its distinguished codes is definable in the structure (M,S). Let namely φ be a function definable in (M,S) and enumerating distinguished codes of elements of S-Def($\mathfrak{A}$) (such a function exists by Lemma C). Let Y_a be a set whose distinguished code is $\varphi(a)$ and let Z_a denote an element of $\mathcal{F}_0$ whose distinguished code is $\delta(a)$ (cf. Lemma E). Define now a function ψ as follows:

$$\psi(x+1) = \min y: y \neq \psi(x) \& Y_y \cap \bigcap_{i \leqslant x} Y_{\psi(i)} \cap \bigcap_{i \leqslant a} Z_i \neq \emptyset \quad \text{for any } a \in M.$$

One can easily see that $\{Y_{\psi(a)}: a\in M\}$ is an ultrafilter containing the set $\mathcal{F}_0$. Define now a function ε such that $\varepsilon(x)$ is a distinguished code of a set $Y_{\psi(x)}$. We put namely (inside M):

$$\varepsilon(x+1) = \min z:\ z \neq \varepsilon(x)\ \&\ z\in C^*\ \&\ (t)(Ey)[S(<(z)_0,<y,(z)_1,\dots$$
$$\dots,(z)_{lh(z)\dot{-}1}>>)\ \&$$
$$\&\ (i)_{<x+1}S(<(\varepsilon(i))_0,<y,(\varepsilon(i))_1,\dots>>)\ \&$$
$$\&\ (i)_{<t+1}S(<(\varepsilon(i))_0,<y(\varepsilon(i))_1,\dots>>).$$

It can be easily seen that ε is definable in (M,S).

Now we prove that for such an $\mathcal{F}$ each set $R_{\mathcal{F}}(X)$ for $X\in$ S-Def($\mathfrak{A}$) is definable in (M,S). Next we use Lemma F and we are done.

Q.E.D.

Take now the ultrafilter $\mathcal{F}$ from Lemma G and build S-($\mathfrak{A}^M/\mathcal{F}$). By Lemma A the structure $(\mathcal{X}_{M_1}\cap M, M, \in)$ where $(\mathcal{X}_{M_1}, M_1, \in)$ is a regular model isomorphic to S-($\mathfrak{A}^M/\mathcal{F}$) is not a model of A_2^- since it contains the set $R_{\mathcal{F}}(X_0)$ such that all other members of $\mathcal{X}_{M_1}\cap M$ are arithmetically definable in $(M, R_{\mathcal{F}}(X_0))$. But this is impossible since we have in A_2^- the full comprehension scheme. Hence Main Theorem is proved.

Q.E.D.

REMARKS. 1. Observe that the model M_1 constructed above is an extension of M (M is even elementarily embeddable in M_1) but it cannot be an end extension. Indeed $M\in\mathcal{X}_M$ and $M\in\mathcal{X}_{M_1}$ (more exactly, the image of M under the canonical embedding of $\mathfrak{A}$ into $\mathfrak{B}$ is an element of $\mathcal{X}_{M_1}$). If now M_1 were an end extension of M then we have had a contradiction with the axiom of induction.

2. Observe that by the construction $\mathcal{X}_M \subsetneq \mathcal{X}_{M_1}\cap M$.

3. The assumption that the satisfaction class S is substitutable can be weakened to the following condition : 1° S is substitutable to the axiom scheme of induction and 2° the comprehension scheme for atomic formulas of the language $L(A_2^-)\cup\{\underline{S}\}$ holds. The second condition is equivalent to the condition that in $\mathfrak{A}$ the comprehension scheme holds also for nonstandard formulas.

4. The assumption of countability of M was not essential. The only thing we need is that M has an A_2^*-expansion with set-universe of cardinality $\leq$ card(M).

5. The very end of the proof of Main Theorem could suggest that it

holds also for models expandable to models of fragments of A_2^- obtained by restricting the comprehension scheme to Σ_n^1 formulas ($n \geq 1$) and denoted by $A_2^-|\Sigma_n^1$. But any model $\mathfrak{A} \models A_2^-|\Sigma_n^1$ having a satisfaction class S such that the comprehension scheme for atomic formulas of the language $L(A_2^-) \cup \{\underline{S}\}$ holds is already a model of the full second order arithmetic A_2^-.

We finish the paper with the following

PROBLEM. Let M be a model of PA such as in The Main Theorem. How many extensions M_1 of M and A_2^--expansions $\mathfrak{B}$ of M_1 do exist such that the trace of $\mathfrak{B}$ on M is not a model of A_2^-?

References

[1] Apt, K.R., Marek, W.: Second order arithmetic and related topics, Annals of Math. Logic 6(1974), 177-239.

[2] Kirby, L.A.S., McAloon, K., Murawski, R.: Indicators, recursive saturation and expandability, Fund. Math. 114(1981), 127-139.

[3] Kotlarski, H.: On Skolem ultrapowers and their nonstandard variant, Zeitschrift f. math. Logik und Grundlagen d. Math. 26 (1980), 227-236.

[4] Krajewski, S.: Nonstandard satisfaction classes, in: Set Theory and Hierarchy Theory, Proc. Bierutowice Conf. 1975, Springer Verlag, 1976, LNM 537, 121-144.

[5] Mostowski, A.: A contribution to teratology, Избранные вопросы алгебры и логики, Наука, Новосибирск 1973, 184-196.

[6] Murawski, R.: On expandability of models of Peano arithmetic I, II, III, Studia Logica 35(1976), 409-419; 35(1976), 421-431; 36(1977), 181-188.

[7] Murawski, R.: Models of Peano arithmetic expandable to models of fragments of second order arithmetic (in Polish), Ph.D. Thesis, Warszawa 1978.

[8] Murawski, R.: Some remarks on the structure of expansions, Zeitschrift f. math. Logik und Grundlagen d. Math. 26(1980), 537-546.

[9] Murawski, R.: Trace expansions of initial segments, Zeitschrift f. math. Logik und Grundlagen d. Math. 30(1984), to appear.

MODEL- AND SUBSTRUCTURE COMPLETE THEORIES OF ORDERED ABELIAN GROUPS

Peter H. Schmitt

Mathematisches Institut
Universität Heidelberg
Im Neuenheimer Feld 294
6900 Heidelberg
W. Germany

Abstract: We give necessary conditions for an arbitrary elementary class of ordered abelian groups to be model-complete, resp. substructure-complete. Ordered abelian groups are considered in a suitable definitional extension of the usual language of ordered groups. We introduce also the concepts of convex model-completeness and convex substructure completeness.

Section 0: Introduction.

Let us first describe the results in the theory of ordered abelian groups we could built on when we started writing this paper. In his pioneering paper [3] Yuri Gurevich associated with every ordered abelian group G for every $n \geq 2$ a coloured chain (i.e. a linear order with additional unary predicates) $Sp_n(G)$, called the n-spine of G. The important fact is, that for any two ordered abelian groups G and H, G is elementarily equivalent to H if and only if for all $n \geq 2$ $Sp_n(G)$ is elementarily equivalent to $Sp_n(H)$. Since coloured chains are in any respect simpler to handle than ordered abelian groups this is a useful tool, e.g. to prove decidability of the first-order theory of ordered abelian groups, which was the main result in [3]. The above transfer theorem was reworked and extended in [4], [10], [11] to allow also the transfer of formulas instead of only sentences from G to $Sp_n(G)$. See the great Transfer-Theorem 1.7 below. As a first application of the great Transfer Theorem we could establish that no complete theory of ordered abelian groups has the independence property, [6].

In the present paper we use the great Transfer Theorem to obtain criteria for model- and substructure completeness of elementary classes $\mathcal{M}$ of ordered abelian groups. Any such elementary class $\mathcal{M}$ can be described as the class of all ordered abelian groups G such that for all $n \geq 2$ $Sp_n(G)$ is a model of T_n, for certain theories T_n in the language of n-spines. The main result (Theorems 3.2 & 3.3 below) reads:

> If for all $n \geq 2$ T_n is modelcomplete (substructurecomplete) then M is modelcomplete (resp. substructurecomplete) in a certain definitional extension of the language of ordered groups.

If G is an ordered subgroup of the ordered abelian group H then $Sp_n(G)$ need not be a substructure of $Sp_n(H)$. Necessary and sufficient conditions for this to be the case are investigated in section 2 below. Forstalling accusations that we oversimplify let us concede that $Sp_n(G)$ can never be a substructure of $Sp_n(H)$ in the strict sense of set theoretic inclusion, but there are natural candiates for embeddings $f_n : Sp_n(G) \to Sp_n(H)$. which for the general pair $G \subseteq H$ need neither be well-defined nor embeddings.

Section 3 proves the main result. Since the definitional extension that has to be used in the main theorem is quite substantial, we consider restrictions on the elementary class M, which allow us to get away with more modest definitional extensions. The various definitional extensions that we are going to consider in this paper are collected in the Appendix. Section 3 also contains hints how earlier partial results on model- and substructure complete classes of ordered abelian groups [7], [8], [12] can be derived from the general result. In section 4 we make use of the observation that when G is a convex subgroup of the ordered abelian group H then $Sp_n(G) \subseteq Sp_n(H)$. We introduce the concepts of convex modelcompleteness and convex substructure completeness and prove, in the notation used above, that modelcompleteness (substructure completeness) of all T_n implies convex modelcompleteness (resp. convex substructure completeness) of M. We close with the conjecture that convex substructure completeness of M coincides with the possiblity to eliminate unbounded quatifiers in the elementary theory of M.

Section 1: Prerequisites.

1.0. The great Transfer Theorem.

We consider ordered abelian groups in the first-order language LOG containing the non-logical symbols +, -, 0, <. Let us remember that every ordered group is torsionfree and its set of convex subgroups is linearly ordered by set theoretic inclusion.

Next we review the crucial definitions and the main results from [11].

The principal concept will be that of an n-spine. Since this appears at first sight rather technical we begin with a brief excursion on how one arrives at this concept.

Suppose we want to give a criterion, when two ordered abelian groups G and H are elementarily equivalent. Let us first consider the very special case that G and H are both lexicographic products (or Hahn-products, as they are sometimes called) over the same ordered index set I, i.e.

$$G = \Pi \{G_i : i \in I\} \quad , \quad H = \Pi \{H_i : i \in I\} .$$

(The elements of G are therefore those functions h in the cartesian product of the family $\{G_i : i \in I\}$ whose support, $\{i \in I : h(i) \neq 0\}$ is well-ordered with respect to the order inherited from I. $0 < h$ in G is defined by $0 < h(i_o)$ where i_o is the minimum of the support of h.) By the Fefermann-Vaught results we know that in this case

$$G_i \equiv H_i \quad \text{for all} \quad i \in I$$

implies $G \equiv H$.

Another simple special case arises when G and H are both archimedean. A. Robinson and E. Zakon proved in [9] that $G \equiv H$ iff

either G and H are both discretely ordered

or G and H are both densely ordered and have the same Szmielew-invariants β_p for each prime p.

We recall that β_p for a torsionfree abelian group G is defined by

$$\beta_p(G) = \dim G/pG .$$

Combining the results of the preceding two paragraphs we may give a criterion for elementary equivalence of the lexicographic products

$$G = \Pi \{G_i : i \in I\} \quad \text{and} \quad H = \Pi \{H_j : j \in J\}$$

over different index sets with G_i, H_j all archimedean. Let I* be an expansion of the ordered set I by adding unary predicates which determine the elementary equivalence type of G_i. To be more precise we add a predicate Dk such that $I^* \models Dk(i)$ iff G_i is discretely ordered and we add predicates $\beta(p,m)$ such that $I^* \models \beta(p,m)(i)$ iff $\beta_p(G_i) \geq m$. Let J* arise in the same way from J. An easy back-and-forth argument shows that

$$I^* \equiv J^* \quad \text{implies} \quad G \equiv H .$$

These examples may guide us in the general case if we remember that any ordered abelian group G can be sliced into archimedean parts in the following way: for every element g of G we define

B(g) = the smallest convex subgroup of G containing g

A(g) = the largest convex subgroup of G not containing g .

The factor group C(g) = B(g)/A(g) is certainly archimedean. The set I(G) = {A(g) : g ∈ G} is linearly ordered by set theoretic inclusion and we view A(g) ∈ I(G) as an index for C(g). It is tempting to conjecture that G is elementarily equivalent to

$$\Pi \{C(g) : A(g) \in I(G)\} .$$

But unfortunately this is far from being true. On obstacle is that the ordered set I(G) is not elementarily definable in G, so we may have G ≡ H but not I(G) ≡ I(H). This is of course due to the fact that the archimedean jump (A(g), B(g)) is not elementarily definable. At this point we remember that Robinson and Zakon determined the elementary class generated by all archimedean groups to be the class of what they called regular abelian groups and they proved their result mentioned above actually for this larger class of ordered abelian groups. This gives us the hint to look for pairs $(A_\infty(g), B_\infty(g))$ such that the factor group $C_\infty(g) = B_\infty(g)/A_\infty(g)$ is regular rather then archimedean. This is still not good enough since regularity is defined by an infinite conjunction of first-order formulas:

G is <u>regular</u> iff G is n-regular for all $n \geq 2$

G is <u>n-regular</u> iff any closed interval [a,b], a,b ∈ G, a < b, which contains at least n elements contains a representative of every congruence class modulo nG.

Now it seems reasonable to abondon the idea that we can determine the elementary equivalence type of G by just looking at one chain of pairs of convex subgroups and we decide to built one such chain for every natural number $n \geq 2$. In detail we define for $n \geq 2$ and g ∈ G:

$B_n(g)$ = the largest convex subgroup C of G such that C/A(g) is n-regular

$A_n(g)$ = the smallest convex subgroup C of G such that B(g)/C is n-regular

$C_n(g) = B_n(g)/A_n(g)$ is certainly n-regular and $A_n(g)$, $B_n(g)$ are elementary definable.

The reader may e.g. try his hand and verify:

$$x \notin A_n(y) \quad \text{iff} \quad |y| \leq |x| \\ \text{or} \\ |x| < |y| \ \& \\ \forall u(|u| < n\cdot|y| \rightarrow \exists v(|u+nv| < n\cdot|x|))$$

here $|g|$ denotes the absolute value of g.

Let $I_n(G)$ be the linearly ordered set $\{A_n(g) : g \in G\}$ where we view $A_n(g)$ as an index for $C_n(g)$. Let $I_n(G)$ arise from $I_n^*(G)$ in the same way as we did obtain I* from I above. We may thus refine our first conjecture to read now:

$$G \equiv H \quad \text{iff} \quad \text{for all } n \geq 2 \quad I_n^*(G) \equiv I_n^*(H) \ .$$

Unfortunately this is still not true in general and we give an example to show what goes wrong:

Let ω be the set of positive integers with its natural order; for $m \in \omega$ let $G_m \simeq Z$ (= the ordered additive group of integers). Let H be $\Pi\ \{G_m : m \in \omega\}$. The element $g_o \in H$ is defined by $g_o(m) = 5$ for all $m \in \omega$. Let finally G be the ordered subgroup of H generated by $\oplus \sum \{G_m : m \in \omega\} \cup \{g_o\}$. ($\oplus \sum \{G_m : m \in \omega\}$ consists by definition of those elements h of H whose support is finite.)

It is not hard to see that for all $n \geq 2$

$$I_n^*(G) \simeq I_n^*(H)$$

but G and H are not elementarily equivalent since in H we have

"for all x: if $x + 5\cdot H$ contains arbitrarily small positive element then $x \in 5\cdot H$"

but in G

"$g_o + 5\cdot G$ contains arbitrarily small positive elements but $g_o \notin 5\cdot G$"

is true.

This leads to the introduction of n-fundaments:

$F_n(g)$ = the largest convex subgroup C of G such that $C \cap (g + nG) = \emptyset$.

In the above example we had

$$F_n^G(g_o) = \{0\} \quad \text{and} \quad F_n^H(g_o) = \emptyset \ .$$

$F_n(g)$ is elementarily definable by:

$$x \in F_n(y) \quad \text{iff} \quad \forall u(|y + n\cdot u| > n\cdot|x|) \ .$$

What we did in the above example with one element g_o we might do with more than one element. This quantitative aspect is measured by the group $\Gamma_n(g)$ defined by:

$$\Gamma_n(g) = \Gamma_{2,n}(g)/\Gamma_{1,n}(g)$$

$$\Gamma_{1,n}(g) = \{h \in G : F_n(h) \subset F_n(g)\}$$

$$\Gamma_{2,n}(g) = \{h \in G : F_n(h) \subseteq F_n(g)\} .$$

Now we are ready to give the definition of an n-spine which are the final versions of what we first tried to approximate by $I_n^*(G)$ above.

The language of spines, LSP, is the first-order language containing as non-logical symbols " < " and unary relation symbols: A, F, Dk, $\alpha(p,k,m)$ for all primes p, $k \geq 1$, $m \geq 0$.

Definition:
For $n \geq 2$ we associate with every ordered abelian group G its n-spine, $Sp_n(G)$, $Sp_n(G)$ is the LSP-structure with universe

$$\{A_n(g) : g \in G\} \cup \{F_n(g) : g \in G\}$$

$C_1 < C_2$	iff	$C_2 \subset C_1$
$A(C)$	iff	$C = A_n(g)$ for some $g \in G$
$F(C)$	iff	$C = F_n(g)$ for some $g \in G$
$Dk(C)$	iff	G/C is discrete
$\alpha(p,k,m)(C)$	iff	$C = F_n(g)$ for some $g \in G$ and $\alpha_{p,k}(\Gamma_n(g)) \geq m$

where $\alpha_{p,k}(H)$ is the Szmielov invariant given by $\dim(p^{k-1}H[p]/p^kH[p])$.

For $n = p^r$ we write $F_{p,r}$ instead of F_n and likewise for A_n, B_n. When necessary we write $F_n^G(g)$ to indicate that $F_n(g)$ is to be formed with respect to the group G and likewise for A_n, B_n.

To complete this explanation of n-spines we return to the situation we started with, i.e. let G be the lexicographic product $\Pi\ \{G_i : i \in I\}$ where all G_i are archimedean. We will describe what $B_n(g)$, $A_n(g)$, $F_n(g)$ are in this special case. For $g \in G$ let $i(g)$ be the minimal element in the support of g and if $g \notin nG$ let $i_n(g)$ be the minimal element i in the support of g such that $g(i) \notin nG_i$.

Then we have:

$B_n(g) = \{h \in G$: for all $i < i(g)$ such that for all j with $i < j < i(g)$ G_j is n-divisible : $h(i) = 0\}$

$A_n(g) = \{h \in G$: for all $i \leq i(g)$ such that for all j with $i \leq j < i(g)$ G_j is n-divisible : $h(i) = 0\}$.

If no G_j, $j \in I$, is n-divisible this reduces to:

$B_n(g) = \{h \in G : \text{for all } i < i(g) : h(i) = 0\}$

$A_n(g) = \{h \in G : \text{for all } i \leq i(g) : h(i) = 0\}$.

If $g \in nG$ then $F_n(g) = \emptyset$. If $g \notin nG$ then:

$$F_n(g) = \{h \in G : \text{for all } i < i_n(g) : h(i) = 0\} .$$

We recall some basic properties of the convex subgroups defined above.

Lemma 1.1.

(1) $A_n(g) = A_n(mg)$ for all integers $m \neq 0$.

(2) $A_n(g + h) \subseteq A_n(g) \cup A_n(h)$.

(3) If $A_n(g) \subset A_n(h)$, then $A_n(g + h) = A_n(h)$.

(4) $F_n(g + nh) = F_n(g)$.

(5) $F_n(g + h) \subseteq F_n(g) \cup F_n(h)$.

(6) If $F_n(g) \subset F_n(h)$, then $F_n(g + h) = F_n(h)$.

(7) If $g \in nG$, then $F_n(g) = \emptyset$.

The following simple properties of n-jumps and n-fundaments will be explicitely used later on

Lemma 1.2.

(1) If $A(g) \subset B(h)$, then $A(g) \subseteq A(h)$.

(2) If $A_n(g) \subset B_n(h)$, then $A_n(g) \subseteq A_n(h)$.

(3) $A_n(g) \subseteq A_n(h)$ iff $A_n(g) = A_n(h)$ or $g < h$.

(4) $h \in A_n(g)$ iff $A_n(h) \subset A_n(g)$.

(5) $h \in F_n(g)$ iff $A_n(h) \subset F_n(g)$.

Proofs of lemmas 1.1 and 1.2 may be found in [10] or [11].

The n-spines $Sp_n(G)$ are definable over G in the sense made precise by the following lemma:

Lemma 1.3.

For every $n \geq 2$, $k \geq 0$ and every sequence of LOG-terms $t_1(\vec{x}),\ldots,t_k(\vec{x})$, every LSP-formula $\chi(y_1,\ldots,y_k,z_1,\ldots,z_k)$ can be translated into a LOG-formula $\varphi(\vec{x})$, such that for all ordered abelian groups G and all $\vec{g} \in G$:

$$Sp_n(g) \models \chi(A_n(t_1(\vec{g})),\ldots,A_n(t_k(\vec{g})),F_n(t_1(\vec{g})),\ldots,F_n(t_k(\vec{g})))$$

iff

$$G \models \varphi(\vec{g}) \ .$$

For a proof see e.g. [11].

Definition:

A LOG-formula is called bounded if it is built up from atomic and negated atomic formulas by the use of $\wedge$, $\vee$, $\exists x(0 \leq x \leq t)$ and $\forall x(0 \leq x \leq t)$, where t is a term not containing x.

The class of bounded existential (bounded universal) LOG-formulas is defined by allowing in addition unrestricted existential (universal) quantification. We can now state the following addendum to Lemma 1.3.

Lemma 1.4.

If the LSP-formula χ in Lemma 1.3 is quantifier-free then we may choose its LOG-translation φ to be bounded.

As an illustration we give the following special cases of Lemma 1.4 explicitely. We shall use the symbol $|x|$ to denote the absolute value of x. It is easily seen that $|.|$ can be eliminated in the following formulas without destroying boundedness.

$$x \notin A_n(y) \quad \leftrightarrow \quad |y| \leq |x| \vee (|x| < |y| \ \& \ \forall u(|u| < n|y| \rightarrow \exists v(|v| \leq n|y| \ \& \ |u+nv| < n|x|))$$

$$x \in F_n(y) \quad \leftrightarrow \quad |x| < |y| \ \& \ \forall z(|z| < n|y| \rightarrow \neg(|y+nz| < n|x|))$$

$$A_n(x) \subseteq A_n(y) \quad \leftrightarrow \quad \forall z(0 \leq z \leq |x|)(z \in A_n(x) \rightarrow z \in A_n(y))$$

We want to add a few remarks on the relationship between fundaments $F_n(g)$ and the lower parts of n-regular jumps, $A_n(h)$. Neither are all $A_n(h)$ equal to some $F_n(g)$ nor is the reverse implication true. But we know:

Lemma 1.5.

Let G be an ordered abelian group, $h \in G$, $n \geq 2$: There is $g \in G$ with $A_n(h) = F_n(g)$ iff for some prime p dividing n $\beta_p(C_n(g)) \geq 1$.

Concerning the question when $F_n(g)$ is equal to some $A_n(h)$ nothing can be said in general, but the following lemma proves sometimes useful.

Lemma 1.6.

(i) $F_n(g) = \bigcap \{A_n(g + nh) : h \in G\}$

(ii) If $Sp_n(G)$ is finite, then for all $g \in G$ there is some $h \in G$ satisfying $F_n(g) = A_n(h)$.

Remark:

One might wonder why the Szmielew-invariants $\beta_p(C_n(g)) = \dim C_n(g)/pC_n(g)$ of the n-regular factors $C_n(g)$ are not incorporated in the n-spine. Well, they in fact are, since we have for any $n \geq 2$, any ordered abelian group G and $h \in G$:

(I) If for all $g \in G$ $A_n(h) \neq F_n(g)$, then for all primes p dividing n:

$$\beta_p(C_n(h)) = 0 .$$

(II) If $A_n(h) = F_n(g)$ then for all primes p dividing n let k be the maximal exponent such that p^k divides n, then

$$\beta_p(C_n(h)) = \alpha_{p,k}(\Gamma(n,F_n(g))) .$$

The language LOG* is the definitional extension of LOG obtained by adding the following unary predicates:

$$D(p,r,i)(x)$$
$$E(n,k)(x)$$
$$x = s(\mathrm{mod}\ A_n(x))$$
$$x < s(\mathrm{mod}\ A_n(x))$$
$$x > s(\mathrm{mod}\ A_n(x))$$

for all $n \geq 2$, all primes p, $r \geq 1$, $0 < i < r$, $0 < k < n$, $s > 0$, defined by:

$$D(p,r,i)(x) \longleftrightarrow \exists y(p^r y = x) \vee \exists y(F_{p,r}(x-p^i y) \subset F_{p,r}(x) = F_{p,r}(y))$$

$E(n,k)(x) \longleftrightarrow \exists y(F_n(x) = A_n(x)$ & "$G/A_n(y)$ is discrete and for the smallest positive element $e + A_n(y)$ in $G/A_n(y)$ we have $F_n(x-ke) \subset A_n(y)$" .

For R one of the symbols =, <, >:

$xRs(\mathrm{mod}\ A_n(x)) \longleftrightarrow$ "$G/A_n(x)$ is discrete and for its smallest positive element $e + A_n(x)$: $x + A_n(x) R (s \cdot e + A_n(x))$" .

It is easily checked that these predicates can be defined by bounded LOG-formulas. We can now state the main result from [11].

Theorem 1.7 (The great Transfer Theorem).

For every LOG-formula $\varphi(\overline{x})$ we find

(i) a natural number $n \geq 2$

(ii) a LSP-formula $\psi_o(y_1,\ldots,y_m,z_1,\ldots,z_r)$

(iii) a quantifierfree LOG*-formula $\psi_1(\overline{x})$

(iv) LOG-terms $t_1(\overline{x}),\ldots,t_m(\overline{x}),s_1(\overline{x}),\ldots,s_r(\overline{x})$

such that for all ordered abelian groups G and all $\overline{g} \in G$:

$$G \models \varphi(\overline{g})$$

iff

$$G \models \psi_1(\overline{g}) \quad \text{and} \quad Sp_n(G) \models \psi_o(C_1,\ldots,C_m,D_1,\ldots,D_r)$$

for $C_i = A_n(t_i(\overline{g}))$, $D_i = F_n(s_i(\overline{g}))$.

Proof: [11, Theorem 3.6].

1.1. Model Theory.

We assume familarity with the notions of substructure- and model-completeness. Ocassionally we will need the following more restricted version of definitional extensions.

Let T be a theory in the language of L and L_1 an extension of L by adding new relation symbols. A theory T_1 is called a Δ_1-definitional extension of T, when T_1 arises from T by adding for every new relation symbols the following two new axioms:

$$\forall\overline{x}(\varphi_R(\overline{x}) \longleftrightarrow R(\overline{x})) \quad , \quad \forall\overline{x}(\psi_R(\overline{x}) \longleftrightarrow R(\overline{x}))$$

where $\varphi_R(x)$ is a universal and $\psi_R(x)$ is an existential L-formula.

Remark:

Let T_1 be a Δ_1-definitional extension of T, $\mathcal{A}$, $\mathcal{B}$ models of T such that $\mathcal{A}$ is an L-substructure of $\mathcal{B}$. Let $\mathcal{A}_1$, $\mathcal{B}_1$ be the unique expansions of $\mathcal{A}$ and $\mathcal{B}$ to models of T_1. Then $\mathcal{A}_1$ is an L_1-substructure of $\mathcal{B}_1$.

Section 2: Embeddings of n-spines.

If H is an ordered divisible abelian group then $Sp_n(H)$ has only $\emptyset$ and $\{0\}$ as its elements. Thus we cannot expect in general when G is an ordered subgroup of some ordered abelian group H that $Sp_n(G)$ is embeddable in $Sp_n(H)$. We shall not investigate this problem in general, but give an answer in the special case when the embedding of $Sp_n(G)$ into $Sp_n(H)$ is effected by the mapping f_n defined by:

$$f_n(A_n^G(g)) = A_n^H(g)$$

$$f_n(F_n^G(g)) = F_n^H(g) \ .$$

Lemma 2.1.

Let G be an ordered subgroup of the ordered abelian group H and $n \geq 2$. Then f_n is an embedding of $Sp_n(G)$ into $Sp_n(H)$ (this asserts in particular that f_n is well-defined)

iff

the following conditions are satisfied for all $g, g' \in G$:

(1) $A_n^G(g) = A_n^G(g')$ iff $A_n^H(g) = A_n^H(g')$

(2) $F_n^G(g) = F_n^G(g')$ iff $F_n^H(g) = F_n^H(g')$

(3) $A_n^G(g) = F_n^G(g')$ iff $A_n^H(g) = F_n^H(g')$

(4) $F_n^G(g) = F_n^H(g) \cap G$

(5) $\exists h \in G(F_n^G(g) = A_n^G(h))$ iff $\exists h \in H(F_n^H(g) = A_n^H(h))$

(6) $\alpha_{p,k}(\Gamma^G(n,g)) = \alpha_{p,k}(\Gamma^H(n,g))$ for all p,k such that p^k divides n

(7) $G/A_n^G(g)$ is discrete iff $H/A_n^H(g)$ if discrete .

Proof. Necessity.
The implications from left to right in (1) - (3) follow since f_n is assumed to be well-defined while the reverse implications are consequences of the assumed injectivity of f_n. (5) - (7) are satisfied since f_n respects the predicates A, $\alpha(p,k,m)$ and Dk of the language LSP. (4) is true since f_n is order preserving using Lemma 1.2 (5).

Sufficiency.
(1) - (3) imply that f_n is well-defined and injective. (5) - (7) assure that f_n respects the predicates A, $\alpha(p,k,m)$ and Dk of LSP. That f_n also respects the predicate F follows from Lemma 1.5. It remains to see that f_n is order-preserving.

Case 1: $A_n^H(g) \subseteq A_n^H(g')$ iff $A_n^H(g) = A_n^H(g')$ or $g < g'$ (Lemma 1.2 (3))
iff $A_n^G(g) = A_n^G(g')$ or $g < g'$ (by (1))
iff $A_n^H(g) \subseteq A_n^G(g')$.

Case 2: $F_n^H(g) \subseteq F_n^H(g') \Rightarrow F_n^H(g) \cap G \subseteq F_n^H(g') \cap G$
$\Rightarrow F_n^G(g) \subseteq F_n^G(g')$ (by (4))

The reverse implication follows since $Sp_n(G)$ is totally ordered.

Case 3: $A_n^H(g) \subseteq F_n^H(g')$ iff $A_n^H(g) = F_n^H(g')$ or $g \in F_n^H(g')$ (Lemma 1.2 (5))
iff $A_n^G(g) = F_n^G(g')$ or $g \in F_n^H(g')$ (by (3),(4))
iff $A_n^G(g) \subseteq F_n^G(g')$.

Remark:
One may wonder why $A_n^G(g) = A_n^H(g) \cap G$ does not appear in the above list. It turns out that this is already a consequence of the assumption that f_n be order-preserving using Lemma 1.2 (4).

One may hope to discover further dependencies among the conditions (1) - (7). Unfortunately the following collection of examples shows that this is not the case.

Example 1 (showing that (1) cannot be ommitted in Lemma 2.1).

Let I be the order theoretic sum $1 + \omega^* + 1 + \omega^*$. We use the following notation for elements of I:

$(0,-\infty)$ for the smallest element ω

$(o,-n)$ for elements in the first copy of ω^*

$(1,-\infty)$ for the greatest element in the part $1 + \omega^* + 1$

$(1,-n)$ for the elements in the second copy of ω^* .

Set $G_{(o,-\infty)} \cong G_{(1,-\infty)} \cong Q$ and for all other $i \in I$ $G_i \cong Z_2$. Let
$H = \sum (G_i : i \in I)$

$$G = \{h \in H : \text{for all } n \in \omega : h(0,-n) = 0\} .$$

The pair $G \subseteq H$ of ordered abelian groups has the following properties:
For $g \in G$, $g \neq 0$:

$$A_2^G(g) = \begin{cases} \{g' \in G: g'(0,-\infty) = g'(1,-\infty) = 0\} \text{ if } g(0,-\infty) \neq 0 \text{ or } g(1,-\infty) \neq \\ \{g' \in G: \text{for all } i \in I\ (i \leq i(g) \rightarrow g'(i) = 0)\} \text{ otherwise.} \end{cases}$$

In the first case we get $\beta_2(C_2^G(g)) = 0$, while in the second case we find $\beta_2(C_2^G(g)) = \omega$.

$$A_2^H(g) = \begin{cases} \{h \in H : h(0,-\infty) = 0\} \text{ if } g(0,-\infty) \neq 0 \\ \{h \in H : \text{for all } n(H(0,-n) = 0) \ \& \ h(0,-\infty) = 0\} \\ \quad \text{if } g(0,-\infty) = 0 \text{ but } g(1,-\infty) \neq 0 \\ \{h \in H : \text{for all } i \in I \ (i \leq i(g) \rightarrow h(i) = 0\} \text{ otherwise }. \end{cases}$$

In the first two cases we get $\beta_2(C_2^H(g)) = 0$ and $\beta_2(C_2^H(g)) = \omega$ in the third case.

Thus we have for $g_1, g_2 \in G$ defined by:

$$g_1(i) = \begin{cases} 1 & \text{if } i = (0,-\infty) \\ 0 & \text{otherwise} \end{cases}$$

$$g_2(i) = \begin{cases} 1 & \text{if } i = (1,-\infty) \\ 0 & \text{otherwise} \end{cases}$$

$$A_2^G(g_1) = A_2^G(g_2) \quad \text{but} \quad A_2^H(g_1) \supsetneqq A_2^H(g_2) \ .$$

On the other hand the conditions (2) - (7) of Lemma 2.1 are true for $G \subseteq H$.

<u>Example 2</u> (showing that condition (2) of Lemma 2.1 cannot be omitted).

Let $I = \omega \cdot 3$ and for all $i \in I$ $G_i \cong Z_2$. Set $H^* = \Pi(G_i : i \in I)$.

$$H_0' = \{h \in H^* : \text{supp}(h) \text{ is finite and for all } i : \omega \leq i < \omega \cdot 3 \rightarrow h(i) = 0\}$$

$$H_1' = \{h \in H^* : \text{supp}(h) \text{ is finite}\} \ .$$

Let $(J_n : n \in \omega)$ be a partition of ω where all J_n are infinite. For $a \in Z_2$, $n \in \omega$ we define elements g_n^a, h_n^a, $k_n^a \in H^*$ by:

$$g_n^a(i) = \begin{cases} a & \text{if for some } j \in J_n \ (i = j \text{ or } i = \omega+j \text{ or } i = \omega \cdot 2+j) \\ 0 & \text{otherwise} \end{cases}$$

$$h_n^a(i) = \begin{cases} a & \text{if } i \in J_n \\ 0 & \text{otherwise} \end{cases}$$

$$k_n^a(i) = \begin{cases} a & \text{if for some } j \in J_n \ (i = \omega + j) \\ 0 & \text{otherwise} \end{cases}$$

Let $C = \{g_n^a : n \in \omega, a \in 2 \cdot Z_2\} \cup \{h_n^a : n \in \omega, a \in Z_2\} \cup \{k_n^a : a \in Z_2, n \in \omega, \text{even}\}$. Finally let

G = the ordered subgroup of H^* generated by
$H'_o \cup \{g_n^a : n \in \omega,\ a \in 2\cdot\mathbb{Z}_2\}$

H = the ordered subgroup of H^* generated by $H'_1 \cup C$.

The pair of groups has the following properties: For $g \in G$, $g \neq 0$:

$$A_2^G = \{g' \in G: \text{for all } i \leq i(g) : g'(i) = 0\}$$

$$A_2^H = \{h \in H : \text{for all } i \leq i(g) : h(i) = 0\}$$

$$C_2^G(g) \simeq C_2^H(g) \simeq Z_2 \ .$$

Now assume $g \notin 2\cdot G$ and we use the notation:
$i_2(g) = \min\{i \in I : g(i) \notin 2\cdot Z_2\}$. If for all $i \in I$ $g(i) \in 2\cdot Z_2$ we set $i_2(g) = \omega\cdot 3$. Note that this may be the case without g being an element of $2\cdot G$.

In calculating $F_2(g)$ we distinguish three cases.

<u>Case 1</u>: $i_2(g) < \omega$.

$$F_2^G(g) = \{g' \in G: \text{for all } i \leq i_2(g) : g'(i) = 0\}$$

$$F_2^H(g) = \{h \in H : \text{for all } i \leq i_2(g) : h(i) = 0\}$$

$$\Gamma^G(2,g) \simeq \Gamma^H(2,g) \simeq Z(2) \ .$$

<u>Case 2</u>: $i_2(g) \geq \omega$.
In this case we have already $i_2(g) = \omega\cdot 3$. Without loss of generality we may assume that $g = \sum (g_n^{a(n)} : n \in K)$ for some finite subset K of ω.

<u>Case 2A</u>: all elements of K are even.

$$F_2^G(g) = \{0\} = F_2^H(g) \quad \text{and} \quad \Gamma^G(2,g) \simeq \Gamma^H(2,g) \simeq \mathbb{Z}(2)^{(\omega)} \ .$$

<u>Case 2B</u>: K contains at least one odd number.

$$F_2^G(g) = \{0\}$$

$$F_2^H(g) = \{h \in H : \text{for all } i < \omega\cdot 2 : h(i) = 0\}$$

$$\Gamma^G(2,g) \simeq \Gamma^H(2,g) \simeq \mathbb{Z}(2)^{(\omega)} \ .$$

Thus we have for the elements $g_o^2, g_1^2 \in G$ defined above:

$$F_2^G(g_o^2) = F_2^G(g_1^2) \quad \text{but} \quad F_2^H(g_o^2) \subsetneq F_2^H(g_1^2) \ .$$

Conditions (1), (3) - (7) are seen to be satisfied.

<u>Example 3</u> (showing that (3) cannot be omitted in Lemma 2.1).

Let $G_1, G_2 \subseteq \mathbb{R}$ be densely ordered groups such that for $i = 1,2$:

$$\beta_p(G_i) = \begin{cases} 0 & \text{if } p \neq 2 \\ \omega & \text{if } p = 2 \end{cases} .$$

Furthermore we choose G_1, G_2 in such a way that there is no nontrivial intersection between G_2 and the divisible hull $\overline{G}_1$ of G_1. Clearly G_1, G_2 are archimedean. The groups G, H will be ordered subgroups of the lexicographic sum $\mathbb{R} \oplus \mathbb{R}$:

$$G = \{(h,h) : h \in G_1\}$$

$$H = \{(h,g) : h \in G_2 \oplus \overline{G}_1,\ g \in G_1\} .$$

These groups have the following properties:

$$G \subseteq H$$

for $g \in G$, $g \neq 0$, $g = (h,h)$: $A_2^G(g) = \{0\}$, $A_2^H(g) = \{(0,g) : g \in G_1\}$

$$\beta_p(C_2^H(g)) = \beta_p(C_2^G(g)) = \begin{cases} \omega & \text{if } p = 2 \\ 0 & \text{if } p \neq 0 \end{cases}$$

$$F_2^H(g) = F_2^G(g) = \begin{cases} \emptyset & \text{if } h \in 2G_1 \\ \{0\} & \text{if } h \notin 2G_1 \end{cases} \quad \text{and } \Gamma^H(2,g) \simeq \Gamma^G(2,g) \simeq \mathbb{Z}(2)^{(\omega)}$$

Thus we have for all $g \in G$ with $g \notin 2G$: $\{0\} = F_2^G(g) = A_2^G(g)$ but $\{0\} = F_2^H(g) \subsetneq A_2^H(g)$. The remaining conditions of Lemma 1.1 are seen to be satisfied by the pair $G \subseteq H$.

<u>Example 4</u> (showing that (4) cannot been omitted in Lemma 2.1).

Let $I = \omega + \omega$ and for all $i \in I$ $G_i \simeq \mathbb{Z}_2$. Denote by H^* the lexicographic product $\Pi(G_i : i \in I)$ and for $a \in Z_2$ let g_a^0 be the following element of H^*:

$$g_a^0(i) = a \quad \text{for all } i \in I$$

G = the ordered subgroup of H^* generated by
$\sum (G_i : i \in I) \cup \{g_a^0 : a \in 2Z_2\}$

H = the ordered subgroup of H^* generated by $G \cup \{g_a^0 : a \in \mathbb{Z}_2\}$.

For $g \in G$, $g \neq 0$ we have:

$$A_2^G(g) = \{g' \in G : \text{for all } i \leq i(g) : g'(i) = 0\}$$

$$A_2^H(g) = \{h \in H : \text{for all } i \leq i(g) : h(i) = 0\}$$

$$C_2^G(g) \simeq C_2^H(g) \simeq \mathbb{Z}_2 .$$

Now assume $g \notin 2G$ and denote by $i_2(g)$ the minimal element of $\{i \in I : g(i) \notin 2G_i\}$ if this set is not empty and $i_2(g) = \omega\cdot 2$ otherwise. There are two cases to be distinguished in calculating $F_2(g)$:

Case 1: $i_2(g) < \omega\cdot 2$

$$F_2^G(g) = \{g' \in G: \text{for all } i \leq i_2(g) : g'(i) = 0\}$$

$$F_2^H(g) = \{h \in H : \text{for all } i \leq i_2(g) : h(i) = 0\}.$$

Case 2: $i_2(g) = \omega\cdot 2$

$$F_2^G(g) = \{g' \in G: \text{for all } i < \omega : g'(i) = 0\}$$

$$F_2^H(g) = \{0\} .$$

In both cases we have $\Gamma^G(2,g) \simeq \Gamma^H(2,g) \simeq \mathbb{Z}(2)$.

Thus condition (4) is violated since we have $F_2^G(g_2^0) \subsetneq F_2^H(g_2^0) \cap G$. But (1) - (3) and (5) - (7) are satisfied.

Example 5 (showing that condition (5) cannot be omitted in Lemma 2.1).

Take $I = \omega + 1$ and for all $i < \omega$ $G_i \simeq \mathbb{Z}_2$ and $G_\omega \simeq \mathbb{Q}$. Let $H^* = \Pi(G_i : i \in I)$. For $a \in \mathbb{Z}_2$ we define $g^a \in H^*$ by:

$$g^a(i) = \begin{cases} a & \text{if } i < \omega \\ 0 & \text{if } i = \omega \end{cases}$$

Take

G = the ordered subgroup of H^* generated by $\sum (G_i : i \in I) \cup \{g^a : a \in 2Z_2\}$

$H = G \oplus G_\omega$.

For $g \in G$, $g \neq 0$ we have:

$$A_2^G(g) = \{g' \in G: \text{for all } i \leq i(g) : g'(i) = 0\}$$

$$A_2^H(g) = \{h \in H : \text{for all } i \leq i(g) : g(i) = 0\}$$

$$C_2^G(g) \simeq C_2^H(g) \simeq Z_2 .$$

Now assume $g \notin 2G$ and let $i_2(g)$ be defined analogously as in Example 4.

Case 1: $i_2(g) < \omega$

$$F_2^G(g) = \{g' \in G: \text{for all } i \leq i_2(g) : g'(i) = 0\}$$

$$F_2^H(g) = \{h \in H : \text{for all } i \leq i_2(g) : h(i) = 0\} .$$

Case 2: $i_2(g) = \omega$

$$F_2^G(g) = F_2^H(g) = \{0\} .$$

In both cases we have $\Gamma^G(2,g) \simeq \Gamma^H(2,g) \simeq \mathbb{Z}(2)$.

For the element $g^2 \in G$ there is no $g \in G$ such that $F_2^G(g^2) = A_2^G(g)$. But the element $g^* \in H$ given by

$$g^*(i) = \begin{cases} 0 & i < \omega \\ 1 & i = \omega \end{cases}$$

satisfies $F_2^H(g^2) = A_2^H(g^*)$. Thus (5) is violated, while it is easily checked that (1) - (4), (6), (7) are true for the pair $G \subseteq H$.

Examples showing that neither (6) nor (7) can be omitted in Lemma 2.1 can be obtained more easily and are left to the reader.

Section 3: Criteria for model- and substructure completeness.

The first result in this section will be a criterion for substructure completeness of arbitrary elementary classes of ordered abelian groups. As was to be expected we have to enlarge the language LOG of ordered groups even beyond LOG*.

For every extension LOG_i (LOG_i^*) of LOG, that we shall consider, we denote by TOG_i (TOG_i^*) the corresponding definitional extension of the theory of ordered abelian groups.

Definition:

LOG_1 is the extension of LOG obtained by adding the following list of new relation symbols for all $n \geq 2$ and all relevant p, k, m.
$EA_n(x,y)$, $EF_n(x,y)$, $AF_n(x,y)$, $F_n(x,y)$, $EAF_n(x)$, $\alpha(n,p,k,m)(x)$, $DK(x)$, $D(p,r,i)$, $M(n)(x)$.

TOG_1 is obtained from the theory of ordered abelian groups by addition of the following definitions:

$$EA_n(x,y) \leftrightarrow A_n(x) = A_n(y) \qquad EF_n(x,y) \leftrightarrow F_n(x) = F_n(y)$$

$$AF_n(x,y) \leftrightarrow F_n(x) = A_n(y) \qquad F_n(x,y) \leftrightarrow x \in F_n(y)$$

$$EAF_n(x) \leftrightarrow \exists y(F_n(x) = A_n(y)) \qquad \alpha(n,p,k,m)(x) \leftrightarrow \alpha_{p,k}(\Gamma(n,x)) \geq m$$

$$Dk(x) \leftrightarrow \text{"}C_n(x) \text{ is discrete"} \qquad M(n)(x) \leftrightarrow x=1 \pmod{A_n(x)}$$

Let LOG_1^* be $LOG^* \cup LOG_1$.

Lemma 3.1.

TOG_1^* is a Δ_1-definitional extension of TOG_1.

Proof: We are asked to give Δ_1-definitions in LOG_1 of $x = k \pmod{A_n(x)}$, $x < k \pmod{A_n(x)}$, $k < x \pmod{A_n(x)}$ and $E(p,r,k)(x)$.

$$x = k(\mathrm{mod}\ A_n(x)) \longleftrightarrow \exists y(M(n)(y)\ \&\ A_n(x-ky) \subset A_n(x))$$
$$\longleftrightarrow \forall y(M(n)(y) \rightarrow A_n(x-ky) \subset A_n(x)$$

where $A_n(x) \subset A_n(y) \longleftrightarrow x < y\ \&\ \neg EA_n(x,y)$

$$x < k(\mathrm{mod}\ A_n(x)) \longleftrightarrow \exists y(M(n)(y)\ \&\ x < k\ \&\ \neg A_n(ky-x) \subset A_n(x))$$
$$\longleftrightarrow \forall y(M(n)(y) \rightarrow x < k\ \&\ \neg A_n(ky-x) \subset A_n(x)) .$$

Similarly we find a Δ_1-definition for $k < x(\mathrm{mod}\ A_n(x))$.

$$E(p,r,k)(x) \longleftrightarrow \exists y(AF_{p,r}(x,y)\ \&\ Dk(y)\ \&\ M(p^r)(y)\ \&\ F_{p,r}(x-ky) \subset F_{p,r}(x))$$
$$\longleftrightarrow EAF_{p,r}(x)\ \&\ \forall y(AF_{p,r}(x,y) \longrightarrow Dk(y))\ \&$$
$$\forall y(AF_{p,r}(x,y)\ \&\ M(p^r)(y) \longrightarrow F_{p,r}(x-ky) \subset F_{pr}(x))$$

where $F_n(x) \subset F_n(y) \longleftrightarrow \exists z(F_n(x-p^r z,y))$

$$\longleftrightarrow \forall z(F_n(z,x) \longrightarrow F_n(z,y))\ \&\ \neg EF_n(x,y) .$$

Theorem 3.2.

For every $n \geq 2$ let T_n be a theory in the language LSP of spines. Let M be the class of all ordered abelian groups G such that for all $n \geq 2$ $Sp_n(G)$ is a model of T_n. If for all $n \geq 2$ T_n is substructure-complete, then M is substructure-complete in the language LOG_1.

Proof: Let $G_1, G_2 \in M$ with a common substructure H (in the language LOG_1). By Lemma 2.1 the mappings $f_{n,j} : Sp_n(H) \longrightarrow Sp_n(G_j)$, $n \geq 2$, $j = 1,2$ defined by

$$f_{n,j}(A_n^H(h)) = A_n^{G_j}(h) \quad , \quad f_{n,j}(F_n^H(h)) = F^{G_j}(h)$$

are embeddings. By assumption we have for all $n \geq 2$:

$$(Sp_n(G_1),\ f_{n,1}(h))_{h\in H} \equiv (Sp_n(G_2),\ f_{n,2}(h))_{h\in H} .$$

By the great Transfer Theorem (1.7) and Lemma 3.1 we get as desired:

$$(G_1,h)_{h\in H} \equiv (G_2,h)_{h\in H} .$$

Remarks:

1. Using Theorem 1.7 for the special case of sentences we see that for every elementary class M of ordered abelian groups there are theories T_n in the language LSP for all $n \geq 2$ such that $G \in M$ iff for all $n \geq 2$ $Sp_n(G) = T_n$.

2. It may happen that a given theory T_n is not substructure-complete in the language LSP but is so in some definitional extension LSP' of LSP. Theorem 3.2 remains true in this case if we consider a definitional extension LOG_1' of LOG_1 such that for every symbol $P(X_1,\dots,X_r,Y_1,\dots,Y_r)$ in LSP' there is a quantifierfree LOG_1'-formula $\varphi_P(x_1,\dots,x_r)$ such that for all ordered abelian groups G and $g_1,\dots,g_r \in G$:

(*) $$Sp_n(G) \models P(A_n(g_1),\dots,A_n(g_r),F_n(g_1),\dots,F_n(g_r))$$

iff

$$G \models \varphi_P(g_1,\dots,g_r) \ .$$

We will consider certain restrictions on the elementary class M which allow us to use more modest extensions of LOG. But first let us state a criterion for modelcompleteness in the most general case.

Theorem 3.3.

Let T_n and M be as in Theorem 3.2. If for all $n \geq 2$ T_n is modelcomplete, then M is modelcomplete in the language LOG_1.

Proof: Analogous to Theorem 3.2.

Remark 3.

What was said in remark 2 above is also true for modelcompleteness in place of substructure completeness. We may this time require (*) only for $G \in M$.

Corollary 3.4.

Let M and $(T_n : n \geq 2)$ be as in Theorem 3.2. This time we assume that for all $G \in M$ and all $n \geq 2$ $Sp_n(G)$ is finite. Then the Theorems 3.2 and 3.3 remain true when we consider the groups in M as LOG_2-strutures.

Definition:

LOG_2 is the language arising from LOG by adding the predicates

$$M(n)(x) \ , \ EA_n(x,y) \ , \ AF_n(x,y) \ , \ Dk(x) \ , \ \beta(n,p,m)(x)$$

for all $n \geq 2$, $m \geq 0$ and all primes p dividing n .

The interpretation of $\beta(n,p,m)(x)$ is given by $G \models \beta(n,p,m)(g)$ iff $\dim C_n(g)/pC_n(g) \geq m$.

Proof of Corollary 3.4.
Let G_{fin} denote the class of all ordered abelian groups G such that $Sp_n(G)$ is finite for all $n \geq 2$. We consider groups in G_{fin} as LOG_2-structures. Since the predicates $EA_n(x,y)$ are in LOG_2 it follows that G_{fin} is closed under substructures. In order to reduce Corollary 3.4 to Theorems 3.2 & 3.3 we ought to give for every predicate P in $LOG_1 \setminus LOG_2$ an existential LOG_2-formula φ_P and an universal formula ψ_P such that for all $G \in G_{fin}$:

$$G \models \forall \bar{x}((P(\bar{x}) \longleftrightarrow \varphi_P(\bar{x})) \ \& \ (P(\bar{x}) \longleftrightarrow \psi_P(\bar{x}))) \ .$$

We call φ_P, ψ_P a Δ_1-definition of P with respect to G_{fin}.
The main simplification lies in the fact that for all G in G_{fin} we have:

$$G \models \forall x \exists y (F_n(x) = A_n(y)) \qquad \text{(by 1.6)} \ .$$

Thus we have in G_{fin} : $EAF_n(x) \longleftrightarrow x = x$.
This can be used to show that the following equivalences are true in G_{fin}:

$$\begin{aligned} F_n(x) = F_n(y) &\longleftrightarrow \exists u,w(F_n(x) = A_n(u) \ \& \ F_n(y) = A_n(w) \ \& \ A_n(u) = A_n(w)) \\ &\longleftrightarrow \forall u,w(F_n(x) = A_n(u) \ \& \ F_n(y) = A_n(w) \longrightarrow A_n(u) = A_n(w)) \\ x \in F_n(y) &\longleftrightarrow \exists u(F_n(y) = A_n(u) \ \& \ x \in A_n(u)) \\ &\longleftrightarrow \forall u(F_n(y) = A_n(u) \longrightarrow x \in A_n(u)) \end{aligned}$$

where $x \in A_n(u) \longleftrightarrow x < u \ \& \ A_n(x) \neq A_n(u)$

$$\begin{aligned} D(p,r,i)(x) &\longleftrightarrow F_{p,r}(x) = A_n(0) \vee [\exists y(A_{p,r}(x+p^r y) = F_{p,r}(x) \\ &\qquad \& \ F_{p,i}(x+p^r y) \neq F_{p,r}(x+p^r y)] \\ &\longleftrightarrow F_{p,r}(x) = A_n(0) \vee [\forall y(A_{p,r}(x+p^r y) = A_{p,r}(x) \longrightarrow \\ &\qquad F_{p,i}(x+p^r y) \neq A_{p,r}(x+p^r y))] . \end{aligned}$$

The formulas on the right side of these equivalences are existential or universal in LOG_2. That $\alpha(n,p,k,m)(x)$ also has a Δ_1-definition in LOG_2 with respect to G_{fin} follows in the same way using the remark on page 7.

Remark 4.
Corollary 3.4 can also be applied "locally" to some fixed n. That is to say, if for a particular n and all $G \in M$ $Sp_n(g)$ is finite, then we may leave out to predicates EAF_n, EF_n, F_n and exchange $\alpha(n,p,r,m)$ for $\beta(n,p,m)$ just for this n.

Corollary 3.5.

Let $\mathcal{M}$ and $(T_n : n \geq 2)$ be as in Theorem 3.3. We assume that for all $G \in \mathcal{M}$ and all $n \geq 2$ $Sp_n(G)$ is finite and for all $g \in G$ and all primes p $\beta_p(C_n(g))$ is also finite. Then the Theorems 3.2 and 3.3 remain true when we consider the groups in $\mathcal{M}$ as LOG_3-structures.

Definition:

The language LOG_3 arises from LOG by adding

$$M(n)(x) \ , \ EA_n(x,y) \ , \ Dk(x) \ , \ D_n(x) \ , \ \beta(n,p,m)(x)$$

for all $n \geq 2$, $m \geq 0$ and all primes p dividing n .

Here $D_n(g)$ holds true for an element g in the ordered abelian group G iff $g = nh$ for some $h \in G$.

Proof of Corollary 3.5.

Let $\mathcal{G}_{ff}$ denote the class of all ordered abelian groups G such that $Sp_n(G)$ is finite for all $n \geq 2$ and for all $g \in G$ and all primes p $\beta_p(C_n(g))$ is finite. We consider the groups in $\mathcal{G}_{ff}$ as LOG_3-structures. Since the predicates $EA_n(x,y)$ and $\beta(n,p,m)(y)$ are in LOG_3 it follows that $\mathcal{G}_{ff}$ is closed under substructures. In order to reduce Corollary 3.5 to Theorems 3.2 and 3.3 we ought to give a Δ_1-definition in LOG_3 of the predicate $AF_n(x,y)$ with respect to $\mathcal{G}_{ff}$. Writing out these definitions would lead to long and messy formulas, so we decided to argue semantically at this point. Let G, H be two ordered abelian groups in $\mathcal{G}_{ff}$ and G a substructure of H with respect to LOG_3. All we have to verify is:

(1) for all $g_o, g_1 \in G$ and $n \geq 2$

$$A_n^G(g_o) = A_n^G(g_1) \quad \text{iff} \quad A_n^H(g_o) = A_n^H(g_1) \ .$$

We may assume in (1) $g = g_o = g_1$, since the following equivalences are true in $\mathcal{G}_{ff}$ (even $\mathcal{G}_{fin}$):

$$A_n(x) = F_n(y) \longleftrightarrow \exists z(A_n(x) = A_n(y+nz) \ \& \ A_n(y+nz) = F_n(y+nz))$$

$$\longleftrightarrow \forall z(A_n(x) = A_n(y+nz) \longrightarrow A_n(y+nz) = F_n(y+nz)) \ .$$

In every ordered abelian group we have

$$A_n(g) = F_n(g) \quad \text{iff} \quad g + A_n(g) \not\subseteq n \cdot C_n(g) \ .$$

This shows that (1) follows from the following assertion:

(2) for all prime powers p^s dividing n and all $g \in G$:

$$g + A_n^G(g) \in p^s C_n^G(g) \quad \text{iff} \quad g + A_n^H(g) \in p^s C_n^H(g) .$$

The following will be the major stepping stone in the proof of (2):

(3) for all prime powers p^s dividing n and all $h \in H$ there are elements $g' \in G$ and $h' \in H$ such that:

$$h - g' = p^s h' \quad \text{and} \quad A_n^H(h) = A_n^H(h') = A_n^H(g') .$$

We prove (2) and (3) by simultaneous induction on the number $q(g)$ (resp. $p(h)$) of elements in $Sp_n(g)$ (resp. $S_p(h)$) that lie strictly above $A_n^G(g)$ (resp. above $A_n^H(h)$).

The greatest element in both $Sp_n(G)$ and $Sp_n(H)$ is the empty set. The second largest element is the trivial subgroup {0} in both cases.

Assume that (2) is true for those $g \in G$ with $q(g) \leq n$ and (3) is true for those $h \in H$ with $p(h) < n$. We will infer from this that (3) is true for all $h \in H$ with $p(n) = n$.

First choose $g \in G$ such that $A_n^H(h) = A_n^H(g)$, which is possible by the finiteness of $Sp_n(H)$. Next let $g_1,\dots,g_r \in G$ be such that $p^{s-1}g_1 + A_n^G(g),\dots,p^{s-1}g_r + A_n^G(g)$ are representatives for a basis of $p^{s-1}C_n^G(g)/p^s C_n^G(g)$. By (2) and the finiteness of $\beta_p(C_n^H(h)) = \beta_p(C_n^G(g))$ $p^{s-1}g_1 + A_n^H(g),\dots,p^{s-1}g_r + A_n^H(g)$ are representatives for a basis of $p^{s-1}C_n^H(h)/p^s C_n^H(h)$. For $s = 1$ this implies that we find a linear combination g' of $g_1,\dots,g_r$ and $h_1 \in H$ such that

$$A_n^H(h-g'-ph_1) = A_n^H(h) \quad \text{and} \quad A_n^H(h_1) = A_n^H(g') = A_n^H(h) .$$

If $p(h) > 1$ we apply the induction hypothesis on (3) to the element $h-g'-ph_1$ and obtain the result. If $p(h) = 1$ then $A_n^H(h-g'-ph_1) = \{0\}$ and we are through. The result for $s > 1$ is obtained by repeated application of the case $s = 1$.

Now we assume (2) for all $g \in G$ with $q(g) < n$ and (3) for all $h \in H$ with $p(h) < n$ and aim to verify (2) for all $g \in G$ with $q(g) = n$.

The implication from left to right is trivial. So we assume $g + A_n^H(g) \in p^s C_n^H(g)$. By definition this means that there is some $h \in H$ such that:

$$A_n^H(g) = A_n^H(h) \quad \text{and} \quad A_n^H(g-p^s h) \subset A_n^H(g) .$$

By (3) applied to $g - p^s h$ there are $g' \in G$, $h' \in H$ with

$$A_n^H(g') = A_n^H(h') = A_n^H(g-p^s h) \quad \text{and} \quad g - g' = p^s(h+h') .$$

Since $D_{p,s}$ is a predicate in LOG_3 this implies that $g - g' \in p^sG$. Since $A_n^G(g-g') = A_n^G(g') \subset A_n^G(g)$ we conclude from this $g + A_n^G(g) \in p^sC_n^G(g)$.

Applications.

I.

M. Kargapolov [8] introduced for a set π of primes the concept of a π-regular ordered abelian group. Translated into our terminology his definition reads:

Definition:
An ordered abelian group G is π-regular iff

(1) $Sp_n(G)$ is finite for all $n \geq 2$.

(2) $\beta_p(C_n(g))$ is finite for all $n \geq 2$, all primes p and all $g \in G$.

(3) for $p \in \pi$ $Sp_n(G)$ contains just one element, i.e. G is p-divisible.

(4) if $\pi \neq \emptyset$ then G is densely ordered.

Kargapolov showed how to axiomatize the elementary theory, Th(G), of a π-regular group G and proved that Th(G) is modelcomplete in the language he considered. By the Great Transfer Theorem it is clear that in order to axiomatize Th(G) we have to specify for each $n \geq 2$:

* the exact (finite) number of elements in $Sp_n(G)$
* the (finite) value $\beta_p(C_n(g))$ for each prime p and each n-regular factor of G
* density or discreteness of the n-regular factors.

Let $T_n(G)$ be the theories of n-spines that arise in this way. Then the class $\mathcal{M}$ of all ordered abelian groups H elementarily equivalent to G coincides with the class of all H such that for all $n \geq 2$ $Sp_n(H)$ is a model of T_n. Since $T_n(G)$ has just one model it is trivially model-complete and by Corollary 3.5 we see that $\mathcal{M}$ is modelcomplete in the language LOG_3. Kargapolov used a different set of additional predicates which we will not describe here, since they are not easily understood and anyhow do not allow a transparent description of what is going on.

II.

V. Weispfenning investigated in [12] ordered abelian groups with finitely many unary predicates $S_o,\dots,S_k$ selecting convex subgroups

$0 = H_o \subseteq H_1 \subseteq \dots \subseteq H_k$ = universe. The main result concerns the case when all factors S_{k+1}/S_k are regular. He shows that a complete theory is obtained by specifying the Szmielev-invariants $\beta_p(S_{k+1}/S_k)$ and whether this factor groups are discrete or not. He also shows that any such complete theory is substructure complete (indeed allows primitiv recursive elimination of quantifiers) when the language is extended by the predicates

$$x \equiv_m^{\kappa} y \quad \longleftrightarrow \quad \exists z(S_k(x-y-mz)$$

and constant symbols e(i) such that e(i) + S_k is the least positive element in S_{k+1}/S_k if this quotient is discrete.

These results can be easily derived from the Great Transfer Theorem and Corollary 3.4. One basically has to show that the predicates of LOG_2 can be defined by quantifier-free formulas in Weispfenning's language. It should be mentioned that [12] also contains results about the case when S_{k+1}/S_k are not regular and results on quantifier elimination in lattice ordered abelian groups.

III.

In [7] B. Jacob studies generalized real closed fields K with an order P of exact level 2m, (K,P). The basic tool in the model theoretic investigation of these structures is a definable valuation ring O(K,P) of K which as it turns out (K,O(K,P)) a henselian valued field with real closed residue class field whose value group G satisfies:

(i) if $p \in S$ then G is p-divisible

(ii) if $p \notin S$ then $\beta_p(G/pG) = 1$

where S is the set of primes dividing m.

In our terminology these groups G can be characterized as those satisfying for all $n \geq 2$:

(1) for $p \notin S$ and all $A_n(g) \in Sp_n(G)$ $\beta_p(C_n(g)) = 0$

(2) for $p \in S$ and $p|n$ there is exactly one $A_n(g) \in Sp_n(G)$ with $\beta_p(C_n(g)) = 1$ for all other $A_n(g') \in Sp_n(G)$ we have $\beta_p(C_n(g')) = 0$.

As a particular consequence we get that no $C_n(g)$ can be discrete.
B. Jacob extends the language of ordered abelian groups to L(S) by adding:

$$D_p(x) \longleftrightarrow \exists y(py = x)$$
$$T_p(x) \longleftrightarrow \exists y(0 < y < x \ \& \ \neg D_p(y))$$

for all $p \in S$.

For p, q ∈ S he uses the abbreviation L(p,q) for

$$\forall x(0 < x \ \&\ \neg D_p(x) \rightarrow \exists y(0 < y < x \ \&\ \neg D_q(x)) \ .$$

To translate these into our terminology let G be a group satisfying (1) & (2). For p ∈ S let $A_n(g_p)$ be the unique element of $Sp_n(G)$ such that $\beta_p(C_n(g_p)) = 1$. Then we observe that for g ∈ G:

$$g \notin pG \quad \text{implies} \quad A_n(g) = A_n(g_p) \ .$$

Furthermore for p, q ∈ S and n such that pq divides n:

$$G \models T(p,q) \quad \text{iff} \quad A_n(g_q) \subseteq A_n(g_p)$$

$$G \models T_p(g) \quad \text{iff} \quad A_n(g_p) \subseteq A_n(g) \ .$$

Now B. Jacob claims that the first-order theory of any ordered abelian group G satisfying (1) & (2) is completely determined by specifying which L(p,q) are true in G and which are not, for p,q ∈ S. ([7], Theorem on page 218). But this is false, since it fails to distinguish between $Z_p + Q$ and Z_p, S = {p}. The sentence $\sigma_p = \exists x(0 < x \ \&\ \forall y(0 < y \longrightarrow D_p(y))$ is true in $Z_p + Q$ but not in Z_p. The theorem becomes true however if in addition all sentences σ_p for p ∈ S that are true in G are added, respectively the negations of σ_p for those p ∈ S for which σ_p is false in G. This follows from the Great Transfer Theorem since T(p,q) and σ_p together allow to determine the Szmielev invariants β_p for all elements in all n-spines.

The second assertion of B. Jacob's theorem, that for groups G satisfying (1) & (2) Th(G) is modelcomplete in L(S') follows from Corollary 2.4. The predicates Dk(x), M(n)(x) of LOG_3 may be dropped in this case since there are no discrete n-regular factors and the remaining predicates of LOG_3 are easily seen to have Δ_1-definitions in L(S) with respect to Th(G).

Concluding remark.

Theorems 3.2 and 3.3 give in general only sufficient conditions for the class M to be modelcomplete resp. substructure complete. In an attempt to prove that these conditions are also necessary, let us say in the case of modelcompleteness, we would have to construct for two given models S_1, S_2 of the theory T_n with $S_1 \subseteq S_2$ two ordered abelian groups G_1, G_2 such that $G_1 \subseteq G_2$ and $Sp_n(G_i) = S_i$ for i = 1,2. This is a problem we could not handle. The situation is even worse; while we are able to axiomatize the elementary theory of {$Sp_n(G)$: G an ordered abelian group} we could not find a full solution to the question: for which countable LSP-structures S do exist ordered abelian groups G with $Sp_n(G) \cong S$? Partial solutions are contained in [10]. For the same reason we could not prove theorems dealing with modelcompletions, prime models or the amalgamation property analogous to theorems 3.2 and 3.3.

Section 4: Convex Theory.

Since convex subgroups play such a crucial role in the study of ordered abelian groups it is worthwhile to investigate the following notions:

Definition:
A class M of ordered abelian groups is convex modelcomplete if for any two groups G, H in M such that G is a convex subgroup of H we have already $G \preceq H$.

Definition:
M is convex substructure complete if for any to groups H_1, H_2 in M with a common convex subgroup G we have:

$$(H_1,g)_{g\in G} \equiv (H_2,g)_{g\in G} \quad .$$

We immediately obtain:

Theorem 4.1.

For every $n \geq 2$, let T_n be a theory in the language LSP of spines. Let M be the class of all ordered abelian groups G such that for all $n \geq 2$: $Sp_n(G)$ is a model of T_n. If for all $n \geq 2$ T_n is substructure complete (modelcomplete) then M is convex substructure complete (resp. convex modelcomplete) in the language LOG.

Proof: We only give the proof for convex modelcompleteness. The case of convex substructure completeness is treated analogously.

Let G, H be in M with G a convex subgroup of H and define as usual the mappings f_n by: $f_n(A_n^G(g)) = A_n^H(g)$, $f_n(F_n^G(g)) = F_n^H(g)$.

Lemma 1.4 guarantees that the assumptions of Lemma 2.1 are satisfied; thus the f_n are well-defined embeddings from $Sp_n(G)$ into $Sp_n(H)$. Modelcompleteness of T_n yields that the f_n are already elementary embeddings and from this we get $G \preceq H$ by the Great Transfer Theorem.

These results are complemented by the following satisfying theorem:

Theorem 4.2.

A theory T of ordered abelian groups is convex model complete iff for every LOG-formula there is a T-equivalent bounded existential LOG-formula.

Proof: One implication is immediate, the other follows from the fact (proved in [2]) that the formulas preserved by convex extensions of T-models are precisely those which are T-equivalent to bounded existential formulas.

Remark:

Theorem 4.2 is of course not restricted to ordered abelian groups, but holds true for any theory where bounded qunatification is defined to be $\forall x(R(x,\bar{y}) \to \varphi(x,\bar{y}))$, resp. $\exists x(R(x,\bar{y}) \,\&\, \varphi(x,\bar{y}))$ for some fixed quantifier-free formula $R(x,\bar{y})$ and convex extensions are replaced by extensions $\mathcal{A} \subseteq \mathcal{B}$ such that for all $\bar{a} \in A$, $b \in B$ $\mathcal{B} \models R(b,\bar{a})$ implies $b \in A$.

We do not know if the following is true:

Conjecture:

A theory of ordered abelian groups, T, is convex substructure complete iff for every LOG-formula there is a T-equivalent bounded LOG-formula.

One implication in the conjectured equivalence is immediate; verfication of the reverse implication runs into the same difficulty as explained in the concluding remarks to section 3. But we know:

Remark 4.3.

Let for all $n \geq 2$ T_n be substructure complete and $\mathcal{M}$ the class of all ordered abelian groups G with $Sp_n(G) \models T_n$ for all $n \geq 2$. Then every LOG-formula $\varphi(\bar{x})$ is in $\mathcal{M}$ equivalent to a LOG-formula $\varphi^*(\bar{x})$ containing at most bounded quantifiers.

Proof: Let ψ_o, ψ_1 be a pair of formulas equivalent to $\varphi(x)$ in the sense of the Great Transfer Theorem 1.7. By assumption we may take ψ_o to be a quantifier-free LSP-formula. Translating ψ_o, ψ_1 back into the language LOG results by Lemma 1.4 in a bounded formula $\varphi^*(x)$ equivalent to $\varphi(x)$ in $\mathcal{M}$.

Let us consider the general problem:

> Given a theory T in a language that contains at least one binary relation symbol $<$ which in every model of T is interpreted as a linear ordering. If T is convex substructure complete, does T then admit elimination of unbounded quantifiers?

We give an example to show that the answer is not always yes:

Let L contain as non-logical symbols binary relations $<$, R and constant symbols c, c_n for $n \in \omega$. Let T be the theory in the language L asserting:

* "$<$ is a dense linear ordering without endpoints"
* $c_0 < c_1 < \ldots < c_n < \qquad < c$
* "there is at most one pair x, y, satisfying Rxy"
* $\forall x,y(Rx,y \longrightarrow (c_n < x < c \;\&\; c < y)$ for all $n \in \omega$.

To prove T substructure complete it suffices to consider two countable T-models $\mathcal{A}$, $\mathcal{B}$ with a common convex substructure $\mathcal{C}$. Without loss of generality we can assume that $\mathcal{A}$, $\mathcal{B}$ are isomorphic if we discard the relation R. If $\neg\exists x,yRxy$ is true in $\mathcal{C}$ the convexity guarantees that $\neg\exists x,yRxy$ is true in both $\mathcal{A}$ and $\mathcal{B}$. If on the other hand $\exists x,yRxy$ hold in $\mathcal{C}$ the isomorphism between $\mathcal{A}$ and $\mathcal{B}$ can easily be arranged to fix the pair of element satisfying R.

If T would admit elimination of unbounded quantifiers, there would in particular be a sentence ψ containing at most bounded quantifiers such that $T \vdash \exists x,yRxy \longleftrightarrow \psi$. Let $\mathcal{A}$ be a model of T, $a,b \in A$ such that $\mathcal{A} \models R(a,b)$. The convex substructure $\mathcal{B}$ of $\mathcal{C}$ with universe $\{a \in A : c < b$ is again a model of T, but $\mathcal{B} \models \neg\exists x,yRxy$.

On the positive side we have:

<u>Theorem 4.4</u>.

Let T be a theory in the language L which contains a binary relation symbol $<$ and an arbitrary number of unary relation symbols such that in every T-model $<$ is a linear order. If for every pair $\mathcal{A}$, $\mathcal{B}$ of T-models with a common one-element substructure $\mathcal{C} = \{c\}$ we have $(\mathcal{A},c) \equiv (\mathcal{B},c)$, then T admits elimination of unbounded quantifiers.

We sketch a proof:

<u>Step 1</u>: For L-formulas $\varphi(x)$ with at most one free variable there is a T-equivalent quantifier-free formula.

Here the standard argument that substructure completeness implies quantifier elimination works since we have to consider only one-element substructures.

Step 2: Every L-formula $\varphi(\bar{x})$ is T-equivalent to a Boolean combination of formulas containing at most two free variables.

This is a consequence of the fact that the disjoint sum operation preserves elementary equivalence.

Step 3: For every L-formula $\varphi(x,y)$ with exactly two free variables there are L-formulas $\varphi_1(x)$, $\varphi_2(y)$, $\chi_{ij}(z_i)$, $\chi'_{ij}(z_i)$ with at most one free variable such that the formula $\varphi(x,y)$ is in T equivalent to:

$$\varphi_1(x) \;\&\; \varphi_2(y) \;\&\; \bigwedge_{1\le i\le n} \exists z_1,\ldots,z_{k_i}(x < z_1 < \ldots < z_{k_i} < y \;\&\; \bigwedge_{1\le j\le k_i} \chi_{ij}(z_i))$$
$$\&\; \bigwedge_{1\le i\le n} \forall z_1,\ldots,z'_{k_i}(x < z_1 < \ldots < z'_{k_i} < y \rightarrow \bigwedge_{1\le j\le k'_i} \chi'_{ij}(z_i))$$

This follows from Theorem 3.10 in M. Rubin's paper "Theories of linear order" in Israel J. of Math. 17 (1974), pp. 392-443.

Appendix.

List of languages:

LOG = language of ordered abelian groups
$+, 0, -, <$

LOG* = LOG plus
$x = k(\mathrm{mod}\ A_n(x))$, $x < k(\mathrm{mod}\ A_n(x))$, $x > k(\mathrm{mod}\ A_n(x))$
for all $n \geq 2$, $k \geq 0$
$D(p,r,i)(x)$ for all primes p, $r \geq 1$, $1 \leq i \leq r$
$E(p,r,k)(x)$ for all primes p, $r \geq 1$, $k \geq 0$

LOG_1 = LOG plus
$M(n)(x)$ for all $n \geq 2$
$D(p,r,i)(x)$ for all primes p, $r \geq 1$, $1 \leq i \leq r$
$EA_n(x,y)$, $EF_n(x,y)$, $AF_n(x,y)$, $F_n(x,y)$, $EAF_n(x)$,
$Dk(x)$ for all $n \geq 2$
$\alpha(n,p,k,m)(x)$ for all $n \geq 2$, $m \geq 0$ all primes p
and $k \geq 1$ such that p^k divides n

LOG_1^* = LOG_1 plus LOG*

LOG_2 = LOG plus

$M(n)(x)$ for all $n \geq 2$

$EA_n(x,y)$, $AF_n(x,y)$, $Dk(x)$ for all $n \geq 2$

$\beta(n,p,m)(x)$ for all primes p, $n \geq 2$, $m \geq 0$

LOG_3 = LOG plus

$M(n)(x)$ for all $n \geq 2$

$EA_n(x,y)$, $Dk(x)$, $D_n(x)$ for all $n \geq 2$

$\beta(n,p,m)(x)$ for all primes p, $n \geq 2$, $m \geq 0$

References.

[1] Feferman, S.: Persistent and invariant formulas for outer extensions. Compositio Math. 20 (1968), 29-52

[2] Feferman, S.: Applications of many-sorted interpolation theorems. Proc. of AMS Symposia in Pure Math. (Tarksi Symposium), vol. 25 (1975), 205-227

[3] Gurevich, Y.: Elementary properties of ordered abelian groups. Translations AMS 46 (1965), 165-192

[4] Gurevich, Y.: The decision problem for some algebraic theories. Doctoral dissertation, Sverdlovsk 1968

[5] Gurevich, Y.: Expanded theory of ordered abelian groups. Annals of Math. Logic 12 (1977), 193-228

[6] Gurevich, Y. and Schmitt, P.H.: The theory of ordered abelian groups does not have the independence property. to appear in Trans. AMS

[7] Jacob, B.: The model theory of generalized real closed fields. J. reine und angew. Math. 323/324 (1981), 213-220

[8] Kargapolov, M.I.: Classification of ordered abelian groups by elementary properties (in Russian). Algebra i Logika 2 (1963), 31-46; MR 27 # 5823

[9] Robinson, A. and Zakon, E.: Elementary properties of ordered abelian groups. Trans. AMS 96 (1960), 222-236

[10] Schmitt, P.H.: Model Theory of ordered abelian groups. Habilitationsschrift, Heidelberg 1982

[11] Schmitt, P.H.: Elementary Properties of ordered abelian groups. to appear

[12] Weispfenning, V.: Elimination of quantifiers for certain ordered and lattice-ordered abelian groups. Bull. Soc. Math. Belg. 33 (1981), 131-155

[13] Zakon, E.: Generalized archimedean groups. Trans. AMS 99 (1961), 21-48

QUANTIFIER ELIMINATION AND DECISION PROCEDURES FOR VALUED FIELDS

Volker Weispfenning
University of Heidelberg

This paper gives an account of model theoretic and algorithmic results in the elementary theory of valued fields, in an approach that uses explicit, primitive recursive quantifier elimination procedures as a unifying principle. We concentrate on positive results concerning quantifier elimination, decision procedures, definable sets, model--completeness, elementary invariants, and prime model extensions. For negative results concerning limitations on quantifier elimination and undecidability, the reader may consult e.g. [Macintyre-McKenna-v.d.Dries 83], [Delon 81], [Becker-Denef-Lipshitz 80] . Our approach covers a good deal of the model theory of valued fields, and provides some significant improvements and new facts. Most notable omissions are the results of Ersov and Delon on Kaplansky fields of characteristic $p > 0$ [Ersov 65, 65-67] , [Delon 81] , and the results of van den Dries and Ersov on fields with several valuations [v.d.Dries 78], [Ersov 80] .

To clarify our viewpoint, let us recall that the model theoretic topics mentioned above can be approached by (at least) two different methods (cf. [Macintyre 77],[Weispfenning a],[Roquette]) : The first ("explicit quantifier elimination") is concerned mainly with syntactical reductions of formulas to formulas of some simple normal form. It dates back to the early work of Skolem, has gained popularity through the work of the Tarski-school, and is also studied by computer scientists under the aspect of feasability. The second ("model-completeness, ultraproducts, saturated structures") was developed by model theorists roughly from 1960 onwards, after earlier pioneering work by Skolem and A.Robinson. It has been very successful in algebraic applications, in particular in field theory. Typical for this method is the combination of model theoretic constructions and compactness arguments with (sometimes deep) results is structural algebra. By contrast, the quantifier elimination method relies on explicit symbolic manipulations (that may be combinatorially involved) in combination with algebraic results of a more elementary nature. So it is closer in spirit to algebraic elimination theory. In comparison, it is fair to say, that in most cases the total complexity of proof (taking both algebraic and logical arguments into account) is about equal.

As far as effectivity is concerned, the second method relies on Gödel' completeness theorem and hence yields -as a rule- only general recursive algorithms, whereas explicit quantifier elimination procedures tend to be primitive recursive. These primitive recursive algorithms can be characterized by having a priori bounds (given by a system of Ackermann functions) on their time complexity, when performed e.g. on a register machine. These bounds are still to large to indicate feasability of the algorithm. On the other hand, most quantifier elimination procedures cannot be expected to run in less than exponential time. So a reasonable goal is to seek for elementary-recursive algorithms, where the time bound is a finite iteration of the exponential function. This has been achieved e.g. for Z-groups (Cooper, Oppen) and for real-closed fields (Collins,Monk,Solovay). Elementary bounds for the Ax-Kochen transfer principle (comparing $\mathbb{Q}_p$ and $\mathbb{F}_p((t))$) and related principles have been computed in [Brown 78]. The primitive recursive procedures presented in this paper may serve as a basis for further research in this direction.

For valued fields, research on decidability and quantifier elimination was carried out mostly in terms of model-completeness, ultraproducts and saturated structures: In 1956, A.Robinson proved model--completeness, completeness and decidability for algebraically closed non-trivially valued fields of fixed pair of characteristics; he also specified prime models. In 1963, Nerode proved the decidability of the diophantine problem in $\mathbb{Q}_p$, and showed that the algebraic part of $\mathbb{Q}_p$ forms a computable field. While these results received only little attention at that time, this changed radically, when Ax and Kochen published their celebrated series of papers [Ax-Kochen 65,66]. For $\mathbb{Q}_p$ and the power series fields F((t)) over a decidable field F of characteristic zero they proved decidability and (relative) quantifier elimination, when a cross-section π is included in the language They also established a transfer of elementary properties from $\mathbb{F}_p((t))$ to $\mathbb{Q}_p$ for large enough prime p . The decidability results for $\mathbb{Q}_p$ and F((t)) (F algebraically closed of characteristic zero or real-closed) were shown independently by Ersov [Ersov 65-67]. Later, Ersov extended the decidability results to Hensel fields of characterisic zero, more general value groups and finite ramification, and to algebraically maximal Kaplansky fields of characterisic $p > 0$. [Ersov 65, 65-67]. [Ziegler 72] and [Baserab 78,79] obtained further refinements of these results. A revised version of [Ax-Kochen 65,66] with simplified proofs and extended results appeared in [Kochen 75]. In 1976, Macintyre showed that quantifier elimination for $\mathbb{Q}_p$ can be obtained without cross-section, when root-predicates are included in the language; as a consequence, definable sets in $\mathbb{Q}_p^n$ have properties comparable to semialgebraic sets in $\mathbb{R}^n$. The result on quantifier elimination was extended in [Cherlin-Dickmann 83] to more general Hensel fields of characteristic zero, and in [Delon 81] to algebraically maximal Kaplansky fields of equal characteristic. A more algebraic treatment of this result for p-adically closed fields appears in [Prestel-Roquette 83]. Macintyre's and Robinson's results have recently found far-reaching applications: Denef uses definable sets of p-adic numbers to prove the rationality of certain Poincaré series [Denef]; Cantor-Roquette and Rumely combine the decidability of algebraically closed valued fields with a local-global principle to derive the decidability of the diophantine problem in algebraic integers (cf. [Roquette]). The study of linear elimination in valued fields was initiated in [v.d.Dries 81] and [Delon 81]. [v.d.Dries b] contains the first results concerning elementary equivalence and elementary extensions for Hensel fields of mixed characteristic and arbitrary ramification.

Primitive recursive methods were introduced into the study of elementary properties of valued fields by [Cohen 69] with a quantifier elimination (using cross-section) and decision method for p-adic fields. His ideas were modified and strongly generalized in [Weispfenning 71, 76]. The results obtained there cover Robinson's theorems on algebraically closed valued fields, as well as a good deal of [Ersov 65, 65-67], [Ziegler 72], [Kochen 75], [Baserab 79]. In [Weispfenning 78] they were applied to prove the primitive recursive decidability of the adele ring and idele group of an algebraic number field. [Brown 78] provides elementary-recursive bounds for the Ax-Kochen transfer principle and related principles.(Weaker results in this direction had been obtained earlier by Ziegler (unpublished) and [Weispfenning 71].)

The <u>plan of this paper</u> is as follows .

<u>Section</u> 1 collects some basic definitions and facts on valued fields together with a few examples to provide a background for the follo-

wing sections.

Section 2 gives a primitive recursive elimination of F-quantifiers (quantifiers ranging over field elements) $Q_1x_1 \dots Q_nx_n(\varphi)$, $Q_i = \exists, \forall$, where $x_1, \dots, x_n$ occur only linearly in φ. The method is uniform for all valued fields, if one includes predicates ε_n in the language, saying that the residue class field has $\geq$ n elements. It exploits again and again the ultrametric triangle inequality and equality ; by means of the approximation theorem it is finally extended to fields with a system of indepenent valuations. Our results generalize the linear elimination for valued fields in [v.d.Dries 81] ; they are also related to the quantifier elimination for ultrametric spaces in [Delon 81] and [Weispfenning b] .

Section 3 presents a primitive recursive quantifier elimination for algebraically closed valued fields of arbitrary characterisitic , following essentially [Weispfenning 71]. It combines the well-known quantifier elimination for divisible ordered abelian groups (elimination of Γ-quantifiers, i.i. quantifiers ranging over the value group Γ and ∞) with the linear elimination of section 2 and a reduction of arbitrary F-quantifiers to quantifiers over zeros of polynomials. Applications include the results on valued fields in [Robinson 56] ; furthermore, a primitive recursive decision procedure, an analogue of Macintyre's theorem on definable sets , prime model extensions, and corresponding results for monically closed valuation rings (cf. [Boffa] , [Weispfenning 82]) .

In section 4 we study Hensel fields of characteristic zero with arbitrary value group, residue class field and arbitrary ramification. To begin with, we find some necessary conditions on the language that have to be satisfied in order to get an F-quantifier elimination. These lead to the statement of our main theorem 4.3 on F-quantifier elimination together with several variants . They extend the author's main results in [Weispfenning 76] and cover the results on quantifier elimination in [Ax-Kochen 66] , [Ziegler 72] , [Macintyre 76] , [Baserab 79] . They are in several aspects more general than the Macintyre-type theorems in [Cherlin-Dickmann 83] and [Delon 81] (for Hensel fields of characteristic zero), but do not quite cover these results due to a mild (but annoying) restriction on the value group. For most applications this restriction is, however, immaterial The results are novel in that

(i) they concern also fields with arbitrary ramification ;

(ii) they give the first explicit, primitive recursive proof of

Macintyre-type quantifier elimination (,where a cross-section is replaced by generalized root-predicates).

Accordingly, applications go in many ways beyond those in [Weispfenning 76] . They concern :

(1) Quantifier elimination and prime model extensions for Hensel fields without cross-section of characteristic (0,0) .

(2) A new proof of the main theorem of [Cherlin-Dickmann 83],i.e. primitive recursive quantifier elimination for real-closed rings. (The results on monically closed valuation rings and real-closed rings have been applies to more general classes of rings and Nullstellensatz-type theorems in [Weispfenning 82] .)

(3) Primitive recursive decision procedures and prime model extensions for p-adically closed fields in the sense [Prestel-Roquette 83] in a language with root-predicates. These cover the model theoretic results in section 5 of [Prestel-Roquette 83] (except 5.3) .

(4) Elementary equivalence and elementary extensions for Hensel fields of mixed characteristic and arbitrary ramification; in particular the corresponding results in [Baserab 79] (3.4.1 , 3.6.1 - 3) and [v.d.Dries b] .

Section 5 is devoted to a proof of the main theorem and its variants. The method of F-quantifier elimination follows partly [Weispfenning 76] in that arbitrary F-quantifiers are replaced by Γ-quantifiers , R-quantifiers (i.e. quantifiers ranging over certain residue class rings), and F-quantifiers ranging over Hensel zeros of polynomials. Due to the lack of a cross-section, the occurences of a bounded F--variable cannot be reduced to linear expression as in [Weispfenning 76] , but only to quotients of powers of linear expressions. This together with the additional generalized root-predicates causes the argument to get more involved; on the other hand, these predicates are necessary in order to eliminate F-quantifiers $\exists x$, where x occurs in powers of linear expressions.

Section 6 indicates some further conclusions that can be drawn from the proofs in section 5 .

I like to thank Jan Denef for stimulating conversations on the subject, and P. Schmitt for bringing lemma 4.1 of [Eklof-Fischer 72] to my attention. Last not least I thank my family for their patience during the preparation of the manuscript, and myself for the many hours of diligent though incompetent typing .

1. SOME BASIC FACTS ON VALUED FIELDS.

In this section, we collect some of the basic definitions, examples and facts concerning valued fields, that we require in the sequel. For a thorough background in valuation theory, we refer the reader e.g. to [Endler 72] .

Let $\mathbb{R}^+$ denote the set of non-negative real numbers. An absolute value (or multiplicative valuation) on a field F is a map $\varphi: F \longrightarrow \mathbb{R}^+$ satisfying

1.1 (i) $\varphi x = 0 \longleftrightarrow x = 0$,

(ii) $\varphi(x \cdot y) = \varphi x \cdot \varphi y$,

(iii) $\varphi(x+y) \leq \varphi x + \varphi y$ (triangle inequality)

So the ordinary absolute value $|\ |$ on the fields $\mathbb{Q}$, $\mathbb{R}$, $\mathbb{C}$ of rational, real, complex numbers, respectively, is an absolute value in the sense of 1.1. An absolute value φ on F is called non-archimedean, if it satisfies the strong (or ultrametric) triangle inequalit

1.1 (iv) $\varphi(x+y) \leq \max(\varphi x, \varphi y)$.

Since $|1+1| > |1|$, the absolute value $|\ |$ on $\mathbb{C}$ is archimedean. Non--archimedean absolute values are constructed conveniently via an (additive, exponential) valuation : Let Γ be an ordered, abelian group , ∞ an extra element on top of Γ , and let F be a field. Then a map $v: F \longrightarrow \Gamma \cup \{\infty\}$ is a valuation on F, if it satisfies

1.2 (i) $v(x) = \infty \longleftrightarrow x = 0$,

(ii) $v(x \cdot y) = v(x) + v(y)$,

(iii) $v(x+y) \geq \min(v(x), v(y))$ (triangle inequality) .

If v is onto $\Gamma \cup \{\infty\}$, then Γ is called the value group of v . Moreover, v is of rank 1 , if Γ is a subgroup of the ordered, additive group of real numbers.

There is a 1-1 correspondence between rank 1 valuations v on F and non-archimedean absolute values φ on F : Fix $1 < a \in \mathbb{R}$; then $v \leftrightarrow \varphi$ via $v(x) = -\log_a \varphi(x)$, $\varphi(x) = a^{-v(x)}$, $-\log_a(0) = \infty$, $a^{-\infty} = 0$. Any rank 1 valuation v defines a metric d on F by $d(x,y) = \varphi(x-y)$ $= a^{-v(x-y)}$. An arbitrary valuation v on F defines a topology τ on F with $\mathbb{U} = \{U_\gamma : \gamma \in \Gamma\}$, $U_\gamma = \{a \in F: v(a) \geq \gamma\}$ as a clopen neighborhood basis of zero. (F, τ) is a totally disconnected, topological field. For $x,y \in F$, x is close to y iff $v(x-y)$ is large in $\Gamma \cup \{\infty\}$.

Any field with absolute value (F,φ) has a <u>completion</u> $(\hat{F},\hat{\varphi}) \supseteq (F,\varphi)$, the completion of $(\mathbb{Q},|\,|)$ being $(\mathbb{R},|\,|)$. If v is a rank 1 valuation on F, then the completion $(\hat{F},\hat{v})$ of (F,v) is defined via the correspondence $v \leftrightarrow \varphi$.

Any field F admits the <u>trivial valuation</u> $v_{triv}: F \to \{0,\infty\}$ defined by $v_{triv}(x) = 0$ for $x \neq 0$. For finite fields, and more generally absolute algebraic fields of positive characteristic this is the only valuation. All other fields have non-trivial valuations, in fact:

1.3 (<u>Extension theorem</u>) Let $F \subseteq F'$ be fields and let v be a valuation on F. Then v has an extension to a valuation v' on F'. In particular, v has an extension to the algebraic closure $\tilde{F}$ of F.

Any map $v: D \to \Gamma \cup \{\infty\}$ defined on an integral domain D and satisfying 1.2 extends in a unique way to a valuation v' of the quotient field F of D; for $a,b \in D$, $b \neq 0$, $v'(a/b) = va - vb$. So in order to define a valuation $v: F \to \Gamma \cup \{\infty\}$, it suffices to determine v on some subdomain D of F with quotient field F . If $v: F \to \Gamma \cup \{\infty\}$ is a valuation with value group Γ , we call (F,Γ,v) a <u>valued field</u>.

<u>Some examples of valued fields</u>. In the following F is a field, F(t) is the rational function field, and F((t)) is the field of formal Laurent series over F.

1.4 $(F(t), \mathbb{Z}, v_\infty)$ with $v_\infty f(t) = -\deg f$ for $f \in F[t]$.

1.5 $(F(t), \mathbb{Z}, v_0)$ with $v_0 f(t) = \operatorname{ord} f$ for $f \in F[t]$, where $\operatorname{ord}(\sum_{n\le i\le m} f_i t^i) = n$ if $f_n \neq 0$.

1.6 $(F((t)), \mathbb{Z}, v_0)$ with $v_0(\sum_{n\le i<\infty} f_i t^i) = n$ if $f_n \neq 0$ ($n \in \mathbb{Z}$).
This is the completion of example 1.5 .

1.7 Let $(F,<)$ be an ordered field with $E = \{a \in F: \exists q \in \mathbb{Q}, -q < a < q\}$ as the set of "<u>finite</u>" elements, $I = \{a \in F: \forall q \in \mathbb{Q}, q > 0 \to -q < a < q\}$ as the set of "<u>infinitesimal</u>" elements. Then E is a subring of F with unique maximal ideal I and $U = E \setminus I$ as multiplicative group of units. Put $\Gamma = F^*/U$, written additively, and define $v: F \to \Gamma \cup \{\infty\}$ by $va = aU$ for $a \neq 0$, $va \le vb$ iff $b=0$ or $b/a \in E$. Then (F,Γ,v) is a valued field.

1.8 For a prime $p \in \mathbb{Z}$, define the p-<u>adic valuation</u> v_p on $\mathbb{Q}$ by $v_p(\sum_{n\le i\le m} a_i p^i) = n$ for $a_n \neq 0$, $0 \le n \le m$ in $\mathbb{Z}$, $0 \le a_i \le p-1$ in $\mathbb{Z}$. Then $(\mathbb{Q}, \mathbb{Z}, v_p)$ is a valued field. The corresponding non-archimedean absolute values φ_p together with $|\,|$ constitute essentially all absolute values on $\mathbb{Q}$.

1.9 The completion of example 1.9 is given by the valued field

$(\mathbb{Q}_p, \mathbb{Z}, v_p)$ of p-adic numbers. The elements of $\mathbb{Q}_p$ can be uniquely expressed as formal infinite sums $a = \sum_{n\leq i<\infty} a_i p^i$ with $0\leq a_i \leq p-1$, $a_i, n \in \mathbb{Z}$; then $v_p a = n$, provided $a_n \neq 0$.

We continue with some properties of valued fields in general.

1.10 The following rules hold in any valued field.
$v(1) = 0$, $v(-a) = va$, $v(a/b) = va - ab$ if $b\neq 0$,
$v(\sum_{1\leq i\leq n} a_i) \geq \min (va_i : 1\leq i\leq n)$; if all the values va_i, $1\leq i\leq n$, are different, then $v(\sum_{1\leq i\leq n} a_i) = \min (va_i : 1\leq i\leq n)$ (triangle equality)

1.11 Put $A = A_v = \{a \in F: va \geq 0\}$, $M = M_v = \{a \in F: va > 0\}$. Then A is a subring of F with quotient field F, M is the unique maximal ideal of A, and $U = A \setminus M$ is the mult plicative group of units of A. A is called the valuation ring of v , and M the valuation ideal of v. For any $0\leq\alpha\in\Gamma$, $I_\alpha = \{a \in F: va > \alpha\}$ is an ideal of A ; $res_\alpha : A \rightarrow R_\alpha = A/I_\alpha$ is the canonical homomorphism onto the residue class ring R_α. For $\alpha = 0$, we put $R_0 = R = F_v$, $res_0 = res$. Then $F_v = A/M$ is a field, the residue class field of (F,Γ,v). For the characteristics of F and F_v, the following cases are possible: either char F = char F_v (comp. ex. 1.6 , where $F((t))_{v_0} = F$) or char F = 0 and char $F_v = p > 0$ (comp. ex. 1.9, where $(\mathbb{Q}_p)_{v_p} = \mathbb{F}_p$, the field of p elements).
In ex. 1.7, F_v is ordered by res a $\leq$ res b iff $b \geq a-i$ for some infinitesimal i; F_v can be embedded as an ordered field into $\mathbb{R}$.

1.12 v can be essentially recovered from its valuation ring $A = A_v$: Define $i: F^*/U \rightarrow \Gamma$ by $va = i(aU)$ for $0\neq a \in F$. Then i is an isomorphism of ordered groups, if F/U is ordered by $aU \leq bU$ iff $b/a \in A$ (comp. ex. 1.7). A subring A with 1 of a field F is a valuation ring of some valuation v on F iff for all $a,b \in A$, a divides b or b divides a.

1.13 Let v be a rank 1 valuation and assume F is complete with respect to v . Then the Newton approximation of a zero of a polynomial in $\mathbb{R}$ can be copied in (F,Γ,v) (cf. [Kochen 75]); as a result , (F,v,Γ) satisfies Hensel's Lemma (HL) :
Let $f(X)\in A[X]$, $a \in A$ with $res f(a) = 0$, $res\, f'(a) \neq 0$; then there is a unique $z \in A$ with $f(z) = 0$, res z = res a , i.e. res a can be lifted to a root of $f(X)$.
From HL one obtains by a suitable linear substitution the more general Newton Lemma (NL) (cf.[Weispfenning 76], 1.2.2)
Let $f(X)\in A[X]$, $a \in A$, $0\leq\alpha\in\Gamma$ with $vf(a) > 2\alpha$, $vf'(a) \leq \alpha$; then

there is a unique $z \in A$ with $f(z) = 0$, $res_{\alpha}(z) = res_{\alpha}(a)$.

(F,Γ,v) is called a <u>Hensel field</u>, if it satisfies HL and hence NL . By the above, any (F,Γ,v), where v is of rank 1, can be extended to a Hensel field $(\hat{F},\hat{\Gamma},\hat{v})$, viz. the completion of (F,Γ,v). Moreover, $(\hat{F},\hat{\Gamma},\hat{v})$ is an <u>immediate extension</u> of (F,Γ,v), i.e. $\hat{\Gamma} = \Gamma$, $\hat{F}_{\hat{v}} = F_v$. A corresponding fact is true for all valued fields: Any valued field (F,Γ,v) has an immediate extension to a Hensel field. Examples 1.6, 1.9 are Hensel fields by virtue of their completeness; ex. 1.7 is a Hensel field, in case $(F,<)$ is real closed, as an application of the intermediate value theorem shows.

1.14 Let Δ be a subset of the value group Γ . A map $\pi : \Delta \to F^{\times}$ is a <u>cross-section of</u> v <u>on</u> Δ , if $v \circ \pi = id_{\Delta}$, and $\pi(\alpha+\beta) = \pi(\alpha)\cdot\pi(\beta)$ for $\alpha, \beta, \alpha+\beta \in \Delta$. Any such π we call a <u>partial cross-section of</u> v; if $\Delta = \Gamma$, then π is a (total) <u>cross-section</u> of v. For any cross-section π of v , $\pi(0) = 1$, $\pi(-\alpha) = \pi(\alpha)^{-1}$; π may be extended to ∞ by setting $\pi(\infty) = 0$. Any valuation with value group $\mathbb{Z}$ has a cross-section (Put $\pi(n) = q^n$ for a fixed $q \in F$ with $v(q) = 1$). This applies in particular to examples 1.4,1.5,1.6,1.8,1.9 . Example 1.7 also has a cross-section, in case F is real closed (Choose a basis X of Γ as $\mathbb{Q}$-vector space, define π on X such that all $\pi(x)$ are positive, and extend π to Γ by positive roots in F). Not all valued fields have cross-sections, but any valued field has an elementary extension with a cross-section (see lemma 4.18) .

The following illustration may help to visualize a valued field (with cross-section) .

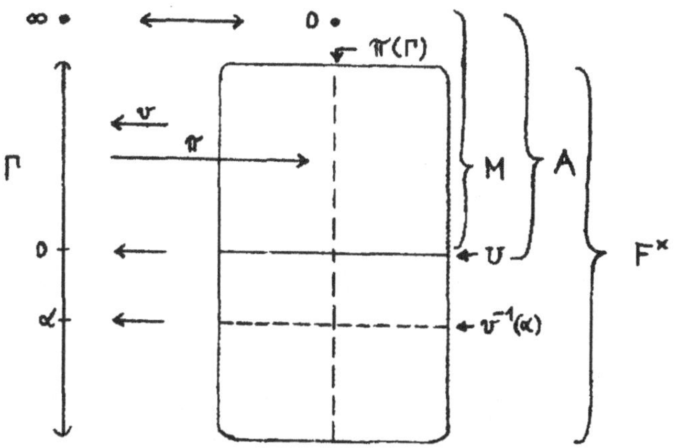

2. LINEAR PROBLEMS IN VALUED FIELDS.

Let $L_F = \{0,1,+,\cdot,{}^{-1}\}$ be the language of fields. Call a formula φ of L_F linear, if all atomic subformulas of φ are of the form $n_1x_1+\dots+n_mx_m = 0$ with $n_i \in \mathbb{Z}$. Then it is easy to specify a procedure (essentially Gauss elimination) that reduces any linear formula φ in L_F equivalently in the theory of infinite fields to a quantifier-free formula φ'.

In the following, we are going to describe a corresponding procedure for arbitrary valued fields; it will show in particular that linear problems in valued fields are primitive recursively reducible to questions on the value group. With the help of the approximation theorem, the method can be extended to fields with an arbitrary system of pairwise independent valuations.

To begin with, we fix a language L for valued fields that exhibits the rôle of the value group. $L = (L_F, L_\Gamma, v)$ is two-sorted, consisting of the language L_F of fields as F-sort , the language $L_\Gamma = \{0,\infty,+,-,<\}$ of ordered abelian groups with an extra element ∞ on top as the Γ-sort , and the operation-symbol v for the valuation. OAG_∞ is the theory of ordered abelian groups with top element ∞ in L_Γ , VF = VF(L) is the theory of valued fields in the language L. We assume by convention that $v: F \to \Gamma \cup \{\infty\}$ is onto, that $0^{-1} = 0$, and that $\infty = -\infty = \alpha + \infty = \infty + \alpha$ for all $\alpha \in \Gamma \cup \{\infty\}$. We denote F-variables by $x, y, \dots$, F-terms by $a, b, \dots$, Γ-variables by $\xi, \eta, \dots$, Γ-terms by $\alpha, \beta, \dots$, and corresponding tuples by $\underline{x}, \underline{y}, \dots, \underline{a}, \underline{b}, \dots, \underline{\xi}, \underline{\eta}, \dots, \underline{\alpha}, \underline{\beta}, \dots$. An F-quantifier (Γ-quantifier) is a quantifier of the form $\exists x$, $\forall x$ ($\exists \xi$, $\forall \xi$).

Let now $\underline{x} = (x_1,\dots,x_n)$ be a tuple of F-variables, a an F-term, α a Γ-term, φ an L'-formula, where $L' = L \cup L'_\Gamma$ and L'_Γ is an extension of L_Γ with $L'_\Gamma \cap L = L_\Gamma$. Then we say " $\underline{x}$ is linear in a " if a is of the form $a = a_1x_1+\dots+a_nx_n + a'$, where x is not in a_i and a'; we say " $\underline{x}$ is linear in α ", if α is of the form $\alpha = \beta(vb_1,\dots,vb_m)$ where $\beta(\xi_1,\dots,\xi_m)$ is an L'_Γ-term and $\underline{x}$ is linear in all b_i; we say " $\underline{x}$ is linear in φ ", if $\underline{x}$ is linear in all terms occuring in φ. Finally, we call φ a linear formula, if the tuple of all F-variables bounded in φ is linear in φ .

In order to eliminate quantifiers in linear formulas, we add to L' nullary predicates $\{E_q\}_{q<\omega}$ and obtain the language L'(E). VF(L'(E)) is then the theory VF(L) together with the following defining axioms for the predicates E_q:

$$E_q \longleftrightarrow \exists x_1 \ldots \exists x_n \bigwedge_{1 \le i < j \le q-1} 0 = v(x_i) = v(x_j) < v(x_i - x_j)$$

So E_q holds in $(F, \Gamma, v) \in VF(L'(E))$ iff $card(F_v) \ge q$.

THEOREM 2.1 There is a primitive recursive procedure assigning to any linear formula φ in L'(E) an F-quantifier-free linear formula φ' in L'(E) such that φ and φ' are equivalent in VF(L'(E)).

For any prime p , the following sets of sentences ,

$char'_p$: $\{p1 = 0\}$,

$char_p$: $\{0 < v(p1) < \infty\}$,

$char_0$: $\{v(n) = 0 : 0 < n < \omega\}$,

specify the characteristic of a valued field :
$char(F, \Gamma, v) = (p,p)$, $(0,p)$, $(0,0)$ iff $(F, \Gamma, v) \models char'_p$, $char_p$, $char_0$, respectively.

COROLLARY 2.2 Let T'_Γ be an extension of OAG_∞ in L'_Γ .

(1) If T'_Γ admits (primitive recursive) quantifier elimination in L'_Γ , then $VF(L'(E)) \cup T'_\Gamma$ admits (primitive recursive) quantifier elimination for linear formulas in L'(E).

(2) Suppose T'_Γ is complete and L'_Γ contains a constant 1'. Then the theories $VF(L'(E)) \cup T'_\Gamma \cup \Phi$ decide (i.e. prove or disprove) all linear sentences in L'(E) , where Φ is one of the following sets of sentences :

$char_0 \cup \{1' = 0\}$,

$char'_p \cup \{1' = 0\} \cup \{E_p n , \neg E_p n+1\}$,

$char'_p \cup \{1' = 0\} \cup \{E_p m : m < \omega\}$,

$char_p \cup \{v(p) = 1'\} \cup \{E_p n , \neg E_p n+1\}$,

$char_p \cup \{v(p) = 1'\} \cup \{E_p m : m < \omega\}$, (p prime, $0 < n < \omega$).

If in addition T'_Γ is primitive recursively decidable, then all these theories have a primitive recursive decision procedure for linear sentences.

The proof of 2.1 is based on the following technical lemma.

LEMMA 2.3 There is a primitive recursive procedure assigning to any L-formula φ of the form

$$\bigwedge_{1 \le i \le n} (v(y_i x - z_i) = \xi_i < \infty \wedge y_i \ne 0)$$

an L-formula φ' and an L(E)-formula φ'' such that :

(i) $\varphi \longleftrightarrow \varphi'$, $\exists x(\varphi) \longleftrightarrow \varphi''$ in VF(L(E)).

(ii) φ', φ'' are quantifier-free and $\underline{z}$ is linear in φ', φ''.

(iii) φ' is a disjunction of formulas of the form

$$\bigwedge_{i\in I} v(ax - c_i) = \alpha_h < \infty \ \wedge \bigwedge_{i,j\in I,\ i\neq j} v(c_i - c_j) = \alpha_i = \alpha_h \wedge \psi ,$$

where the z_i are not in a, x is not in a, c_i, α_i, ψ ; ψ is quantifier-free, $h \in I \subseteq \{1,\dots,n\}$.

PROOF of 2.1. By induction on the number of F-quantifiers in φ , it suffices to treat the case $\exists x(\psi)$, where ψ is F-quantifier-free, and say x, $\underline{z}$ are linear in ψ. $\exists x(\psi)$ is equivalent in VF(L) to a formula

$$\exists \xi_1 \dots \exists \xi_n (\exists x(\bigwedge_{1\leq i\leq n} v(a_i x - b_i) = \xi_i < \infty \ \wedge \ a_i \neq 0) \wedge \psi') \vee \psi'',$$

where $\underline{z}$ is not in a_i, x is not in a_i, b_i, ψ', ψ'', and $\underline{z}$ is linear in b_i, ψ', ψ''. The quantifier $\exists x$ can now be eliminated by lemma 2.3.

PROOF of 2.3. With $a = \prod y_i$, $a_i = \prod_{j\neq i} y_j$, $b_i = z_i a_i$, $\alpha_i = \xi_i + v(a_i)$, φ is equivalent to

$$\bigwedge_{1\leq i\leq n} (v(ax - b_i) = \alpha_i < \infty \ \wedge \ a \neq 0).$$

Next, we adjoin to φ a disjunction over the possible orderings of the α_i; this reduces φ to a disjunction of formulas of the form $\varphi_1 \wedge \psi_1$, where

$$\varphi_1 := a \neq 0 \wedge \bigwedge_{i\in I_1} v(ax - b_i) = \alpha_i = \alpha_h , \quad h \in I_1 \subseteq \{1,\dots,n\} ,$$

$$\psi_1 := \bigwedge_{i\notin I_1} v(b_i - b_h) = \alpha_i < \alpha_h .$$

φ_1 reduces to a disjunction of formulas of the form $\varphi_2 \wedge \psi_2$, where

$$\varphi_2 := a \neq 0 \wedge \bigwedge_{i\in I_2} v(ax-b_i) = \alpha_h \wedge \bigwedge_{i,j\in I_2, i\neq j} v(b_i - b_j) = \alpha_i = \alpha_h < \infty ,$$

$$\psi_2 := \bigwedge_{i\in I_1 - I_2} \bigvee_{j\in I_2} v(b_j - b_i) = \alpha_h , \quad I_2 \subseteq I_1 .$$

This gives the construction of φ'. Notice that $\exists x(\varphi_2)$ is equivalent in VF(L(E)) to $E_{q+1} \wedge \bigwedge_{i,j\in I_2, i\neq j} v(b_i - b_j) = \alpha_i = \alpha_h < \infty \wedge a \neq 0$, where $q = \mathrm{card}(I_2)$. This shows how to construct φ''.

Theorem 2.1 and corollary 2.2 have a useful variant for a slightly more restricted class of linear formulas: Assume that L'_Γ extends L_Γ only by additional constant-symbols. Then a linear formula $\varphi \in L'(E)$ is <u>strictly linear</u> if all Γ-terms in φ contain no Γ-operation-symbol "-" and at most one Γ-variable. The well-known quantifier elimination procedures for dense and for discrete orderings can now be adapted to prove the following lemma.

LEMMA 2.4 Let $DeOG_\infty$ be the theory of non-zero, dense, ordered abelian groups with top element ∞ in L_Γ, and let $DtOG_\infty$ be the theory of discretely ordered, abelian groups with top element ∞ and smallest positive element 1' in $L'_\Gamma = L_\Gamma \cup \{1'\}$. Then $DeOG_\infty$ and $DtOG_\infty$ admit primitive recursive quantifier elimination for strictly linear formulas in L_Γ and L'_Γ, respectively.

By way of contrast, the weakest extension theories of $DeOG_\infty$ and $DtOG_\infty$ in L_Γ and L'_Γ that admit quantifier elimination are the theories DOG_∞ and ZG_∞ of non-zero, divisible, ordered abelian groups and of Z-groups with top element ∞, respectively.

The proof of 2.1 shows that this theorem holds for strictly linear formulas as well. Together with lemma 2.4, this yields the following generalization of theorem 4 in [v.d.Dries 81].

COROLLARY 2.5 (i) $VF(L(E)) \cup DeOG_\infty$ and $VF(L'(E)) \cup DtOG_\infty$ admit primitive recursive quantifier elimination for strictly linear formulas.

(ii) Let Φ be one of the sets of sentences in 2.2. Then the theories $VF(L'(E)) \cup DeOG_\infty \cup \Phi$, $VF(L'(E)) \cup DtOG_\infty \cup \Phi$ decide in a primitive recursive way all strictly linear sentences in L'(E).

In the rest of this section, we extend theorem 2.1 and corollary 2.2 to fields with several, pairwise independent valuations.

Let L'_h be L' with v, Γ replaced by v_h, Γ_h for $h < \omega$, and assume that the languages L'_{Γ_h} are pairwise disjoint. Then $L'(h^*) = \bigcup \{L'_h : h < h^*\}$ for $h^* \leq \omega$ is a suitable language for fields with finitely or countably many valuations $\{v_h\}_{h<h^*}$. Let the predicates E^h_q be defined as E_q with v replaced by v_h, let $L'(h^*,E)$ be $L'(h^*) \cup \{E^h_q\}_{q<\omega, h<h^*}$, and let $VF(L'(h^*,E))$ be the class of all fields with pairwise independent valuations $\{v_h\}_{h<h^*}$ in $L'(h^*,E)$. Then $VF(L'(h^*,E))$ is an elementary class. The definitions concerning linearity of tuples and formulas carry over in an obvious way to the language $L'(h^*,E)$. The envisaged extension of theorem 2.1 can now be stated as follows.

THEOREM 2.6 There is a primitive recursive procedure assigning to any linear formula φ in $L'(h^*,E)$ an F-quantifier-free linear formula φ' in $L'(h^*,E)$ such that φ and φ' are equivalent in $VF(L'(h^*,E))$.

PROOF. As in the proof of 2.1, it suffices to eliminate the quantifier in formulas φ of the form

$$\exists x\Big(\bigwedge_{h<h'}\ \bigwedge_{1\le i\le n_h} v_h(a_{ih}x - b_{ih}) = \xi_{ih} < \infty \ \wedge\ a_{ih} \neq 0\ \Big),$$

where x is not in a_{ih}, b_{ih}, and $\underline{z}$ is linear in b_{ih} and does not occur in a_{ih}, $h' \le h^*$, $h' < \omega$. By the approximation theorem we get

$$\exists x\Big(\bigwedge_{h<h'} \exists \xi_h (v_h(x - x_h) = \xi_h < \infty \ \wedge \bigwedge_{1\le i\le n_h} \xi_{ih} < \xi_h\ )\Big).$$

So φ is equivalent in $VF(L'(h^*,E))$ to

$$\bigwedge_{h<h'} \exists x_h\Big(\bigwedge_{1\le i\le n_h} v_h(a_{ih}x_h - b_{ih}) = \xi_{ih} < \infty \ \wedge\ a_{ih} \neq 0\ \Big).$$

The remaining quantifiers can now be eliminated as in the proof of 2.1

To formulate a counterpart of 2.2 for fields with infinitely many valuations, we need the following concepts: Let T'_h be theories extending OAG_∞ in L'_h . Then we say $\{T'_{\Gamma_h}\}_{h<h^*}$ are uniformly primitive recursively decidable, if the set $\{(h,\varphi) : h<h^*,\ \varphi\in L'_{\Gamma_h},\ T'_{\Gamma_h}\models\varphi\}$ is primitive recursive. We say $\{T'_{\Gamma_h}\}_{h<h^*}$ admit uniform, primitive recursive quantifier elimination, if there is a primitive recursive procedure assigning to any pair (h,φ) with $h<h^*$, $\varphi\in L'_{\Gamma_h}$ a quantifier-free formula φ'_h in L'_{Γ_h} such that $T'_{\Gamma_h} \models \varphi \leftrightarrow \varphi'_h$.

COROLLARY 2.7 Let T'_{Γ_h} be extensions of OAG_∞ in L'_{Γ_h} for $h<h^*$.

(1) If $\{T'_{\Gamma_h}\}_{h<h^*}$ admit (uniform, primitive recursive) quantifier elimination , then $VF(L'(h^*,E)) \cup \bigcup_{h<h^*} T'_{\Gamma_h}$ admits primitive recursive quantifier elimination for linear formulas in $L'(h^*,E)$.

(2) Suppose T'_{Γ_h} are complete and L'_{Γ_h} contains a constant 1_h for $h<h^*$ For any $h<h^*$, let Φ_h be one of the sets of sentences in 2.2 reformulated for v_h in place of v. Then $VF(L'(h^*,E)) \cup \bigcup_{h<h^*} T'_{\Gamma_h} \cup \bigcup_{h<h^*} \Phi_h$ decides (i.e. proves or disproves) all linear sentences in $L'(h^*,E)$. If in addition the theories $\{T'_{\Gamma_h}\}_{h<h^*}$ are uniformly primitive recursively decidable, then all these theories have a primitive recursive decision procedure for linear sentences.

It should now be clear, how to state and prove variants of 2.6 and 2.7 for strictly linear formulas. We leave the details to the reader.

3. ALGEBRAICALLY CLOSED VALUED FIELDS.

Let the languages L, $L'_\Gamma \cap L = L_\Gamma$, $L' = L \cup L'_\Gamma$ be as in section 2. We let ACVF(L') denote the theory of algebraically closed, nontrivially valued fields in L'. Recall that for any $(F,\Gamma,v) \models$ ACVF(L), Γ is nontrivial and divisible and F_v is an algebraically closed (and hence infinite) field. So all the predicates E_q of section 2 are true in ACVF(L), and hence can be dispensed with. Our main result shows that in ACVF(L') the quantifier elimination for linear formulas of theorem 2.1 can be extended to arbitrary formulas.

THEOREM 3.1 There is a primitive recursive procedure assigning to any L'-formula φ an F-quantifier-free L'-formula φ' such that φ and φ' are equivalent in ACVF(L').

The well-known primitive recursive quantifier elimination procedures for nontrivial, divisible, ordered abelian groups extends easily to the theory DOG_∞ of such groups with top element ∞. Together with 3.1 we may thus conclude:

THEOREM 3.2 ACVF(L) admits primitive recursive quantifier elimination.

COROLLARY 3.3 (i) ACVF(L) is primitive recursively decidable.
(ii) For any prime p, the theories ACVF(L) $\cup$ $char'_p$, ACVF(L) $\cup$ $char_p$, ACVF(L) $\cup$ $char_0$ are complete and primitive recursively decidable.

These results may also be rephrased in terms of valuation rings: Let $L_D = \{0,1,+,-,\cdot,D\}$ be the language of rings with a binary predicate D. Let VD be the theory of valuation domains in L_D with $D(x,y) \longleftrightarrow \exists z(xz = y)$ as defining axiom for D. Call $R \models$ VD monically closed if every monic polynomial in R[X] has a root in R. Let MCVR be the theory of monically closed valuation domains that are not fields, in L_D. Then any L_D-formula φ has a coding φ^* in L, and any quantifier-free L-formula ψ without free Γ-variables has a quantifier-free coding $\hat{\psi}$ in L_D in the following sense: For $(F,\Gamma,v) \models$ VF(L) with valuation ring A, and $\underline{a}$ in A, $A \models \varphi(\underline{a})$ iff $(F,\Gamma,v) \models \varphi^*(\underline{a})$ and $(F,\Gamma,v) \models \psi(\underline{a})$ iff $A \models \hat{\psi}(\underline{a})$.
Using this coding, we may conclude:

<u>COROLLARY</u> 3.4 (i) MCVR admits primitive recursive quantifier elimination and is primitive rcursively decidable.
(ii) For all primes p, the theories MCVR $\cup$ $\{p1=0\}$, MCVR $\cup$ $\{p1 \neq 0, \neg D(p,1)\}$, MCVR $\cup$ $\{D(n,1) : 0<n<\omega\}$ are complete and primitive recursively decidable.

(The model-completeness of MCVR is observed also in [Boffa].)

Another application of 3.2 concerns <u>definable relations</u> in algebraically closed valued fields: If $(F,\Gamma,v) \models ACVF(L)$, $\varphi(x_1 \ldots x_n, y_1 \ldots y_m)$ is an L-formula, $b_1,\ldots,b_m \in F$, then $\varphi^F(\underline{b}) = \{\underline{a} \in F^n : (F,\Gamma,v) \models \varphi(\underline{a},\underline{b})\}$. Sets of this form are called <u>definable</u> (<u>with parameters</u>) in (F,Γ,v). The following is a simpler counterpart to Macintyre's result for p-adically closed fields ([Macintyre 76]).

<u>COROLLARY</u> 3.5 Let $(F,\Gamma,v) \models ACVF(L)$.
(i) Any definable set $M \subseteq F^n$ is a finite union of sets of the form $Z \cap \mathcal{O}$, where Z is the zero-set of a polynomial ideal in $F[X_1 \ldots X_n]$ and $\mathcal{O}$ is open in F^n.
(ii) Any infinite definable set $M \subseteq F$ has a nonempty interior.

<u>PROOF</u>. (ii) follows from (i), since for n=1 any zero-set Z is finite or equals F. To prove (i), assume $M = \varphi^F(\underline{b})$ for some L-formula $\varphi(\underline{x},\underline{y})$. By 3.2, we may assume that φ is quantifier-free and in disjunctive normal form. So M is a finite union of sets $\psi^F(\underline{b})$, where $\psi(\underline{x},\underline{y})$ is of the form

$$\bigwedge_{1\leq i\leq l} f_i(\underline{x},\underline{y})=0 \wedge g(\underline{x},\underline{y})\neq 0 \wedge \bigwedge_{1\leq i\leq l'} vh_i(\underline{x},\underline{y}) \; r \; vk_i(\underline{x},\underline{y}) \quad ,$$

where f_i, g, h_i, k_i are polynomials with integer coefficients and r is one of the relations $=$, $<$. The claim is now obvious from the fact that $(g(\underline{x},\underline{b}) \neq 0)^F$ is open and $(vh_i(\underline{x},\underline{b}) \; r \; vk_i(\underline{x},\underline{b}))^F$ is clopen in F^n.

Our final application concerns prime model extensions.

<u>COROLLARY</u> 3.6 (i)([Robinson 56]) Any $(F,\Gamma,v) \models VF(L)$ has a prime model extension $(\bar{F},\bar{\Gamma},\bar{v}) \models ACVF(L)$ (i.e. $(\bar{F},\bar{\Gamma},\bar{v})$ embeds over (F,Γ,v) into any $(F',\Gamma',v') \models ACVF(L)$ extending (F,Γ,v)).
(ii) $(\bar{F},\bar{\Gamma},\bar{v})$ is unique up to an isomorphism over (F,Γ,v).
(iii) If v is nontrivial on F, then $\bar{F}$ is the algebraic closure of F.

<u>PROOF</u>. If v is nontrivial on F, $\bar{F}$ is the algebraic closure of F and

$\overline{v}$ is any extension of v to $\overline{F}$ with value group $\overline{\Gamma}$, then (i) and (ii) are a consequence of the following well-known model theoretic fact:

FACT 3.7 Let T ba a theory that admits quantifier elimination in a language L. Let $M \subseteq M' \models T$ such that any $b \in M'$ is algebraic over M (i.e. $M' \models \varphi(b,\underline{a})$ for some tuple $\underline{a}$ of elements of M, where $\varphi(y,\underline{x})$ is an L-formula such that $\varphi^{M'}(\underline{a})$ is finite). Then M' is a minimal prime model extension of M and any other prime model extension M'' of M is isomorphic to M' over M.

If v is trivial on F, we extend $(F,\{0\}, v_{triv})$ to $(F(t), \mathbb{Z}, v_0)$, where $v_0(t) = 1$, and continue as in the first case. The uniqueness follows from the fact that for any $(F',\Gamma',v') \supseteq (F,\{0\},v_{triv})$, any $b \in F'$ with $v(b) > 0$ is trancendental over F.

The rest of this section is devoted to a proof of theorem 3.1.

By induction on the number of F-quantifiers in an L-formula, it suffices to eliminate the F-quantifier $\exists x$ in formulas $\exists x(\varphi)$, with F-quantifier-free φ . φ is equivalent to a formula $\varphi'(va_1,\dots,va_n)$ with an L'_Γ-formula $\varphi'(\eta_1,\dots,\eta_n)$ (notice that $a=0$ iff $va=\infty$). Each a_i can be rewritten - modulo a case distinction by a disjunction of conjunctions of formulas $b=0$ and $c\neq 0$,where the operation $^{-1}$ does not occur in b and c - as quotients of formal polynomials in x. (We use the following conventions concerning polynomials: A (formal) polynomial in x of (formal) degree $n = \deg(f)$ is an F-term f(x) of the form $\sum_{0\le i\le n} f_i x^i$,where f_i are F-terms that do not contain the variable x ; f is monic if $f_n = 1$. We denote polynomials by f(x), g(x), h(x),...; the corresponding tuple of coeficients is then $\underline{f} = (f_0,\dots,f_n)$, $\underline{g}$, $\underline{h}$,... . f'(x) is the (formal) derivative of f(x), i.e. $f'(x) = \sum_{1\le i\le n} i f_i x^{i-1}$. $T(\underline{f},a)$ is the tuple of coefficients of the Taylor expansion of f(x) at a; i.e. $T(\underline{f},a) = (f_0^*,\dots,f_n^*)$ with $f_k^* = \sum_{k\le i\le n} \binom{i}{k} f_i a^{i-k}$, and the equation $f(x) = f^*(x-a)$ is true in all fields. If f(x), g(x) are polynomials in x of degree n , m , then $Q_{f,g}$, $R_{f,g}$ are the tuples of coefficients of the quotient and the remainder of f(x) , when (formally) divided by g(x) ; i.e. for $Q_{f,g} = \underline{q}$, $R_{f,g} = \underline{r}$, q(x) and r(x) are polynomials of degree max(0, n-m) and max(0, m-1) , and the formula $g_n \neq 0 \rightarrow f(x) = q(x)g(x) + r(x)$ is true in all fields.)

In this way, we arrive at an equivalent formula

$$\exists\xi_1\dots\exists\xi_n(\exists x(\bigwedge_{1\le i\le n} vf_i(x) = \xi_i) \wedge \varphi''(\xi_1,\dots,\xi_n)) ,$$

where $f_i(x)$ are polynomials in x and φ'' is an L'_Γ-formula. So it suffices to treat formulas of the form $\exists x(\bigwedge_{1\le i\le n} vf_i(x)=\xi_i)$. Moreover, we may assume that the f_i are monic.

In the following, it will be convenient to enlarge L' by n-ary Γ-operation-symbols $\min_n$ with defining axioms
$\min_n(\xi_1,\dots,\xi_n) = \xi \leftrightarrow \bigvee_{1\le i\le n} (\xi = \xi_i \wedge \bigwedge_{1\le j\le n} \xi \le \xi_j)$.
Since these operations can be eliminated in formulas without introducing new quantifiers, they do not affect the quantifier elimination procedure. If $(\alpha_1,\dots,\alpha_n) = \underline{\alpha}$, we write $\min \underline{\alpha}$ for $\min_n(\alpha_1,\dots,\alpha_n)$; if $f(x)$ is a polynomial, $\min \underline{vf}$ stands for $\min_{n+1}(vf_0,\dots,vf_n)$, and $\min \underline{vfa}$ for $\min_{n+1}(vf_0,v(f_1a),\dots,v(f_na^n))$.

<u>LEMMA</u> 3.8 Let f(x) be monic of degree n. Then the following equivalences hold in ACVF(L).

(i) $va = \alpha < \infty \rightarrow (\exists x(v(x-a)>\alpha \wedge f(x)=0) \leftrightarrow vf(a) > \min \underline{vfa})$.

(ii) $\alpha<\infty \rightarrow (\exists x(vx=\alpha \wedge f(x)=0) \leftrightarrow \bigvee_{0\le i<j\le n} vf_i+i\alpha = vf_j+j\alpha = \min (vf_h+h\alpha : 0\le h\le n))$.

<u>PROOF</u>.(i) Let $g(x) = \sum (f_ia^i)x^i = \sum g_ix^i$. Then it suffices to show that $va = \alpha < \infty$ implies :
$\exists x(v(x-1) > 0 \wedge g(x) = 0) \leftrightarrow vg(1) > \min \underline{vg}$.
$\Rightarrow$: The right hand side implies $vg(1) = v(g(x) - g(1) \ge \min \{v(g_i(x^i-1)): 1\le i\le n\} > \min \underline{vg}$, since $v(x^i-1) = v(x-1) + v(1+\dots+x^{i-1}) > 0$.
$\Leftarrow$: We argue in some fixed $(F,\Gamma,v) \in$ ACVF for $g(X) \in F[X]$ by contraposition. Since F_v is infinite, there is $b\in F$ with $vb=0 \wedge vg(b)= \min \underline{vg}$. Let $g^*(X) = g(X-b) = g_n\cdot(X-c_1)\cdot\dots\cdot(X-c_n)$ with $c_i \in F$. Then $\neg\exists x(v(x-1) > 0 \wedge g(x) = 0)$ means that $v(c_i-b-1) \le 0$ for $1\le i\le n$. Hence $v(c_i-b-1) \le vc_i$, and so $vg(1) = vg^*(1+b) = vg_n + \sum v(1+b-c_i) \le vg_n + \sum vc_i = vg^*_0 = vg(b) = \min \underline{vg}$.

(ii) We argue again in some fixed $(F,\Gamma,v) \in$ ACVF for $\alpha\in\Gamma$, $f(X)\in F[X]$. $\Rightarrow$: $\infty = vf(x) > \min \underline{vfx}$ implies that for some $0\le i<j\le n$, $vf_ix^i = vf_jx^j$.
$\Leftarrow$: Choose $a\in F$ with $va=\alpha$ and put $g(X) = \sum(f_ia^i)X^i$. If for $0\le i<j\le n$, $vf_i+i\alpha = vf_j+j\alpha = \min (vf_h+h\alpha : 0\le h\le n)$, then $vg_i = vg_j = \min \underline{vg} < \infty$. Choose $b\in F$ with $vb = \min \underline{vg}$, and put $h(X) = b^{-1}g(X)$. Then res h(X) has a nontrivial zero in F_v, and so there exists $c\in F$ with $vc = 0 \wedge vg(c) > \min \underline{vg}$. By (i), there exists $d\in F$ with $g(d) = 0 \wedge v(d-c) > 0$. So $f(ad) = 0$ and $v(ad) = va = \alpha$.

<u>LEMMA</u> 3.9 ACVF(L') admits primitive recursive quantifier elimination for formulas of the form

(i) $\exists x(\bigwedge_{1\le i\le n} v(x - a_i) = \alpha_i)$

(ii) $\exists x(\ f(x) = 0 \wedge \bigwedge_{1\le i\le n} v(x - a_i) = \alpha_i)$,

where x is not in a_i, α_i , and f(x) is monic of degree m.

<u>PROOF</u>. (i) is an immediate consequence of 2.3. The quantifier elimination for formulas of type (ii) reduces by 2.3 to formulas of the form

(iii) $\exists x(\ f(x) = 0 \wedge \bigwedge_{1\le i\le n'} v(x - a_i') = \alpha < \infty \wedge \bigwedge_{1\le i<j\le n'} v(a_i' - a_j') = \alpha)$,

where x is not in a_i', α . In case n'= 1, let $\underline{f}^* = T(\underline{f}, a_1')$; then (iii) is equivalent to $\exists x'(\ f^*(x') = 0 \wedge vx' = \alpha < \infty\ )$. For this formula, 3.8(ii) provides a quantifier elimination. Next assume n'> 1, and put $b_i = a_i' - a_1'$. Then (iii) is equivalent to

$\exists x'(\ f^*(x') = 0 \wedge vx' = \alpha < \infty \wedge \bigwedge_{2\le i\le n'} v(x' - b_i) = \alpha) \wedge \bigwedge_{2\le i<j\le n'} vb_i = v(b_i - b_j) = \alpha$.

By lemma 3.8(i) , this is equivalent to (iv) :

$\exists y(\ vf^*(y) > \min v\underline{f^*y} \wedge vy = \alpha < \infty \wedge \bigwedge_{2\le i\le n'} v(y - b_i) = \alpha\ ) \wedge$
$\bigwedge_{2\le i<j\le n'} vb_i = v(b_i - b_j) = \alpha$.

Let $g(x) = \sum (f_i^* b_2^i) x^i = \sum g_i x^i$, $c_i = b_i b_2^{-1}$, h(x) the formal polynomial corresponding to $(x - c_2)\cdot\ldots\cdot(x - c_{n'})$. Then (iv) is equivalent to

$\exists z(\ vg_m < \infty \wedge vg(z) > \min v\underline{g} \wedge vz = 0 \wedge vh(z) = 0 = \min v\underline{h}\ ) \wedge$
$\bigwedge_{2\le i<j\le n'} vb_i = v(b_i - b_j) = \alpha$.

This formula reduces to a disjunction of formulas of the form (v) :

$\exists z(\ vg^*(z) > 0 = \min v\underline{g}^* = vg^*_{m^*} \wedge vz = 0 \wedge vh(z) = 0 = \min v\underline{h}\ ) \wedge \Psi$,

where Ψ is quantifier-free and does not contain z , deg $g^*(z) = m^*$. Let k(z) be the formal polynomial corresponding to $g^*(z)^{n'-1}$, and put $\underline{r} = R(k,h)$. Then a simple instance of Hilbert's Nullstellensatz applied to the residue class fields shows that (v) is equivalent in ACVF(L') to $\min v\underline{r} = 0 = \min v\underline{g}^* = vg^*_{m^*} = \min vh\ ) \wedge \Psi$.

For the next lemma, we introduce the following notation: Ft($\underline{w}$) is the set of F-terms containing F-variables from the tuple $\underline{w}$ only. $Z(\underline{z},\underline{w})$ denotes any formula of the form $\bigwedge_{1\le i\le n} g_i(z_i) = 0$, where $g_i(z_i)$ are monic polynomials in z_i with coefficients $g_{ij} \in Ft(\underline{w}, z_1, \ldots, z_{i-1})$ – (" <u>z is a tuple of zeros of polynomials with parameters w</u> "). We put deg $Z(\underline{z},\underline{w})$ = max (deg $g_i(z_i)$: $1\le i\le n$) .

<u>LEMMA</u> 3.10 There is a primitive recursive procedure assigning to any formula $\varphi := \bigwedge_{1\le i\le n} vf_i(x) = \alpha_i$, where α_i are L'_Γ-terms , $f_i(x)$ are monic with coefficients in $Ft(\underline{w})$, a formula $\varphi' :=$
$\exists \underline{\xi}(\exists \underline{z}(Z(\underline{z},\underline{w}) \wedge \bigwedge_{1\le i\le n'} v(x - a_i) = \alpha'_i) \wedge \Psi)$, where $a_i \in Ft(\underline{z})$, α'_i are L'_Γ-terms, Ψ is a quantifier-free L'_Γ-formula, and deg $Z(\underline{z},\underline{w})$ $\le$ max (deg $f_i(x)$: $1\le i\le n$) , such that φ and φ' are equivalent in ACVF(L') .

<u>PROOF</u> by induction on $k = \Sigma(\deg f_i : \deg f_i > 1 , 1\le i\le n)$.
The case $k = 0$ is trivial. If $k > 0$, choose $f_j(x)$ with $\deg f_j > 1$, and put $f^* = T(\underline{f_j},z')$, $f^*(x) = f_0^* + x\cdot g(x)$, $h(x) = g(x-z')$.
Then φ is equivalent to

$$\exists\eta\exists\zeta\exists z'(f_j(z') = 0 \wedge v(x-z') = \eta \wedge vh(x) = \zeta \wedge \bigwedge_{1\le i\le n,\ i\ne j} vf_i(x) = \alpha_i \wedge \eta + \zeta = \alpha_j) .$$

The induction assumption can now be applied to finish the proof.

<u>COROLLARY</u> 3.11 Let φ be as in 3.10. Then $\exists x(\varphi)$ can be reduced equivalently in ACVF(L') to a formula of the form
$\exists \underline{\xi}\ \exists \underline{z}(Z(\underline{z},\underline{w}) \wedge \Psi)$ with $Z(\underline{z},\underline{w})$ as in 3.10 , Ψ quantifier-free

<u>PROOF.</u> Combine 3.10 with 3.9(i) .

The elimination of the quantifier $\exists x$ for formulas $\exists x(\varphi)$ as in 3.11 looks like a bad deal : To eliminate $\exists x$, a tuple of new F--quantifiers $\exists \underline{z}$ had to be introduced. Nevertheless, the additional condition on $\underline{z}$ expressed be the formula $Z(\underline{z},\underline{w})$ will make sure that the elimination of the quantifiers $\exists \underline{z}$ will come to an end after finitely many steps.

<u>LEMMA</u> 3.12 ACVF(L') admits primitive recursive F-quantifier elimination for formulas φ of the form $\exists \underline{z}(Z(\underline{z},\underline{w}) \wedge \Psi(\underline{z},\underline{w}))$, where Ψ is F-quantifier-free.

<u>PROOF.</u> Let $n = \deg Z(\underline{z},\underline{w})$ and let $\underline{z} = (z_1,\dots,z_m)$. We proceed by induction on $\omega\cdot n + m$: The case $n = m = 0$ is trivial.
$n > 0$, $m=1$: Then φ is of the form $\exists z(g(z) = 0 \wedge \Psi(\underline{z},\underline{w}))$,where the coefficients of $g(z)$ are in $Ft(\underline{w})$. By the reduction at the beginning of the proof of theorem 3.1, we may assume that $\Psi(\underline{z},\underline{w})$ is of the form $\bigwedge_{1\le i\le n} vf_i(z) = \alpha_i$, where α_i are L'_Γ-terms

and $f_i(z)$ are monic. Moreover, by the division algorithm for formal polynomials, we may assume that $\deg f_i(z) < \deg g(z) = n$. Then by 3.10, φ reduces to a formula

$$\exists \underline{\xi}(\exists \underline{z}'(Z(\underline{z}',\underline{w}) \wedge \exists z(g(z)=0 \wedge \bigwedge_{1\le i\le n'} v(z-a_i) = \alpha_i')) \wedge \Psi'),$$

where $\deg Z(\underline{z}',\underline{w}) < n$, $a_i \in Ft(\underline{z}')$, α_i' are L_Γ'-terms and Ψ' is an L_Γ'-formula. Next, the quantifier $\exists z$ can be eliminated by 3.9(ii), and the induction assumtion completes the F-quantifier elimination for the resulting formula.
The case $n > 0$, $m > 1$ follows immediately from the case $n > 0$, $m=1$ and the induction assumption.

The proof of theorem 3.1 is now complete by 3.11 and 3.12. It should be clear that the F-quantifier elimination procedure outlined above is indeed primitive recursive. Moreover, it avoids a source of high complexity occuring frequently in quantifier elimination procedures - the formation of disjunctive normal forms as preparation for the elimination of an existential quantifier.

4. HENSEL FIELDS OF CHARACTERISTIC ZERO.

In the previous section we gave a quantifier elimination procedure for the theory ACVF of algebraically closed nontrivially valued fields (of arbitrary characteristic) in the language L of valued fields. Let now HF(L') be the theory of Hensel fields (F,Γ,v) such that F has characteristic zero in a language $L' \supseteq L$. We want to find a "small" extension L' of L in which HF(L') admits elimination of F-quantifiers.

Let us remark to begin with that HF(L) does not admit quantifier elimination. For, any quantifier elimination for HF(L) yields a quantifier elimination for the theory of all value groups Γ of models (F,Γ,v) of HF(L) in the language of ordered groups; but this is the theory OAG of all ordered abelian groups, a contradiction.

Next, we show that HF(L) does not even admit F-quantifier elimination: The element 2 satisfies the same F-quantifier-free L-formulas in $(\mathbb{Q}((t)), \mathbb{Z}, v_0)$ and in $(\mathbb{R}((t)), \mathbb{Z}, v_0)$, where $v_0 \restriction \mathbb{R}$ is trivial and $v_0(t) = 1$; on the other hand 2 has a root in $\mathbb{R}((t))$ and no root in $\mathbb{Q}((t))$. In fact, this shows that even the theory $HF_0(L) = HF(L) \cup$

$char_0$ does not admit F-quantifier elimination. To remove this problem for $HF_0(L)$, we make quantifiers ranging over the residue class fields explicit in our language so that F-quantifier elimination refers only to "true" F-quantifiers and not to quantifiers over the residue class field "in disguise".

Accordingly, we extend L by a new sort, the R_0-sort, consisting of the language $L_R = \{0,1,+,-,\cdot\}$ of rings, and a new operation-symbol res_0 from the F-sort to the R_0-sort. Call the resulting language $L(R_0)$. We let $HF_0(L(R_0))$ be the theory of all $(F,\Gamma,v) \models HF_0(L)$, where the R_0-sort is interpreted as ranging over the residue class field F_v and res_0 as the canonical homomorphism $res_0 : A \to F_v$ extended to F by the convention that $res_0(a) = 0$ for $va < 0$. We denote R_0-variables by $r,s,t,r',\ldots$, R_0-terms by $d,e,d',\ldots$. Then the problem considered above disappears, since $\exists r(\ r^2 = res_0(x)\ )$ is an F-quantifier-free $L(R_0)$-formula.

But there are more obstacles to an F-quantifier elimination in $L(R_0)$: In $(\mathbb{Q}((t)), \mathbb{Z}, v_0)$ the elements t^2 and $2t^2$ satisfy the same F-quantifier-free $L(R_0)$-formulas, but $2t^2$ has no square root in $\mathbb{Q}((t))$. One way to overcome this difficulty is to introduce an operation-symbol π (from the Γ-sort to the F-sort) for a cross-section $\pi : \Gamma \cup \{\infty\} \to F$ into the language (cf. [Weispfenning 76]). This has the disadvantage that quantifier-free formulas can be quite complicated, inhibiting e.g. a perspicuous description of definable sets (cf. [Macintyre 76]). In our present treatment we rather try to avoid cross-sections. Instead, one may want to add predicates $\{W'_n\}_{1<n<\omega}$ with defining axioms $W'_n(x) \leftrightarrow \exists z(\ z^n = x)$ to $L(R_0)$. But even in this extended language $L(R_0,W')$, $HF_0(L(R_0,W')$ does not admit F-quantifier elimination: Consider $(K((t)), \mathbb{Z}, v_0)$ with $K = \mathbb{C}((u))$, and let $a = ut^3$, $b = u^2t^3$. Then a and b satisfy the same F-quantifier-free formulas in $L(R_0,W')$, but $(K((t), \mathbb{Z}, v_0) \models \Psi_3(a,u) \wedge \neg\Psi_3(b,u)$, where $\Psi_n(x,r)$ is the formula

$$\exists z(\ res_0(\ xz^n) = r \neq 0\ ) \quad \text{for} \quad 1 \leq n < \omega \quad .$$

This suggests a slightly different extension of $L(R_0)$: Let $L(R_0,W)$ be obtained from $L(R_0)$ by adding new binary predicates $\{W_n\}_{1\leq n<\omega}$, with defining axioms $W_n(x,r) \leftrightarrow \Psi_n(x,r)$ for $1 \leq n < \omega$. Then a theorem of F. Delon([Delon], thm. 2.1) says that $HF_0(L(R_0,W))$ does indeed admit elimination of F-quantifiers. We shall derive an explicit, primitive recursive version of this result under a mild additional restriction on the value groups involved (see 4.2 below).

Next, we consider the general case of Hensel fields of characteristic zero with residue class fields of arbitrary characteristic. Here another obstacle for F-quantifier elimination comes up: Let $(\mathbb{Q}_p, \mathbb{Z}, v_p)$ be the valued field of p-adic numbers, let t be transcendental over $\mathbb{Q}_p$ and extend v_p to $\mathbb{Q}_p(t)$ by $v_p f(t) = \min(vf_i : 0 \leq i \leq n)$ for a polynomial $f(t) \in \mathbb{Q}_p[t]$ of degree n. Let $(F, \mathbb{Z}, v)$ be the completion of $(\mathbb{Q}_p(t), \mathbb{Z}, v_p)$. Then $a = t^p$ and $b = t^p + p$ satisfy the same F-quantifier-free formulas in $L(R_0)$, but a has a p-th root and b has no p-th root in F. So the theory of $(F, \mathbb{Z}, v)$ does not admit F-quantifier elimination in $L(R_0)$. By varying this example, it becomes plausible that there is no chance for an F-quantifier elimination for HF even in $L(R_0, W)$.

The solution to this problem is to extend $L(R_0)$ in such a way that quantifiers over certain residue rings $R_\alpha = A/I_\alpha$, $\alpha > 0$, are admitted in F-quantifier-free formulas. For this purpose, we introduce a new Γ-constant 1_Γ (or 1 for short) and specify a function $w: \omega \to \omega$ with $w(n) \leq n$ for $0 \leq n < \omega$. Then HF_w is the class of all $(F, \Gamma, v) \models HF$ with $0 \leq 1_\Gamma < \infty$ and $v(i) \leq w(n) \cdot 1_\Gamma$ for $1 \leq i \leq n$. Roughly speaking, the language L(R) is now obtained from L by adding to L infinitely many new sorts, the R_k-sorts, $k < \omega$, together with suitable operation-symbols to describe explicitly properties of the residue class rings $R_k = R_\alpha$ for $\alpha = k \cdot 1_\Gamma$. L(R,W) is then an extension of L(R) by appropriate generalizations of the predicates W_n considered above, and $HF_w(L(R,W)) = HF_w(R,W)$ is the class of all $(F, \Gamma, v) \models HF_w(L)$, regarded as L(R,W)-structures.

We are going to prove an F-quantifier elimination theorem of moderate generality in this language. Consider the following condition on the value group Γ.

4.1 $\Gamma / \hat{\mathbb{Z}}$ is divisible, where $\hat{\mathbb{Z}}$ is the convex hull of $\mathbb{Z} = \mathbb{Z} \cdot 1_\Gamma$ in Γ.

This can be rephrased by the following set of sentences in the Γ-sort: $\{ \forall \xi \exists \eta (\xi \equiv_n \eta \wedge 0 \leq \eta < n 1_\Gamma) : 1 < n < \omega \}$, where $\alpha \equiv_n \beta$ stands for $\exists \zeta (\alpha + n\zeta = \beta)$. 4.1 is certainly satisfied in case $\mathbb{Z}$ is cofinal in Γ. More generally, it holds if for some convex subgroup Δ of Γ with $1_\Gamma \notin \Delta$, Γ/Δ is regular. (An ordered, abelian group Γ is regular, if it is elementarily equivalent to an archimedean ordered group; equivalently, if Γ is a Z-group or Γ is dense regular. Γ is a Z-group, if Γ has a smallest positive element 1' and $\Gamma/\mathbb{Z} \cdot 1'$ is divisible; Γ is dense regular, if for all $0 < n < \omega$, $n\Gamma$ is dense in Γ, i.e. if $\Gamma \models \forall \xi \forall \eta (0 < \xi \to \exists \zeta (0 < \zeta < \xi \wedge \zeta \equiv_n \eta))$. Both the <u>class</u> ZG <u>of</u> Z-<u>groups</u> and the <u>class</u> DR <u>of dense regular groups</u> are elementary classes.) A specific example of this kind is $\Gamma = \mathbb{Q} \oplus \Gamma' \oplus \mathbb{Z}^n$ (lexicographical sum) with $\Gamma' = \mathbb{Z}\cdot 1 + \mathbb{Z}\sqrt{2}$ and $1_\Gamma = 1$.

We let $HF'_w(R,W)$ be the theory $HF_w(R,W) \cup \{4.1\}$.

<u>THEOREM</u> 4.2 $HF'_w(R,W)$ admits primitive recursive F-quantifier elimination. In other words, there is a primitive recursive procedure assigning to every formula φ in $L(R,W)$ an F-quantifier-free formula φ' such that φ and φ' are equivalent in $HF'_w(R,W)$.

This theorem will be a consequence of our main theorem 4.3 . The latter requires still another extension of language. To make this rather complicated setup more transparent, we shall first give a precise description of the language $L(R,W)$ and the theory $HF_w(R,W)$ and afterwards introduce the full framework for 4.3 .

$L(R)$ is obtained from the two-sorted language L as follows :

(1) The Γ-sort is augmented by the new constant 1_Γ; the F-sort is augmented by a new constant q .

(2) For $k<\omega$, we let L_{R_k} be a copy of the ring language $\{0,1,+,-,\cdot\}$ together with a constant q_k . L_{R_k} will be the R_k-<u>sort</u> of $L(R)$.

(3) For all $k<\omega$, res_k is a unary operation-symbol from the F-sort to the R_k-sort and v_k is a unary operation-symbol from the R_k-sort to the Γ-sort. For $h,k,l<\omega$, $h+k \leq l$, $res^l_{k,h}$ is a unary operation-symbol from the R_l-sort to the R_k-sort.

$L(R)$ is then the language $L \cup \bigcup_{k<\omega} L_{R_k} \cup \{1_\Gamma, q, \{res_k\}_{k<\omega}, \{v_k\}_{k<\omega},$ $\{res^l_{k,h}\}_{h,k,l<\omega,\ h+k\leq l}\}$. So $L(R)$ has infinitely many sorts and extends $L(R_0)$. As in $L(R_0)$, we denote R_k-variables by $r,s,t,r',\dots$, R_k-terms by $d,e,d',\dots$. If there is no danger of confusion, we drop the subscripts of the symbols 1_Γ, v_k , q_k .

Next let $w:\omega \to \omega$ be fixed. Then $L(R,W) = L_w(R,W)$ is obtained from $L(R)$ by adding binary relation-symbols $W_{n,k,h}$ with first argument in the F-sort and second argument in the R_k-sort ,for $n,k,h<\omega$, $n \geq 1$, $k \geq h+2w(n)$.

$HF_w(R,W)$ is the theory of all $(F,\Gamma,v) \models HF_w(L)$, regarded as $L(R,W)$--structures with the following interpretation of the new symbols:

(1) $0 \leq 1_\Gamma < \infty$, $v(q) = 1_\Gamma$;

(2) The R_k-sort ranges over the residue-class rings $R_k = A/I_k$, where $I_k = \{a \in F: va > k1_\Gamma\}$, res_k is the canonical homomorphism from the valuation ring A onto R_k , extended to F by the

convention that $res_k(a) = 0$ for $va < 0$. $q_k = res_k(q)$.
$v_k : R_k \to \Gamma \cup \{\infty\}$ is defined by $v_k(res_k(a)) = va$, if $0 \leq va \leq k1_\Gamma$, and $v_k(res_k(a)) = \infty$ otherwise .
$res^1_{k,h} : R_1 \to R_k$ is defined by $res^1_{k,h}(res_1(a)) = res_k(q^{-h}a)$, if $h1_\Gamma \leq va$, and $res^1_{k,h}(res_1(a)) = 0$ otherwise .

(3) $W_{n,k,h}(x,r)$ is defined by $\exists z(res_k(xz^n) = r \wedge v_k(r) \leq h1_\Gamma)$. The choice of $k \geq h + 2w(n)$ is made in order to guarantee that in $HF_w(R,W)$, $W_{n,k,h}(x, res_k(y))$ is equivalent to $\exists z(xz^n = y)$ (see 5.3(iii)).

This completes the description of $L_w(R,W)$ and $HF_w(R,W)$.

The following extension of the language $L(R,W)$ is performed in order to relax condition 4.1 .

Let $C = \{c_1, \ldots, c_m\}$ be a new finite set of F-constants, let $C_\Gamma = \{ c_{1\Gamma}, \ldots, c_{m\Gamma} \}$ be a corresponding set of Γ-constants, put $L_w(R,W,C) = L_w(R,W) \cup C \cup C_\Gamma$, and let $HF'_w(R,W,C)$ be the theory of all $(F,\Gamma,v) \models HF_w(R,W)$ with the following interpretation of the new constants:

(1) $0 < c_{i\Gamma} = vc_i < \infty$.

(2) If Δ is the subgroup $\langle C_\Gamma \rangle$ of Γ generated by C_Γ, then $\Delta \cap \mathbb{Z} = \{0\}$ and $\Gamma / \langle \hat{\mathbb{Z}} \cup C_\Gamma \rangle$ is divisible.

Condition (2) can be expressed by the following sentences :

$-n1_\Gamma < \sum m_i c_i < n1_\Gamma \to m_i c_i = 0$ for $0 < n < \omega$, $m_i \in \mathbb{Z}$;

$(\bigvee_{\check{c} \in \check{C}_n} \xi \equiv_n \eta + v(\check{c}) \wedge 0 \leq \eta < n1_\Gamma)$,where $\check{C}_n$ is the set of all F-terms of the form $\prod_i c_i^{j_i}$ with $0 \leq j_i < n$.

As with theorem 3.1, we phrase our main theorem in a fairly general extension language L' of $L(R,W,C) = L_w(R,W,C)$. Let L'_Γ be an extension of L_Γ , let L'_{R_k} be an extension of L_{R_k} for $k < \omega$, and assume that $L' \cap L(R,W,C) = L(R,W,C)$, $L'_{R_k} \cap L(R,W,C) = L(R,W,C)$. Then we put

$$L' = L(R,W,C) \cup L'_\Gamma \cup \bigcup_{k<\omega} L'_{R_k} .$$

MAIN THEOREM 4.3 Let L' be as above. There is a primitive recursive procedure assigning to every L'-formula φ an F-quantifier-free L'-formula φ' such that φ and φ' are equivalent in $HF'_w(R,W,C)$.

Specializing $C = \emptyset$, this yields theorem 4.2 even in an extended language L'. Under a mild assumption on the language L', theorem 4.3 can be sharpened by specifying also the quantifier-structure of the formulas φ' obtained in 4.3 : Define binary R_k-relations $\widetilde{\approx}_{n,k}$ (or $\approx_n$, for short) by $r \widetilde{\approx}_{n,k} s \leftrightarrow (r=s=0) \vee (r\neq 0 \wedge s\neq 0 \wedge \exists t(rt^n= s \vee st^n= r)$. In any model (F,Γ,v) of $HF_w(R)$, these are equivalence relations on the rings R_k .

SUPPLEMENT 4.4 Suppose all the relations $\widetilde{\approx}_{n,k}$ and their complements $\widetilde{\not\approx}_{n,k}$ are existentially definable without F-quantifiers in L'. Then for a prenex L'-formula φ , the corresponding formula φ' in theorem 4.3 can be taken prenex of the same prefix-type as φ.

Next, we state several useful variants of the main theorem. The first one deals with F-quantifier elimination in languages, where the generalized root-predicates $W_{n,k,h}$ are replaced by the ordinary root-predicates W'_n .

Let $L(R,W',C,E)$ be obtained from $L(R,W,C)$ by dropping all the predicates $W_{n,k,h}$ and adding unary F-predicates W'_n and finite sets E_n of F-constants for $1\leq n<\omega$. Let $HF'_w(R,C)$ be obtained from $HF'_w(R,W,C)$ by 'forgetting' the predicates $W_{n,k,h}$; let $HF'_w(R,W',C,E)$ be the theory of all $(F,\Gamma,v) \models HF'_w(R,C)$ regarded as $L(R,W',C,E)$-structures and satisfying the following axioms on W'_n , $\check{e}\in E_n$:

(1) $W'_n(x) \leftrightarrow \exists z(x=z^n)$.

(2) $0\leq v\check{e}\leq 1_\Gamma$ for all $\check{e} \in E_n$, $1\leq n< \omega$.

(3) $\forall x(0\leq vx\leq 1_\Gamma \rightarrow \bigvee_{\check{e}\in E_n} \exists z(xz^n= \check{e})$, $1\leq n<\omega$.

VARIANT 4.5 Theorem 4.3 and supplement 4.4 hold for $L(R,W',C,E)$ and $HF'_w(R,W',C,E)$ in place of $L(R,W,C)$ and $HF'_w(R,W,C)$.

This result is in many respects more general than the quantifier elimination theorem in [Cherlin-Dickmann 83] (theorem 5) for unramified Hensel fields of characteristic zero ; due to our restriction on the value group, it does, however, not quite cover their result.

The next variant avoids the constants in C and C_Γ at the expense of introducing a partial cross section π for the valuation v . Our intention is , to have π defined on a subgroup Δ of Γ such that $\Delta \cap \hat{\mathbb{Z}} = \mathbb{Z}$ and $\Gamma/(\Delta+ \hat{\mathbb{Z}})$ is divisible . Accordingly, we extend $L(R,W)$ by a unary operation-symbol π from the Γ-sort to the F-sort to $L(R,W,\pi)$, and let $HF'_w(R,W,\pi)$ be the theory of all models (F,Γ,v)

of $HF_w(R,W)$, regarded as $L(R,W,\pi)$-structures and satisfying the following axioms:

4.6 (i) $\pi(\infty) = 0$,

(ii) $\pi(\xi) \neq 0 \wedge \pi(\eta) \neq 0 \rightarrow (\pi(\xi+\eta) \neq 0 \wedge \pi(\xi+\eta) = \pi(\xi)\cdot\pi(\eta))$,

(iii) $\pi(0) = 1$, $\pi(1_\Gamma) = q$, $0 < \xi < 1_\Gamma \rightarrow \pi(\xi) = 0$,

(iv) $\pi(\xi) \neq 0 \quad \pi(-\xi) = (\pi(\xi))^{-1}$,

(v) $\forall\xi\exists\eta\exists\zeta(\xi<\infty \rightarrow \pi(\eta) \neq 0 \wedge 0 \leq \zeta \leq 1_\Gamma \wedge \xi \equiv_n \eta + \zeta)$.

<u>VARIANT</u> 4.7 Theorem 4.3 and supplement 4.4 hold for $L(R,W,\pi)$ and $HF_w'(R,W,\pi)$ in place of $L(R,W,C)$ and $HF_w'(R,W,C)$.

Variants 4.5 and 4.7 may be combined in an obvious manner to yield :

<u>VARIANT</u> 4.8 Theorem 4.3 and supplement 4.4 hold for $L(R,W',\pi,E)$ and $HF_w'(R,W',\pi,E)$ in place of $L(R,W,C)$ and $HF_w'(R,W,C)$.

Finally, we may require that the subgroup Δ , on which π is defined (i.e. $\neq 0$) is 'large' in Γ in the sense that every $\gamma \in \Gamma$ has a distance $\leq 1_\Gamma$ to some $\delta \in \Delta$. Call such a partial cross section π for v '<u>almost total</u>' . Then neither the predicates $W_{n,k,h}$ nor the predicates W_n' are required for an F-quantifier elimination : Let $L(R,\pi) = L(R) \cup \{\pi\}$, and let $HF_w''(R,\pi)$ be the theory of all $(F,\Gamma,v) \models HF_w(R)$, regarded as $L(R,\pi)$-structures and satisfying 4.6 (i) - (iv) and

4.6 (vi) $\forall\xi(\xi<\infty \rightarrow \exists\eta(0 \leq \xi-\eta \leq 1_\Gamma \wedge \pi(\eta) \neq 0)$.

(This readily implies 4.6(v).) Then we get:

<u>VARIANT</u> 4.9 Theorem 4.3 and supplement 4.4 hold for $L(R,\pi)$ and $HF_w''(R,\pi)$ in place of $L(R,W,C)$ and $HF_w'(R,W,C)$.

This variant generalizes the main F-quantifier elimination result for Hensel fields with cross section in [Weispfenning 76] .

<u>The proof of the main theorem, its supplement and its variants will be given in section</u> 5 . In this section, we shall be concerned with <u>applications of these results.</u> While F-quantifier-free formulas are quite simple in comparison with arbitrary formulas, they are not yet simple enough for most applications. One reason is that in such a formula properties of the value group and the various residue class rings are still interconnected via the maps v_k , $res^1_{k,h}$, and in languages with partial cross section also via π and res_k .

This prevents a straightforward separation' of bound Γ-variables and bound R_k-variables for various $k<\omega$. Another problem for the transfer of (absolute) quantifier elimination from the Γ-sort and R_k-sorts to an **extension theory** of $HF_w'(R,W,C)$ is the joint occurence of F-terms and R_k-terms in the predicates $W_{n,k,h}$. In this respect, the theories $HF_w'(R,W',C,E)$ and $HF_w''(R,\pi)$ behave much better, and hence will be the basis of our results on absolute quantifier elimination.

There is, however, one result that can be derived directly from F--quantifier elimination; this concerns certain properties of definable sets, proved for p-adic fields by Macintyre (cf. 3.5 and [Macintyre 76]).

THEOREM 4.10 Let (F,Γ,v) be a model of $HF_w'(R,W,C)$ or of $HF_w'(R,W',C,E)$ in a language L' as in 4.3 or 4.5, respectively.

(i) Any definable set $M\subseteq F^n$ is a finite union of sets of the **form** $Z\cap\mathcal{O}$, where Z is the zero-set of a polynomial ideal in $F[X_1,\dots,X_n]$ and $\mathcal{O}$ is open in F^n.

(ii) Any infinite definable set $M\subseteq F$ has a non-empty interior.

PROOF. As in 3.5, it suffices to prove (i): An induction on the complexity of F-quantifier-free formulas shows that any such formula $\varphi(x_1,\dots,x_n,\underline{y})$ is equivalent to a disjunction of formulas

$\bigwedge_{i\in I} a_i=0 \quad \bigwedge_{1\le j\le m} b_j\ne 0 \ \wedge\ \vartheta(b_1,\dots b_m)$,where

$\vartheta(z_1,\dots,z_m)$ is an F-quantifier-free formula containing z_i only in terms vz_i, $res_{k_i}z_i$ and in formulas $W_{n_i,k_i,h_i}(z_i,d_i)$ or $W_n'(z_i)$, respectively. So by 5.2, $\underline{z}$ is k-invariant in $\vartheta(\underline{z})$ for some $k<\omega$. We may assume that $a_i(\underline{x},\underline{y})$ are formal polynomials in $\underline{x}$ and that $b_j(\underline{x},\underline{y})$ are quotients of formal polynomials in $\underline{x}$. So for all models (F,Γ,v) of the theories in question and all parameters $\underline{c}$ in F,

$\bigwedge_{1\le j\le m} b_j(\underline{x},\underline{c})\ne 0 \wedge \vartheta(b_1(\underline{x},\underline{c}),\dots,b_m(\underline{x},\underline{c}))$ defines an open set in F^n, by continuity of the $b_j(\underline{x},\underline{c})$. This proves the theorem.

Next, we turn to Hensel fields with residue class field of characteristic zero, where the F-quantifier elimination can be applied in a fairly direct manner.

Let $HF_0'(R_0,W,C)$, $HF_0'(R_0,W',C,E)$, $HF_0'(R_0,W,\pi)$, $HF_0''(R_0,\pi)$ be obtained from the corresponding theories $HF_w'(R,W,C)$, $HF_w'(R,W',C,E)$, $HF_w'(R,W,\pi)$, $HF_w''(R,\pi)$ by the following operations :

(1) Add the axioms $v(n)=0$ for all $0<n<\omega$.

(2) Put $w \equiv 0$, $1_\Gamma = 0$, $q \quad 1$, $q_0 = 1$.
(3) Replace $\varphi(v_0(d))$ by $(d=0 \wedge \varphi(\infty)) \vee (d \neq 0 \wedge \varphi(0))$ in a formula $\varphi(\xi)$.

So, in these theories all R_k-sorts with $k > 0$ have disappeared. **The main** theorem 4.3 together with supplement 4.4 and variants 4.5, 4.7, 4.9 specialize in an obvious manner for these theories. For $HF_0'(R_0,W',C,E)$ and $HF_0''(R,\pi)$, we can in addition prove a 'separation of bound variables' :

LEMMA 4.11 Let L' be as in 4.5 (4.9) with $L(R,W',C,E)$ ($L(R,\pi)$) specialized to $L(R_0,W',C,E)$ ($L(R_0,\pi)$). In $HF_0'(R_0,W',C,E)$ ($HF_0''(R_0,\pi)$) any F-quantifier-free L'-formula φ is equivalent to a disjunction φ' of formulas of the form $\psi(\underline{a}) \wedge \varrho(\underline{va}) \wedge \sigma(\underline{res_0 a})$, where $\psi(\underline{x})$ is quantifier-free , $\varrho(\underline{\xi})$ is an L'_Γ-formula, and $\sigma(\underline{r})$ is an L'_{R_0}-formula . Moreover the assignment $\varphi \longmapsto \varphi'$ is primitive recursive, and for prenex φ, φ and φ' have the same prefix type.

PROOF. As in [Weispfenning 76] , lemma 3.2 .

Combining variants 4.5 and 4.9 with lemma 4.10 , we get the following transfer of quantifier elimination.

THEOREM 4.12 Let L' be as in 4.10 , let T_Γ be a theory in L'_Γ and let T_{R_0} be a theory in L'_{R_0} . If T_Γ and T_{R_0} admit (primitive recursive) quantifier elimination, then $HF_0'(R_0,W',C,E) \cup T_\Gamma \cup T_{R_0}$ ($HF_0''(R_0,\pi) \cup T_\Gamma \cup T_{R_0}$) admits (primitive recursive) quantifier elimination in L' .

(For $HF_0''(R_0,\pi) \cup T_\Gamma \cup T_{R_0}$ this is cor. 3.5(iii) in [Weispfenning 76] .)

EXAMPLES 4.13 The following theories T_Γ , T_{R_0} satisfy the hypothesis of theorem 4.12 , when L'_Γ , L'_{R_0} , C_Γ , E are as specified below . (Notice that in a model (F,Γ,v) of $HF_0'(R_0,W',C,E)$, E is a set of representatives for U/U^n , where U is the group of units of the valuation ring A.)

(i) $T = DvOG_\infty$ (theory of non-trivial, divisible ordered abelian groups with extra element ∞ on top) in L_Γ , $C = \emptyset$.
(ii) $T = ZG_\infty$ (theory of Z-groups with smallest positive element 1' and extra element ∞ on top) in $L_\Gamma \cup \{1', \{\equiv_n\}_{n<\omega}\}$, $C_\Gamma = \{1'\}$.

(iii) $T = Z^nG_\infty$ (theory of structures $G \cup \{\infty\}$,where G is elementarily equivalent to the lexicographical product $\mathbb{Z}1_1 \oplus \ldots \oplus \mathbb{Z}1_n$) in $L_\Gamma \cup \{1_1,\ldots,1_n \{\equiv_n\}_{n<\omega}\}$, $C_\Gamma = \{1_1,\ldots,1_n\}$. Similarly for $Z^nG^*_\infty$ (theory of structures $G \cup \{\infty\}$,where G is elementarily equivalent to $\mathbb{Z}1_1 \oplus \ldots \oplus \mathbb{Z}1_n \oplus \mathbb{Q}$) , cf. [Weispfenning 80], cor. 2.14 .

(iv) $T = DR_\infty(C_\Gamma)$ (theory of structures $G \cup \{\infty\}$,where G is dense regular and G/pG is generated by C_Γ + pG for any prime p) in $L_\Gamma \cup C_\Gamma$.

(v) T_{R_0} = ACF (theory of algebraically closed fields) in L_F with $E = \{1\}$.

(vi) T_{R_0} = ACVF (theory of algebraically closed non-trivially valued fields) in L_{VF} with $E = \{1\}$.

(vii) T_{R_0} = RCF (theory of real-closed ordered fields) in L_{OF} with $E = \{+1,-1\}$.

A <u>real-closed valued field</u> is a real-closed ordered field $(F,<)$ with a distinguished convex subring A such that $1 \in A \neq F$. Let RCVF be the theory of real-closed valued fields in the language $L_{OF}(A) = \{0,1,+,-,\cdot,^{-1}, <, A\}$.

<u>COROLLARY</u> 4.14 RCVF admits primitive recursive quantifier elimination , is complete and primitive recursively decidable .

<u>PROOF</u>. It is easy to verify the following facts about real-closed valued fields $(F,<,A)$ (cf. 1.7 , [Cherlin-Dickmann 83]) : A is a valuation ring ; the associated valuation has a divisible, non-trivial value group and a real-closed residue class field F_v , which is ordered by $res_0 a < res_0 b \leftrightarrow a<b \wedge res_0 a \neq res_0 b$. (F,Γ,v) is a Hensel field and hence a model of $HF_0'(R_0,W',\emptyset,\{+1,-1\})$. Any $L_{OF}(A)$-formula φ translates uniformly into a $L(R_0)$-formula φ_1 (replace $a>0$ by $\exists x(x^2=a)$, $A(a)$ by $va \geq 0$). φ_1 is equivalent in $HF_0'(R_0,W',\emptyset,\{+1,-1\}) \cup DvOG_\infty \cup RCF$ to a quantifier-free formula φ_2 . Replacing in φ_2, $W_n'(a)$ by $a=a$ for odd n , and by $a \geq 0$ for even n ; $va \leq vb$ by $A(b/a)$; $res_0 a = 0$ by $\neg A(a) \vee (A(a) \wedge \neg A(a^{-1}))$; $res_0 a > 0$ by $a>0 \wedge \neg A(a^{-1})$, we obtain a quantifier--free $L_{OF}(A)$-formula φ_3 such that $RCVF \models \varphi \leftrightarrow \varphi_3$. The rest is now obvious .

This corollary can be rephrased in terms of certain ordered rings with divisibility in the same way as this was done with theorem 3.2 in corollary 3.4 :

A <u>real-closed ring</u> is an ordered integral domain which is not a field and satisfies the intermediate value theorem for polynomials. Let RCR be the theory of real-closed rings in the language $L_{OR}(D) = \{0,1,+,-,\cdot,<,D\}$ with $D(x,y) \leftrightarrow \exists z(xz = y)$.

<u>COROLLARY</u> 4.15 ([Cherlin-Dickmann 83]) RCR admits primitive recursive quantifier elimination, is complete and primitive recursively decidable.

<u>PROOF</u>. For any $(F,<,A) \models$ RCVF , $(A,<,D)$ is a real-closed ring with quotient field F ; conversely, if $(A,<,D) \models$ RCR and F is the quotient field of A, then $(F,<,A) \models$ RCVF. This establishes a 1-1 correspondence $(A,<,D) \leftrightarrow (F,<,A)$. An $L_{OR}(D)$-formula $\varphi(\underline{x})$ translates into an $L_{OF}(A)$-formula $\varphi_1(\underline{x})$ by replacing $D(x,y)$ by $A(y/x)$ and restricting quantifiers to A . A quantifier-free $L_{OF}(A)$-formula translates into a quantifier-free $L_{OR}(D)$-formula by replacing $A(b/a)$ by $D(a,b)$, $b/a = 0$ by $b=0 \wedge a \neq 0$, $b/a > 0$ by $a \cdot b > 0$. So the quantifier elimination procedure for RCVF yields a corresponding procedure for RCR . The rest is obvious .

We return for a moment to the general theory $HF_0^!(R_0,W',C,E)$ to prove a <u>theorem</u> <u>on</u> <u>prime</u> <u>model</u> <u>extensions</u>.

<u>THEOREM</u> 4.16 Let L', T_Γ , T_{R_0} be as in 4.12 , and assume in addition

(i) If $\Gamma_1 \subseteq \Gamma_2 \models T_\Gamma$ and $\Gamma_1 \setminus \{\infty\}$ is a pure subgroup of $\Gamma_2 \setminus \{\infty\}$, then $\Gamma_1 \models T$.

(ii) If $K_1 \subseteq K_2 \models T_{R_0}$ and K_1 is algebraically closed in K_2 , then $K_1 \models T_{R_0}$.

Then any L'-substructure of an L'-model of $HF_0^!(R_0,W',C,E)$ has a unique, minimal prime model extension to an L'-model of $HF_0^!(R_0,W',C,E)$.

<u>PROOF</u>. Let $(F,\Gamma,v) \subseteq (F',\Gamma',v') \models HF_0^!(R_0,W',C,E)$, and let $\tilde{F}$ be the algebraic closure of F in F'. Then by Hensel's lemma, the value group $\tilde{\Gamma}$ of $\tilde{F}$ under $\tilde{v} = v'|\tilde{F}$ is the divisible hull of Γ in Γ' and the residue class field $\tilde{F}_{\tilde{v}}$ of $(\tilde{F},\tilde{\Gamma},\tilde{v})$ is the algebraic closure of F_v in $F'_{v'}$; moreover, $(\tilde{F},\tilde{\Gamma},\tilde{v})$ is a Hensel field and satisfies the relevant axioms on C, C_Γ, E . So by (i),(ii), $(\tilde{F},\tilde{\Gamma},\tilde{v}) \models HF_0^!(R_0,W',C,E) \cup T_\Gamma \cup T_{R_0}$, and $(\tilde{F},\tilde{\Gamma},\tilde{v})$ is algebraic over (F,Γ,v) in the model theoretic sense. So fact 3.7 yields the theorem.

This theorem may be applied e.g. to $T_\Gamma \in \{DvOG_\infty , ZG_\infty\}$ and $T_{R_0} \in \{ACF, ACVF, RCF\}$. We list some typical applications.

COROLLARY 4.17 (i) Every substructure $(F,<,A)$ of a real-closed valued field with $A \neq F$ has a unique, minimal prime model extension to a real closed valued field.
(ii) Every substructure $(A,<,D)$ of a real-closed ring such that A is not a field has a unique, minimal prime model extension to a real--closed ring. (For a detailed discussion of prime model extensions for real-closed rings see [Th. Becker 83].)
(iii) Let $(F,\Gamma,v) \subseteq (F_i,\Gamma_i,v_i) \models HF_0(R_0)$, assume that Γ, Γ_i have the same smallest positive element $1'$, Γ_i are Z-groups, F_{iv_i} are algebraically closed extensions of F_v, F_i are algebraic extensions of F for $i = 1,2$, and that $F \cap F_1^n = F \cap F_2^n$ for all $0 < n < \omega$. Then (F_1,Γ_1,v_1) and (F_2,Γ_2,v_2) are isomorphic over (F,Γ,v) .

For a transfer of model-completeness, completeness and (primitive recursive) decidability from T and T_{R_0} to $HF_0''(R_0,\pi) \cup T_\Gamma \cup T_{R_0}$ and $HF(R_0) \cup T_\Gamma \cup T_{R_0}$, we refer the reader to [Weispfenning 76] . We add only one remark : For the first theory, the transfer is a direct consquence of 4.9 and 4.11. To reduce the case of the second theory (without cross-section) to the first, one may use - instead of the arguments in [Weispfenning 76],section 4 - more conveniently the following lemma by Ziegler:

LEMMA 4.18([Ziegler 72] , comp. also [Kochen 75]) Let $(F_1,\Gamma_1,v_1) \subseteq (F_2,\Gamma_2,v_2)$ be valued fields in some extension language L' of L, and assume Γ_1 is a pure subgroup of Γ_2. Then there exist valued fields with cross-section such that $(F_1',\Gamma_1',v_1',\pi_1) \subseteq (F_2',\Gamma_2',v_2',\pi_2)$ as $L'(\pi)$-structures and such that (F_i',Γ_i',v_i') is an elementary extension of (F_i,Γ_i,v_i) for $i=1,2$.

PROOF. Combine a compactness argument with unions of elementary chains .

Next, we consider Hensel fields of characteristic zero with finite residue class rings R_k ($k < \omega$), and in particular p-adically closed fields in the sense of [Prestel-Roquette 83] .

Let $L(W',C)$ be the language obtained from $L(R,W',C,\emptyset)$ by dropping all R_k-sorts $(k < \omega)$; let $L(W',C,D)$ result from $L(W',C)$ by adding a new finite set D of F-constants. For any prime p , we let $HF_p'(W',C,D)$ be the theory of all Hensel fields (F,Γ,v) in $L(W',C,D)$ such that
(i) $v(p1) = 1_\Gamma > 0$ (so F_v has characteristic p) ;
(ii) W_n' satisfy the defining axioms (1), preceding 4.5 ;

(iii) C, C_Γ satisfy the axioms (1),(2), preceding 4.3 ;

(iv) Let $A=A_v$ be the valuation ring of v, let $res':A \to A/pA = R'$ be the canonical homomorphism. Then $\{1\} \cup \{res'(\check{d}) : \check{d} \in D\}$ is a basis of R' as vector space over the field $F_p = \mathbb{Z}/\mathbb{Z}_p$.

This has the following consequences : If for $0<k<\omega$, $res_k':A \to A/p^kA = R_k'$ is canonical , then $R_k' = \{res_k'(a) : a \in \check{D}_k\}$, where $\check{D}_k = \{m1 + \sum_{\check{d}\in D} m_{\check{d}}\cdot\check{d} : 0 \leq m, m_{\check{d}} < p^k\}$, and so R_k' is finite. R_k is a homomorphic image of R_{k+1}' , and so $R_k = \{res_k(a) : a \in \check{D}_{k+1}\}$; in particular, if 1_Γ is the smallest positive element of Γ , then $R_k = R_{k+1}'$.

This theory admits F-quantifier elimination and 'separation of bound variables' :

THEOREM 4.19 Let $L_\Gamma' \supseteq L_\Gamma$ with $L_\Gamma' \cap L(W',C,D) = L(W',C,D)$, and put $L' = L(W',C,D) \cup L_\Gamma'$. Then there is a primitive recursive procedure assigning to every L'-formula φ an F-quantifier-free L'-formula φ' such that $HF_p'(W',C,D) \models \varphi \leftrightarrow \varphi'$. Moreover, φ' may be taken as a disjunction of formulas of the form $\Psi(\underline{a}) \wedge \varrho(\underline{va})$, where $\underline{a}$ is a tuple of F-terms, $\Psi(\underline{x})$ is quantifier-free and $\varrho(\underline{\xi})$ is an L_Γ'-formula ; for prenex φ, φ and φ' have the same prefix type .

PROOF. Use variant 4.5 for $HF_w'(R,W',C,E)$ with $w(n) = \max\{v_p(i)1 : 1 \leq i \leq n\}$ and $E_n = \check{D}_2$ $(n<\omega)$, to get F-quantifier elimination in $L(R,W',C,D)$. Then eliminate all R_k-sorts by means of the constant terms in $\check{D}_{k+1}$. Finally, the separation of bound variables is obtained as in 4.11 .

COROLLARY 4.20 Let L' be as in 4.19, and let T_Γ be a theory in L_Γ' . If T admits (primitive recursive) quantifier elimination , so does $HF_p'(W',C,D) \cup T_\Gamma$.

The theory pCF(D) of p-adically closed fields (with parameters in D) is obtained from $HF_p'(W',C,D)$ in L' by the following specialization : We fix $L_\Gamma' = L_\Gamma \cup \{1_\Gamma , \{\equiv_n\}_{n<\omega}\}$, $C = \emptyset$, and put $pCF(D) = HF_p'(W',\emptyset,D) \cup ZG_\infty$. Then from 4.20, we get :

THEOREM 4.21 pCF(D) admits primitive recursive quantifier elimination in the language $L' = L(W',D) \cup \{1_\Gamma , \{\equiv_n\}_{n<\omega}\}$. Moreover, for formulas without free Γ-variables, the predicates $\{\equiv_n\}$ are not needed in the quantifier elimination.(This generalizes [Macintyre 76] thm.1).

PROOF. ZG_∞ admits primitive recursive quantifier elimination in the language $L_\Gamma \cup \{1', \{\equiv_n\}\}$, where $1'$ is a constant for the smallest positive element of the group. A slight modification of such a quantifier elimination procedure yields the result also with the following weaker axioms on $1'$: $0<1'$, $(0<\xi \rightarrow 1' \leq e\xi)$,where $0<e<\omega$ is fixed. Taking $1' = 1_\Gamma$, $e = \mathrm{card}(D)$, the first statement follows now from 4.20 . To prove the second statement, we remark that in pCF(D) , $b \neq 0 \wedge c \neq 0 \wedge vb \equiv_n vc$ is equivalent to $\bigvee_{0 \leq m < p^k,\, a \in \check{D}_k} (W'_n((m1+a)b/c) \wedge v(m1+a) = 0)$, for $k = 2 \cdot v_p(n)$.

<u>COROLLARY</u> 4.22 (cf.[Weispfenning 76],4.10, 4.11 ,[Prestel-Roquette 83], cor. 5.3) (i) The maximal sets Φ of formulas of the form $v(m+a) \leq v(m'+a')$, $W'_n(m''+a'')$ $(0 \leq m,m' < p$, $0 \leq m'' < p^n$, $a,a' \in \check{D}_1$, $a'' \in \check{D}_n)$ that are consistent with pCF(D) form elementary invariants for models of pCF(D) .

(ii) pCF(D) and all its completions are primitive recursively decidable .

<u>PROOF</u>. (i) By 4.21 , it suffices to decide atomic sentences of the form $v(m+a) \leq v(m'+a')$ and $W'_n(m''+a'')$ with $0 \leq m,m',m'' < p^k$, $a,a',a'' \in \check{D}_k$ for arbitrary $k < \omega$. In view of the axioms for $\check{d} \in D$, this can be reduced to a decision about the formulas mentioned in (i).
(ii) Notice that all the sets Φ in (i) are finite and that there are only finitely many such sets. This together with (i) proves (ii).

Specializing further to <u>unramified</u> p-<u>adically</u> <u>closed</u> <u>fields</u>, we get the following result: Let $pCF_f(d)$ be $pCF(D) \cup \{\forall\xi(0<\xi \rightarrow 1_\Gamma \leq \xi)\}$, where D has been replaced by a single F-constant d with an axiom saying that $|F_v| = p^f$, $g(d) = 0$, and resg(x) is irreducible over $\mathbb{F}_p$ for a specific monic $g(x) \in \mathbb{Z}[x]$ of degree f . Then 4.21 yields :

<u>COROLLARY</u> 4.23 (cf. [Weispfenning 76], 4.10,4.11 ,[Prestel-Roquette 8 cor. 5.4) $pCF_f(d)$ admits primitive recursive quantifier elimination , is complete and primitive recursively decidable .

Concerning <u>prime</u> <u>model</u> <u>extensions</u> <u>for</u> p-<u>adically</u> <u>closed</u> <u>fields</u>, a slight variant of the proof of 4.16 , together with 4.21, proves a corresponding result :

<u>COROLLARY</u> 4.24 Any substructure (F,Γ,v) of a model of pCF(D) has a unique, minimal prime model extension to a model of pCF(D). More concretely, if $(F,\Gamma,v) \subseteq (F_i,\Gamma_i,v_i)$ as L-structures, $(F_i,\Gamma_i,v_i) \models$

pCF(D), F_i are algebraic extensions of F for i=1,2 , and $F \cap F_1^n = F \cap F_2^n$ for all $0 < n < \omega$, then the (F_i, Γ_i, v_i) are isomorphic over (F, Γ, v) (cf. [Prestel-Roquette 83], cor. 3.11).

We conclude this section with applications of F-quantifier elimination to Hensel fields of characteristic zero with residue class field of characteristic $p > 0$, and arbitrary ramification. In other words, we want to make no assumtion on the size of the interval $[0, 1_\Gamma]$ in the value group Γ , where $1_\Gamma = vp$. The crucial problem is again, to find a language that allows 'separation of bound variables' in F--quantifier-free formulas. We restrict our attention for the time being to Hensel fields with an almost total cross-section , i.e. to models of the theory $HF_w''(R,\pi)$. Recall that in a model (F,Γ,v,π) of this theory , π is a cross-section on a discrete subgroup Δ of Γ with smallest positive element 1_Γ such that for every $\gamma \in \Gamma$ there is a $\delta \in \Delta$ with $0 \leq \gamma - \delta \leq 1_\Gamma$, and π is zero outside Δ . For a fixed prime p ,we let $HF_p''(R,\pi) = HF_w''(R,\pi) \cup \{vp = 1_\Gamma > 0\}$ with $w(n) = \max\{v_p(i) \cdot 1_\Gamma : 1 \leq i \leq n\}$.

Let $(F,\Gamma,v,\pi) \models HF_p''(R,\pi)$. Regarding Δ as 'generalized integers', we may define an 'integer part' operation [ ] on the Γ-sort by $[\infty] = \infty$, $[\alpha]$ = the unique $\delta \in \Delta$ with $0 \leq \alpha - \delta < 1_\Gamma$, for $\alpha < \infty$. In other words, [ ] is axiomatized by :

4.25 $\quad [\infty] = \infty \ , \ \xi < \infty \rightarrow (0 \leq \xi - [\xi] < 1_\Gamma \wedge \pi(\xi) \neq 0)$.

We let $HF_p''(R,\pi,[\,])$ be the theory of all $(F,\Gamma,v,\pi,[\,])$ in $L(R,\pi,[\,]) = L(R,\pi) \cup \{[\,]\}$, such that $(F,\Gamma,v,\pi) \models HF_p''(R,\pi)$ and [ ] satisfies 4.25.

LEMMA 4.26 Any F-quantifier-free formula φ without free Γ-variables and free R_l-variables ($l < \omega$) is equivalent in $HF_p''(R,\pi,[\,])$ to a disjunction φ' of formulas of the form

$\rho([va_1],\ldots,[va_m]) \wedge \sigma(res_k a_1,\ldots,res_k a_m, res_k(a_1/\pi[va_1]),\ldots, res_k(a_m/\pi[va_m]))$, where k is fixed , $\rho(\xi_1,\ldots,\xi_m)$ is an L_Γ-formula with all quantifiers restricted to Δ (i.e. of the form $\exists\eta(\pi(\eta)\neq 0 \wedge \ldots)$), $\sigma(r_1,\ldots,r_m,s_1,\ldots,s_m)$ an L_{R_k}-formula . Moreover, the assignment $\varphi \mapsto \varphi'$ is primitive recursive, and for prenex φ, φ and φ' have the same prefix type .

PROOF. Let φ be given. By the procedure in [Weispfenning 76], lemma 3.1 , we may assume that in φ no Γ-variable occurs in the scope of a π-operation-symbol (comp. the proof of 4.7 in section 5). To simplify the notation, we write $\Delta(\xi)$ for $\pi(\xi) \neq 0$ and $(\exists\xi\in\Delta)\psi$ for $\exists\xi(\Delta(\xi) \wedge \psi)$, and call such a quantifier 'restricted to Δ'.

To restrict all Γ-quantifiers in φ to Δ , we replace any such quantifier $\exists\xi\,\Psi(\xi)$ by $(\exists\xi'\in\Delta)(\exists r)(\,vr<1_\Gamma \wedge \Psi(\xi'+vr))$, where r is a R_1-variable. In the resulting formula φ_1 , we may write any atomic Γ-subformula in the form $\alpha+\beta \overset{=}{<} \alpha'+\beta'$, where α, α' contain no symbols v , v_1 , and hence no F-term or R_1-term , and β, β' are of the form va or v_1d . Replacing va by $[va]$ + $v_1res_1(a/\pi[va])$, we arrive at the following form for atomic subformulas of φ_1 :

(*) $\gamma+\delta \overset{=}{<} \gamma'+\delta'$, where γ, γ' are built up from 1_Γ , ∞, Γ-variables and terms $[va]$ by means of + , - ; δ, δ' are of the form $v_1(d)$ for an R_1-term d .

Since, provably $0 \leq v_1d \leq 11_\Gamma \vee v_1d = \infty$, and $\Delta(\gamma) \vee \gamma=\infty$, (*) can be rewritten as a boolean combination of atomic formulas involving only terms of the form γ, γ' <u>or</u> terms of the form δ, δ'; so these new atomic formulas do not contain Γ-variables and R_1-variables simultaneously. Next, atomic formulas of the form $v_1d \leq v_1d'$ are replaced by $\exists r(\,dr \leq d'\,)$, and afterwards the bound Γ-variables and bound R_1-varables are separated. Finally, the various R_1-sorts occuring in the resulting formula can be coded into a single R_k-sort, where k is taken greater or equal to all l occuring in the formula.

A combination of this lemma with variant 4.9 of the main theorem and its supplement yields now the following results on $HF_p''(R,\pi,[\,])$:

<u>THEOREM</u> 4.27 Let T be a theory in a definable extension L'_Γ of L_Γ , let T_{R_k} be a theory in a definable extension L'_{R_k} of L_{R_k} for $k<\omega$, and let $T^\Delta = \{\Psi^\Delta : \Psi\in T\}$ be T relativized to $\Delta(\xi)$ $(\leftrightarrow \pi(\xi)\neq 0)$ Put $L' = L(R,\pi) \cup L'_\Gamma \cup \bigcup_{k<\omega} L'_{R_k}$.

(i) If T and all T_{R_k} admit (primitive recursive) quantifier elimination , then $HF_p''(R,\pi,[\,]) \cup T^\Delta \cup \bigcup_{k<\omega} T_{R_k}$ admits (primitive recursive) quantifier elimination for L'-formulas without free Γ-variables and free R_k-variables .

(ii) If T and all T_{R_k} are complete (and primitive recursively decidable) , so is $HF_p''(R,\pi,[\,]) \cup T^\Delta \cup \bigcup_{k<\omega} T_{R_k}$.

(iii) If $(F_i,\Gamma_i,v_i,\pi_i,[\,]_i)$ are models of $HF_p''(R,\pi,[\,])$ with a commom substructure $(F,\Gamma,v,\pi,[\,])$ such that Δ_i are elementarily (existentially) equivalent over Δ and $R_k(F_i)$ are elementarily (existentially) equivalent over $R_k(F)$ for $k<\omega$, then $(F_i,\Gamma_i,v_i,\pi_i,[\,]_i)$ are elementarily (existentially) equivalent over $(F,\Gamma,v,\pi,[\,])$.

(iv) If T and all T_{R_k} are model-complete, so is $HF''_p(R,\pi,[\,]) \cup T^{\Delta} \cup \bigcup_{k<\omega} T_{R_k}$.

REMARK 4.28 In 4.27 (ii),(iii),(iv) the distinguished operation may be deleted, since it is existentially and universally definable.

Our next goal is to carry over some of these properties to the corresponding Hensel fields without an almost total cross-section. Let $HF_p(R) = HF_w(R) \cup \{vp = 1_\Gamma > 0\}$ with $w(n) = \max(\, v_p(i) : 1 \le i \le n\,)$ be the theory of all Hensel fields of characteristic zero with residue class field of characteristic $p > 0$ in $L(R)$. In the next lemma, we show that the assumption of the existence of an almost total cross--section is no severe restriction on the models of $HF_p(R)$, since $HF''_p(R,\pi)$ and hence $HF''_p(R,\pi,[\,])$ are conservative extensions of $HF_p(R)$. If $(F,\Gamma,v) \models HF_p(R)$, we call a subgroup Δ of Γ a large discrete subgroup of Γ, if Δ is discrete with smallest positive element 1_Γ and every $\gamma \in \Gamma$ has a distance $< 1_\Gamma$ to an element $\delta \in \Delta$. In particular, for any model (F,Γ,v,π) of $HF''_p(R,\pi)$, $\{\gamma \in \Gamma : \pi(\gamma) \neq 0\}$ is a large discrete subgroup of Γ.

LEMMA 4.29 (i) Every model (F,Γ,v) of $HF_p(R)$ has an elementary extension (F',Γ',v'), where Γ' contains a large discrete subgroup.
(ii) Let $(F_0,\Gamma_0,v_0) \subseteq (F_1,\Gamma_1,v_1)$ be models of $HF_p(R)$ such that Γ_1 is an elementary extension of Γ_0. Then there exist structures $(F'_0,\Gamma'_0,\Delta'_0,v'_0) \subseteq (F'_1,\Gamma'_1,\Delta'_1,v'_1)$ such that Δ'_i is a large discrete subgroup of Γ'_i, (F'_i,Γ'_i,v'_i) is an elementary extension of (F_i,Γ_i,v_i) for $i=1,2$, and (Γ'_1,Δ'_1) is an elementary extension of (Γ'_0,Δ'_0).
(iii) Let (F_i,Γ_i,v_i) ($i=1,2$) be as in (ii). Then there exist models $(F''_0,\Gamma''_0,v''_0,\pi''_0) \subseteq (F''_1,\Gamma''_1,v''_1,\pi''_1)$ of $HF''_p(R,\pi)$ such that (F''_i,Γ''_i,v''_i) is an elementary extension of (F_i,Γ_i,v_i) for $i=1,2$, and (Γ''_1,Δ''_1) is an elementary extension of (Γ''_0,Δ''_0) for $\Delta''_i = \{\gamma \in \Gamma''_i : \pi''_i(\gamma) \neq 0\}$.

PROOF. (i) Put $\Delta = \mathbb{Z} = \mathbb{Z}\cdot 1_\Gamma$ and form an $\aleph_1$-saturated elementary extension (F',Γ',Δ',v') of (F,Γ,Δ,v) Then the convex hull $\hat{\Delta}'$ of Δ' in Γ' is definable in (Γ',Δ') and pure in Γ'. So by [Eklof-Fisher 72], lemma 4.1, $\hat{\Delta}'$ is a direct summand of Γ', and so Γ' can be written as a lexicographical sum $\Gamma' = \Gamma'' \oplus \hat{\Delta}'$. Then $\Delta'' = \Gamma'' + \Delta'$ is the lexicographical sum of Γ'' and Δ' and hence by construction a large discrete subgroup of Γ' .

(ii) From (i), we get $(F_0',\Gamma_0',\Delta_0',v_0')$ such that (F_0',Γ_0',v_0') is an elementary extension of (F_0,Γ_0,v_0) and Δ_0' is a large discrete subgroup of Γ_0' . By compactness, (F_1,Γ_1,v_1) has an elementary extension (F_2,Γ_2,v_2) extending (F_0',Γ_0',v_0') , such that Γ_2 is an elementary extension of Γ_0' ($\Gamma_2 \succeq \Gamma_0'$). By compactness , we find $(\Gamma_2',\Delta_2') \succeq (\Gamma_0',\Delta_0')$ with $\Gamma_2' \succeq \Gamma_2$. Again by compactness, we find $(F_3,\Gamma_3,v_3) \succeq (F_2,\Gamma_2,v_2)$ with $\Gamma_3 \succeq \Gamma_2'$. Continuing in this way , we get elementary chains (Γ_i',Δ_i') $(2 \leq i < \omega)$, (F_i,Γ_i,v_i) $(2 \leq i < \omega)$ with $\Gamma_i \preceq \Gamma_i' \preceq \Gamma_{i+1}$. Passing to the limit, we obtain the desired structure $(F_1',\Gamma_1',\Delta_1',v_1')$.

(iii) Let $(F_i',\Gamma_i',\Delta_i',v_i')$ be as in (ii), and let $(F_i'',\Gamma_i'',\Delta_i'',v_i'',\pi_i)$ be obtained from these structures by Ziegler's lemma 4.18. The proof of this lemma shows that we may assume $(\Gamma_1'',\Delta_1'') \succeq (\Gamma_0'',\Delta_0'')$. Modifying π_i to π_i'' , by $\pi_i'' | \Delta_i'' = \pi_i$, $\pi_i'' | \Gamma_i'' \setminus \Delta_i'' = 0$, we obtain $(F_i'',\Gamma_i'',v_i'',\pi_i'')$ satisfying (iii) .

A direct application of 4.29 and 4.28 allows us now to transfer 4.27 (ii),(iii) (modified), and (iv) to $HF_p(R)$:

<u>THEOREM</u> 4.30 Let T , T_{R_k} , L' be as in 4.27 .

(i) If T and all T_{R_k} are complete (and primitive recursively decidable), so is $HF_p(R) \cup T \cup \bigcup_{k<\omega} T_{R_k}$.

(ii) If $(F,\Gamma,v) \subseteq (F',\Gamma',v')$ are models of $HF_p(R)$ such that Γ' is an elementary extension of Γ and $R_k(F')$ is an elementary extension of $R_k(F)$ for $k < \omega$, then (F',Γ',v') is an elementary extension of (F,Γ,v) .

(iii) If T and all T_{R_k} are model-complete, so is $HF_p(R) \cup T \cup \bigcup_{k<\omega} T_{R_k}$.

<u>REMARK</u>. The transfer of completeness in 4.30(i)(without parenthesis), and of elementary extensions in 4.30(ii) for Hensel fields with arbitrary ramification was first established in [v.d.Dries b] by a reduction to the equal characteristic case, using non-standard arguments.

Due to the problem of separation of bound variables in F-quantifier--free formulas, we have no transfer of quantifier elimination corresponding to 4.27(i) in a language with root-predicates W_n' in place of an almost total cross-section π . This problem can be overcome, when we restrict our attention to <u>finitely ramified Hensel fields (with arbitrary residue class rings)</u>: Let $HF_{p,e}'(R,W',C,E) = HF_w'(R,W',C,E)$

$\cup \{vp = 1_\Gamma > 0\} \cup \{\forall \xi(0 < \xi \rightarrow 1_\Gamma \leq e\xi)\}$ for $0 < e < \omega$, with $w(n) = \max\{v_p(i) : 1 \leq i \leq n\}$. Call a theory T in $L'_\Gamma \supseteq L_\Gamma$ normal if every L'_Γ-formula $\Psi(\xi)$ is equivalent in T to a formula $\Psi'(\underline{e\xi})$, and the assignment $\Psi \mapsto \Psi'$ is primitive recursive. In particular, T is normal if T contains a primitive recursive set of defining axioms for every relation-symbol in $L'_\Gamma \setminus L_\Gamma$.

THEOREM 4.31 Let L'_Γ, L'_{R_k}, L' be as in 4.5, let T_Γ be a normal theory in L'_Γ and let T_{R_k} be a theory in L'_{R_k} for $k < \omega$. It T_Γ and all T_{R_k} admit (primitive recursive) quantifier elimination, then $HF'_{p,e}(R, W'^k, C, E) \cup T_\Gamma \bigcup_{k<\omega} T_{R_k}$ admits (primitive recursive) quantifier elimination for L'-formulas without free R_k-variables.

PROOF. In view of 4.5, it suffices to separate bound variables in F-quantifier-free formulas. To this end, we eliminate R_k-terms from atomic Γ-formulas : $\Psi(v_{k_1}d_1, \ldots, v_{k_m}d_m) \wedge \bigwedge_{1 \leq j \leq m} d_j \neq 0$ is equivalent to $\bigvee_{0 \leq i_1 \leq k_1} \cdots \bigvee_{0 \leq i_m \leq k_m} [\bigwedge_{1 \leq j \leq m} \exists r(d_j^e r = p^{i_j}) \wedge \neg \exists s(d_j^{e+1} s = p^{i_j}) \wedge \Psi'(i_1 1_\Gamma, \ldots, i_m 1_\Gamma)]$. After this step, Γ-variables and R_k-variables can be separated as in 4.11, and finally, various R_k-sorts can be coded into a single one.

Applications of the results 4.27(ii), 4.30(i) on a transfer of primitive recursive decidability hinge on the possibility of finding (infinite) decidable residue class rings R_k. For finitely ramified Hensel fields a further recursive reduction of sentences to sentences about the value group and the residue class field only has been established for certain cases in [Ziegler 72]. For unramified Hensel fields with perfect residue class field there is in fact a primitive recursive reduction of L_{R_k}-sentences to L_{R_0}-sentences: One represents elements of R_k in the form $\sum_{0 \leq i \leq n} \{r_i\} p^i$, where $\{r_i\}$ is the Teichmüller representative of $r_i \in R_0$ in R_k. Then addition, multiplication and equality can be coded uniformly in terms of the r_i (cf. [Greenberg 69], chapter 6). Together with 4.30(i), this yields:

COROLLARY 4.32 Let T, T_{R_0}, L' be as in 4.30. If T and T_{R_0} are complete (and primitive recursively decidable), so is $HF_p(R) \cup T \cup T_{R_0} \cup \{\forall \xi(0 < \xi \rightarrow 1_\Gamma \leq \xi), \forall r \exists s(s^p = r)\}$ (with R_0-variables r,s).

With $T = ZG_\infty$, $T_{R_0} = ACF_p$ this shows in particular :

COROLLARY 4.33 (cf. [Kochen 75]) The theory of the maximal unramified extension $\check{\mathbb{Q}}_p$ of $\mathbb{Q}_p$ is primitive recursively decidable.

5. PROOF OF THE MAIN THEOREM AND ITS VARIANTS .

Recall the outline of the proof of theorem 3.1: In order to eliminate F-quantifiers in algebraically closed valued fields, it was sufficient to treat formulas $\exists x\, \varphi$,where φ was F-quantifier-free; φ could be further reduced to the form

$$(*) \quad \bigwedge_{1\le i\le n} vf_i(x) = \xi_i \quad \text{with monic polynomials } f_i(x)\ .$$

By introducing tuples of zeros of polynomials via additional F-quantifiers, the f_i could be reduced to linear polynomials. At this point, the quantifier $\exists x$ in (*) was eliminated using lemma 2.3. The same procedure was then applied successively to the quantifiers over zeros of polynomials introduced earlier, resulting in formulas of the form

$$(**) \quad \exists z(\ g(z) = 0 \wedge \bigwedge_{1\le i\le n} v(x-a_i) = \alpha_i\ )\ .$$

The condition $g(z) = 0$ was equivalently replaced by a condition saying that z is an approximate zero of g. In this way (**) was reduced to a formula essentially about the (algebraically closed) residue class field, where the quantifier $\exists z$ could be eliminated by a simple instance of Hilbert's Nullstellensatz. The fact that the procedure came to an end after finitely many steps was due to a decrease of the degrees of the polynomials involved, obtained by the division algorithm for polynomials.

For the proof of the main theorem 4.3 , we shall roughly proceed along the same lines: Tuples of zeros will be replaced by tuples of Hensel-zeros (i.e. zeros obtained by an application of Hensel's lemma), and linear polynomials by quotients of powers of linear polynomials. The predicates $W_{n,l,h}$ are used to eliminate the quantifier $\exists x$ in formulas $\exists x\, \varphi$,where x occurs only in quotients of powers of linear polynomials in x. The essential fact underlying the reduction of arbitrary F-quantifier-free formulas to formulas of this special type is the decomposition lemma 5.14 . It is based on ideas of P.J. Cohen and taken over verbatim from [Weispfenning 76] .

We retain the notation and conventions introduced in section 3.

To begin with, we introduce relations $\underset{k}{\sim}$ for $1\le k<\omega$. $x \underset{k}{\sim} y$ is defined by $(x=y=0 \vee v(x-y) > vx + k1 = vy + k1$. In any valued field (F,Γ,v) with $0\le 1<\infty$, $\underset{k}{\sim}$ is an equivalence relation on F compatible with multiplication and division (but not with addition):

LEMMA 5.1 The following rules hold in any valued field with $0 \leq 1 < \infty$.

(i) $x \underset{k}{\sim} x$,

(ii) $x \underset{k}{\sim} y \rightarrow y \underset{k}{\sim} x$,

(iii) $x \underset{k}{\sim} y \wedge y \underset{k}{\sim} z \rightarrow x \underset{k}{\sim} z$,

(iv) $x \underset{k}{\sim} y \rightarrow x \underset{h}{\sim} y$ for $0 \leq h \leq k < \omega$,

(v) $x \underset{k}{\sim} y \rightarrow vx = vy \wedge res_k x = res_k y$,

(vi) $x \underset{k}{\sim} x' \rightarrow x^{-1} \underset{k}{\sim} x'^{-1}$,

(vii) $x \underset{k}{\sim} x' \wedge y \underset{k}{\sim} y' \rightarrow xy \underset{k}{\sim} x'y'$,

(viii) $v(x-y) \leq vy + k1 \wedge x \underset{k}{\sim} x' \rightarrow v(x'-y) \leq vy + k1$,

(ix) $v(x-y) \leq vy + k1 \wedge x \underset{2k}{\sim} x' \rightarrow x'-y \underset{k}{\sim} x-y$.

PROOF. Easy.

If $\underline{x}$, $\underline{y}$ are n-tuples of variables $(x_1,\ldots,x_n)$, $(y_1,\ldots,y_n)$, we write $\underline{x} \underset{k}{\sim} \underline{y}$ for $\bigwedge_{1 \leq i \leq n} x_i \underset{k}{\sim} y_i$. We call $\underline{x}$ k-<u>invariant</u> in a formula $\varphi(\underline{x})$, if $\underline{x} \underset{k}{\sim} \underline{y}$ implies $\varphi(\underline{x}) \leftrightarrow \varphi(\underline{y})$ in the theory $HF'_w(R,W,C)$. The following formulas $Bd_k(f(x))$: $vf(x) \leq \min v\underline{fx} + k1$ ("<u>f(x) is bounded by k</u>") will be of special importance in the sequel. From 5.1, we get the following facts on k-invariance:

LEMMA 5.2

(i) x is k-invariant in $vx = \xi$, $res_k x = r$, $W_{n,k,h}(x,r)$, $Bd_k(f(x))$.

(ii) (x,y) is k-invariant in $res_k(x/y) = r$, $W_{n,k,h}(x/y,r)$.

The next lemma collects some properties of the maps res_l and the relations $W_{n,l,h}$.

LEMMA 5.3 The following hold in $HF'_w(R,W,C)$.

(i) $0 \leq v(xy) \wedge 0 \leq vx \leq k \rightarrow.$
$res_l(xy) = t \leftrightarrow \exists r \exists s(res_{l+k}(x)=r \wedge res_{l+k}(q^k y)=s \wedge t=res^{l+k}_{l,k}(rs))$

(ii) $-k \leq vx \leq 0 \rightarrow.$
$res_l(x^{-1})=t \leftrightarrow \exists r \exists s(res_{l+k}(q^k x)=r \wedge rs=q^k \wedge res^{l+k}_{l,0}(s)=t)$

(iii) $res_l(y)=t \rightarrow (W_{n,l,h}(x,t) \leftrightarrow (\exists z(xz^n=y) \wedge 0 \leq vy \leq h))$

(iv) $W_{n,l,h}(x,t) \rightarrow W_{n,l',h'}(x,t)$ for $h \leq h'$.

(v) $W_{n,l,h}(x,t) \rightarrow (W_{n,l,h}(x,s) \leftrightarrow t \underset{n}{\approx} s \wedge vs \leq h)$
for $l \geq 2h + 2w(n)$.

(vi) $W_{n,l,h}(x,s) \wedge W_{n,l,h}(y,t) \rightarrow$
$(W_{n,l,h}(xy,t_1) \leftrightarrow st \underset{n}{\approx} t_1 \wedge vt_1 \leq h$ for $l \geq 2h + 2w(n)$.

(vii) $W_{n,l,h}(x,t) \leftrightarrow W_{n,l,2h}(q^h x, q^h t)$ for $l \geq 2h + 2w(n)$.

(viii) $W_{n,l,h}(x,t) \rightarrow (W_{n,l,h}(q^n x^{-1}, s) \leftrightarrow$
$\exists t_1(t_1 t = q^h \wedge t_1 \underset{n}{\approx} s \wedge vt_1 \leq h \wedge vs \leq h))$
for $l \geq 2h + 2w(n)$.

(ix) $0 \leq v(xy) \wedge 0 \leq vx \leq k \rightarrow (W_{n,l,h}(xy,t) \leftrightarrow$
$\exists r \exists s(res_{l'}(x) = r \wedge W_{n,l',h+k}(q^k y, s) \wedge res^{l'}_{l,k}(rs) =$
$t)$ for $l' \geq h+k + 2w(n)$.

(x) $\exists \eta(0 \leq vx + n\eta \leq h) \rightarrow \exists r\, W_{n,l,h}(x,r)$.

(xi) $\bigvee_{\check{c} \in \check{C}_n} \bigvee_{0 \leq i \leq n} \exists r\, W_{n,l,1}(q^i \check{c} x, r)$.

<u>PROOF</u>. (i),(ii), as well as (iv) - (ix) are straightforward, once (iii) is established: $\Leftarrow$ is obvious ; for $\Rightarrow$, let $res_l(xz^n) = t = res_l(y)$, $vt \leq h$. Then $0 \leq vy \leq h$ and $v(xz^n - y) > l$, and so $v(1 - xz^n/y) > l-h$. Let $f(x') = x'^n - xz^n/y$. Then $vf(1) > 2w(n)$, $vf'(1) \leq w(n)$, and so by the Newton lemma , there exists z' with $f(z') = 0$. Thus $z'^n = xz^n/y$, and so $x(z/z')^n = y$.

We are now in a position to find a simple normal form for formulas in which a given F-variable occurs only in quotients of powers of linear polynomials.

<u>LEMMA</u> 5.4 Let φ be F-quantifier-free and suppose x occurs in φ only in terms vb, $res_l b$ and formulas $W_{n,l,h}(b,d)$, where b is of the form $c(x-a)^m/(x-a')^{m'}$ and a,a',c,d do not contain x . Then φ is equivalent in $HF'_w(R,W,C)$ to a disjunction φ' of formulas of the form

(i) $\exists \xi \exists s(v(x-a) = \xi < \infty \wedge W_{n,l,1}(q^i \check{c}(x-a), s) \wedge \psi)$,

(ii) $\exists \xi \exists r \exists s(v(x-a) = \xi < \infty \wedge c \neq 0 \wedge res_l(c(x-a)^m) = r \wedge$
$W_{n,l,1}(q^i \check{c}(x-a), s) \wedge \psi)$, (iii) $x = a \wedge \psi$,

where x is not in c, a, ψ , $\check{c} \in \check{C}_n$, ψ is F-quantifier-free.
Moreover, the assignment $\varphi \mapsto \varphi'$ is primitive recursive .

<u>PROOF</u>. Using 5.3(vi),(viii),(xi), $\varphi(x)$ may be reduced to a disjunction of formulas of the form $(x = a_i \wedge \varphi(a_i))$ and

φ_1: $\exists \underline{\xi} \exists \underline{s} \exists \underline{t}(\bigwedge_{i \in I} v(x-a_i) = \xi_i < \infty \wedge \bigwedge_{i \in I_1, j \in I_2} res_l(c_i(x-a_i)^{m_i}/(x-a_j)^{m_j}) = s_{ij} \wedge \bigwedge_{i \in I_3} W_{n,l,h}(q^{j_i} \check{c}_i(x-a_i), t_i) \wedge \psi_1)$, where $I_1, I_2, I_3 \subseteq I$, ψ_1 is F-quantifier-free, $\check{c}_i \in \check{C}_n$, x is not in ψ_1, c_i, a_i. Next, we adjoin to φ_1 the true disjunction $\bigvee_{\emptyset \neq J \subseteq I} \bigvee_{g \in J} \rho_{J,g}$, where $\rho_{J,g}$ is the formula

$\bigwedge_{i \in J} \xi_g - 1 \leq \xi_i \leq \xi_g \wedge \bigwedge_{i \in I \setminus J} \xi_i < \xi_g - 1$.

For fixed J, g, we put $\varphi_1 \wedge \rho_{J,g} = \varphi_2$, $a = a_g$, $\xi = \xi_g$, $t = t_g$, $b_i = a_i - a_g$. Then for $i \in I \setminus J$, $(x-a_i) = (x-a) - b_i \underset{1}{\sim} b_i$, and so by 5.2, we may substitute b_i for $(x-a_i)$ in φ_2. Next, we adjoin the true disjunction $\bigvee_{J' \subseteq J} \sigma_{J'}$, where $\sigma_{J'}$ is the formula $\bigwedge_{i \in J'} \xi - 1 \leq vb_i \leq \xi + 1 \wedge \bigwedge_{i \in J \setminus J'} \xi + 1 < vb_i$. For fixed J' and $i \in J \setminus J'$, $(x-a_i) = (x-a) - b_i \underset{1}{\sim} (x-a)$; so we may substitute $(x-a)$ for $(x-a_i)$ in $\sigma_{J'} \wedge \varphi_2$ for $i \in J \setminus J'$.

As a consequence, it suffices to treat formulas of the form φ_3 :

$$v(x-a) = \xi < \infty \wedge \bigwedge_{i \in I} \xi - 1 \leq v((x-a)-b_i) = \xi_i \leq \xi \wedge \bigwedge_{i \in I} \xi - 1 \leq vb_i \leq \xi + 1 \wedge$$
$$\bigwedge_{i \in I_1} res_1(c_i(x-a)^{\pm m_i}) = s_i \wedge W_{n,1,h}(q^j\check{c}(x-a), t) \wedge$$
$$\bigwedge_{i \in I_2, j \in I_3} res_1(c_i((x-a)-b_i)^{m_i}/((x-a)-b_j)^{m_j}) = s_{ij} \wedge$$
$$\bigwedge_{i \in I_4} W_{n,1,h}(q^{j_i}\check{c}_i((x-a)-b_i), t_i) ,$$

with $I_1, \dots, I_4 \subseteq I$, $I_1 \cap (I_2 \cup I_3 \cup I_4) = \emptyset$.

Next, we notice that $q^{j_i}\check{c}_i((x-a)-b_i) = q^{j_i - 1}c_i(x-a)[q^1((x-a)-b_i)/(x-a)]$ with $0 \leq v[q^1((x-a)-b_i)/(x-a))] \leq 1$; furthermore

$c_i((x-a)-b_i)^{m_i}/((x-a)-b_j)^{m_j} = q^{21(m_j - m_i)} c_i b_i^{m_i} b_j^{-m_j}$

$(q^{21}b_i^{-1}(x-a) - q^{21})^{m_i} / (q^{21}b_j^{-1}(x-a) - q^{21})^{m_j}$ with

$0 \leq q^{21}b_i^{-1}(x-a) - q^{21}$, $q^{21}b_j^{-1}(x-a) - q^{21} \leq 31$; finally

$v((x-a) - b_i) = \xi_i$ is equivalent to $(\xi_i = vb_i < \xi) \vee$

$(\xi_i = \xi < vb_i) \vee (\xi = vb_i \wedge res^1_{0,0} res_1(b_i^{-1}(x-a) - 1) \neq 0)$.

So, 5.3 (i),(ii),(ix) can be applied to φ_3 to reduce the problem to formulas of the form φ_4 :

$$v(x-a) = \xi < \infty \wedge \bigwedge_{i \in I} res_1(c_i(x-a)^{m_i}) = s_i \neq 0 \wedge W_{n,1,h}(q^j\check{c}(x-a), t) ,$$

with $0 < m_i < \omega$, $\check{c} \in \check{C}_n$.

If in φ_4, $I = \emptyset$, we are done ; otherwise, it remains to reduce the conjunction $\bigwedge_{i\in I} \mu_i$, with μ_i : $res_1(c_i(x-a)^{m_i}) = s_i$ to a single formula μ of the same form. To this end, let $d = gcd(m_i : i \in I)$, $m_i = m_i' d$, $d = \sum_{I_1} q_i m_i - \sum_{I_2} q_i m_i$ for $0=m_i'$, $q_i < \omega$, $I = I_1 \dot\cup I_2$, $c = \prod_{I_1} c_i^{q_i} / \prod_{I_2} c_i^{q_i}$, $q' = \sum_I q_i$, $l' = l + 4q'$. Then

$\bigwedge_I \mu_i$ is equivalent to $\exists r \{\exists s_i' : i \in I\} \{\exists r_i' : i \in I\}$

$$\left[res_{l'}(cq^{q'}(x-a)^d) = r \wedge \bigwedge_{i\in I} res_{l'}(q^{q'} c_i c^{-m_i'}) = r_i' \wedge res^{l'}_{l,0}(s_i') = s_i \wedge \prod_{I_1} s_i'^{q_i} \cdot r = \prod_{I_2} s_i'^{q_i} \wedge \bigwedge_I r^{m_i'} \cdot r_i' = q^{2q'} \cdot s_i' \right.$$

COROLLARY 5.5 Let φ, x be as in 5.4 . Then $\exists x \varphi$ is equivalent in $HF'_w(R,W,C)$ to an F-quantifier-free formula φ_1. Moreover φ_1 can be obtained from φ by a primitive recursive procedure.

PROOF. By 5.4, we may assume that φ is of the form

(i) $v(x-a) = \xi < \infty \wedge W_{n,1,1}(q^i c(x-a), s)$ or

(ii) $v(x-a) = \xi < \infty \wedge res_1(c(x-a)^m) = r \wedge vr < 1 \wedge W_{n,1,1}(q^i \check{c}(x-a), s)$,

with x not in a, c, $\check{c} \in \check{C}_n$.

So by 5.3(iii), $\exists x \varphi$ is equivalent to

$$\exists \eta(\xi + i + v(c) + n\eta = vs \leq 1) \quad \text{in case (i), and to}$$

$$vc + m\xi = vr \wedge \exists r' \exists s' \exists t(W_{nm,1;nm+1}(c/(q^i c)^m, res^{l'}_{1,0}(t)) \wedge t \cdot s'^m = q^{nm} r' \wedge res^{l'}_{1,0}(r') = r \wedge res^{l'}_{1,0}(s') = s \wedge vr < 1 \wedge vs \leq 1)$$

in case (ii) .

Our next goal is to extend 5.4 and 5.5 to F-quantifier-free formulas φ, where x occurs only in polynomials that are known to be bounded by some natural number k .

LEMMA 5.6 Let $0 \leq k, n < \omega$ .There is a primitive recursive procedure assigning to any formula $\varphi(y)$, such that y is k-invariant in φ and $\varphi(y)$ implies $y \neq 0$, and any polynomial $f(x)$ of degree n , formulas $\psi_h(\xi)$ and polynomials $\hat{f}_h(x)$, $0 \leq h \leq n$, such that

(1) ψ_h is quantifier-free and does not contain x, deg $\hat{f}_h \leq n$;

(2) $\varphi(f(x))$ is equivalent in $HF'_w(R,W,C)$ to

$\bigvee_{0\le h\le n} (\Psi_h(vx) \wedge 0\le \min v\underline{\hat{f}_h x} \le k \wedge f_h \ne 0 \wedge \varphi(f_h q^{-k} x^h \hat{f}_h(x)))$

PROOF. Put $\hat{f}_h(x) = \sum_{h\le i\le n} q^k f_h^{-1} f_i x^{i-h}$, and let $\Psi_h(vx)$ be a suitable equivalent of

$\bigwedge_{0\le i<h} v(f_i x^i) > v(f_h x^h) + k \le \min v\underline{fx} + 2k < \infty$. Then $\Psi_h(vx) \wedge f(x) \ne 0$ implies $f(x) \underset{k}{\sim} f_h q^{-k} x^h \hat{f}_h(x)$.

LEMMA 5.7 The following hold in $HF'_w(R,W,C)$.

(i) $0 \le \min v\underline{fx} \le k \longrightarrow$

$(\neg)Bd_k(f(x)) \leftrightarrow \exists r_0 \ldots \exists r_n (\bigwedge_{0\le i\le n} res_{2k}(f_i x^i) = r_i \wedge$
$v(\sum_{0\le i\le n} r_i) \underset{(>)}{\le} \min(vr_i : 0\le i\le n) + k$.

(ii) $0 \le \min v\underline{fx} \le k \wedge Bd_k(f(x)) \wedge c \ne 0 \longrightarrow$

$v(cx^h f(x)) = \eta < \infty \leftrightarrow \exists\xi\exists r_0 \ldots \exists r_n (\bigwedge_{0\le i\le n} res_{2k}(f_i x^i) = r_i \wedge$
$vx = \xi \wedge v(\sum_{0\le i\le n} r_i) + vc + h\xi = \eta < \infty)$.

PROOF. Obvious.

LEMMA 5.8 Let φ be a formula of the form

$\bigwedge_{j\in J} f_j(x) = 0 \quad \bigwedge_{i\in I} vf_i(x) = \xi_i < \infty \wedge \bigwedge_{i\in I_1, j\in I_2} res_k(f_i(x)/f_j(x)) = s_{ij} \ne 0$
$\wedge \bigwedge_{i\in I_3} W_{n_i,k,h}(f_i(x), t_i)$,

where $f_i(x)$ are polynomials in x, $J \cap I = \emptyset$, $I_1, I_2, I_3 \subseteq I$. Let a_i be F-terms not containing x, let $\underline{f}_i^* = T(\underline{f}_i, a_i)$ (i.e. the coefficients of the Taylor expansion of $f(x)$ at a_i) for $i \in I \dot\cup J$, and let $\tilde{\varphi}$ be the formula $\bigwedge_{i\in I\cup J} Bd_k(f_i^*(x-a_i)) \wedge \varphi$). Then

(1) $\tilde{\varphi}$ is equivalent in $HF'_w(R,W,C)$ to a disjunction φ' of formulas of the form 5.4(i),(ii),(iii).

(2) $\exists x\, \tilde{\varphi}$ is equivalent in $HF'_w(R,W,C)$ to an F-quantifier-free formula φ''.

Moreover, φ', φ'' can be constructed from $\tilde{\varphi}$ by a primitive recursive procedure.

PROOF. (2) follows from (1) by 5.5. To prove (1), notice that $Bd_k(f_i^*(x-a_i))$ implies $f_i(x) = 0 \leftrightarrow (x = a_i \wedge f_i(a_i) = 0)$. So we may assume that $J = \emptyset$. Let $\tilde{\varphi}_1$ be obtained from $\tilde{\varphi}$ by replacing $f_i(x)$ by $f_i^*(x-a_i)$ in φ for all $i \in I$, and write $\tilde{\varphi}_1 = \tilde{\varphi}_2(\{f_i^*(x-a_i) : i \in I\})$.

Then the tuple $\{y_i : i \in I\}$ is k-invariant in $\tilde{\varphi}_2(\{y_i : i \in I\})$, and $\tilde{\varphi}_2(\{y_i : i \in I\})$ implies $\bigwedge_{i \in I} y_i \neq 0$. So by 5.6, we can replace $\tilde{\varphi}_1$ equivalently by a disjunction of formulas $\tilde{\varphi}_3$ of the form

$$\exists \eta \Big[\bigwedge_{i \in I} (v(x-a_i) = \eta_i < \infty \wedge 0 \leq \min \underline{vg_i(x-a_i)} \leq k \wedge c_i \neq 0 \wedge Bd_k(g_i(x-a_i)) \wedge vc_i + m_i \eta + vg_i(x-a_i) = \xi_i < \infty) \wedge$$
$$\bigwedge_{i \in I_1, j \in I_2} res_k [(c_i(x-a_i)^{m_i} / c_j(x-a_j)^{m_j}) \cdot (g_i(x-a_i)/g_j(x-a_j))] = s_{ij} \neq 0 \wedge$$
$$\bigwedge_{i \in I_3} W_{n_i,k,h}(c_i(x-a_i)^{m_i} g_i(x-a_i), t_i) \wedge \psi \Big] ,$$

where ψ is F-quantifier-free, x is not in c_i, ψ; $g_i(y)$ polynomials in y. At this point, we apply 5.7 and 5.3(i),(ii),(ix) to reduce $\tilde{\varphi}_3$ equivalently to an F-quantifier-free formula φ satisfying the hypothesis of 5.4. Then lemma 5.4 completes the proof.

Hensel's lemma (HL) assert the existence of certain zeros of polynomials; in order to describe these zeros, we introduce the concept of a <u>Hensel zero</u>: Let $HZ_k(f(z))$ ("z <u>is a Hensel zero of</u> f(x) <u>of order</u> k") be the formula
$\min v\underline{f} < \infty \wedge z \neq 0 \wedge f(z) = 0 \wedge Bd_k(f'(z))$.
Then NL may be rephrased as follows.

<u>LEMMA</u> 5.9 (cf.[Weispfenning 76], lemma 2.3) Let f(x) be of degree n and let $0 \leq k < \omega$. Then $HF_w(L) \models Bd_k(f'(x)) \wedge \neg Bd_{2(k+w(n))}(f(x)) \longrightarrow \exists z(HZ_k(f(z)) \wedge z \underset{k+w(n)}{\sim} x)$.

As a consequence, we get:

<u>LEMMA</u> 5.10 Let z be k'-invariant in $\varphi(z)$, $k \leq k'$. Then $HF_w'(R,W,C) \models \exists z(HZ_k(f(z)) \wedge \varphi(z)) \leftrightarrow \exists x(Bd_k(f'(x)) \wedge \neg Bd_{2(k'+w(n))}(f(x)) \wedge \varphi(x))$.

<u>LEMMA</u> 5.11 Let $\varphi(x)$ satisfy the hypothesis of 5.4 and let x be l-invariant in $\varphi(x)$; let a be an F-term not containing z and let $0 \leq k \leq l < \omega$. Then φ' : $\exists z(HZ_k(f(z)) \wedge \varphi(z-a))$ is equivalent in $HF_w'(R,W,C)$ to an F-quantifier-free formula φ'', and φ'' can be obtained from φ' by a primitive recursive procedure.

<u>PROOF</u>. Lemma 2.6 in [Weispfenning 76] says that for $k' = k+l+w(n)$,
$HZ_k(f(z)) \wedge z \underset{k'}{\sim} a \rightarrow z-a \underset{l}{\sim} f(a)/f'(a)$. So
$\exists z(HZ_k(f(z)) \wedge \varphi(z-a))$ is equivalent to
$\exists z(HZ_k(f(z)) \wedge z \underset{k'}{\sim} a \wedge \varphi(f(a)/f'(a))) \vee \varphi_1$, where φ_1 is the

formula $\exists z(HZ_k(f(z)) \wedge v(z-a) \leq va + k' \wedge \varphi(z-a))$. By 5.1(viii), (ix), z is $2k'$-invariant in $v(z-a) \leq va + k' \wedge \varphi(z-a)$. So by 5.10, φ_1 is equivalent to

$$\exists x(Bd_k(f'(x) \wedge \neg Bd_{2(k'+w(n))}(f(x)) \wedge v(x-a) \leq va+k' \wedge \varphi(x-a)) .$$

By 5.6, this is equivalent to a formula φ_2 of the form

$\exists \xi \exists x(Bd_k(g_1(x)) \wedge \neg Bd_{k''}(g_2(x)) \wedge \bigwedge_{i=1,2}(0 \leq \min v\underline{g_i x} \leq k'' \wedge vx = \xi \wedge v(x-a) \leq va + k' \wedge \varphi(x-a) \wedge \psi)$, where g_1, g_2 are polynomials in x, $k'' = 2(k' + w(n))$, ψ is F-quantifier-free and does not contain x. Lemma 5.7(i) can now be applied to reduce φ_2 to a formula $\exists \xi \exists x(\varphi_3)$, where $\varphi_3(x)$ satisfies the hypothesis of 5.4. So by 5.5, φ_2 is equivalent to an F-quantifier-free formula.

COROLLARY 5.12 Let $\varphi(x)$, $\tilde{\varphi}(x)$ be as in 5.8. Then $\tilde{\varphi}_1$: $\exists x(HZ_k(f(x)) \wedge \tilde{\varphi}(x))$ is equivalent in $HF_w'(R,W,C)$ to an F-quantifier-free formula φ'. φ' can be obtained from $\tilde{\varphi}_1$ by a primitive recursive procedure.

PROOF. Combine 5.8 (1) with 5.11.

In order to eliminate an F-quantifier $\exists x \psi$ in front of an arbitrary F-quantifier-free formula ψ, we need two more steps:

(1) Reduction of ψ to a formula φ as in 5.8.
(2) Construction of suitable Taylorexpansions $f^*(x-a)$ of a given polynomial $f(x)$ such that $f^*(x-a)$ is provably bounded by some natural number k.

The first step is treated in the next lemma; the second is accomplished by the decomposition lemma in [Weispfenning 76].

LEMMA 5.13 Let $\varphi(x)$ be F-quantifier-free. Then $\varphi(x)$ is equivalent in $HF_w'(R,W,C)$ to a formula φ_1 of the form

$$\exists \underline{\xi} \exists \underline{s} \exists \underline{t} \Big[\bigwedge_{i \in J} f_j(x) = 0 \wedge \bigwedge_{i \in I} vf_i(x) = \xi_i < \infty \wedge \bigwedge_{i \in I_1, j \in I_2} res_k(f_i(x)/f_j(x)) = s_{ij} \neq 0 \wedge \bigwedge_{i \in I_3} W_{n_i,k,h}(f_i(x),t_i) \wedge \psi \Big] ,$$

where $f_i(x)$ are polynomials in x, $J \cap I = \emptyset$, $I_1, I_2, I_3 \subseteq I$, ψ is F-quantifier-free and does not contain x. φ_1 is obtained from φ by a primitive recursive procedure.

PROOF. $\varphi(x)$ contains x in terms $v(a(x))$, $res_l(a(x))$ and formulas $W_{n,l,h}(a(x),d)$, where d may contain x in subterms of the form $res_k(a'(x))$. So, $\varphi(x)$ can be rewritten in the form φ' :

$\exists \underline{\xi} \exists \underline{s} \exists \underline{t} (\bigwedge_{j \in J_1} a_j(x) = \xi_j \wedge \bigwedge_{j \in J_2} res_{k_j}(a_j(x)) = s_j \wedge \varphi'')$,

where φ'' contains x only in subformulas of the form $W_{n_j,k_j,h_j}(a_j(x),t_j)$. Next, the terms $a_j(x)$ can be rewritten - modulo case distinctions - as quotients $f_j(x)/g_j(x)$ of polynomials in x. Using the operations $res^k_{k_j,0}$ and 5.3(iv) , we may assume that $k_j = k$, $h_j = h$ for all j . Finally, φ' obtains the desired form by applications of 5.3(v),(vi),(viii) .

The following decomposition lemma is taken from [Weispfenning 76] (lemma 2.7) . It is stated and proved there for Hensel fields with weak cross-section π ; but both the statement and the proof do not involve π . We employ the following notation : $HZP_k(\underline{y}, z_1, \dots, z_n)$ ("$(z_1,\dots,z_n)$ <u>is a tuple of Hensel zeros of order k in the parameters</u> $\underline{y}$ ") denotes a formula of the form $\bigwedge_{1 \le i \le m} HZ_{k_i}(g_i(z_i))$, where $g_i(z_i)$ are polynomials in z_i of degree m_i with coefficients $\underline{g}_i \subseteq Ft(\underline{y} \cup \{z_1,\dots,z_{i-1}\})$, $k = \max(k_i : 1 \le i \le m)$. We put $\deg HZP_k(\underline{y},\underline{z}) = \max(m_i : 1 \le i \le m)$.

<u>DECOMPOSITION LEMMA</u> 5.14 For every polynomial f(x) of degree n there exists a finite set D(f(x)) of formulas of the form

(i) $Bd_k(f(x))$,

(ii) $\exists \underline{z} (HZP_{k'}(\underline{f},\underline{z}) \wedge Bd_m(f^*(x - \Sigma \underline{z})))$,

where $\underline{f}^* = T(\underline{f}, \Sigma \underline{z})$, $length(\underline{z}) = n$, $k,k',m \le 3^{n-1}w(n)$, $\deg HZP_{k'}(\underline{f},\underline{z}) \le n$, such that $HF_w(R) \models \bigvee \{\varphi : \varphi \in D(f(x))\}$. Moreover, D is constructed primitive recursively from f(x) .

<u>COROLLARY</u> 5.15 Let $\varphi(x,\underline{y})$ be F-quantifier-free .

(1) φ is equivalent in $HF'_w(R,W,C)$ to a disjunction φ' of formulas of the form $\exists \underline{\xi} \exists \underline{s} (\vartheta(x,\underline{y}) \wedge \psi)$ and $\exists \underline{\xi} \exists \underline{s} \exists \underline{z} (HZP_k(\underline{y},\underline{z}) \wedge \vartheta(x,\underline{y},\underline{z}) \wedge \psi)$, where ψ is as in 5.13 , ϑ is of the form 5.4(i),(ii),(iii) .

(2) $\exists x \varphi$ is equivalent in $HF'_w(R,W,C)$ to an F-quantifier-free formula φ'' .

Moreover, φ', φ'' can be computed from φ by a primitive recursive procedure. If, in addition, φ is as in 5.8, then $\deg HZP_k(\underline{y},\underline{z})$ can be taken smaller or equal to $\max(\deg f_i(x) : i \in I \cup J)$.

<u>PROOF</u>. (2) follows from (1) and 5.5 . To prove (1), we may assume that φ is of the form φ_1 described in 5.13. Successive applications

of 5.14 to all the polynomials $f_i(x)$, $i \in J \cup I$, yield an equivalent disjunction of formulas of the form $\exists \underline{\xi} \exists \underline{s} \exists \underline{t}(\tilde{\varphi} \wedge \psi)$ and $\exists \underline{\xi} \exists \underline{s} \exists \underline{t} \exists \underline{z}(\mathrm{HZP}_{k'}(\underline{y},\underline{z}) \wedge \tilde{\varphi} \wedge \psi)$,where $\tilde{\varphi}$ is as in 5.8, ψ is as in 5.13. Now, 5.8(1) yields the desired result.

COROLLARY 5.16 Let $\varphi(x,\underline{y})$ be F-quantifier-free and let $f(x)$ be a polynomial of degree n . Then $\exists x(\mathrm{HZ}_k(f(x)) \wedge \varphi)$ is equivalent in $\mathrm{HF}'_w(R,W,C)$ to a disjunction φ' of formulas of the form $\psi_1(\underline{y})$ and $\exists \underline{z}(\mathrm{HZP}_{k'}(\underline{y},\underline{z}) \wedge \psi_2(\underline{y},\underline{z}))$,where ψ_1, ψ_2 are F-quantifier--free and $\deg \mathrm{HZP}_{k'}(\underline{y},\underline{z}) < n$.

PROOF. We may assume that φ is of the form φ_1 described in 5.13. Moreover, by the division algorithm for formal polynomials, we may assume that $\deg f_i(x) < \deg f(x) = n$ for all $i \in J \cup I$. The claim follows now from 5.15 (1) and 5.11 .

To complete the proof of the main theorem, it suffices now to eliminate the F-quantifiers in formulas of the form $\exists \underline{z}(\mathrm{HZP}_{k'}(\underline{y},\underline{z}) \wedge \varphi)$, where φ is F-quantifier-free. This is done as in the proof of 3.12, by induction on $\omega \cdot n + m$, where $n = \deg \mathrm{HZP}_{k'}(\underline{y},\underline{z})$, $m = \mathrm{length}(\underline{z})$, using corollary 5.16 .

The rest of this section is devoted to proofs of supplement 4.4 and variants 4.5 , 4.7 , 4.9 of the main theorem .

PROOF of 4.4. In the proof of 4.3, we have eliminated an existential F-quantifier $\exists x\, \psi$, introducing in the elimination procedure only existential Γ-quantifiers and R_k-quantifiers - provided one disrerards the quantifiers hidden in the relations $\underset{n,k}{\approx}$ and their negations . So, if these relations are also F-quantifier-free and existential in L' , then the induction on the number of F-quantifiers in φ shows that φ' has the same prefix-type as φ (but of course a lot more quantifiers).

PROOF of 4.5. $\exists z(x = z^n)$ is definable in $\mathrm{HF}'_w(R,W,C)$ by $W_{n,k,1}(x,1)$ with $k = 1 + 2w(n)$. Conversely, the defining formula $\exists z(\mathrm{res}_k(xz^n) = r \wedge vr \leq h)$ of $W_{n,k,h}(x,r)$ has for $k \geq 2h + 2w(n)$ an F-quantifier-free equivalent in $\mathrm{HF}'_w(R,W',C,E)$, viz. the formula

$\bigvee_{\check{e}\in E_n} \bigvee_{0\leq i\leq h} [vr\leq h \wedge r \underset{n}{\approx} res_k(q^i\check{e}) \wedge W'_n(x/q^i\check{e})]$. (Use 5.3 .)

PROOF of 4.7. We indicate the changes necessary in the proof of the main theorem 4.3 : 5.3(xi) has to be replaced by

5.3(xi)' : $\exists\xi\exists r(\pi(\xi) \neq 0 \wedge W_{n,1,1}(\pi(\xi)x,r))$.

In 5.4 , the formulas (i) , (ii) have to be replaced by

(i)' $\exists\xi\exists\eta\exists s(v(x-a)=\xi<\infty \wedge \pi(\eta)\neq 0 \wedge W_{n,1,1}(\pi(\eta)(x-a),s) \wedge \Psi)$,

(ii)' $\exists\xi\exists\eta\exists r\exists s(v(x-a)=\xi<\infty \wedge \pi(\eta)\neq 0 \wedge c\neq 0 \wedge res_1(c(x-a)^m)= r \wedge$
$W_{n,1,1}(\pi(\eta)(x-a),s) \wedge \Psi)$.

The proof of 5.4 and 5.5 , as well as the statement of 5.8 and 5.15 has to be adapted accordingly. The only delicate point comes up in the proof of 5.13 , whose statement remains unaltered : In presence of the operation-symbol π , Γ-variables and R_k-variables may enter into F-terms, by being in the scope of a symbol π . These occurences are eliminated by the procedure described in [Weispfenning 76], lemma 3.1 . The essential point is that the subgroup Δ of Γ , on which π is non-trivial , is discrete with smallest positive element 1_Γ . Subsequently, occurences of the distinguished F-variable x in the scope of a symbol π are eliminated as in [Weispfenning 76] , lemma 2.9 , by introducing new existentially quantified Γ-variables.

PROOF of 4.9. We have to show that the defining formula $\exists z(res_k(xz^n) = r \wedge vr \leq h)$ of $W_{n,k,h}(x,r)$ is equivalent in $HF''_w(R,\pi)$ to an F-quantifier-free formula φ' . By 5.3 , φ' can be taken as

$(\pi(\eta)\neq 0 \wedge 0\leq vx + n\eta \leq (n+h)1_\Gamma \wedge \exists s\exists t(res_{k+n+2h}(x\cdot\pi(n\eta))s^n =$

$t\cdot q^{(n+h)} \wedge res^{k+n+2h}_{k,0}(t) = r))$.

6. CONCLUDING REMARKS .

The method of proof presented in section 5 has a potential that goes beyond the applications in section 4. In the following, we indicate some further conclusions that can be drawn from this method.

(1) Quantifier elimination without root-predicates and cross-section.

THEOREM 6.1 Let L'_Γ be an extension of L_Γ with $L'_\Gamma \cap L(R) = L(R)$, let L'_{R_k} be an extension of L_{R_k} with $L'_{R_k} \cap L(R) = L(R)$, and put $L' = L(R) \cup L'_\Gamma \bigcup_{k<\omega} L'_{R_k}$. Let T be an extension of HF(R) in L' such that all the defining formulas $\Psi_{n,k,h}(x,r)$ for $W_{n,k,h}(x,r)$ are equivalent in T to F-quantifier-free formulas $\Psi'_{n,k,h}(x,r)$ (and the assignment $\Psi_{n,k,h} \mapsto \Psi'_{n,k,h}$ is primitive recursive) . Then T admits (primitive recursive) elimination of F-quantifiers in L' .

PROOF. The sets of constants C, C_Γ with the corresponding axioms on the value group (preceeding 4.3) were introduced solely to cope with the predicates $W_{n,k,h}$ in the F-quantifier elimination. If all $W_{n,k,h}$ are F-quantifier-free definable they are superfluous in the proof of 4.3 .

This theorem can be applied e.g. to Hensel fields with radicable residue class fields of characteristic zero.

COROLLARY 6.2 Let $HF_0(R_0)$ be the theory of Hensel fields with residue class field of characteristic zero in $L(R_0)$. Let L' be as in 6.1 with L(R) replaced by $L(R_0)$.

(i) $HF_0(R_0) \cup \{\forall r \exists s (s^n = r) : 1 \le n < \omega\}$ admits primitive recursive elimination of F-quantifiers in L'.

(ii) If T_Γ is a theory in L'_Γ , T_{R_0} is a theory in L'_{R_0} , and T_Γ , T_{R_0} admit (primitive recursive) quantifier elimination, then $HF_0(R_0) \cup \{\forall r \exists s (s^n = r) : 1 \le n < \omega\}$ admits (primitive recursive) quantifier elimination in L' .

This improves V , theorem 6 in [Ziegler 72] .

PROOF. In the given theory, $\Psi_{n,0,0}(x,r)$ is equivalent to $vx \equiv_n 0 \wedge r \neq 0$.

(2) Replacing the residue class rings R_k $(k<\omega)$ by a single 'large' residue class ring $\overline{R}$.

If one wants to avoid the somewhat cumbersome manipulations of indices k for the various R_k-sorts involved in the proof of the main theorem, one may obtain a weaker but more streamlined result as follows: Drop all R_k-sorts, including res_k , $res^1_{k,h}$, q_k , v_k ; furthermore q , 1_Γ, and introduce instead

(i) a unary Γ-predicate Δ for a convex subgroup of the value group;

(ii) a new R-sort consisting of the ring language and operation symbols $\overline{res}$, $\overline{v}$.

Let $HF(\Delta,\overline{R})$ be the theory of Hensel fields (F,Γ,Δ,v) with distinguishe convex subgroup Δ of Γ , in the language $L(\Delta,\overline{R})$, where the $\overline{R}$-sort ranges over the residue class ring $\overline{R} = A/I$ for the prime ideal $I = \{a\in F : \text{for all } \delta\in\Delta,\ va>\delta\}$ of the valuation ring $A = A_v$, $\overline{res}: A \longrightarrow \overline{R}$ is canonical , $\overline{res} \mid F\setminus A = 0$, and $\overline{v}:\overline{R} \longrightarrow \Gamma\cup\{\infty\}$ is defined by $\overline{v}(0) = \infty$, $\overline{v}\ \overline{res}(a) = va$ for $\overline{res}(a) \neq 0$. Let $HF'(\Delta,\overline{R},\overline{W})$ be the theory of all $(F,\Gamma,\Delta,v) \models HF(\Delta,\overline{R})$ in $L(\Delta,\overline{R},\overline{W}) = L(\Delta,\overline{R}) \cup \{\overline{W}_n : 0<n<\omega\}$, where

(i) $v(n1)\in\Delta$ for all $0<n<\omega$;

(ii) Γ/Δ is divisible ;

(iii) $\overline{W}_n(x,r)$ is defined by $\exists z(\ \overline{res}(xz^n) = r \wedge r \neq 0\ )$.

Then the main theorem has the following counterpart.

THEOREM 6.3 Let L' be as in 4.3 with L(R,W,C) replaced by $L(\Delta,\overline{R},\overline{W})$. Then $HF'(\Delta,\overline{R},\overline{W})$ admits primitive recursive elimination of F-quantifiers in L' .

For the proof, it suffices to replace in section 5 the R_k-sorts by the $\overline{R}$-sort and corresponding concepts as $\underset{k}{\sim}$, $\underset{n,k}{\approx}$, Bd_k , HZ_k in the obvious manner (e.g. $\overline{Bd}(f(x)) \longleftrightarrow \Delta(\ vf(x) - \min \underline{vfx}\ )$).

This theorem may be viewed as an effective 'decomposition' of the valuation v .

(3) Definable Skolem functions and definable functions in p-adically closed fields.

The normal form for formulas with a distinguished F-variable x obtained in 5.15 can be used :

(i) to compute in a primitive recursive manner definable Skolem

functions for p-adically closed fields (in the sense of [Prestel-Roquette 83]) and real-closed rings , generalizing the results in [v.d.Dries a] ;
(ii) to reprove Denef's representation of definable functions from $\mathbb{Q}_p$ into $\mathbb{Z}$ ([Denef], 6.3) and of polynomials $f(\underline{x},t)$ by means of definable functions in $\mathbb{Q}_p$ ([Denef], 7.3) in an effective way.

These topics will be dealt with in a separate note.

(4) The main theorem and its variants are valid for languages L', where L'_Γ includes quantification over convex subgroups of the value group. This in combination with Gurevic's results on ordered abelian groups has been applied in [Weispfenning c] to prove the decidability of some classes of fields with quantification over certain valuation subrings.

REFERENCES.

J.Ax-S.Kochen 65 , Diophantine problems over local fields I,II , Amer. J. Math. 87, 605-648 .

----- " ----- 66 , Diophantine problems over local fields III, Annals of Math. 83, 437-456 .

S.Baserab 78 , Some model theory for henselian valued fields, J. of Algebra 55, 191-212 .

----"---- 79 , A model theoretic transfer theorem for henselian valued fields , Crelle's Journal 311/312 , 1-30 .

J.Becker-J.Denef-L.Lipshitz 80 , Further remarks on the elementary theory of formal power series , in Model Theory of Algebra and Arithmetic, Proc. Karpacz 1979, Springer LNM vol. 834 .

Th.Becker 83 , Real closed rings and ordered valuation rings , Zeitschr. f. Math. Logik u. G. M. 29, 417-425 .

M.Boffa , Unpublished manuscript.

S.S. Brown 78, Bounds on transfer principles for algebraically closed and complete discretely valued fields , Memoirs AMS , vol. 204 .

G.Cherlin-M.Dickmann 83 , Real-closed rings II. Model Theory , Ann. of pure and appl. Logic 25, 213-231 .

P.J.Cohen 69 , Decision procedures for real and p-adic fields , Comm. pure and appl. Math. 22, 131-153 .

F. Delon 81 , Quelques propriétés des corps valués en théorie des modèles , Thèse , Paris .

J. Denef , The rationality of the Poincaré series associated to the p-adic points on a variety (Second version) , preprint .

L.van den Dries 78 , Model theory of Fields , thesis, Utrecht .

---- " ---- 81 , Quantifier elimination for linear formulas over ordered and valued fields, Bull.Soc.Math. Belg. 23 , 19-32 .

L.van den Dries a , Algebraic theories with definable Skolem functions , preprint .

----- " ----- b , Elementary invariants for henselian valuation rings of mixed characteristic, and relative versions , manuscript , Jan. 1983 .

P.Eklof-E.Fischer 72 , The elementary theory of abelian groups , Ann. math. Logic 4 , 115-171 .

O.Endler 72 , Valuation Theory , Springer , Berlin-Heidelberg.

Ju.Ersov 65-67, On the elementary theory of maximal valued fields (russian), Algebra i Logika I : 4, 31-69 , II : 5 , 8-40 , III : 6 , 31-73 .

---"--- 65 , On the elementary theory of maximal normed fields , Sov. Math. Doklady 6 , 1390-1393 .

---"--- 80 , Multiply valued fields, Sov. Math. Doklady 22, 63-66.

M.J.Greenberg 69 , Lectures on forms in many variables , Benjamin , New York .

S.Kochen 75 , The model theory of local fields , Logic Conf. Kiel 1974 , Springer LNM , vol. 499 .

A.Macintyre 76, On definable sets of p-adic numbers , J. Symb. Logic 41 , 605-610 .

---- " ---- 77, Model-completeness , in Handbook of math. Logic , North-Holland , Amsterdam , 139-180 .

A.Macintyre-K.McKenna-L.v.d.Dries 83 , Elimination of quantifiers in algebraic structures, Adv. in Math. 47 , 74-87 .

A.Nerode 63 , A decision method for p-adic integral zeros of diophantine equations, Bull. AMS 69 , 513-517 .

A.Prestel-P.Roquette 83 , Formally p-adic fields , Springer LNM, vol. 1050 .

A.Robinson 56 , Complete Theories , North-Holland , Amsterdam .

P.Roquette , Some tendencies in contemporary algebra , to appear .

V.Weispfenning 71 , Elementary theories of valued fields , Dissertation , Universität Heidelberg .

---- " ---- 76 , On the elementary theory of Hensel fields , Ann. math. Logic 10 , 59-93 .

---- " ---- 78 , Model theory of lattice products , Habilitationsschrift, Universität Heidelberg

---- " ---- 81 , Quantifier elimination for certain ordered and lattice-ordered abelian groups , Bull. Soc. Math. Belg. 23 , 131-156 .

---- " ---- 82 , Valuation rings and boolean products , Proc. Conf. F.N.R.S., Brussels .

---- " ---- a , Aspects of quantifier elimination in algebra , to appear in Proc. Conf. Univ. Alg. , Darmstadt 1983 .

---- " ---- b , Quantifier elimination for ultrametric spaces , Abstract , Table ronde de logique , Paris 1983 .

---- " ---- c , Some decidable second-order field theories , Abstract , Table ronde de logique , Paris 1983 .

M.Ziegler 72 , Die elementare Theorie henselscher Körper , Dissertation , Universität Köln .

On $\underset{\sim}{\Sigma}^1_3$

Philip Welch
Mathematical Institute,
University of Oxford.

Contents: We show, in ZF+DC, under the assumption that every real has a sharp, that if the second uniform indiscernible u_2 equals $\aleph_2$ then in fact every real has a dagger. By previously known results this implies, as a contrapositive statement Theorem 1 below on the decomposition of $\underset{\sim}{\Sigma}^1_3$ sets.

Introduction

It was known to Sierpinski that every $\underset{\sim}{\Sigma}^1_1$, or analytic set, and every $\underset{\sim}{\Sigma}^1_2$ set was the union of $\aleph_1$ Borel sets. Shoenfield's proof of the absoluteness of Σ^1_2 predicates between L and V shows that every Σ^1_2 set of reals (reals here identified with $N = {}^{\omega}\omega$) is the projection of a tree, in L, on $\omega \times \omega_1$ (see definitions 1 & 2 below). It turned out that performing such decompositions of such sets hinges on the possibility of representing it as the projection of some tree on $\omega \times \lambda$, for some λ, or to put it differently of defining a scale into some λ on it. Martin showed that an assumption about the size of our set-theoretical universe, namely that every inner model of the form L(x) has a closed and unbounded class of generating indiscernibles (which contain every uncountable cardinal) provides for scales for every $\Pi^1_2(x)$, and hence by Prop.2 every $\Sigma^1_3(x)$, set of reals (Theorem 2 below.)

If as usual we abbreviate this statement as saying "$x^{\#}$ exists" his result implies the following propositions concerning the decomposition of $\underset{\sim}{\Sigma}^1_3$ sets (all results are in the theory ZF+DC unless "AC" for the full axiom of choice is indicated.)

(1) ($\forall a \subseteq \omega (a^{\#}$ exists) Every $\underset{\sim}{\Sigma}^1_3$ wellfounded relation $\subseteq N^2$ has length $< \aleph_{\omega+1}$ and thus $\underset{\sim}{\delta}^1_3 < \aleph_{\omega+1}$.

(Here $\underset{\sim}{\delta}^1_3$ = the supremum of lengths of prewellorderings of N in the class $\underset{\sim}{\Delta}^1_3$.) A result of Solovay's on these classes of indiscernibles yields:

(2) (AC + $\forall a \leq \omega (a^{\#}$ exists)) Every $\underset{\sim}{\Sigma}^1_3$ wellfounded relation $\subseteq N^2$ has length $< \aleph_3$ and $\underset{\sim}{\delta}^1_3 < \aleph_3$.

If we define $\underset{\sim}{B}_\kappa$= the smallest boolean algebra containing all closed sets and closed under unions of length $< \kappa$ (thus $\underset{\sim}{B}_{\aleph_1}$ = Borel) then Martin's result also yielded:

(3) a) ($\forall a \in N(a^{\#}$ exists)) Every $\underset{\sim}{\Sigma}^1_3$ set is the union of $\aleph_{\omega+1}$ sets in $\underset{\sim}{B}_{\aleph_{\omega+1}}$.

b) (AC + $\forall a \in N(a^{\#}$ exists)) Every $\underset{\sim}{\Sigma}^1_3$ set is the union of $\aleph_2$ Borel sets.

This paper shows that by making some additional assumptions about the universe of sets we can produce some better bounds (Theorem 1). These assumptions are of a restrictive nature - that the universe is not too large. We assume that if there's an inner model of the form $\langle L(\mu), \varepsilon, \mu \rangle$ where in the latter structure μ is a normal measure on some cardinal κ say, then there is no such c.u.b. class of indiscernibles above κ (or equivalently there is no elementary embedding $j: L(\mu) \rightarrow L(\mu)$ with $j \upharpoonright \kappa+1 = id \upharpoonright \kappa+1$.) This is abbreviated as $\sim 0^+$. $\sim b^+$ is the analogous statement for models of the form $\langle L(b,\mu), \varepsilon, b, \mu \rangle$ for $b \in N$.

<u>Theorem</u> <u>1</u> ($\forall a \in N(a^{\#}$ exists) $\to \exists b \in N(b^{+}$ doesn't exist))

a) Every $\underset{\sim}{\Sigma}^1_3$ wellfounded relation of reals has length $< \aleph_2$ and $\underset{\sim}{\delta}^1_3 \leq \aleph_2$.

b) Every $\underset{\sim}{\Sigma}^1_3$ set is the union of $\aleph_1$ Borel sets.

Notice that the improvements go in two directions. The use of AC is eliminated and we drop a cardinal even over the previous use of AC.

For $a \subsetneq \tau$, let I^a be the class of Silver indiscernibles for $L(a)$ given by $a^{\#}$ if such exists (thinking now of reals as subsets of ω, I^a must contain every uncountable cardinal$>\tau$.) Suppose every bounded in τ set of ordinals has a sharp. The we may form

$$C^{\tau} = \bigcap_{a \subseteq \gamma < \tau} I^a ; \qquad C^{\tau} = < u^{\tau}_i : i \in On >$$

say. Solovay defined C^{ω_1} (= C hereafter); $C = <u_i: i \in On>$. C is thus a closed unbounded class, which contains all cardinals, the "uniform" indiscernibles.

Theorem 1 (as do (1)-(3) above) depend completely on the following theorem of Martin's. A consequence of it is that Π^1_2 sets can be represented as projections of trees on $\omega \times u_{\omega}$.

<u>Theorem</u> <u>2</u> (Martin see (DST,8H.9)) ($\forall a \in N$ $a^{\#}$ exists)

Every $\Pi^1_2(a)$ set admits a $\Delta^1_3(a)$ scale into u_{ω}.

and the task thereafter is to compute the u_n's. We know that u_1 must be $\aleph_1, u_2 \leq \aleph_2$ trivially, and (Solovay) $cf(u_{\xi+1}) = cf(u_2)$. So, using AC $u_{\omega} < \aleph_3$. Without AC we only know $u_n \leq \aleph_n$ (and indeed $u_n = \aleph_n$ in ZF+DC+AD and thus $\aleph_3, \aleph_4, \ldots$ are all singular.) Clearly CH + AC imply u_2 (and hence $u_n, \ldots u_{\omega}$) $< \aleph_2$.

The substance of this paper is to show under the assumptions of Theorem 1 simply that $u_{\omega} < \aleph_2$.

The tool for this is a lemma (lemma 5 below) that appears as a central one in Jensen's proof of the absoluteness of Σ^1_3 predicates between V and K, the Core Model, (if we are taking $\sim 0^{\dagger}$ in the above) for which the reader may refer to (CM) or (SAK).

Interpreted this means that Π^1_2 (and thus Σ^1_3) sets can be represented as projections of trees in K.

Set now and for the rest of the paper $\tau = \aleph_1$ the ordinal that we identify with the first uncountable cardinal of V. This lemma shows that for any real a we have a class J^a, a closed and unbounded class of ordinals. J^a is definable from a "mouse" in K, (which can be regarded as a bounded subset of τ) J^a has the further property that the indiscernibles in I^a occur with a regular "periodicity" amongst the enumeration of the ordinals in J^a. This period is countable (in V) and using this fact we can show $u_\omega < \aleph_2$ and indeed the following is true, that under the hypothesis $\sim 0^{\dagger}$ for example the computation of $u_2^{\tau}, \ldots u_\omega^{\tau}$ ($= u_2^V, \ldots u_\omega^V$) yields the <u>same</u> ordinals irrespective of whether it is effected in K or in V, where K is the Core Model. Since K is a model of ZF+GCH we obtain that $u^V < \tau^{+K}$. Having said all this the demonstrations of these facts consists of stringing together a few observations; there are no difficult proofs.

Some definitions and theorems from descriptive set theory.

We give some definitions and results which will be very familiar to anyone versed in current descriptive set theory, so they can easily omit this. We follow (DST) and all these results are there and in (KM) to which one may refer for more information.

A <u>tree</u> T on $\omega \times \lambda$ ($\lambda \in On$) is a set of finitesequences of the form $((n_0,\xi_0),(n_1,\xi_1),\ldots(n_k,\xi_k))$ $n_i \in \omega, \xi_i \in \lambda$, closed under initial segments with $\emptyset$ at the "top", ie the tree grows downwards.

<u>Definition</u> 1 T(a) ($a \in {}^{\omega}\omega$) is the projection of T by a, and is the following

$$T(a) = \{ (\xi_0,\ldots,\xi_k) \mid ((a(0),\xi_0),(a(1),\xi_1)\ldots(a(k),\xi_k)) \in T \}.$$

<u>Definition</u> 2 For T a tree on $\omega \times \lambda$

$$p(T) = \{ a \in N \mid T(a) \text{ is not wellfounded } \}$$

We don't need to go into the definition of scales - the reader need only know the following:

<u>Proposition</u> 1 (KM,6B-1) Suppose $A \subseteq N$ is a pointset and admits a κ-scale, then there is a canonically associated tree on $\omega \times \kappa$ such that

$$a \in T \equiv T(a) \text{ is not wellfounded}$$
$$\equiv \exists f\ (a,f) \in (T) \quad (f \in N)$$

where by (T) we mean the set of all "branches" through a tree T.

Knowing that a set $A \subseteq N^2$ admits a κ-scale we can use

<u>Proposition</u> 2 (KM,6c-1) If $A \subseteq N^2$ admits a κ-scale then $\exists^N$ admits a κ -scale.

($\exists^N$ is the projection of the set A onto one axis) That scales on sets give bounds on wellfounded relation follows from the Kunen-Martin theorem:

Theorem 3 (Kunen,Martin see (KM,7A-1),(DST(2G-2))

"Let $\underset{=}{\leq} \subseteq N^2$ be a wellfounded relation and assume $\leq$, as a pointset, admits a κ-scale, then $|\underline{\leq}| < \kappa^+$"

where $|\underline{\leq}|$ is the rank or length of the wellfounded relation and κ^+ means the next cardinal $> \kappa$.

Again knowing that a set has a scale, we can go via the canonical tree to get an estimate of the complexity of its construction using the following:

Theorem 4 (Sierpinski for $\kappa = \omega$, (KM,8A-2) or (DST 2F-2))

Let T be a tree on $\omega \times \kappa$ and $A = p(T)$. A is then the union and intersection of κ^+ sets in $\underline{B}_{\kappa^+}$.

Proof of Theorem 1

We assume from now on that all reals have sharps, and for ease of writing that 0^{+} (loosely, a remarkable character for the theory ZF + $V = L^{\mu}$, where $V = L^{\mu}$ says that V is the constructible closure of a normal measure μ on some ordinal, or any of the usual equivalents) does not exist. All of the proof relativises to K^{a} if $a^{\dagger}$ does not exist, where K^{a} is the core model built up from a-mice. For all these notions see (CM). We quote the results as we need them.

Lemma 3 $\forall a \in {}^{\omega}\omega\ (K \not\subseteq L(a))$

Proof: Suppose it were. Then the existence of $a^{\#}$ implies there is an elementary embedding of $L(a) \to L(a)$, and thus of $K \to K$. But the latter implies the existence of an inner model with a measurable cardinal (CM,16.21). All this in $L(a^{\#})$. Thus $L(a^{\#}) \models$ " There exists an inner model with a measure"

But then in $L(a^{\#\#})$ we have an embedding of $L(a^{\#}) \to L(a^{\#})$ at a point above the measurable of $L(a^{\#})$'s L^{μ} and thus by (CM,13.19) $0^{\dagger}$ exists in $L(a^{\#\#})$ and in V.

QED

We use the notation of (CM) when considering mice. Given a mouse M we let

$$\langle\langle M_{\alpha}\rangle_{\alpha \in On}, \langle\pi_{\alpha\beta}\rangle_{\alpha \leq \beta \in On}, \langle\kappa_{\alpha}\rangle_{\alpha \in On}\rangle$$

be its mouse iteration, calling M_{α} its <u>α'th iterate</u> and κ_{α} is referred to as the <u>α'th iteration point</u>.

For M,N Core mice (CM,10.18), we use the <* order to compare them (< in (CM 15.7)). This is a well order.

Definition 3 M <* N if there exist $\alpha, \beta \in On$ such that $M_{\alpha} \in N_{\beta}$.

<u>Definition</u> 4 M is <u>sharplike</u> ((CM,15.5) or in the terminology of (SAK 1.4), <u>critical</u>) if $M \models$ "$\exists y(y = H_{\kappa_0})$" where κ_0 is the measurable cardinal of M.

Then we have

<u>Lemma</u> 4 (SAK,1.4),(CM,15.6) Let W be an inner model of ZF. Suppose $K \not\subseteq W$. Let M be the $<^*$-least mouse with $M \not\subseteq W$. Then M is sharplike and

$$K^W = \bigcup_{i \in On} H_{\kappa_i}^{M_i}.$$

This is due to Jensen, as is:

<u>Lemma</u> 5 (SAK,3.9),(CM,21.22) Suppose $a \in {}^{\omega}\omega$, $a^{\#}$ exists and $K \not\subseteq L(a)$. Let $\langle \tau_i^a \mid i \in On \rangle$ enumerate I^a. Then there exists i_0, α, β such that $\tau_{i_0} = \kappa_\alpha$, and indeed $\tau_{i_0+j} = \kappa_{\alpha+\beta j}$ $(j \in On)$ where κ_α is the α'th iteration point of the $<^*$-least mouse not in L(a).

The last lemma is a key one in proving the Σ_3^1-absoluteness or "correctness" of K assuming $\forall a(a^{\#}$ exists) + $\sim 0^{\dagger}$ and it is this lemma that provides our results. We remark that lemmas 4 & 5 are theorems of ZF alone.

Notice now that Lemma 5, being a theorem of ZF is true in $L(a^{\#})$ and hence the "least such ordinals i_0, α, β " are thus definable in $L(a^{\#})$ and hence are countable ordinals of $L(a^{\#\#})$ and thus of V.

<u>Remark</u> We can actually do away with the assumption of $a^{\#}$ in this last lemma. Also a better computation, not requiring the existence of $a^{\#\#}$ shows that i_0, α, β can be taken $\leq \omega_1^{L(a^{\#})}$.

The import of Lemma 5 is that for any real a the indiscernibles τ_i^a periodically appear amongst the κ_α^M where M is the $\leq^*$-least mouse $\notin L(a)$, and they commence doing so below $\aleph_1$ (of V).

Consider now the following set B_M for some fixed countable-in-V sharplike core mouse M.

$$B_M = \{a \in N \mid M \text{ is the } \leq^*\text{- least mouse not in } L(a)\}$$

Clearly $\omega_1 = \kappa_{\omega_1}^M \in \bigcap_{a \in B_M} I^a$.

But by the "countable periodicity" established by Lemma 5 and the remarks following it

$$\kappa_{\omega_1 \cdot (1+j)} \in \bigcap_{a \in B_M} I^a \qquad \text{also, for } j \in On.$$

So define

$$C_M = \{\kappa^M_{\omega_1 \cdot (1+j)} \mid j \in On\}, M \text{ a countable sharplike core mouse}$$

Notice now that this is a definable class of K since if τ is the ordinal that we identify with $\aleph_1$ of V :

$$C_M = \{\kappa^M_{\tau\cdot(1+j)} \mid j \in On\}, M \in K_\tau$$

Thus C_M is a definable class of K, definable from τ and M and thus is clearly c.u.b. in On.

Now form:

$$C^* = \bigcap_{\substack{\text{sharplike, core mice } M \\ M \in H_{\aleph_1}}} C_M.$$

again a definable class of K, definable from τ. Clearly

<u>Claim</u> $C^* \subseteq C$

Proof: Let $a \in N$ be arbitrary. Let M as above be the $\leq^*$-least mouse $\notin L(a)$. Then by design $C^* \subseteq C_M \subseteq I^a$.

QED

The following is also clear:

Lemma 6 $C = C^*$

Proof: We need to show $C \subseteq C^*$. Suppose not and let

$$u \in C - (C^* \cup \{\omega_1\}).$$

Then $\exists M \in K_{\omega_1}$ such that $u \neq \kappa^M_{\omega_1 \cdot (1+j)}$ for any $j \in On$.

Let $f \subseteq \omega$ such that $M \in L(f)$ and is countable in $L(f)$. Then $u \in I^f$ and $u = \tau^f_u$.

But $\{ \kappa^M_\alpha \mid \alpha \in On \} \supseteq I^f$ and $\tau^f_\alpha = \kappa^M_{\tau^f_\alpha}$, thus

$$u = \tau^f_u = \kappa^M_u \quad \text{and } u = \omega_1 . (1+u). \text{ Contradiction.}$$

QED

Remark It is not hard to see that if we set

$$Q = Q^{\omega_1} = \{ N_{\omega_1} \mid N \in K_{\omega_1}, \ N \text{ a core mouse} \}$$

$$= J^F_\theta \quad \text{say,}$$

where F is the cub filter on $\aleph_1$, for some θ and if we let

$< \lambda_\alpha \mid \alpha \in On >$ be the iteration points of Q_α then

$$\{ \lambda_\alpha \mid \alpha \in On \} \subseteq C \quad \text{(we may show equality here)}$$

Thus if we let $\underline{B}$ = smallest admissible set containing $K_{\aleph_1}$,

we then have $Q \in \underline{B} = L_{\tau^*}(K_{\aleph_1})$ say

$$u_i < \tau^* \quad (i < \tau^*) \quad \text{since } Q_i \in \underline{B}, \text{for such } i$$

$$u_{\tau^*} = \tau^*$$

Thus

Lemma 7 $u_2, \ldots u_\omega$ are less than the height of the first admissible set containing $K_{\aleph_1}$.

So in particular u_ω is a small ordinal compared to $\aleph_1^{+K}$ let alone $\aleph_2$. Indeed Martin/Kechris have shown:

Proposition 8 $\forall a(a^{\#}$ exists) implies $u_2 = \underset{\sim}{\delta}^1_2$.

Proof of Theorem 1a) We simply apply Proposition 2 and the Kunen-Martin Theorem 3 . Every $\Sigma^1_3(a)$ set admits an u_ω -scale. But $u_\omega < \aleph_2$. Thus if a $\underset{\sim}{\Sigma}^1_3$ set has a scale into this ordinal if it's a well-founded relation, its length is $< \aleph_2$. Trivially then $\underset{\sim}{\delta}^1_3 \leq \aleph_2$.

Proof of b)

Since in ZF+DC $\aleph_1$ is regular, and further that the canonical tree obtained from the u_ω -scale e-njoyed by any particular $\underset{\sim}{\Sigma}^1_3$ set A, on $\omega \times u_\omega$ can be replaced by one on $\omega \times \aleph_1$, since $\bar{u}_\omega = \aleph_1$ (using Prop.1 to get from the scale to such a tree). We reason as follows:

We have a tree on $\omega \times \aleph_1$ such that

$$a \in A \equiv T(a) \text{ is not wellfounded.}$$

But $cf(\aleph_1) > \omega$ so there exists $\xi < \omega_1$ such that

$T^\xi(a)$ is not well founded where

$$T^\xi = \{ ((n_0,\xi_0)),\ldots,(n_k,\xi_k)) \in T \mid 0 \leq i \leq k,\ \xi_i \leq \xi \}.$$

Thus

$$a \in A \equiv \exists \xi < \omega_1 \quad T^\xi(a) \text{ is not well-founded} \qquad (*)$$

Now T^ξ is a tree on $\omega \times \xi$ and ξ is countable .

We just now note that the standard decomposition of a set of the form p(S) where S is a tree on $\omega \times X$, with X countable, into $\aleph_1$ Borel sets is uniformly definable in S. We may then, given T, uniformly in ξ define a function $f(\alpha,\xi)$ to be the α'th Borel set inthe decomposition of $p(T^\xi)$ for $\alpha,\xi < \omega_1$. Thus $A = \bigcup_{\xi < \omega_1} \bigcup_{\alpha < \omega_1} f(\alpha,\xi)$ as required.

QED Theorem 1

(CM) "The Core Model" Tony Dodd. L.M.S. Lecture Note Series No.61. Cambridge University Press.

(KM) "Notes on the Theory of Scales" in "Cabal Seminar 76-77" Springer Lecture Note Series No.689, by A.Kechris and Y.Moschovakis.

(DST) "Descriptive Set Theory" Y.Moschovakis. North Holland Publishing Company. 1980.

(SAK) "Some applications of the Core Model" H.-D.Donder, B. Koppelberg, and R.B.Jensen, in "Set Theory and Model Theory" Springer Lecture Note Series No.872.

* The author would like to gratefully acknowledge that this research was made possible through his holding of an S.E.R.C. Research Fellowship.

Vol. 952: Combinatorial Mathematics IX. Proceedings, 1981. Edited by E. Billington, S. Oates-Williams, and A.P. Street. XI, 443 pages. 1982.

Vol. 953: Iterative Solution of Nonlinear Systems of Equations. Proceedings, 1982. Edited by R. Ansorge, Th. Meis, and W. Törnig. VII, 202 pages. 1982.

Vol. 954: S.G. Pandit, S.G. Deo, Differential Systems Involving Impulses. VII, 102 pages. 1982.

Vol. 955: G. Gierz, Bundles of Topological Vector Spaces and Their Duality. IV, 296 pages. 1982.

Vol. 956: Group Actions and Vector Fields. Proceedings, 1981. Edited by J.B. Carrell. V, 144 pages. 1982.

Vol. 957: Differential Equations. Proceedings, 1981. Edited by D.G. de Figueiredo. VIII, 301 pages. 1982.

Vol. 958: F.R. Beyl, J. Tappe, Group Extensions, Representations, and the Schur Multiplicator. IV, 278 pages. 1982.

Vol. 959: Géométrie Algébrique Réelle et Formes Quadratiques, Proceedings, 1981. Edité par J.-L. Colliot-Thélène, M. Coste, L. Mahé, et M.-F. Roy. X, 458 pages. 1982.

Vol. 960: Multigrid Methods. Proceedings, 1981. Edited by W. Hackbusch and U. Trottenberg. VII, 652 pages. 1982.

Vol. 961: Algebraic Geometry. Proceedings, 1981. Edited by J.M. Aroca, R. Buchweitz, M. Giusti, and M. Merle. X, 500 pages. 1982.

Vol. 962: Category Theory. Proceedings, 1981. Edited by K.H. Kamps, D. Pumplün, and W. Tholen. XV, 322 pages. 1982.

Vol. 963: R. Nottrot, Optimal Processes on Manifolds. VI, 124 pages. 1982.

Vol. 964: Ordinary and Partial Differential Equations. Proceedings, 1982. Edited by W.N. Everitt and B.D. Sleeman. XVIII, 726 pages. 1982.

Vol. 965: Topics in Numerical Analysis. Proceedings, 1981. Edited by P.R. Turner. IX, 202 pages. 1982.

Vol. 966: Algebraic K-Theory. Proceedings, 1980, Part I. Edited by R.K. Dennis. VIII, 407 pages. 1982.

Vol. 967: Algebraic K-Theory. Proceedings, 1980. Part II. VIII, 409 pages. 1982.

Vol. 968: Numerical Integration of Differential Equations and Large Linear Systems. Proceedings, 1980. Edited by J. Hinze. VI, 412 pages. 1982.

Vol. 969: Combinatorial Theory. Proceedings, 1982. Edited by D. Jungnickel and K. Vedder. V, 326 pages. 1982.

Vol. 970: Twistor Geometry and Non-Linear Systems. Proceedings, 1980. Edited by H.-D. Doebner and T.D. Palev. V, 216 pages. 1982.

Vol. 971: Kleinian Groups and Related Topics. Proceedings, 1981. Edited by D.M. Gallo and R.M. Porter. V, 117 pages. 1983.

Vol. 972: Nonlinear Filtering and Stochastic Control. Proceedings, 1981. Edited by S.K. Mitter and A. Moro. VIII, 297 pages. 1983.

Vol. 973: Matrix Pencils. Proceedings, 1982. Edited by B. Kågström and A. Ruhe. XI, 293 pages. 1983.

Vol. 974: A. Draux, Polynômes Orthogonaux Formels – Applications. VI, 625 pages. 1983.

Vol. 975: Radical Banach Algebras and Automatic Continuity. Proceedings, 1981. Edited by J.M. Bachar, W.G. Bade, P.C. Curtis Jr., H.G. Dales and M.P. Thomas. VIII, 470 pages. 1983.

Vol. 976: X. Fernique, P.W. Millar, D.W. Stroock, M. Weber, Ecole d'Eté de Probabilités de Saint-Flour XI – 1981. Edited by P.L. Hennequin. XI, 465 pages. 1983.

Vol. 977: T. Parthasarathy, On Global Univalence Theorems. VIII, 106 pages. 1983.

Vol. 978: J. Ławrynowicz, J. Krzyż, Quasiconformal Mappings in the Plane. VI, 177 pages. 1983.

Vol. 979: Mathematical Theories of Optimization. Proceedings, 1981. Edited by J.P. Cecconi and T. Zolezzi. V, 268 pages. 1983.

Vol. 980: L. Breen. Fonctions thêta et théorème du cube. XIII, 115 pages. 1983.

Vol. 981: Value Distribution Theory. Proceedings, 1981. Edited by I. Laine and S. Rickman. VIII, 245 pages. 1983.

Vol. 982: Stability Problems for Stochastic Models. Proceedings, 1982. Edited by V. V. Kalashnikov and V. M. Zolotarev. XVII, 295 pages. 1983.

Vol. 983: Nonstandard Analysis-Recent Developments. Edited by A.E. Hurd. V, 213 pages. 1983.

Vol. 984: A. Bove, J. E. Lewis, C. Parenti, Propagation of Singularities for Fuchsian Operators. IV, 161 pages. 1983.

Vol. 985: Asymptotic Analysis II. Edited by F. Verhulst. VI, 497 pages. 1983.

Vol. 986: Séminaire de Probabilités XVII 1981/82. Proceedings. Edited by J. Azéma and M. Yor. V, 512 pages. 1983.

Vol. 987: C. J. Bushnell, A. Fröhlich, Gauss Sums and p-adic Division Algebras. XI, 187 pages. 1983.

Vol. 988: J. Schwermer, Kohomologie arithmetisch definierter Gruppen und Eisensteinreihen. III, 170 pages. 1983.

Vol. 989: A.B. Mingarelli, Volterra-Stieltjes Integral Equations and Generalized Ordinary Differential Expressions. XIV, 318 pages. 1983.

Vol. 990: Probability in Banach Spaces IV. Proceedings, 1982. Edited by A. Beck and K. Jacobs. V, 234 pages. 1983.

Vol. 991: Banach Space Theory and its Applications. Proceedings, 1981. Edited by A. Pietsch, N. Popa and I. Singer. X, 302 pages. 1983.

Vol. 992: Harmonic Analysis, Proceedings, 1982. Edited by G. Mauceri, F. Ricci and G. Weiss. X, 449 pages. 1983.

Vol. 993: R. D. Bourgin, Geometric Aspects of Convex Sets with the Radon-Nikodým Property. XII, 474 pages. 1983.

Vol. 994: J.-L. Journé, Calderón-Zygmund Operators, Pseudo-Differential Operators and the Cauchy Integral of Calderón. VI, 129 pages. 1983.

Vol. 995: Banach Spaces, Harmonic Analysis, and Probability Theory. Proceedings, 1980–1981. Edited by R.C. Blei and S.J. Sidney. V, 173 pages. 1983.

Vol. 996: Invariant Theory. Proceedings, 1982. Edited by F. Gherardelli. V, 159 pages. 1983.

Vol. 997: Algebraic Geometry – Open Problems. Edited by C. Ciliberto, F. Ghione and F. Orecchia. VIII, 411 pages. 1983.

Vol. 998: Recent Developments in the Algebraic, Analytical, and Topological Theory of Semigroups. Proceedings, 1981. Edited by K.H. Hofmann, H. Jürgensen and H. J. Weinert. VI, 486 pages. 1983.

Vol. 999: C. Preston, Iterates of Maps on an Interval. VII, 205 pages. 1983.

Vol. 1000: H. Hopf, Differential Geometry in the Large. VII, 184 pages. 1983.

Vol. 1001: D.A. Hejhal, The Selberg Trace Formula for PSL(2, ℝ). Volume 2. VIII, 806 pages. 1983.

Vol. 1002: A. Edrei, E.B. Saff, R.S. Varga, Zeros of Sections of Power Series. VIII, 115 pages. 1983.

Vol. 1003: J. Schmets, Spaces of Vector-Valued Continuous Functions. VI, 117 pages. 1983.

Vol. 1004: Universal Algebra and Lattice Theory. Proceedings, 1982. Edited by R.S. Freese and O.C. Garcia. VI, 308 pages. 1983.

Vol. 1005: Numerical Methods. Proceedings, 1982. Edited by V. Pereyra and A. Reinoza. V, 296 pages. 1983.

Vol. 1006: Abelian Group Theory. Proceedings, 1982/83. Edited by R. Göbel, L. Lady and A. Mader. XVI, 771 pages. 1983.

Vol. 1007: Geometric Dynamics. Proceedings, 1981. Edited by J. Palis Jr. IX, 827 pages. 1983.

Vol. 1008: Algebraic Geometry. Proceedings, 1981. Edited by J. Dolgachev. V, 138 pages. 1983.

Vol. 1009: T. A. Chapman, Controlled Simple Homotopy Theory and Applications. III, 94 pages. 1983.

Vol. 1010: J.-E. Dies, Chaînes de Markov sur les permutations. IX, 226 pages. 1983.

Vol. 1011: J. M. Sigal. Scattering Theory for Many-Body Quantum Mechanical Systems. IV, 132 pages. 1983.

Vol. 1012: S. Kantorovitz, Spectral Theory of Banach Space Operators. V, 179 pages. 1983.

Vol. 1013: Complex Analysis – Fifth Romanian-Finnish Seminar. Part 1. Proceedings, 1981. Edited by C. Andreian Cazacu, N. Boboc, M. Jurchescu and I. Suciu. XX, 393 pages. 1983.

Vol. 1014: Complex Analysis – Fifth Romanian-Finnish Seminar. Part 2. Proceedings, 1981. Edited by C. Andreian Cazacu, N. Boboc, M. Jurchescu and I. Suciu. XX, 334 pages. 1983.

Vol. 1015: Equations différentielles et systèmes de Pfaff dans le champ complexe – II. Seminar. Edited by R. Gérard et J. P. Ramis. V, 411 pages. 1983.

Vol. 1016: Algebraic Geometry. Proceedings, 1982. Edited by M. Raynaud and T. Shioda. VIII, 528 pages. 1983.

Vol. 1017: Equadiff 82. Proceedings, 1982. Edited by H. W. Knobloch and K. Schmitt. XXIII, 666 pages. 1983.

Vol. 1018: Graph Theory, Łagów 1981. Proceedings, 1981. Edited by M. Borowiecki, J. W. Kennedy and M. M. Sysło. X, 289 pages. 1983.

Vol. 1019: Cabal Seminar 79–81. Proceedings, 1979–81. Edited by A. S. Kechris, D. A. Martin and Y. N. Moschovakis. V, 284 pages. 1983.

Vol. 1020: Non Commutative Harmonic Analysis and Lie Groups. Proceedings, 1982. Edited by J. Carmona and M. Vergne. V, 187 pages. 1983.

Vol. 1021: Probability Theory and Mathematical Statistics. Proceedings, 1982. Edited by K. Itô and J.V. Prokhorov. VIII, 747 pages. 1983.

Vol. 1022: G. Gentili, S. Salamon and J.-P. Vigué. Geometry Seminar "Luigi Bianchi", 1982. Edited by E. Vesentini. VI, 177 pages. 1983.

Vol. 1023: S. McAdam, Asymptotic Prime Divisors. IX, 118 pages. 1983.

Vol. 1024: Lie Group Representations I. Proceedings, 1982–1983. Edited by R. Herb, R. Lipsman and J. Rosenberg. IX, 369 pages. 1983.

Vol. 1025: D. Tanré, Homotopie Rationnelle: Modèles de Chen, Quillen, Sullivan. X, 211 pages. 1983.

Vol. 1026: W. Plesken, Group Rings of Finite Groups Over p-adic Integers. V, 151 pages. 1983.

Vol. 1027: M. Hasumi, Hardy Classes on Infinitely Connected Riemann Surfaces. XII, 280 pages. 1983.

Vol. 1028: Séminaire d'Analyse P. Lelong – P. Dolbeault – H. Skoda. Années 1981/1983. Edité par P. Lelong, P. Dolbeault et H. Skoda. VIII, 328 pages. 1983.

Vol. 1029: Séminaire d'Algèbre Paul Dubreil et Marie-Paule Malliavin. Proceedings, 1982. Edité par M.-P. Malliavin. V, 339 pages. 1983.

Vol. 1030: U. Christian, Selberg's Zeta-, L-, and Eisensteinseries. XII, 196 pages. 1983.

Vol. 1031: Dynamics and Processes. Proceedings, 1981. Edited by Ph. Blanchard and L. Streit. IX, 213 pages. 1983.

Vol. 1032: Ordinary Differential Equations and Operators. Proceedings, 1982. Edited by W. N. Everitt and R. T. Lewis. XV, 521 pages. 1983.

Vol. 1033: Measure Theory and its Applications. Proceedings, 1982. Edited by J. M. Belley, J. Dubois and P. Morales. XV, 317 pages. 1983.

Vol. 1034: J. Musielak, Orlicz Spaces and Modular Spaces. V, 222 pages. 1983.

Vol. 1035: The Mathematics and Physics of Disordered Media. Proceedings, 1983. Edited by B. D. Hughes and B. W. Ninham. VII, 432 pages. 1983.

Vol. 1036: Combinatorial Mathematics X. Proceedings, 1982. Edited by L. R. A. Casse. XI, 419 pages. 1983.

Vol. 1037: Non-linear Partial Differential Operators and Quantizatio Procedures. Proceedings, 1981. Edited by S. I. Andersson an H.-D. Doebner. VII, 334 pages. 1983.

Vol. 1038: F. Borceux, G. Van den Bossche, Algebra in a Locali Topos with Applications to Ring Theory. IX, 240 pages. 1983.

Vol. 1039: Analytic Functions, Błażejewko 1982. Proceedings. Edite by J. Ławrynowicz. X, 494 pages. 1983

Vol. 1040: A. Good, Local Analysis of Selberg's Trace Formula. II 128 pages. 1983.

Vol. 1041: Lie Group Representations II. Proceedings 1982–1983 Edited by R. Herb, S. Kudla, R. Lipsman and J. Rosenberg. IX 340 pages. 1984.

Vol. 1042: A. Gut, K. D. Schmidt, Amarts and Set Function Processes III, 258 pages. 1983.

Vol. 1043. Linear and Complex Analysis Problem Book. Edited by V. P. Havin, S. V. Hruščëv and N. K. Nikol'skii. XVIII, 721 pages. 1984.

Vol. 1044: E. Gekeler, Discretization Methods for Stable Initial Valu Problems. VIII, 201 pages. 1984.

Vol. 1045: Differential Geometry. Proceedings, 1982. Edited b A. M. Naveira. VIII, 194 pages. 1984.

Vol. 1046: Algebraic K-Theory, Number Theory, Geometry and Analysis. Proceedings, 1982. Edited by A. Bak. IX, 464 pages. 1984.

Vol. 1047: Fluid Dynamics. Seminar, 1982. Edited by H. Beirão d Veiga. VII, 193 pages. 1984.

Vol. 1048: Kinetic Theories and the Boltzmann Equation. Semina 1981. Edited by C. Cercignani. VII, 248 pages. 1984.

Vol. 1049: B. Iochum, Cônes autopolaires et algèbres de Jordar VI, 247 pages. 1984.

Vol. 1050: A. Prestel, P. Roquette, Formally p-adic Fields. V, 167 pages 1984.

Vol. 1051: Algebraic Topology, Aarhus 1982. Proceedings. Edited b I. Madsen and B. Oliver. X, 665 pages. 1984.

Vol. 1052: Number Theory. Seminar, 1982. Edited by D. V. Chudnovsky, G. V. Chudnovsky, H. Cohn and M. B. Nathanson. V, 30 pages. 1984.

Vol. 1053: P. Hilton, Nilpotente Gruppen und nilpotente Räume V, 221 pages. 1984.

Vol. 1054: V. Thomée, Galerkin Finite Element Methods for Paraboli Problems. VII, 237 pages. 1984.

Vol. 1055: Quantum Probability and Applications to the Quantu Theory of Irreversible Processes. Proceedings, 1982. Edited b L. Accardi, A. Frigerio and V. Gorini. VI, 411 pages. 1984.

Vol. 1056: Algebraic Geometry. Bucharest 1982. Proceedings. 198 Edited by L. Bădescu and D. Popescu. VII, 380 pages. 1984.

Vol. 1057: Bifurcation Theory and Applications. Seminar, 1983. Edite by L. Salvadori. VII, 233 pages. 1984.

Vol. 1058: B. Aulbach, Continuous and Discrete Dynamics nea Manifolds of Equilibria. IX, 142 pages. 1984.

Vol. 1059: Séminaire de Probabilités XVIII, 1982/83. Proceeding Edité par J. Azéma et M. Yor. IV, 518 pages. 1984.

Vol. 1060: Topology. Proceedings. 1982. Edited by L. D. Faddeev an A. A. Mal'cev. VI, 389 pages. 1984.

Vol. 1061: Séminaire de Théorie du Potentiel. Paris, No. 7. Proceedings. Directeurs: M. Brelot, G. Choquet et J. Deny. Rédacteurs F. Hirsch et G. Mokobodzki. IV, 281 pages. 1984.